NATURALLY FRACTURED RESERVOIRS
Second Edition

NATURALLY FRACTURED RESERVOIRS

Second Edition

Roberto Aguilera, Ph.D.

PennWell Books

PENNWELL PUBLISHING COMPANY
TULSA, OKLAHOMA

Copyright © 1995 by
PennWell Publishing Company
1421 South Sheridan/P.O. Box 1260
Tulsa, Oklahoma 74101

Library of Congress Cataloging-in-Publication Data

Aguilera, Roberto.
    Naturally fractured reservoirs / Roberto Aguilera. -- 2nd ed.
      p. cm.
    Includes bibliographical references (p.) and index.
    ISBN 0-87814-449-8
    1. Petroleum engineering.   2. Oil reservoir engineering.
  I. Title.
  TN871.A335 1995
  622'.3382--dc20   95-18164
                                          CIP

Printed in the United States of America

DEDICATION

Dedicated to my wife, María Ester; my daughter, María Silvia, and, my sons, Roberto Federico and Carlos Gustavo.

Acknowledgements

I wish to thank the Society of Petroleum Engineers, the Petroleum Society of CIM, the Society of Professional Well Log Analysts, the American Association of Petroleum Geologists, Western Atlas, Schlumberger, Halliburton, Gulf Publishing Co., PennWell Publishing Co., and *Petroleum Engineer* for permission to draw material from their publications. In addition, I express my gratitude to the various authors and organizations that have published material on the subject of naturally fractured reservoirs.

Although I am the only person responsible for the final form of this book, I would like to thank Dr. H.K. van Poollen and Dr. G.R. Pickett, r.i.p., for their help and encouragement during the development of some of the concepts and techniques presented here. Mr. Alonso Marin, r.i.p., provided valuable help and guidance during my early years as a petroleum engineer.

During the last 16 years I have been teaching courses on the subject of naturally fractured reservoirs all over the world. I thank my students for their questions and suggestions that have led to this second edition of my book.

Since 1984 I have been privileged to cooperate with AAPG presenting a course on fractured reservoir analysis with Dr. David Stearns, Dr. Melvin Friedman, and Dr. Ronald Nelson. Chapter 1 of this book dealing with geologic aspects reflects many of the geological techniques I have learned with Dave, Melvin and Ron. However, the responsibility for the way in which these techniques are presented is only mine.

Last, but not least, I wish to thank my wife, María Ester; my daughter María Silvia, and my sons, Roberto Federico and Carlos Gustavo, for their patience and understanding during the long evenings, weekends and holidays needed for the preparation of this book.

Preface

Since the publication of the first edition of my book on naturally fractured reservoirs in 1980, the science and art of evaluating these types of reservoirs has advanced at a rapid pace.

The first edition included in a single book many subjects that are usually discussed in separate volumes. There were nine chapters of the first edition that included geologic aspects, drilling and completion methods, log interpretation, well testing, fractured shales, primary and secondary recovery, numerical simulation, case histories, and economic evaluations.

Initially I attempted to include the same nine chapters in this second edition. However, I quickly realized that due to the many advances in the field, this was not going to be feasible. Consequently, this second edition includes only six chapters dealing with geologic aspects, drilling and completion, formation evaluation by well log analysis, tight gas reservoirs, case histories, and economic evaluations and reserves. Presently I am working on a second volume covering other aspects of naturally fractured reservoirs.

Chapter 1 deals with geologic aspects of naturally fractured reservoirs, reasons for generations of fractures including tectonic, regional, contractional and surface-related fractures, migration and accumulation of petroleum, direct and indirect sources of information, fractured reservoirs in various lithologies, how to avoid walking away from a commercial fractured reservoir due to an improper evaluation, and the importance of in-situ stresses on the study of naturally fractured reservoirs.

Chapter 2 reviews some important drilling and completion concepts for deviated holes. Since most fractures at depths of interest are vertical to subvertical, deviated and horizontal wells probably stand better chances of finding hydrocarbons than vertical wells. The advantages and disadvantages of open-hole vs. perforated completions are reviewed. Key elements associated with hydraulic fracturing and acidizing of naturally fractured reservoirs are also discussed.

Chapter 3, which focuses on formation evaluation by well log analysis, examines the use of many conventional and specialized well log curves in the qualitative and quantitative evaluation of naturally fractured reservoirs. The importance of electric and sonic imaging tools is discussed in detail. Special techniques dealing with the porosity exponent m and the water saturation exponent n allow quantitative estimates of porosity and water saturation in matrix, fractures and the combined matrix-fractures system. The effect of lithology variations and shaliness is reviewed as well as logging of horizontal wells. The uncertainty of calculating hydrocarbons-in-place in fracture media and the effect of miscallibrated logs is also analyzed.

Chapter 4 is concerned with tight gas reservoirs where production is possible in many cases thanks to the presence of natural fractures. Many of these reservoirs are also multi-layered adding a great deal of complexity to the evaluation. The chapter covers both fractured shales and tight gas sands. Geographical distribution is discussed. Various methods of well log interpretation, well test analysis and performance forecasts are reviewed in detail.

Chapter 5 presents case histories of naturally fractured reservoirs around the world. Giant, modest, and non-commercial oil and gas reservoirs are reviewed, highlighting the most important features associated with each one of them. Fractured reservoirs in sandstones, carbonates, cherts, shales, basement and tight gas sandstones are considered in this chapter. The case histories include the effect on recoveries of aquifers, various injection schemes (water, gas, polymers, CO_2, steam), subsidence, rapid pressure decline, strong gravity segregation with counterflow, fold and fault-related fractures, vertical communication through fractures, slanted and horizontal wells, and retrograde condensation.

Chapter 6 discusses some of the most important aspects of economic analysis. The optimum equilibrium between well spacing, maximum efficient rate and economic recovery is reviewed. Emphasis is placed in the economic analysis of acceleration projects as recovery from most naturally fractured reservoirs is directly related to actual acceleration projects. The effect of directional and horizontal wells on costs and recoveries is reviewed. The chapter includes reserves definitions as provided by various organizations and some guidelines based on my experience for estimating oil and gas reserves in naturally fractured reservoirs.

Contents

CHAPTER 1

Geologic Aspects

Many of present-day producing naturally fractured reservoirs have been accidentally discovered when looking for some other type of reservoir. Some years ago McNaughton and Garb (1975) estimated that ultimate recovery from producing fractured reservoirs would surpass 40 billion stock tank barrels of oil (STBO). Today I firmly believe that this figure was very conservative. I am convinced that there are significant volumes of hydrocarbons that have been left behind pipe as undiscovered, or behind plugged and abandoned wells or because of vertical wells that have not intercepted vertical fractures.

Figure 1–1 shows the location of some important naturally fractured reservoirs. They are found all over the world, in all types of lithologies and throughout the geologic stratigraphic column. This is demonstrated in Chapter 5 dealing with Case Histories.

WHAT IS A NATURAL FRACTURE?

A natural fracture is a macroscopic planar discontinuity that results from stresses that exceed the rupture strength of the rock (Stearns, 1990).

Another definition provided by Nelson (1985) is as follows:

"A reservoir fracture is a naturally occurring macroscopic planar discontinuity in rock due to deformation or physical diagenesis."

WHAT IS A NATURALLY FRACTURED RESERVOIR?

A naturally fractured reservoir is a reservoir which contains fractures created by mother nature. These natural fractures can have a positive or a negative effect on fluid flow. Open uncemented or partially mineralized fractures might have, for example, a positive effect on oil flow but a negative effect on water or gas flow due to coning effects. Totally mineralized natural fractures might create permeability barriers to all types of flow. This in turn might generate small compartments within the reservoir that can lead to uneconomic or marginal recoveries.

In my opinion all reservoirs contain a certain amount of natural fracturing. However, from a geologic and a reservoir engineering point of view, I regard as naturally fractured reservoirs only those where the fractures have an effect, either positive or negative, on fluid flow as suggested by Nelson (1985).

REQUIREMENTS FOR HYDROCARBON ACCUMULATION

In general, a petroleum reservoir consists of source rock, reservoir rock, seal rock, trap, and fluid content.

Source rock, or source environment, is believed to be responsible for the origin of petroleum. Most geologists believe that the origin of petroleum is organic, related mainly to vegetables which were altered by pressure, temperature, and bacteria. Some geologist (Hunt et al, 1992), however, believe that the origin of petroleum is igneous and indicate that oil rises from depth in granitic shield terrains of the world.

FIGURE 1-1 Location of some oil and gas fields producing from naturally fractured reservoirs. Some of the reservoirs are very prolific, some produce at more modest rates but still are economic, some are marginal, and some have tested at very low rates. Naturally fractured reservoirs are found all over the world in all types of lithologies and throughout the stratigraphic column all the way from the pre-Cambrian to the Miocene.

FIGURE 1-1 CODE
Location of some oil and gas fields producing from naturally fractured rocks.

1. Carbonate oil and gas reservoirs, Monkman area, Sukunka and Bullmoose fields, British Columbia, Canada (Tertiary)

   Carbonate gas reservoirs, Beaver River Field, British Columbia, Canada (Devonian)
2. Shale (Second White Specks) oil reservoirs, Alberta, Canada (Cretaceous)
3. Carbonate oil reservoir, Norman wells, NW Territories, Canada (Devonian)
4. Carbonate oil reservoirs, Weyburn Unit & Midale fields, Saskatchewan, Canada (Mississippian)
5. Carbonate oil reservoirs, Trenton, Ontario, Canada (Ordovician)
6. Carbonate gas reservoir, St. Flavien field, Quebec, Canada (Ordovician)
7. Monterey shale oil reservoirs offshore California (Miocene)
8. Basement oil reservoirs and Monterey shale onshore California
9. Carbonate oil reservoirs, Great Canyon & Bacon Flat fields (Devonian)
10. Sandstone oil reservoirs, Altamont & Blue Bell fields (Devonian)
11. Bakken shale oil reservoirs, Montana, North Dakota (Mississippian)

    Carbonate oil reservoirs, Cabin Creek and Pennel fields, Montana (Ordovician, Silurian)

    Carbonate oil reservoir, Killdeer field, North Dakota (Ordovician)

    Carbonate oil reservoir, Little Knife field, North Dakota (Mississippian)

    Carbonate oil reservoir, Sanish pool, North Dakota (Devonian)
12. Mancos shale oil reservoir, Colorado
13. Basement oil reservoirs, Orth field, Kansas (Pre-Cambrian)
14. Carbonate Karst oil reservoir, Oklahoma City field & Cottonwood Creek, Oklahoma (Ordovician)

    Mississippian lime oil reservoirs, Oklahoma (Mississppian)

    Meteorite impact oil reservoirs, Ames structure, Oklahoma
15. Niobrara chalk tight oil production, Silo field, Oklahoma
16. Coalbed methane reservoirs, San Juan Basin, New Mexico
17. Austin chalk oil reservoirs, Texas (Upper Cretaceous)

    Sandstone oil reservoirs, Spraberry, Texas (Permian)

18. Tight gas reservoirs, Cotton Valley Basin, Texas
19. Black shale gas reservoirs, Appalachian Basin (Devonian)
20. Carbonate oil reservoir, Bagre field, Mexico
21. Carbonate oil reservoirs, Reforma area, Sitio Grande and Cactus fields, Mexico (Cretaceous)
22. Carbonate oil reservoirs, Rubelsanto and Chinaja fields, Guatemala, (Triassic)
23. Sandstone oil and condensate reservoirs, Andes foothills, Colombia (Tertiary)
24. Sandstone Mogollon oil reservoirs, onshore and offshore, Peru (Cretaceous)
25. Sandstone gas and condensate reservoirs, Bolivia (Devonian)
26. Sandstone gas and condensate reservoirs, Argentina (Devonian)
27. Carbonate oil reservoirs, Puesto Rojas area, Argentina
    Carbonate oil reservoirs, Lindero Atravesado, Quintuco fm. Argentina (Cretaceous)
    Carbonate oil reservoirs, Pampa Paluaco (Argentina)
28. Sandstone oil reservoirs, Chile (Cretaceous)
29. Sandstone oil reservoir, offshore Brazil (Cretaceous)
30. Calcarenite oil reservoirs, offshore Brazil (Cretaceous)
31. Carbonate oil reservoirs, La Paz and Mara Fields, Venezuela (Cretaceous)
32. Carbonate (chalk) oil reservoirs, North Sea, Ekofisk field (Cretaceous, Paleozoic)
33. Carbonate oil reservoir, Lacq field, France (Upper Cretaceous)
    Carbonate gas reservoir, Lacq gas field, France (Cretaceous)
    Carbonate gas reservoir, Meillon gas field, France (Jurassic)
34. Carbonate oil reservoir, Casablanca field, offshore Spain (Jurassic)
35. Carbonate oil reservoir, Cavone field, Italy (Mesozoic)
36. Basement oil reservoir, Sidi Fili and Baton fields, Morocco
37. Carbonate oil reservoir, on-shore, Tunisia
38. Basement quartzite oil reservoir, Amal field, Libya (Cambrian)
39. Carbonate oil reservoir, Alamein field, Egypt (Cretaceous)
40. Carbonate oil reservoirs, Syria
41. Carbonate oil reservoir, Raman field, Turkey
42. Carbonate oil reservoir, Ain Zalah field, Iraq (Cretaceous)
    Carbonate oil reservoir, Kirkuk field, Iraq (Eocene, Oligocene)
43. Carbonate Asmari oil reservoirs, Iran (Oligocene, Miocene)
44. Basement oil reservoirs, Yemen
45. Carbonate oil reservoirs, Dukhan field, Oman (Jurassic)
46. Carbonate oil reservoirs, Dhurnal and Meyal fields, Pakistan (Tertiary)
47. Carbonate offshore oil reservoirs, India
48. Carbonate gas reservoir, Dachigangjing field & many others, China (Carboniferous)
49. Basement offshore oil reservoirs, Vietnam
50. Carbonate offshore gas reservoirs, Indonesia (Eocene)
51. Carbonate oil reservoirs, Nido A & B fields, Philippines
52. Sandstone gas reservoir, Palm Valley field, Australia (Ordovician)
53. Carbonate oil reservoir, Waihapa field, New Zealand (Oligocene)
54. Siliceous shale and chert oil and gas reservoir, Ayukawa field, Japan (Miocene)
55. Kouybychev region, Karabulak, Achaluki, Zamanjul, Russia

It is difficult to prove that oil actually came from a definite source. However, it is believed that the source rock is usually near the hydrocarbon reservoir, i.e., that petroleum was formed within that particular area. Source rock is difficult to identify because it usually contains no visible hydrocarbons. Snider (1934) indicates that the main source rock is shale, followed by limestone.

Reservoir rock is provided by porous and permeable beds. Precise determination of matrix and fracture porosity is important for accurate calculations of hydrocarbon-in-place. Matrix and fracture permeabilities are important parameters in calculating flow capacities.

Igneous, sedimentary, or metamorphic rocks can make an acceptable reservoir rock. However, most of the world's hydrocarbon accumulations occur in sandstones and carbonate rocks.

Seal rock confines hydrocarbons in the reservoir rock because of its extremely low level of permeability. Usually, seals have some plasticity, which allows them to deform rather than fracture during earth crust movements. The most important seal is shale, followed by carbonate rocks and evaporites.

A trap is formed by impervious material which surrounds the reservoir rock above a certain level. The trap holds the hydrocarbons in the reservoir. Traps are formed by a variety of structural and stratigraphic features. Landes (1959) provides a simplified classification of oil and gas traps:

I. Structural traps
   a. dry synclines
   b. anticlines
   c. salt-cored structures
   d. hydrodynamic
   e. fault
II. Stratigraphic traps
   a. Varying permeability caused by sedimentation.
   b. Varying permeability caused by ground water.
   c. Varying permeability caused by truncation and sealing.

The combination of structural and stratigraphic features is not uncommon. Natural fractures can be found in all of the above traps.

Fluid contents is the water and hydrocarbon that occupy the porous beds.

POROSITY

Porosity represents the void space in a rock. It can be quantified by dividing the void space by the bulk volume of the rock. In general, porosity can be classified as primary and secondary.

Primary Porosity

Primary porosity is established when the sediment is first deposited. Thus, it is an inherent, original characteristic of the rock.

For example, a sandstone rock usually has primary porosity. The value of primary porosity depends on many factors, including its arrangement and distribution, cementation and degree of interconnection among the voids.

Therefore, it is necessary to distinguish between total primary porosity and effective porosity. Total primary porosity is the ratio between the total primary void spaces and the bulk volume of the rock. Effective primary porosity is the ratio between the interconnected void space and the bulk volume of the rock. The commercial point of view is interested in effective porosity.

Graton and Fraser (1935) have evaluated porosity of a systematic packing of spheres (Figure 1–2A). For example, in the cubic or wide-packed system, porosity can be evaluated by first considering the volume of the spheres (or sand-grain volume).

$$Sphere\ volume = \frac{4\pi r^3}{3} \tag{1-1}$$

where r is the radius of the sphere. The unit cell presented in the lower part of Figure 1–2A is a cube with each side equal to 2r, or:

$$Bulk\ volume = (2r)^3 = 8r^3 \tag{1-2}$$

By definition, porosity (ϕ) is equal to the void space divided by the bulk volume of the rock, or:

$$\phi = \frac{void\ space}{bulk\ volume} = \frac{bulk\ volume - sphere\ volume}{bulk\ volume} \tag{1-3}$$

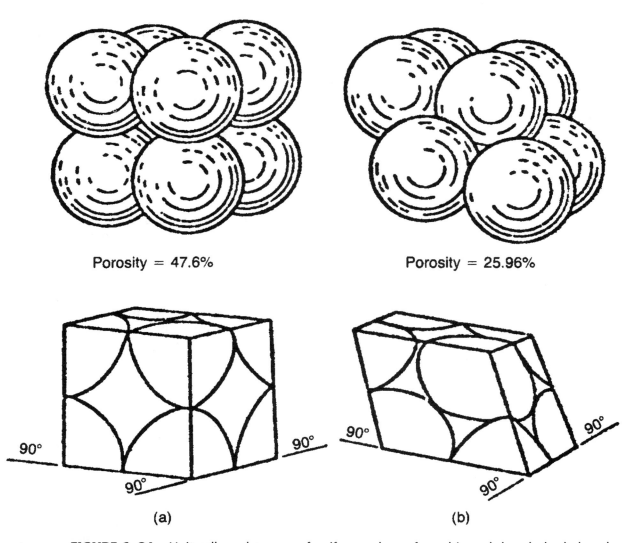

Porosity = 47.6%   Porosity = 25.96%

(a)   (b)

FIGURE 1–2A Unit cells and groups of uniform spheres for cubic and rhombohedral packing. (a) Cubic or wide-packed; (b) rhombohedral, or close-packed (after Fraser & Graton, 1935)

For cubic packing:

$$\phi = \frac{8r^3 - (4\pi r^3/3)}{8r^3} = 0.476 \ or \ 47.6\% \tag{1-4}$$

Notice in Equation 1–4 that porosity for the cubic or wide-packed system is only a function of packing and is independent of the radius of the sphere as these radii cancel out.

For the rhombohedral or closed-packed system, Graton and Fraser (1935) found a porosity of 25.96%. In general, primary porosities are lower than the above theoretical values due to cementation, irregularity of grain size, and shaliness. For example, the porosity of a clean, average sandstone is about 20%.

Secondary Porosity

Also known as induced porosity, secondary porosity is the result of geologic processes after the deposition of sedimentary rock and has no direct relation to the form of the sedimentary particles. Most reservoirs with secondary porosity are either limestones or dolomites. However, naturally fractured reservoirs are found sometimes in other types of lithologies including sandstones, shales, anhydrites, igneous metamorphic rocks and coal seams.

In general, secondary porosity is due to solution, recrystallization, dolomitization, and fractures.

Secondary porosity by solution can be generated by percolating acid waters which dissolve mostly limestones and dolomites, thus improving their porosities.

Dolomitization also improves porosity of carbonates. The equation describing dolomitization can be written as:

$$\begin{array}{cc} \text{limestone} & \text{dolomite} \\ 2\ CaCO_3 + Mg\ CL_2 \rightarrow & CaMg(CO_3)_2 + CaCl_2 \end{array}$$

Since porosity is equal to void space divided by bulk volume, fracture porosity can be attached to single point properties or total bulk properties. As such fracture-porosity is strongly scale-dependent.

Fracture porosity (ϕ_f) attached to single point properties is equal to the void space within the fractures divided by bulk volume of the fracture. As such, ϕ_f is a large number, sometimes close to 100%.

Fracture porosity (ϕ_2) attached to bulk properties is equal to void space within the fractures divided by total bulk volume. As such, ϕ_2 is usually a small number, in many cases less than 1%. Keep in mind that ϕ_2 is strongly scale-dependent. For example, if we have a 1-foot drilling break, the value of ϕ_2 at that particular location within that single foot is 100%. However the value of ϕ_2 for the whole reservoir might be less than 1%.

Figure 1–2B shows typical fracture networks considered by Reiss (1980). Fracture porosity for each one of the models is calculated as follows:

Model	ϕ_2
Sheets or stratum	w/D
Match sticks and cubes with 2 open fractures (horizontal fracture is closed)	2w/D
Cubes	3w/D

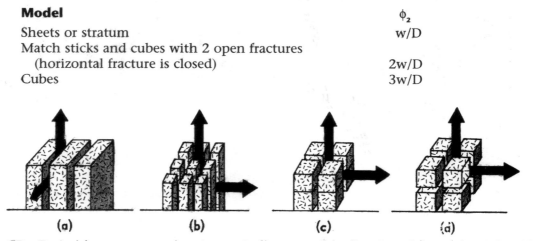

(a)	(b)	(c)	(d)

FIGURE 1–2B Typical fracture networks—Arrows indicate possible direction of flow (after Reiss, 1980)

where w = fracture width and D is distance between fractures. Following are some values of fracture porosity reported in the literature (adapted from Hensel, 1989):

Source	Formation Information	ϕ_2 (%)
Pittman (1979)	General statement	1
Stearns and Friedman (1972)	Austin chalk, Texas	0.2
Confidential study	Monterey, California	0.01 to 1.1
Weber and Bakker (1981)	South African karst zone	1 to 2
Bergosh and Lord (1987)	CT scan examples	1.53 to 2.57
Bergosh and Lord (1987)	Epoxy injection examples	1.81 to 9.64
Nelson (1985)	Amal field, Libya	1.7
Davidson and Snowdon (1978)	Beaver River, British Columbia	0.05 to 5
Kelton (1950)	Ellenburger, Texas	0.23 to 1.04
Pickett and Reynolds (1969)	Mississipian Lime, Oklahoma	0.5
Sahuquet and Ferrier (1982)	Lacq Superieur, France	0.5

DUCTILITY AND BRITTLENESS

Fractures are usually formed in rocks which are brittle. After analyzing the deformation characteristics of rocks in the laboratory, Griggs and Handin (1960) found that rock with very low ductility was characterized by a nearly linear stress-strain relationship, as shown in stage 1 of Figure 1–3. Strain is the deformation caused by stress. It may be dilation which

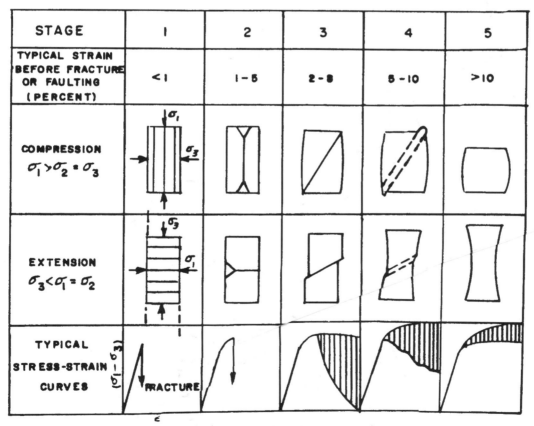

σ_1, σ_2, σ_3, ARE MAXIMUM, INTERMEDIATE, AND MINIMUM PRINCIPAL STRESSES, RESPECTIVELY.

FIGURE 1-3 Generalized spectrum of deformation characteristics of rocks as measured and observed in the laboratory (after Griggs and Handin, 1960)

is a change in volume, or distortion which is a change in form or both. As ductility is increased (stages 2 to 4), the stress-strain curve departs from linearity; however, some fractures are still formed. For very ductile rocks (stage 5), there is a large, permanent strain and a lack of fracturing.

A sample under directed forces as in Figure 1–3 goes through different types of deformation. During the approximately straight line portion in the stress-strain diagram there is *elastic deformation*, i.e., if the stress is withdrawn, the body returns to its original size and shape. The *elastic limit* is a limiting stress, above which, the sample does not return to its original size and shape. Hooke's law, which states that strain is proportional to stress, applies below the elastic limit. All five stages in Figure 1–3 show elastic deformation at the beginning of the experiments.

Plastic deformation occurs when the stress exceeds the elastic limit, i.e., when the curve in the stress-strain diagram departs from the approximate linearity. If the stress is withdrawn during plastic deformation the body returns only partially to its original shape. Stage 1 in Figure 1–3 does not show plastic deformation. Stages 2–5 show plastic deformation.

Rupture or generation of one or more fractures might occur when there is a continued increase in stress as shown in stages 1 and 2 of Figure 1–3.

In some cases the body deforms elastically and plastically as stress increases, but a point is reached after which progressively less stress is necessary to continue deformation. This high point in the stress-strain diagram is called *ultimate strength*. It is a function of many variables including confining pressure, rock type and temperature. Some key equations related to failure by rupture are presented in Chapter 2 (Equations 2–18 through 2–33).

Handin (1963) ascertained that ductility is affected by rock type, temperature, and net overburden. Figure 1–4 shows how ductilities vary as a function of depth for various lithologies. Quartzite is less ductile (more brittle), followed by dolomite, sandstone and limestone.

These findings were corroborated by Stearns (1967) who measured the relative frequency of fractures in various lithologies. He found the highest degree of fracturing was present in quartzite, followed by dolomite, quartzitic sandstone, calcite cemented sandstone and limestone (Figure 1–5A).

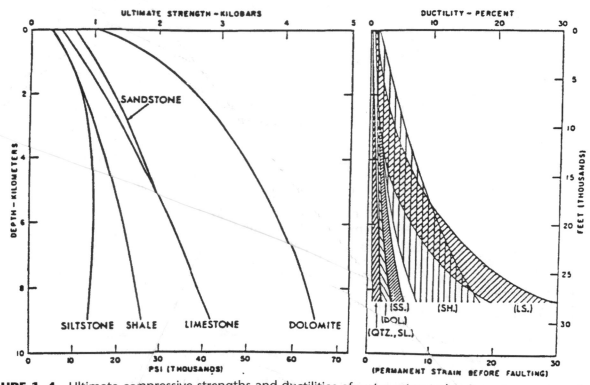

FIGURE 1–4 Ultimate compressive strengths and ductilities of water-saturated sedimentary rocks as function of depth. Effects of confining pressure, normal pore pressure, and temperature (30°C/km) are included (after Handin et al., 1963)

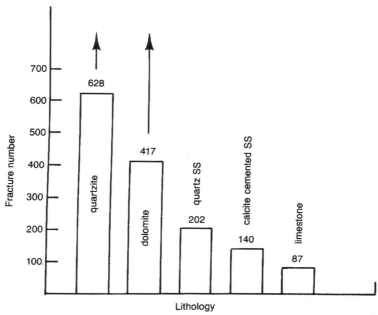

FIGURE 1–5A Average fracture number for several common rock types naturally deformed in the same physical environment (after Stearns, 1967)

Stearns' findings were extended by Sinclair in 1980. He found that, other things being equal, the degree of fracturing was larger in intervals with fine grain size as shown on Figure 1–5B.

Evaluating the value of fracture porosity with a satisfactory level of certainty is difficult even with whole core analysis, because cores usually break along the natural fractures planes. However, it is important to quantify fracture porosity as precisely as possible because fracture systems do provide important storage capacity in some reservoirs.

Important hydrocarbon reservoirs are found in fractured limestones, dolomites, cherts, shales, siltstone, sandstone, coals, igneous, and metamorphic rocks.

Fracture widths are usually very small, varying from paper-thin to 6 mm and more. The other two dimensions are also variable. I have examined cores with partially mineralized fractures exhibiting widths of more than 1 inch. The partial mineralization might still allow significant fluid flow and might act as a natural proppant agent that helps to maintain the fractures open as the reservoir is depleted.

REASONS FOR GENERATION OF FRACTURES

Fracture generation is attributed to various causes including (Landes, 1959):

- Diastrophism as in the case of folding and faulting. The faulting tends to generate cracks along the line of fault, which in turn produce a zone of dilatance. The dilatancy effect is probably responsible for a large part of the migration and accumulation of petroleum in fracture reservoirs (Mead, 1925; McNaughton, 1953; Thomas, 1986).
- Deep erosion of the overburden that permits the upper parts to expand, uplift, and fracture through planes of weakness.
- Volume, shrinkage, as in the case of shales that lose water, cooling of igneous rocks, and desiccation of sedimentary rocks.
- Paleokarstification and solution collapse as suggested by core studies of the Brown zone at the Healdton and Cottonwood Creek fields in Oklahoma (Lynch and Al-Sheib, 1991).
- Fluid pressure release when pore fluid pressure approaches the lithostatic pressure, as in the case of geopressured sedimentary strata (Hubbert and Wills, 1957; Capuano, 1993).
- Meteorite impact that can lead to complex, extensively brecciated, fractured reservoirs (Shoemaker, 1960; Kuykendall et al, 1994).

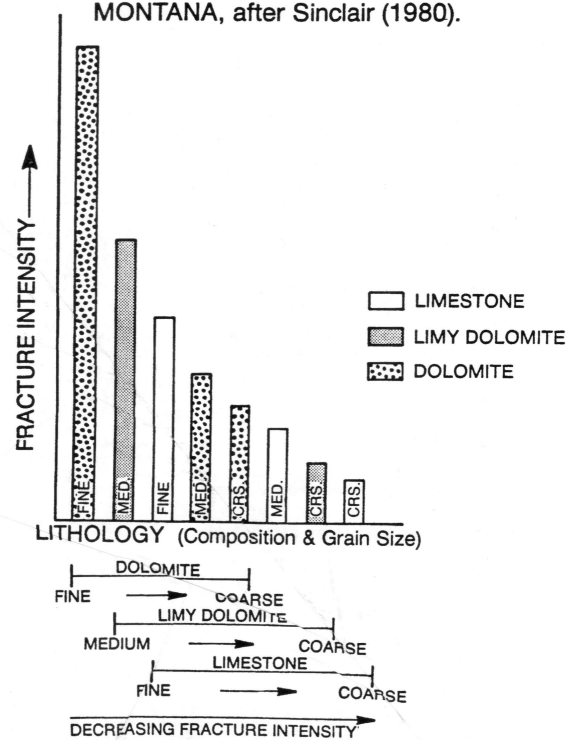

FIGURE 1–5B Fracture intensity as a function of composition and grain size (after Sinclair, 1980)

STORAGE CAPACITIES

Developing naturally fractured reservoirs has led to numerous economic failures. Initial high oil rates have led engineers in many instances to overestimate production forecasts of wells. Reservoir engineers usually make two key assumptions: (1) the fractures have a negligible storage capacity and are only channels of high permeability that allow fluids to flow; and (2) the matrix has an important storage capacity, but a very small permeability.

The first assumption has led to many fiascos in developing naturally fractured reservoirs. In fact, many reservoirs that produce at high initial rates decline drastically after a short period of time. This occurs because the producible oil has been stored in the fracture system. Consequently, it is important to estimate oil-in-place with reasonable accuracy within the fracture system.

The second assumption must be considered carefully. If the permeability of the matrix is very low, then the oil bleed-off from the matrix into the fractures might be very slow and only the oil originally within the fractures will be produced in a reasonable span of time. If the matrix has a reasonable permeability, then the storage capacity of the matrix becomes of paramount importance.

Other parameters that play an important role on how quickly oil moves from the matrix into the fractures include matrix porosity, total matrix compressibility, fracture spacing or distance between fractures, and oil viscosity. Because of its low viscosity, gas movement from the matrix into the fractures is faster than oil movement.

It is important to visualize that the storage capacity of naturally fractured reservoirs varies extensively, depending on the degree of fracturing in the formation and the value of primary porosity. The greater the value of primary porosity, the greater the success possibilities of naturally fractured reservoirs.

Figure 1–6 shows schematic sketches of porosity distribution in fractured reservoir rocks. The storage capacity in the matrix porosity of Figure 1–6A is large compared with storage capacity in the fractures. For this case, it can be seen in the lower part of Figure 1–6A that a very small percentage of the total porosity is made out of fractures. In general, this situation would tend to occur in reservoirs where the matrix porosity is rather high (larger than 10 up to more than 35%). Consequently, conventional exploration methods could probably be applied to locate these kind of reservoirs.

If the matrix has some permeability so as to allow flow into the wellbore, Type A reservoirs can be considered equivalent to what Nelson (1985) has called "fracture permeability assist" reservoirs, i.e., reservoirs where the fractures contribute permeability to an already producible reservoir.

The earth pulsates an average of 10 to 15 in., four times a day, under the gravitational attraction of the moon and the sun (Pirson, 1967). This triggers numerous small earthquakes

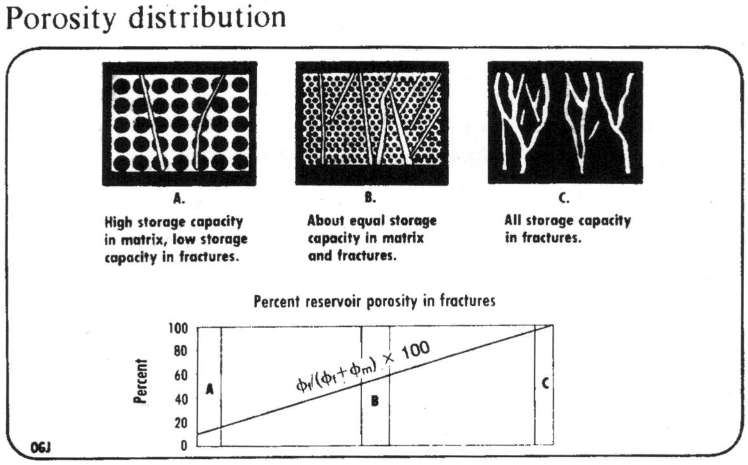

Porosity distribution

FIGURE 1–6 Schematic sketches showing porosity distribution in fractured reservoir rocks (after McNaughton & Garb, 1975)

daily that may generate natural fractures. It is very likely then that some of the so-called conventional high porosity reservoirs might be naturally fractured, and they would fall into Classification A in Figure 1–6. This has been discovered accidentally in some sandstone reservoirs with excellent well to well correlations when there has been an almost instantaneous water breakthrough following initiation of a water injection project.

Figure 1–6B shows a schematic of rock with about the same storage capacity in fracture and matrix porosity. In this case, the matrix has rather low porosities and the fractures provide the essential permeability.

The Type B reservoirs have been subdivided into B-I and B-II based on the characteristics of the matrix system. If the matrix has low but effective porosity, if it shows permeability to oil (or gas), if capillary pressures suggest good pore geometry, then the matrix will contribute effectively to the storage capacity of the reservoir.

The B-I Type is an ideal combination of porosities and permeabilities which has facilitated production of over 100 million stbo from some individual wells in Iran (McNaughton and Garb, 1975). Figure 1–7A shows schematic sketches of relative permeability and capillary pressure curves for Type B-I reservoirs.

The relative permeability curves for the fractures are shown as straight lines with 45° angles. This assumes that the fracture system is approximately equivalent to a bundle of tubes, where the irreducible water and residual oil saturations are equal to zero. Similar types of relative permeability curves for the fractures have been used in numerical simulation studies with good level of success.

The relative permeability curve of the matrix indicates an irreducible water saturation (S_{wmi}) of about 30% and a residual oil saturation of 20%. The capillary pressure curves suggest that the matrix by itself is a good reservoir rock and that the oil saturation within the fractures approaches 100%.

Figure 1–7B shows schematic sketches of relative permeability and capillary pressure for Type B-II reservoirs. In this case the matrix system is not a good reservoir rock as shown by the capillary pressure curve, even if there is some matrix porosity. Consequently, the fractures have only a fraction of the total porosity, but they might have nearly 100% of the hydrocarbon storage capacity.

Conventional log interpretation in this type of rock might lead to very high values of water saturation due to large amounts of water in the matrix. This is dangerous because based on conventional water saturation cutoffs potential hydrocarbon fractured intervals might be bypassed. To avoid this problem, it is better to resort to specialized methods that allow calculation of fracture, matrix and total water saturation using unconventional values of the petrophysical parameters m and n, as discussed in Chapter 3.

Distinction between B Type reservoirs is extremely important as it might help to avoid economic fiascos. As an example, several wells in the Austin Chalk of Texas produce at high rates which decline drastically after a short period of time. This occurs because the producible oil has been stored in the fracture system, in spite that the average matrix porosity ranges between 3% and 7%. The Austin chalk could then be classified in several areas as a B-II Type reservoir.

Figure 1–6C shows the schematic of a rock where the matrix porosity is zero and all the storage capacity is due to fractures. In this case, the fractures provide the essential porosity and permeability. Reservoirs of this type are generally characterized by initially high production rates that decline to uneconomic limits in a short period of time.

However, there are exceptions. For example, the Edison and Mountain View fields in the San Joaquin Valley of California and the El Segundo, Wilmington, and Playa del Rey fields in the Los Angeles Basin produced above 15,000 bo/d from fractured pre-Cretaceous basement schist. The storage in the basement rock of the La Paz-Mara oil fields in western Venezuela is in the fracture system. This field produced at one point in time over 80,000 bo/d from the basement reservoir. Matrix porosity contributes very little, if at all to the overall reservoir capacity of the Osage and Meramec limestones in the Eastern Anadarko basin. In these limestones essentially all the oil is within the fracture system (Harp, 1966). The Amal field in Libya (tectonic fractures) produces from a Cambrian quartzite which has a fracture porosity of about 1.7%, an area of approximately 100,000 acres, and reserves in

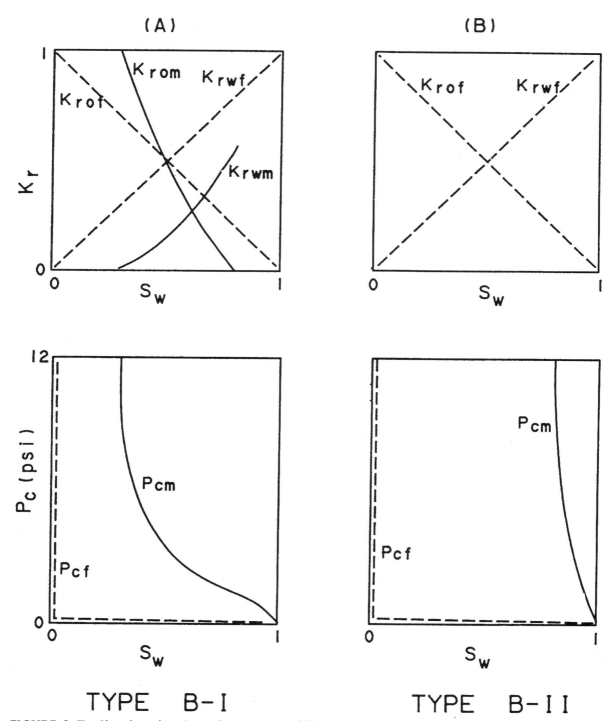

(A)

(B)

TYPE B-I

TYPE B-II

FIGURE 1–7 Sketches showing relative permeability (k$_r$) and capillary pressure (P$_c$) curves for Type B-I and Type B-II porosities. Subscripts o, w, m and f stand for oil, water, matrix, and fractures, respectively (after Aguilera, 1983)

the order of 1,700 MMstbo. The Big Sandy field in west Virginia and Kentucky produces from regional fractures that cut many structures. Reserves from Devonian shales at Big Sandy field are estimated at 3 Tscf. The reservoir area is estimated at 650,000 acres (Bagwal and Ryan, 1976.)

In summary, there is probably enough evidence to banish the generalized assumption that the storage capacity of a fractured system is negligible compared to the storage capacity of the matrix.

FRACTURE MORPHOLOGY

It relates to the form of natural fractures including open, deformed, mineral-filled and vuggy fractures (Nelson, 1985).

Open Fractures. They are uncemented and do not contain any kind of secondary mineralization. Fracture width is very small, probably the size of one pore, but increases permeability significantly parallel to the fracture. (See Example 1–3). On the other hand it has negligible effect on permeability perpendicular to the fracture. Porosity of open fractures is very small, usually a fraction of a percent, although there are exceptions.

Deformed Fractures. Included in-here are gouge-filled fractures and slickenside fractures. The gouge is provided by the finely abraded material resulting from grinding or sliding motion. This drastically reduces permeability. A slickenside is the result of frictional sliding along a fracture or fault plane. This generates a polished or striated surface that might increase permeability parallel to the fracture but drastically reduce permeability perpendicular to the fracture. Slickensides are thus a great cause of strong anistropy.

Mineral-filled Fractures. These fractures are cemented by secondary mineralization. Usual filling materials include quartz and calcite. These types of fractures might be formidable permeability barriers. On the other hand, partial secondary mineralization might have positive effects on hydrocarbon recoveries because it might act as a "natural proppant agent" that eliminates or reduces the closing of the fracture as the reservoir is depleted.

Vuggy Fractures. They can provide significant porosities and permeabilities. Due to the round shape of the vugs this type of fractures probably do not close as the reservoir is depleted. These types of fractures are the result of percolating acid waters through fractures which can lead to the development of Karst and very prolific reservoirs. Karst is an altered region made up of porous limestone containing deep fissures and sinkholes characterized by underground caves and streams.

PERMEABILITY

Permeability is a property of the porous medium and is a measure of the capacity of the medium to transmit fluids. Reservoirs can have primary and secondary permeability. Primary permeability is also referred to as matrix permeability by reservoir engineers. Secondary permeability can be either by fractures or solution vugs.

Matrix Permeability

Matrix permeability can be evaluated with the use of Darcy's law:

$$v = -\frac{k}{\mu} \times \frac{dp}{dl} \tag{1-5}$$

where:
$\quad v \quad = \quad$ apparent flow velocity, cm/sec
$\quad \mu \quad = \quad$ viscosity of the flowing fluid, cp
$\quad dp/dl \quad = \quad$ potential gradient in the flow direction, atm/cm
$\quad k \quad = \quad$ permeability of the rock, darcys

Darcy's law applies under the following conditions: (1) steady-state flow, (2) horizontal and linear flow, (3) laminar flow, (4) isothermal conditions, (5) constant viscosity, and (6) pore space 100% saturated with flowing fluid.

For the case of linear, incompressible fluid flow, permeability can be calculated from the equation:

$$k = v\frac{\mu L}{\Delta p} = \frac{q\mu L}{A\,\Delta p} \tag{1-5A}$$

where: $\quad q \quad = \quad$ flow rate, cm^3/sec
$\qquad\quad\; A \quad = \quad$ area, cm^2

Since the darcy unit is too large for most cases, permeability is usually expressed as one thousandth of a darcy, or a millidarcy (md).

Darcy's law in petroleum field units is expressed as:

$$q = \frac{0.001127 k A \Delta p}{\mu L} \qquad (1\text{--}6)$$

where: $\quad q \quad = \quad$ flow rate, b/d
$\qquad\quad\; k \quad = \quad$ permeability, md
$\qquad\quad\; A \quad = \quad$ area, sq ft
$\qquad\quad\; \Delta p \quad = \quad$ pressure differential, psi
$\qquad\quad\; \mu \quad = \quad$ viscosity, cp
$\qquad\quad\; L \quad = \quad$ distance, ft

Solution Vug Permeability

In some carbonate reservoirs, the percolation of acid waters can improve porosity and permeability by dissolution of the matrix. Poiseuille's law for capillary flow and Darcy's law for flow of liquids in permeable beds can be combined to estimate permeability in solution channels. Craft and Hawkins (1959) made a comprehensive study of this problem, modeled in the following discussion.

Assume a capillary tube with the following characteristics:

$\quad$ L $\;=\;$ length of capillary, cm
$\quad$ r $\;=\;$ inside radius, cm
$\quad$ A $\;=\;$ area, cm^2

A fluid of viscosity μ in poises is flowing in laminar or viscous flow under a pressure drop equal to p_1–p_2 dynes/sq cm. As the fluid wets the walls of the capillary, the velocity at the walls is considered to be zero and the velocity at the center of the tube is a maximum. The viscous force can be expressed by:

$$F = \mu A \frac{dv}{dx} \qquad (1\text{--}7)$$

where dv/dx is cm/sec/cm. The phase area of a capillary equals $2\pi rL$. Consequently, for a cylinder the previous equation can be written as:

$$F = \mu(2\pi rL)\frac{dv}{dr} \qquad (1\text{--}8)$$

If the fluid is not accelerating, the driving forces plus the viscous forces must equal zero. The driving force is equal to the pressure differential $(p_2$–$p_1)$ times the cross-section area πr^2. Consequently,

$$\pi r^2(p_1 - p_2) + \mu(2\pi rL)\frac{dv}{dr} = 0 \qquad (1\text{--}9)$$

and

$$dv = \frac{-(p_1 - p_2)rdr}{2\mu L} \qquad (1\text{--}10)$$

Integrating,

$$\int dv = \frac{-(p_1 - p_2)}{2\mu L}\int rdr \qquad (1\text{--}11)$$

$$v = \frac{-(p_1 - p_2)r^2}{4\mu L} + C \qquad (1\text{--}12)$$

The integration constant (C) is evaluated from the previous equation by taking v = 0 at r = r_o, or

$$0 = \frac{-(p_1 - p_2)r_o^2}{4\mu L} + C \tag{1-13}$$

Consequently,

$$C = \frac{(p_1 - p_2)r_o^2}{4\mu L} \tag{1-14}$$

Inserting Equation 1–14 into Equation 1–12 leads to

$$v = \frac{(r_o^2 - r^2)(p_1 - p_2)}{4\mu L}$$

The previous equation indicates that the liquid velocity in a capillary varies parabolically and reaches a maximum at the center of the tube. Velocity at the walls is zero.

The flow rate (dq) through an element (dr) is dq = vdA, where the area dA equals $2\pi r dr$, or

$$q = \int_o^q dq = \int_o^{r_o} vdA \tag{1-15}$$

and,

$$q = \int_o^{r_o} \frac{(r_o^2 - r^2)(p_1 - p_2)}{4\mu L} 2\pi r dr \tag{1-16}$$

The integration leads to Poiseuille's law for viscous flow of liquids through capillary tubes:

$$q = \frac{\pi r_o^4 (p_1 - p_2)}{8\mu L} \tag{1-17}$$

Darcy's law for steady state linear flow of compressible fluids can be written as:

$$q = \frac{9.86 \times 10^{-9} kA(p_1 - p_2)}{\mu L} \tag{1-18}$$

where:
A = area available to flow = πr^2, sq cm
k = permeability, darcys

Combination of Equation 1–17 and 18 leads to

$$k = \frac{Ar_o^2(p_1 - p_2)}{8\mu L} \times \frac{\mu L}{9.86 \times 10^{-9} A(p_1 - p_2)}$$

and,

$$k = 12.7 \times 10^6 r_o^2 \; darcys \tag{1-19}$$

where r_o is in centimeters.

If the inside radius (r_o) is in inches rather than centimeters, the permeability is given by:

$$k = 80 \times 10^6 r_o^2 = 20 \times 10^6 D^2 \; darcys \tag{1-20}$$

where D is the capillary diameter in inches. Therefore, the permeability of a circular opening 0.0001 in. in diameter is 0.20 darcys or 200 md. The permeability of a circular opening 0.01 in. in diameter is 2,000 darcys or 2,000,000 md.

Average permeability from a vug-matrix system can be obtained from the relationship:

$$k_{av} = \frac{k_v N\pi r^2 + k_b(A - N\pi r^2)}{A} \tag{1-21}$$

where
k_v = solution vug permeability, darcys
N = number of solution channels per section
A = cross-sectional area, in.^{2}
k_b = matrix block permeability, darcys
r = solution channel radius, in.

Example 1–1.   Consider a cube of limestone rock, 1 ft on each side with a matrix permeability of 1 md. The rock contains 20 solution channels, each one having a diameter of 0.05 in. What is the average permeability of the system if the channels lie in the direction of flow? What percentage of the fluid is stored in the matrix and what percentage in solution channels if the matrix porosity is 2%?

The permeability of each solution channel is obtained from the equation:

$$k_v = 20 \times 10^6 \times 0.05^2 = 50{,}000 \ darcys/channel$$

The average permeability of the system is obtained from the equation:

$$k_{av} = \frac{(50{,}000 \times 20 \times \pi \times .025^2) + 0.001(144 - 20 \times \pi \times .025^2)}{144}$$

$$k_{av} = \frac{1963.50 + 0.14}{144} = 13.64 \ darcys$$

The pore volume of the 20 solution channels is obtained from:

$$\pi r^2 \ NL = \pi \times 0.025^2 \times 20 \times 12 = 0.47 \ in.^3/ft^3$$

The pore volume of matrix rock is obtained from:

$$\phi_b \ (V_{rock} - V_{channels}) = 0.02 \ (12^3 - 0.47) = 34.55 \ in.^3/ft^3$$

The percentage of fluids stored in the matrix is:

$$\frac{34.55 - 0.47}{34.55} = 98.6\%$$

And the percentage of fluids stored in the solution channels is:

$$100 - 98.6 = 1.4\%$$

If the channels lie *perpendicular* to the direction of flow the permeability of the system would be approximately the same as the matrix permeability. The percentage of fluids stored in the solution channels and matrix, however, would be the same.

Example 1–2.   Calculate the flow rate through a cube, 1 ft on each side with a matrix permeability of 1 md (Amyx et al, 1960). If there is a solution channel 0.05 in. in diameter, calculate the flow rate through the solution channel and the combined flow rate. Assume $\Delta p = 1$ and $\mu = 1$.

Darcy's law (Equation 1–6) can determine the flow rate in the matrix, or

$$q = 1.127 \frac{kA\Delta p}{\mu L} = 1.127 \times \frac{0.001 \times 1 \times 1}{1 \times 1} = 0.001127 \ b/d$$

The vug permeability per solution channel is calculated from Equation 1–20 as follows:

$$K_v = 20 \times 10^6 \ D^2 = 20 \times 10^6 \times (0.05 \ in)^2 = 50{,}000 \ darcys$$

The area per solution channel is:

$$A = \frac{\pi 0.05^2}{4(144)} = 1.3635 \times 10^{-5} \ ft^2/solution \ channel$$

and the rate through the solution channel is:

$$q = \frac{1.127 \times 50000 \times 1.3635 \times 10^{-5} \times 1}{1 \times 1} = 0.768422 \ b/d$$

Consequently, the combined rate is 0.001127 + 0.768422 = 0.769549 b/d. The importance of the solution channel can be better appreciated in terms of percentage increase in flow rate which for this case is 68,183%.

Fracture Permeability

The presence of unhealed, uncemented, open fractures greatly increases the permeability of a rock. It is possible to estimate fracture permeability and flow rates through open fractures by following a development similar to the one presented for solution vugs.

Assume a fracture with a width equal to w_o, a length equal to L, and a lateral extent of the fracture equal to h. For this system the cross-section area open to flow is equal to $w_o h$. The driving force on the fracture is the pressure differential (p_1-p_2) acting on the area w_oh, or (p_1-p_2) w h dynes. The viscous forces are given by:

$$F = \mu A \frac{dv}{dw} \tag{1-22}$$

where A is the area equal to h L. If the liquid is not accelerating, the driving force plus the viscous force must equal zero, or:

$$(p_1 - p_2)wh + \mu hL \frac{dv}{dw} = 0 \tag{1-23}$$

separating variables and integrating,

$$(p_1 - p_2) \int w dw = \mu L \int dv \tag{1-24}$$

and,

$$(p_1 - p_2) \frac{w^2}{2} = -\mu L v + C \tag{1-25}$$

The integration constant may be evaluated at $v = 0$ and $w = w_o/2$, or

$$(p_1 - p_2) \frac{(w_o/2)^2}{2} = C \tag{1-26}$$

and,

$$(p_1 - p_2) \frac{w_o^2}{8} = C \tag{1-27}$$

Inserting Equation 1–27 into Equation 1–25 leads to:

$$(p_1 - p_2) \frac{w^2}{2} = -\mu L v + (p_1 - p_2) \frac{w_o^2}{8} \tag{1-28}$$

and,

$$(p_1 - p_2)\left(\frac{w^2}{2} - \frac{w_o^2}{8}\right) = -\mu L v \tag{1-29}$$

Consequently,

$$\frac{(p_1 - p_2)}{\mu L}\left(\frac{w_o^2}{8} - \frac{w^2}{2}\right) = v \tag{1-30}$$

The flow rate (dq) through an element (dw) equals vdA, where the area (dA) is given by 2hdw. Consequently,

$$q = \int_o^q dq = \int_o^{w_o} v dA \tag{1-31}$$

and,

$$q = \int_o^{w_o} \frac{(p_1 - p_2)}{\mu L}\left(\frac{w_o^2}{8} - \frac{w^2}{2}\right) 2h dw \tag{1-32}$$

Integrating,

$$q = \frac{w_o A(p_1 - p_2)}{12\mu L} \tag{1-33}$$

The previous equation can be combined with Darcy's law (Equation 1–18) to obtain a relationship for fracture permeability attached to single point properties (k_f) as follows:

$$k_f = \frac{w_o^2 A(p_1 - p_2)}{12\mu L} \times \frac{\mu L}{9.86 \times 10^{-9} A(p_1 - p_2)} \qquad (1\text{–}34)$$

and,

$$k_f = 8.35 \times 10^6 w_o^2 \; darcys \qquad (1\text{–}35)$$

where w_o is in centimeters. If the fracture width (w_o) is in inches rather than in centimeters, the permeability is given by:

$$k_f = 54 \times 10^6 w_o^2 \; darcys \qquad (1\text{–}36)$$

Consequently, the intrinsic permeability attached to single point properties of a fracture 0.01 in. thick would be 5,400 darcys or 5,400,000 md. These extremely high values of permeability clearly indicate the importance of fractures on production of tight reservoirs which otherwise would be noncommercial, even in the presence of high hydrocarbon saturations.

Example 1–3. Calculate the average permeability of a rock which contains three fractures, each one 0.01 in. wide. Dimensions of the rock are 1 ft × 1 ft × 1 ft. Matrix permeability is 1 millidarcy.

$$K_{av} = \frac{0.001[144 - 3(12 \times 0.01)] + [(54 \times 10^6 \times 0.01^2)3(12 \times 0.01)]}{144}$$

$$K_{av} = 13.51 \; darcys = 13,510 \; md$$

This is an average permeability parallel to the direction of the fractures. The average permeability perpendicular to the direction of the fractures, however, would be approximately equal to the matrix permeability if the fracture does not contain any secondary minerals.

Fracture permeability (k_f) from equation 1–36 is attached to single point properties. It can be extended to fracture permeability (k_2) attached to bulk properties of the system for one set of parallel fractures by using the equation:

$$k_2 = \frac{k_f w_o}{D} \qquad (1\text{–}37)$$

where D is distance between fractures.

Example 1–4. Calculate fracture permeabilities, k_f and k_2, for one set of parallel fractures assuming a fracture spacing (D) of 1 cm, and a fracture width of 5×10^{-4} cm. If the matrix permeability is 0.01 md, what is the amount of total permeability due to fractures?

$$k_f = 8.35 \times 10^6 \times (5 \times 10^{-4})^2 = 2.088 \; darcys = 2088 \; md$$

$$k_2 = 2.088 \times \frac{(5 \times 10^{-4})}{1} = 0.00104 \; darcys = 1.04 \; md$$

If we assume a total permeability of 0.01 + 1.04 = 1.05 md, the amount of total permeability due to fractures is 1.04/1.05 = 99%.

Equation 1–37 assumes one set of parallel fractures. Equation 1–37 can be extended for calculations of fracture permeability, k_2, in the case of match sticks and cubes models by using the following equations:

$$Cubes, \; k_2 = \frac{2}{3} \frac{k_f w_o^2}{D} \qquad (1\text{–}38)$$

$$Match \; sticks, \; k_2 = \frac{1}{2} \frac{k_f w_o^2}{D} \qquad (1\text{–}39)$$

MECHANICAL BEHAVIOR OF ROCK

Mechanical properties of rocks are determined by the interaction of intrinsic and environmental properties (Friedman, 1976–1994).

Intrinsic properties include composition of load-bearing framework, grain size, matrix porosity, matrix permeability, bed thickness, and previously existing mechanical discontinuities.

Environmental properties include effective pressure, i.e., the difference between confining and pore fluid pressure, temperature, time (strain rate), differential stress and pore fluid composition.

Intrinsic Properties. The effect of *composition* of the framework was illustrated in Figure 1-5A which showed, for the same physical environment, a larger degree of fracturing in quartzite followed by dolomite, quartzitic sandstone, calcite cemented sandstone, and limestone.

The effect of *grain size* on fracture abundance was studied by Sinclair (1980) at Teton anticline (Montana). He found a larger degree of fracturing in fine dolomite, followed by coarse dolomite, medium limestone, coarse limy dolomite and coarse limestone (Figure 1–5B).

The effect of *matrix porosity* and *matrix permeability* has been studied by various investigators who have found that the degree of fracturing increases with decreasing matrix porosity and permeability.

The effect of *bed thickness* was studied by McQuillan (1985) who showed data indicating that thinner beds contained closer spaced fractures. An interesting empirical observation is that in some cases the plot of fracture density in a Cartesian scale vs. bed thickness in a logarithmic scale results in a straight line as shown on Figure 1–8. Data for generation of these straight lines come from the Asmari limestone Kuh-E Pabdeh Gurpi, Kuh-E Asmari and Kuh-E Pahn in Southwestern Iran.

Other investigators (Narr and Lerche, 1984), however, indicate that fracture spacing is commonly observed to be a direct, linear function of the thickness of the fractured bed (Figure 1–9A and B). Figure 1–10 illustrate the effect of bed thickness on fracture intensity. The photo was taken in the Teton anticline of Northwestern Montana.

Previously existing mechanic discontinuities might have an effect on degree of fracturing. For example, it has been noticed in some areas that the largest fracturing occurs near faults. As we move away from the fault fracture porosity decreases.

Environmental Properties. Effective or net *confining pressure* (p_k), i.e., the difference between confining and pore fluid pressure, plays an important role in generation of fractures. Figure 1–11 shows the effect of effective pressure on the behavior of Solenhofen limestone under compression (Robertson, 1955) for seven different experiments. At a net confining pressure of 1 kg/cm^2 the sample behaved elastically until it reached a compressive stress of 2800 kg/cm^2 when it fractured.

Experiments run at net confining pressures of 300 and 700 kg/cm^2 showed elastic deformation followed by short periods of plastic deformation and finally the samples fractured at compressive stresses of about 4500 and 3700 kg/cm^2, respectively.

The specimens tested at p_k's of 1000, 2000, 3000, and 4000 kg/cm^2 started to deform elastically at about 4000 kg/cm^2 and then deformed plastically. Note that in these cases there was no fracturing probably because the tests were not run for a large period of time. These experiments allow us to conclude that rock strength increases with net confining pressure. Furthermore, they point out that rocks that might fracture rather easily near the surface might undergo large plastic deformations at bigger depths.

Temperature increases the probability of plastic deformation. Experiments carried out by Griggs (1951) on Yule Marble cylindrical specimens cut perpendicular to foliation under a net confining pressure of 10,000 atmospheres led to this result. He also found that if temperature is held constant but one sample is dry and the other wet, the elastic limit will be larger in the dry specimen. This leads to the conclusion that near the surface where temperatures are low, plastic deformation is smaller than at greater depths where temperature increases.

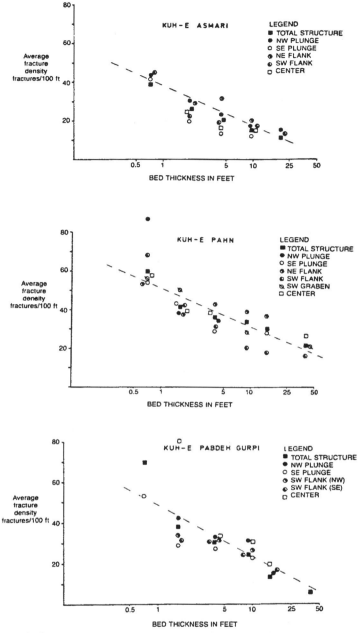

FIGURE 1–8 Tectonic fracture spacing vs. bed thickness plot for carbonate rocks on three outcropping folds in the Middle East (from McQuillan, 1973)

The effect of *time* including creeps, strain rate, and viscosity has been studied by Griggs (1951). He found in a 400-day experiment using Solenhofen limestone under a compressive stress of 1400 kg/cm², that most of the shortening occurred at very early times. The strength of Solenhofen limestone at room conditions is 2560 kg/cm². Consequently, the experiment was run at about 55% of the strength value. The following table summarizes results of the experiment:

Time, days	Shortening, %
1	0.006
10	0.011
100	0.016
400	0.019

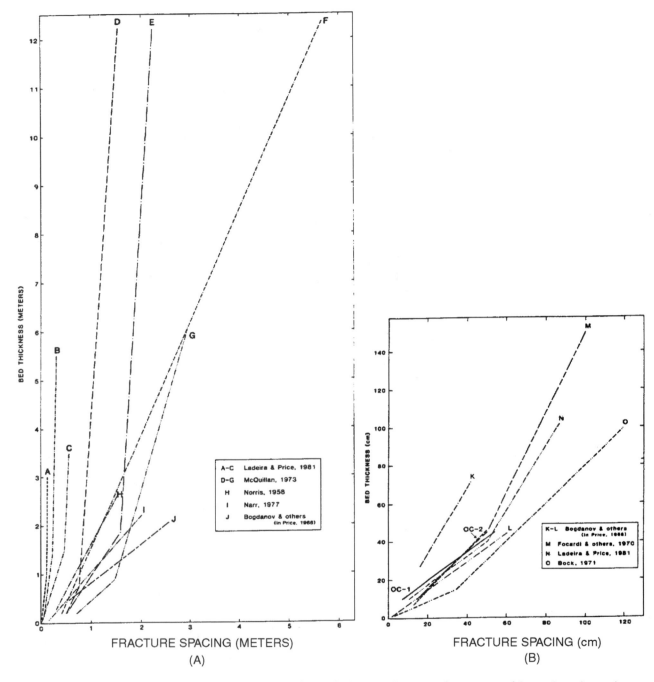

FIGURE 1-9 Linear approximations of bed thickness/joint spacing trends measured in various investigations. (A) Trends from data sets having bed thicknesses greater than 2 m (6.6 ft). (B) Trends from data sets having bed thicknesses less than 2 m (6.6 ft) (after Narr and Lerche, 1984)

The above data allow construction of a creep curve (Figure 1–12). Creep is defined as a slow continuous deformation with the passage of time. Deformation that occurs very quickly is called instantaneous deformation. It ends at a shortening of about 0.008% in Figure 1–12. This is followed by a primary deformation between a shortening of approximately 0.008 and 0.015%, and ends with a secondary creep above 0.015% shortening. An ideal creep curve would also include a tertiary deformation bending upwards followed by rupture.

Strain rate (ë), i.e., the ratio between strain and time is an important environmental property. Figure 1–13 taken from Friedman's AAPG Fracture Reservoir Analysis

FIGURE 1–10 Bed thickness effect on fracture spacing, Teton anticline, Northwestern Montana.

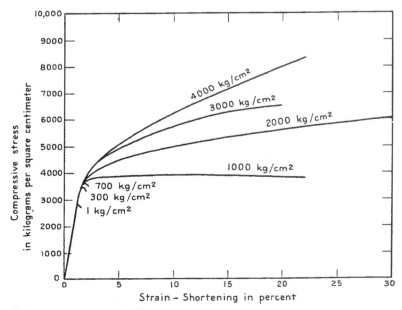

FIGURE 1–11 Effect of confining pressure on behavior of Solenhofen Limestone under compression (after E. Robertson, 1955)

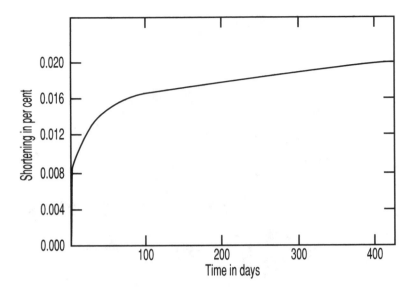

FIGURE 1–12 Creep curve for Solenhofen Limestone under a stress of 1400 kg/cm² (after D.T. Griggs, 1951)

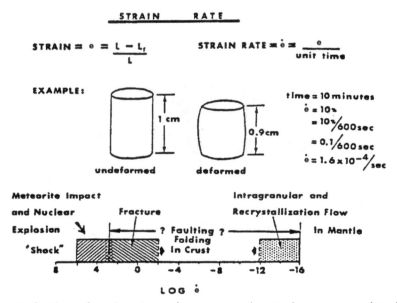

FIGURE 1–13 Definition of strain rate and representative strain rates associated with various events (after Friedman, 1985)

(1982–1994) notes illustrates the concept of strain rate and gives representative strain rates associated with various events including meteorite impact and nuclear explosion shock (10^6 – 10^3/sec), fracturing (10^3 – 10^{-2}/sec), folding in crust (10^{-2} – 10^{-12}/sec), and intergranular and recrystallization flow (10^{-12} – 10^{-16}/sec).

FRACTURES CLASSIFICATION

From an *experimental* point of view fractures can be classified as shear, extension and tensile fractures (Stearns and Friedman 1972, Nelson, 1985).

Shear Fractures. They are the result of stresses that tend to slide one part of the rock past the adjacent part, i.e., they involve movement parallel with the plane of fracture. Planes B and C in Figure 1–14 represent shear fractures.

POTENTIAL FRACTURES DEVELOPED IN LABORATORY EXPERIMENTS

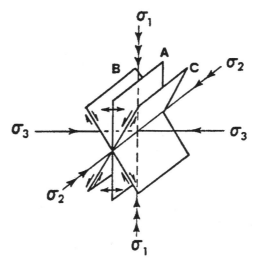

FIGURE 1–14 Potential fractures developed in laboratory experiments (after Nelson, 1985)

Extension Fractures. Those in which the two matrix walls move away from each other. Plane A in Figure 1–14 represents an extension fracture. It is parallel to the plane containing the greatest principal (σ_1) and intermediate principal (σ_2) stress axes and perpendicular to the least principal stress axis (σ_3).

Tension Fractures. They are similar to extension fractures because in both cases the walls displace perpendicularly to and away from the fracture plane. However, extension fractures are formed with an assumed positive compressive minimum principal stress component as opposed to tension fractures which are formed with an assumed negative tensile minimum principal stress component.

For any triaxial stress state, there might be two shear-fracture orientations and one extension-fracture orientation. The angle of the biggest principal stress (σ_1) bisects the acute angle between shear fractures. The extension fractures are normal to σ_3 and the line of interception of the fractures is parallel to σ_2.

From a *geologic* point of view natural fractures can be classified as follows (Stearns, 1972; Nelson, 1985):

1. Tectonic fractures (due to surface forces)
2. Regional fractures (probably due to surface forces)
3. Contractional fractures (due to body forces)
4. Surface related fractures (due to body forces)

Tectonic Fractures. "Those whose origin can, on the basis of orientation, distribution and morphology, be attributed to or be associated with, a local tectonic event. As such, they are developed by the application of surface or external forces" (Nelson, 1985) and can be related to diastrophism as in the case of folding and faulting.

Fractures associated with faults can be assigned the same stress scenario that was present when the faulting occurred. Consequently, the shear fractures can be considered miniatures of the fault, and its orientation, as well as the orientation of the extension fractures, can be determined based on the attitude of the fault. In some cases, logs might allow estimates of lateral distance to the fault.

Fractures associated with folds are genetically related to the folding process rather than to regional stresses that caused the folding. Stearns and Friedman (1972) have shown that common fracture geometries are formed by compression and extension of beds when σ_2 is either perpendicular or parallel to bedding. This indicates that the three principal stresses can have four different orientations as shown on Figure 1–15.

Set 1 shows Type 1 fractures where σ_1 is parallel to dip, and σ_1 and σ_3 are in the bedding plane. In set 2 (Type 2 fractures), σ_1 and σ_3 remain in the bedding plane but σ_1 is parallel to strike. Set 3 (Type 3 fractures) shows a σ_1 perpendicular to the bedding plane and σ_2 parallel to strike. Set 4 (Type 4 fractures) shows σ_2 parallel to strike and σ_3 perpendicular to the bedding plane.

The four orientations of the three principal stresses generate 12 potential fracture planes, including two shear and one extension for each orientation. However, Figure 1–15 shows that sets 2 and 3 have a common extension fracture plane. Furthermore, bedding is parallel to the extension fracture plane for set 4. The final product is that there are 10 potential orientations for fracture planes plus an eleventh which is the common conjugate slip plane to bedding.

Stearns (1982–1994) points out that these eleven potential orientations have led some field geologist to conclude that fractures are not consistent in folds. However, he emphasizes that after seeing the total fracture pattern on a fold, it is possible to make an interpretation of the partial patterns on lesser exposed structures.

The Teton anticline in Northwestern Montana, close to Great Falls provides a good example of the four orientation sets (Schematic in Figure 1–16). The flanks are almost perfect dip slopes. Lack of vegetation permits complete examination of large areas. Climbing of the anticline is possible without gaining or losing more than 10–15 ft of stratigraphic section. The fractures are very pervasive in scale and can grade down to small shear fractures.

Another good example of tectonic fractures in outcrops is found in the Phosphoric limestone, Ferris mountains of eastern Wyoming.

Although outcrops provide powerful information, they are usually suspect, because weathering and stress relaxation may have enhanced the fracture development. Friedman and McKiernan (1994) have used Formation Microscanner (FMS) and outcrop data of the Austin chalk to show that only one set of nearly vertical strongly oriented N.E. striking fractures occur in the subsurface whereas at least two sets are developed at the outcrop. This suggests that fracture patterns might be significantly simpler in the subsurface than at the outcrop. Their study further suggests that the outcrop spacing frequency distribution curve acts as a minimum-spacing boundary curve.

Tectonic fractures are the most important with respect to hydrocarbon production around the world. One example of a reservoir producing from Tectonic fractures is provided by the Palm Valley gas field of Central Australia (Milne and Barr, 1990).

Regional Fractures. They have been defined by Nelson, Stearns and Friedman (1982–1994) as "those which are developed over large areas of the earth's crust with relatively little change in orientation, show no evidence of offset across the fracture plane, and are always perpendicular to major bedding surfaces". These fractures seem to be unrelated to local structures, are probably due to surface forces, tend to follow an orthogonal pattern, and seem to be omnipresent.

The change in orientation can be as small as 10° or 15° over 100 miles. Because there is no offset across the fracture plane, they are very conducive to fluid flow. Good examples of regional fractures are found in the Uinta Basin, approximately one fourth of the Colorado Plateau and the entire Michigan basin.

Various reasons have been offered to try to explain the existence of regional fractures, including (1) regional up-lift (Price, 1966), (2) fatigue due to low level cyclic stress differentials, and (3) formation of the fractures soon after sedimentation, due to prolongation of fractures in the beds below.

According to Stearns (1982–1994), the first explanation is the best offered. However, our knowledge about small regional stress differential sources is so limited at this time, that the

SET 1

$\sigma_1\,\sigma_3$ in bedding plane

σ_1 parallel to dip

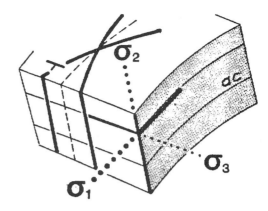

SET 2

$\sigma_1\,\sigma_3$ in bedding plane

σ_1 parallel to strike

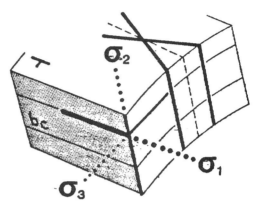

SET 3

σ_1 perpendicular to bedding plane

σ_2 parallel to strike

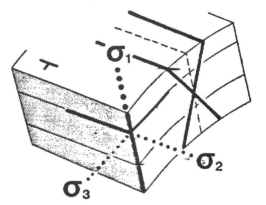

SET 4

σ_2 parallel to strike

σ_3 perpendicular to bedding plane

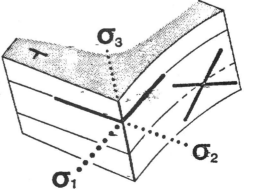

FIGURE 1–15 Fracture sets related to folding (after Stearns, 1985)

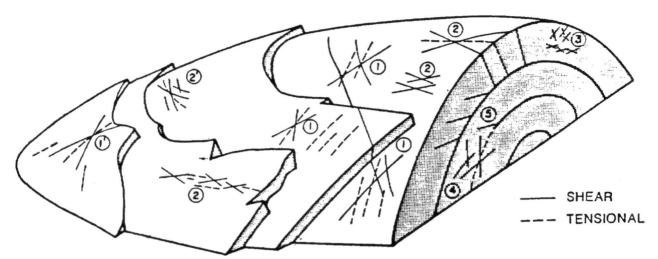

— SHEAR

--- TENSIONAL

FIGURE 1-16 Schematics of most common fractures associated with folds (after Stearns and Friedman, 1972)

subject is open to questions. The second potential reason is interesting but is based upon sources that are not clearly understood at this time, such as lunar tides. The third argument is probably weak as in some areas an unfractured sandstone is underlain and overlain by beds that contain fractures of the same orientation.

Some examples of reservoirs producing from regional fractures include the Austin chalk, the Big Sandy field of eastern Kentucky and West Virginia which produces from Devonian Shales (Bagwal and Ryan, 1976), the Spraberry field of West Texas which produces from fractured sandstones and Siltstones (Wilkinson, 1953), and the Altamont-Blue Bell field of Utah which produces from fractured sandstones (Baker, 1972). In some cases excellent production can be obtained from regional fractures superimposed over local tectonic fractures.

Contractional Fractures. Those which are the result of either (1) desiccation, (2) Syneresis, (3) thermal gradients, and (4) mineral phase changes. Each one of these can be either a tensile or extension fracture associated with a general bulk volume reduction throughout the rock (Nelson, 1985).

They are initiated by body rather than surface forces, i.e., they are started by forces internal to the body rather than external to it as in the case of tectonic fractures.

Desiccation Fractures (mud cracks) are tensile and are due to volume shrinkage due to loss of water in subaerial drying. They dip steeply with respect to bedding and tend to be wedge-shaped and filled with later material. They are usually developed in clay-rich environments and have the form of cuspate polygons of several nested sizes. Kahle and Floyd (1971) has discussed these kind of fractures found in sedimentary structures in Cayugan (Silurian) tidal flat carbonates in northwestern Ohio. The same type of fractures have been found by Netoff (1971) in sandstones near Boulder, Colorado. These fractures do not appear to be very important for hydrocarbon production.

Syneresis Fractures are the result of a chemical process which involves volume reduction due to subaqueous or subsurface dewatering brought about by a salinity change or a pore chemistry change. They can be either tension or extension fractures. Sometimes they have been called "chicken-wire" fractures due to their three-dimensional polygonal shape. They tend to be closely and regularly spaced and are commonly isotropic in three dimensions. Syneresis fractures can occur in shales, fine to course-grained sandstones, siltstones, limestones and dolomites. Picard (1966) has reported on oriented-linear shrinkage syneresis cracks in the Eocene Green River Formation, Raven Ridge Area, Uinta Basin, Utah. An example of production from syneresis fractures is provided by the Panoma field, Council Grove formation, Kansas (For log analysis see Figure 3–64).

Thermal Contractional Fractures are due to cooling of rock and their generation is dependent on the formation of a thermal gradient across the rock (Thirumalai, 1969).

They can be either extension or tension fractures depending upon the depth of burial. They are usually presented in columnar jointing in fine-grained igneous rocks. They are important from the point of view of hydrocarbons production in igneous rocks. For example, hydrocarbons and water were produced from a fracture basalt offshore Salt Lake City, Utah. Sometimes this thermal contractional fractures might be superimposed on local tectonic fractures due to faulting (Nelson, 1985).

Mineral-Phase Change Fractures are due to volume reductions resulting from mineral phase change. They can be either extension or tension fractures and their geometry can be very irregular. One example of this type of fracture is provided by dolomitization, where the phase change from calcite to dolomite produces a molar volume reduction of approximately 13%. The same approximate reduction occurs in the phase change from montmorillonite to illite.

Surface Related Fractures. They are developed as a result of the application of body forces. This group includes fractures developed during unloading, weathering in general, and creation of free surfaces or unsupported boundaries (Nelson, 1985).

Unloading fractures occur, for example, in quarrying operations. Weathering fractures are related to diverse processes of mechanical and chemical weathering such as freeze-thaw cycles, small scale collapse and subsidence, mineral alterations and diagenesis. Fractures due to the creation of free surfaces or unsupported boundaries tend to be parallel to canyon walls. They can be extension or tension fractures and are generated principally by gravitational forces. These fractures do not appear important from the point of view of petroleum production.

Stylolites. These fractures are initiated at planar stress concentration within a rock. These concentrations can occur at lithologic boundaries, clay seams and low porosity zones. Figure 1–17 shows an example of a stylolite associated with induced unloading fractures, tension gashes and extension natural fractures (Nelson, 1981).

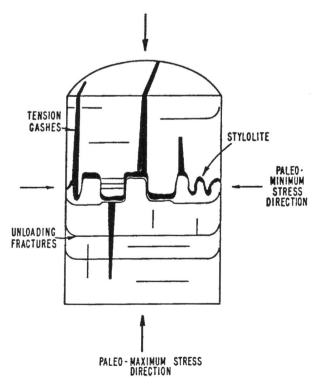

FIGURE 1–17 Schematic diagram showing geometric relation of stylolites, tension gashes, unloading fractures, and the paleo-state of stress that caused them (from Nelson, 1981)

Stylolites can become a problem in some reservoirs, as they reduce permeability perpendicular to the seam and porosity in the near vicinity.

MIGRATION AND ACCUMULATION

One reasonable explanation for petroleum migration and accumulation in naturally fractured reservoirs is provided by the theory of dilatancy. The principle of this theory is explained with the use of Figure 1–18, as in the case of earthquakes. Figure 1–18A shows a fault under tectonic stresses. In Figure 1–18B the stresses have built up sufficiently to fracture the rock.

Then, fluids start moving into the dilatant zone, due to the vacuum produced by the fractures. In Figure 1–18C the fluids have already filled the fractures. In Figure 1–18D a displacement and earthquake occurs. As certain seismic velocities decrease in stages B and C, it is possible to predict the occurrence of an earthquake within reasonable time limits.

McNaughton and Garb (1975) have analyzed the possibility of the same sequence of events for the migration and accumulation of petroleum in naturally fractured reservoirs. Upon the breaking of brittle rock by tectonic stresses, oil, water, or gas migrate toward the zone of dilatancy due to the vacuum produced by the fractures. The geological requirement for this hydrocarbon migration is a source rock contiguous to the brittle rock. According to this theory the fractures were formed after the generation of petroleum.

Figure 1–19 depicts the evolution of a basement reservoir according to the theory of dilatancy. In stage A, the basement metamorphic rocks are fractured, generating pressure gradients across the unconformity. The vacuum produced by the fractures causes oil and water to migrate downward into the fractures. In stage B, continued fracturing caused by pulsatory tectonic movements establishes cell-like sizable oil reservoirs below an impermeable capping. This capping may be formed by deposition of calcite in fractures of the metamorphic rock.

Thomas (1986) indicates that there is good evidence to support McNaughton's (1953) dilatancy theory along the Sooner trend in Central Oklahoma where oil "can and does migrate downward several hundred feet along vertical fractures." In this case oil production is obtained from the Meramec-Osage Mississippian lime only when it is conformably overlain by the Chester shale (source rock). When the Chester is truncated the Meramec Osage changes sharply from oil-bearing to water-bearing (Figure 1–20).

Dilatancy model fluid flow

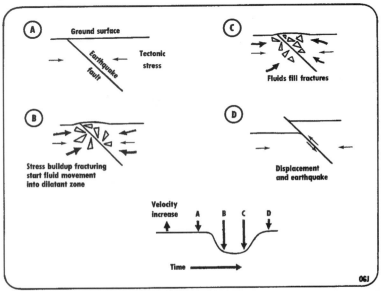

FIGURE 1–18 Fluid flow according to dilatancy model (after Kanamori, 1974)

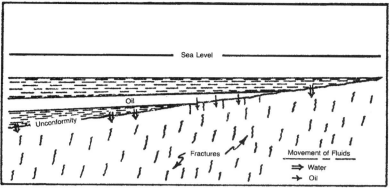

Stage A. Fracturing of metamorphic rocks establishes pressure gradients across the unconformity. Oil, gas, and water move into the dilated basement complex.

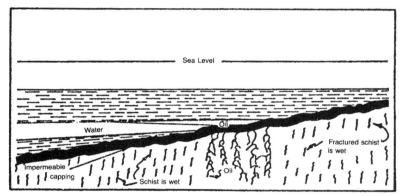

Stage B. Continued fracturing, followed by influent seepage and cementation, has established discrete, cell-like reservoirs below impermeable capping. Fractured schists bordering the reservoirs are wet.

FIGURE 1–19 Schematic sketches showing evolution of basement oil pool according to dilatancy hypotheses (after McNaughton, 1953)

Section across eastern edge*

FIGURE 1–20 Section across eastern edge of Sooner trend (Oklahoma) (after Thomas, 1986)

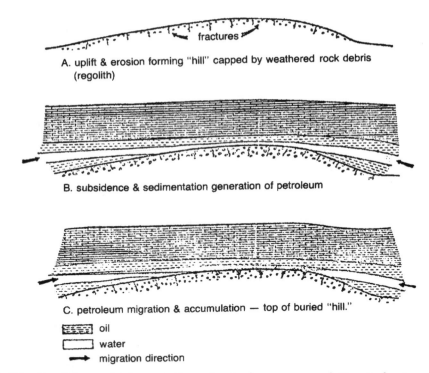

A. uplift & erosion forming "hill" capped by weathered rock debris (regolith)

B. subsidence & sedimentation generation of petroleum

C. petroleum migration & accumulation — top of buried "hill."

oil
water
migration direction

FIGURE 1–21 Possible geologic evolution of petroleum accumulation in fractures formed before time of petroleum migration (after McNaughton and Garb, 1975)

Another theory for explaining the migration and accumulation of petroleum in fractured rocks is depicted in Figure 1–21. According to this theory the fractures were formed before the migration of petroleum.

Figure 1–21A shows how deep erosion of the overburden permitted uplift of the rock, generating fractures in the uplifted "hill." In Figure 1–21B coarse rock debris accumulated in the lower slopes of the hill. The hill became deeply buried with continuous subsidence and sedimentation. In Figure 1–21C petroleum is generated and migrates up the crest of the hill, indicated by the arrow. This petroleum is trapped in both the sediments and the fractured basement.

Another theory indicates that in some reservoirs the oil may enter the reservoir by upward migration along fractures from some deeper bed. This appears to be the case of the Ain Zalah field in Iraq (Daniel, 1954).

SOURCES OF INFORMATION

Sound reservoir engineering studies should use as a base a combination of direct and indirect sources of information. Direct sources of information include cores, drill cuttings and downhole cameras. Indirect sources of information include all types of well logs (including mud logs), well testing data, inflatable packers and production history. These types of information can be mapped in many different ways and combined with reservoir engineering techniques that lead to estimates of hydrocarbons-in-place and recoveries under different depletion strategies.

Direct Sources of Information

Core analysis, oriented cores, drill cuttings, and downhole cameras provide direct sources for evaluating fracture reservoirs.

Core Analysis

It is not unusual to have very low core recoveries from intervals that are intensively fractured. In fact, it has been found in many naturally fractured reservoirs that there is a good correlation between poor core recovery and good production intervals. Because of this the natural fractures that we observe in cores are not the best developed. This leads to the conclusion that fracture porosities and fracture permeabilities from cores are conservative. The "big, most important" fractures that might provide important hydrocarbon storage are rarely seen in cores. In cases like these all we see is rubble.

The successful study of a naturally fractured core must start at the well site as indicated by Bergosh et al (1985), and Skopec (1994). The laboratory must be selected carefully, following meetings with laboratory personnel, and an inspection of the facilities where the experiments will be conducted (Sprunt, 1994).

Core retrieval must be handled very carefully due to the fragile nature of the fractured core. If the operating company suspects the presence of natural fractures, it should advise the service company well in advance to make sure that a specialized person in handling fracture cores is sent to the wellsite. Disruption of the fractured core might be minimized by using core barrels with inner liners such as PVC, and aluminum or wire mesh. In case of some disruption the core must be properly fitted together and marked with scribe lines to make sure that it is accurately laid out in the laboratory for fracture analysis (Bergosh et al, 1985). Consolidated rocks should never be frozen because this might produce irreversible structural damage to the core (Skopec, 1994).

The geologist should label and document the core thoroughly, including information such as if the fracture was generated during coring, removal of the barrel or by a geologist's rock hammer.

The core should be preserved as best as possible, preferentially by wrapping it in plastic and placing it into ziplock bags. This precaution will help to prevent loss of reservoir fluids and/or core dehydration.

Natural vs. Induced Fractures

Cores provide an important tool for direct examination of fractures. However, it is important to distinguish whether fractures are natural or artificially induced (Figure 1–22). Sangree (1969) suggests various criteria for differentiating natural from artificially induced fractures in cores.

The fracture is probably natural if:

1. Cementation is observed along the fracture surface. (Be careful that crystals on fracture surface are not halite deposited by evaporation of core fluids or other materials deposited from drilling fluids). In general, any fracture surface which appears to be a fresh break (i.e., unweathered and free of mineralization) should not be counted as a natural fracture unless there is some special supporting evidence.
2. Fracture is enclosed in core. One end (penetrating) or both ends (enclosed) of the fracture occur in the core.
3. Parallel sets of fractures are observed in a single core.
4. Slickensides (friction grooves) are observed on fracture. Unfortunately, drilling-induced slickensides are not uncommon, particularly in semiplastic shales or marls drilled at shallow depths. This criterion should be used with care.

The fracture is probably artificially induced if:

1. An uncemented vertical fracture angles in abruptly from the core edge in the down hole direction (Figure 1–22). Watch out for this type—it is most probably induced during drilling or pulling cores. Drilling-induced fractures commonly split the core into equal halves, often with a slight rotation about the core axis.
2. Fractures are conchoidal or very irregular. Natural fractures tend to be relatively plane. An exception occurs in highly porous, coarse-textured rocks where natural fracture surfaces may be quite irregular.

Fracture types compared

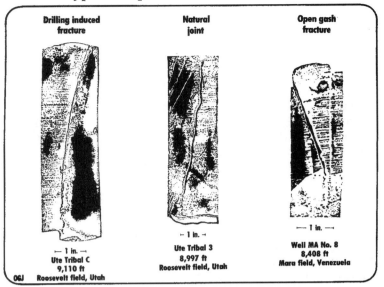

FIGURE 1–22 Comparison of drilling induced fracture, natural joint, and open-gash fracture (after Sangree, 1969)

Some other criteria for distinguishing a natural fracture include the presence of oil staining and/or asphaltic material in the fracture and stylolites which grade into fractures.

Whole Cores

Whole core analysis provides a valuable tool when the core does not break along the plane of fracture. A method developed by Kelton (1950) for analyzing fractured and vugular limestones reveals that fractures and vugs may provide an important part of the storage capacity in reservoirs with double-porosity systems. Table 1–1 shows a comparison of matrix and whole core data in the Ellenburger formation, Fullerton field, for four groups of samples.

Group 1 consisted of 14 samples with limited fracture development, while Group 2 provided 13 samples. Group 3 used 15 samples with more development of intergranular porosity. Group 4 consisted of four samples which had intergranular and/or highly vugular porosity. Notice from Table 1–1 the importance that secondary porosity can have on storage capacity. This is highlighted in the last line of Table 1–1 which shows the parti-

TABLE 1–1 Matrix vs. Whole Core Data, Ellenburger, Fullerton Field (after Kelton, 1950)

Group	1	2	3	4
Gas, % bulk	0.08	0.91	0.63	3.72
Oil, % bulk	0.06	0.06	0.88	1.84
Water, % bulk	2.07	1.65	1.66	3.27
% pore	94	63	52	39
Matrix porosity, % bulk	1.98	1.58	2.56	7.92
% total pore	90	60	81	94
Total porosity, % bulk	2.21	2.62	3.17	8.40
Fracture porosity, % bulk	0.23	1.04	0.61	0.48
k_{max}, md	10	409	23	94
$k_{90°}$, md	0.6	1.2	10	38
Matrix k, md	0.3	0.2	0.3	3.7
Partitioning coefficient, v	0.1	0.39	0.19	0.05

Locke-Bliss method

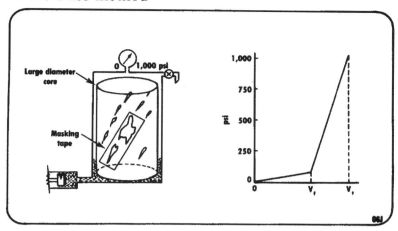

FIGURE 1–23 Lock-Bliss method of porosity partitioning (after Pirson, 1962)

tioning coefficient, i.e., the ratio of secondary porosity divided by total porosity. Also remarkable are the differences between matrix and whole core (fractures plus vugs plus matrix) permeabilities.

Locke and Bliss (1950) presented a technique which allows the direct determination of matrix and fracture porosity. This technique has been used successfully by Pirson (1962) to evaluate two-porosity systems. The method consists essentially of masking the fractures (and/or vugs) with adhesive tape before submerging the core into a water pressuring chamber. Water is injected at controlled volumes and the pressures are recorded. The result is shown in Figure 1–23.

Between zero and V_f the water invades the coarse porosity; consequently, no drastic pressure increase is noticeable. When the coarse porosity is water-saturated, the water starts penetrating the matrix porosity and the pressure increase is more pronounced. The total pore volume of the core is the total water injected V_t.

The breaking point in Figure 1–23 represents the partitioning coefficient, which has been defined by Pirson as:

$$v = \frac{V_f}{V_t} = \frac{V_f}{V_f + V_b \phi_b} \qquad (1\text{–}40)$$

where: V_f = fracture volume
V_b = matrix bulk volume
ϕ_b = matrix porosity attached to bulk volume of only the matrix (excluding the bulk volume of the fractures), fraction

Full diameter whole cores have an average size of 4 in. in diameter and 6 in. in length, as opposed to core plugs which typically have one inch in diameter and two inches in length. This indicates that bulk volumes of whole cores and plugs are 75.4 and 1.57 cubic inches, respectively, or a volume of the whole core which is 48 times as large as the plug volume. This larger volume gives better probabilities of analyzing samples which contain natural fractures, allows a more accurate quantification of the fracture system, and permits determination of permeability anisotropies by carrying out multi-directional permeability measurements. As discussed previously, however, I believe that fracture porosities and fracture permeabilities extracted from core data are, in general, too conservative.

Water and oil saturations in whole cores can be determined with the use of the solvent extraction process and the CO_2-toluene cleaner. The method consist of removing the unbound water to determine water saturation. A limitation is that it might take days or weeks to completely remove the unbound water. The oil saturation is determined by placing the whole core in the CO_2-toluene cleaner unit. Initially, CO_2 gas and toluene solvent are injected into the core at high temperatures and pressures. Then there is a slow depressurization. The

CO_2 reduces the oil viscosity and the toluene cleans the sample. Oil saturation is determined after various cleaning cycles and drying of the core by knowing the bulk and grain volume of the sample, total weight loss, volume of extracted water, and oil density.

One advantage of the previous procedure is that the same whole core sample can be used for determination of oil and water saturation, porosity, and permeability. On the other hand, the conventional plug used for determination of oil and water saturation is not the same one used for measuring porosity and permeability. Thus adjacent plugs must be used for these analyses.

Bergosh et al (1985, 1987) have reported on the use of some new fracture analysis techniques which have evolved from standard methods published in the API Publication #40 (1960), dealing with recommended practices for core-analysis procedures.

The technique involves the use of a Computerized Tomographic (CT) Scanner, which consists of a rotating x-ray source and detector. This device circles the core which lays horizontally in a gantry table. The table permits incremental advance of the core to ensure constant and thorough examination of each core interval. The x-ray data is converted into an image on a CRT screen. A valuable feature of this device is its ability to view the core without removing it from the liner or the core preservation material. It helps to characterize fracture widths greater than 0.5 mm for open and partially open fractures, tortuosity, distance between fractures, inter-connectedness and invasion by drilling mud while avoiding disruption and destruction of the fractured core.

Another technique reported by Bergosh et al (1985) consists of multi-directional permeability measurements (including three-dimensional permeabilities), which can be carried out routinely at 300 psi net confining pressure and room temperature.

Whenever possible, laboratory experiments should be carried out at simulated conditions of net overburden pressure and temperature. This will provide more realistic measurements of porosity and permeability and the ability to determine fracture compressibility. Closing of the fractures as the net confining pressure increases can be analyzed and the importance of partial mineralization as a propping agent can be assessed.

In certain reservoirs where the matrix permeability is very low (< 1 md) dynamic displacement techniques for calculating permeability might affect the sample due to various reasons including the presence of sensitive clay minerals and/or siliceous fine within the pore structure. In these cases, a pulse decay technique (Amaefule et al, 1986) might provide more reliable permeabilities. The pulse decay is dependant on sample permeability, dimensions of the sample and reservoir, and fluid characteristics. The pulse decay permits thus rapid determination of k_o and k_w with minimal pore volumes throughout at low velocities. Amaefule et al. (1986) found that in spite of the low quality of some rocks, they are still stress sensitive. It is necessary then that these permeabilities be determined at in-situ stress conditions.

Pulse Decay Profile Permeater

The pressure-decay profile permeater (PDPP) provides a nondestructive permeability measurement with a useful range from 0.001 md to 20,000 md (Jones, 1992; Georgi and Jones, 1992). A probe is sealed against the flat surface of a slabbed core sample. Gas is released from a tank of known volume and pressure. By recording the pressure decay as a function of time, the permeability of the core at the position of the probe can be calculated. The time required for low permeability measurement is about 20 seconds and the uncertainty of the measurements at room conditions of pressure and temperature is less than 5%. These permeability measurements are at zero confining stress. They should be corrected to reservoir conditions.

Care must be taken to avoid placing the probe over fractures when measurements are made for the matrix. Fracture permeabilities can be measured by placing the probe directly over the fractures. The fracture permeabilities measured with the PDPP instrument are not necessarily accurate since the flow pattern is no longer hemispherical in shape, an assumption used in the PDPP permeability calculation. The fracture permeability must therefore be substantiated with values determined from pressure transient analysis.

Data from the PDPP can be used to micro-simulate the naturally fractured core (Au and Aguilera, 1994). The micro-simulation is an additional tool that helps to answer the question—Is the matrix going to contribute to hydrocarbon production?

Oriented Cores

One of the methods available for obtaining oriented cores is the Christensen-Hugel approach, in which three grooves are cut continuously into the core as it enters the barrel. There is a reference grove with a known relation to the oriented lug which appears on the survey instrument's photograph of the compass (Rowley et al, 1971). A multishot instrument allows to take three photographs per minute, one per minute or one every two minutes depending on the clock used. This multishot device is part of the orienting system and can be located with a non-magnetic core barrel tube. The following are some of the outer barrel outside diameters, typical bit sizes and core sizes:

Outer Barrel O.D. (In.)	Typical Bit Size (In.)	Core Size (In.)
NXC3	2–31/32	1–7/8
4–1/8	5–1/2	2–1/8
4–3/4	6–1/8	2–5/8
5–3/4	6–5/8	3–1/2
6–1/4	7–1/2	4
6–3/4	8–9/16	4
7–5/8	8–7/8	5–1/4

Figure 1–24 shows a goniometer which helps to calculate dip direction, dip angle, and strike. With this information, core plugs taken from the parent core can be oriented.

A new method developed for determining core orientation measurements is the computerized goniometer. In this approach the sample is placed on the device with the orienting scribe mark to the north. Then three points are digitized on the planar feature to be quantified. Based on some additional geological data, the computer transforms this planar attitude to an in-situ orientation, from which dip and strike can be calculated (Bergosh et al, 1985). Figure 1–25 shows a very useful plot which compares directional permeability and fracture orientation as determined from a Monterey core using methods discussed above.

The first row shows directional permeability of each 30° segment. Other rows show fracture orientation for open, closed, partially open and healed fractures. The number of fractures along the true fracture orientation is represented by the letter n. A summary of results is presented at the bottom of Figure 1–25.

A three-dimensional visual core diagram of petrophysical properties and their relation to fracture type is presented on Figure 1–26. Note the open, closed and healed fractures, strike, permeabilities and values of matrix and fracture porosities.

Core data should be assembled into a core fracture log for better application. Core Laboratories, Inc. fracture log, for example, includes a track with oil saturation; core porosity; stylolite height and frequency per foot; lithology; fracture length, frequency and condition (open, partially open, closed, broken, mineralized); fracture density per foot; rock hardness; fracture dip angle; fluorescence; and rubble zones. Similar fracture logs are provided by other service companies.

Although oriented cores are very powerful it is important to keep in mind the error associated with the technique. Nelson (1987) has shown that the error in the coring and surveying procedure is in the order of 5°, the error provided by the mechanical goniometer is ± 2.75° (the computerized goniometer reduces this error), the reproducibility error in using the goniometer is about 2° for one analyst and 4° for multiple analysts. Overall oriented core errors are in the order of 11°.

Core-Scale Pressure Transients

This pulse-decay test technique was developed by Kamath et al (1992), who setup a laboratory that captures several hundred pressure readings per second as a response to an

FIGURE 1–24 Goniometer (courtesy of Christensen)

initial disturbance in a core. This permits to determine permeabilities of homogeneous cores, matrix and fractured properties of naturally fractures cores, and the individual segment properties of a butted core sample.

Drill Cuttings

These can detect natural fractures in only a few instances. However, natural fractures may not be preserved in cuttings due to breakage along fractures. Consequently, the reservoir might be naturally fractured even if the cuttings do not show any fractures.

Downhole Cameras

Cameras can obtain direct information regarding bed boundaries, faulting, fractures, hole size, and hole shape (Fons, 1960). They also provide valuable information during fishing operations.

A compass may be added below the camera to take photographs which help determine borehole deviation from the vertical axis and orientation of the fractures intersecting the well bore. In this sense, downhole cameras and oriented cores provide similar information.

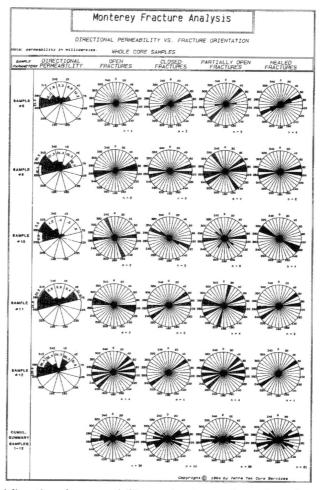

FIGURE 1–25 Multidirectional permeability vs. fracture orientation plot.

The method applies equally to dry or gas-filed wells. One of the cameras uses 16 mm rolled film and can take 1000 pictures in a single trip. After each photo is taken, the film automatically rolls to the next position.

Operations are restricted to temperatures below 200°F and pressures below 4000 psi. The entire tool is 4 1/2 in. in diameter and 4 ft long. The usual problems of photography, like clean lenses and focus depth, are associated with this type of technique. Figure 1–27 is a photograph of a fracture obtained with a downhole camera.

Borehole video cameras have been shown to be successful for the analysis of naturally fractured reservoirs in horizontal wells (Overbey et al, 1989). In this case the camera is pushed through the wellbore while attached to the drill string. Recording of the wellbore can be made going in the hole and coming out. Low pressure reservoirs, preferentially air-drilled, appear as prime candidates for video technology.

Indirect Sources of Information

Indirect sources of information for evaluating naturally fractured reservoirs include drilling history, log analysis, well testing, inflatable packers, and production history.

Drilling history provides valuable information regarding mud losses and penetration rates. Mud losses might be associated with natural fractures, vugs, underground caverns, or induced fractures. Penetration rates can be increased considerably while drilling all types of secondary porosity (see Chapter 2 and mud logging in Chapter 3).

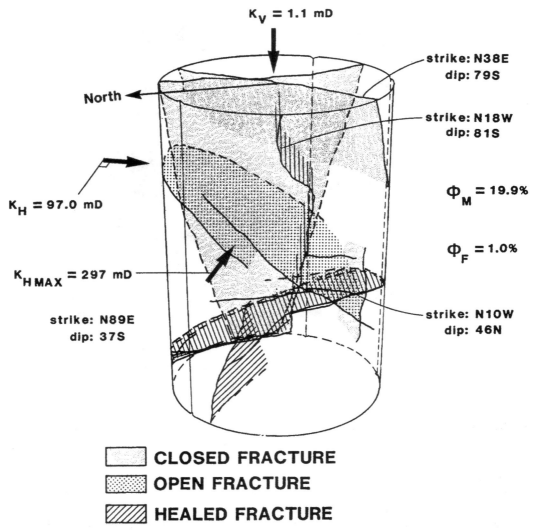

SAMPLE # 76
DEPTH : 11959'

$K_V = 1.1$ mD

North

strike: N38E
dip: 79S

strike: N18W
dip: 81S

$\Phi_M = 19.9\%$

$\Phi_F = 1.0\%$

$K_H = 97.0$ mD

$K_{H MAX} = 297$ mD

strike: N89E
dip: 37S

strike: N10W
dip: 46N

CLOSED FRACTURE
OPEN FRACTURE
HEALED FRACTURE

FIGURE 1–26 Three-dimensional core diagram of petrophysical properties and their relation to fracture type (after Bergosh et al, 1985)

Log analysis is a powerful tool for detecting and evaluating naturally fractured reservoirs. There are logs which in some cases, are run specifically to locate fractures. For example, sonic amplitude logs, variable intensity logs, the borehole televiewer, dipmeter logs, and formation microscanners (FMS and FMI) have been used successfully in detecting fractures. Conventional logs can be used in some instances for quantitative analysis of naturally fractured reservoirs (see Chapter 3).

Well testing is also a powerful tool for evaluating fractured reservoirs. Much progress has been made on the subject of pressure analysis of fracture media. Most pressure time curves of odd shapes can now be analyzed by straightforward analytical procedures and/or numerical techniques. In some instances, it is possible to evaluate parameters such as fracture, matrix, and combined permeability and/or porosity, and distance between fractures from well testing.

Inflatable packers obtain impressions of the borehole on the pliable material of the packer. Fraser and Pettit (1961) reported the results of a field test using inflatable packers to determine the type and orientation of the fractures. Care must be exercised when work-

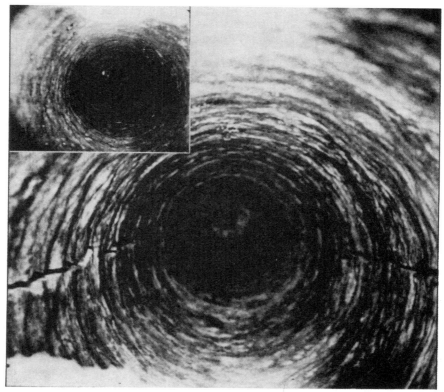

FIGURE 1–27 Photograph of fracture obtained with a borehole camera (after Fons, 1960)

ing with these devices, since the packer can be accidentally over-inflated and blown out in enlarged boreholes.

Figure 1–28 shows the photo of a borehole impression obtained with an inflatable packer. The impression of the fractures is recorded by a special rubber element when the packer is inflated to a certain pressure and maintained under these conditions for a certain length of time.

An example reported by Fraser and Pettit (1961) from a well in Howard Glasscock field, Howard County, Texas included an inflation pressure of 1300 psi and letting the packer

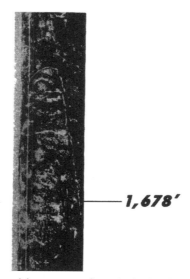

1,678'

FIGURE 1–28 Impression: top of fracture at borehole (Well 2) (after Anderson and Stahl, 1967)

seated for approximately 13 hours. Prior to unseating, the orientation was taken with a standard directional survey. Then the packers were pulled from the hole. A vertical fracture was impressed with an orientation of N 73°E.

Production history provides qualitative information regarding the presence of natural fractures. If, for example, the matrix permeability of a sandstone reservoir is less than 0.1 md, its porosity is 5%, and well rates exceed 1000 bo/d, then it is clear that the well performance is a function of fracture permeability. In some cases, early water or gas breakthroughs in secondary recovery projects indicate the presence of fractures and their directions. This late knowledge about the presence of natural fractures must be avoided in practice by using all possible means to determine fractured trends before a secondary recovery project is started.

MAPPING FRACTURED TRENDS

Aerial Photography. It has been used successfully in some areas to map fracture trends. Alpay (1969) has reported a field application of aerial photography in eight reservoirs of West Texas. He found that, in general, a good match was obtained between predominant fractured trends determined from aerial photography and subsurface trends determined from reservoir performance.

The rationale supporting aerial photography indicates that surface fracture trends conserve essentially the same direction with depth in tectonically undisturbed areas. Consequently, surface-fracture trends can provide valuable information with respect to potential fracture direction in underground hydrocarbon reservoirs. One advantage of aerial photography is that large areas (and consequently large fractured trends) can be evaluated quickly and rather inexpensively.

Figure 1–29 shows surface fracture traces determined from aerial photography in West Texas. Also shown is the area where engineering experience is available, and a rosette which summarizes the surface fracture traces on a field-wide basis. The interior of the rosette indicates three prominent directions of the surface fractures: (1) a N25–35E with a mode at N30E, (2) a N60–75E with a mode at N65E, and (3) a N55–75E with twin modes at N60W and N70W.

The exterior part of the rosette indicates the preferential direction of water channelling in the area of engineering experience. The agreement between direction determined from the photo study and direction determined from water channelling (Figures 1–30 and 1–31) is good.

Mapping fractured trends from aerial photography must be done by experts, since fences, power lines, roads, pipelines, etc., may be mistaken for natural fractured trends. To avoid these possible mistakes, it is better to find areas which have no unnatural features.

If, in addition to the delineation of fractured trends, it is necessary to know fracture morphology, small-scale fracture densities, and determination of total fracture geometries, Stearns and Friedman (1972) recommend that extensive ground control be used in conjunction with the aerial photography interpretation.

Fracture Intensity Indices. Indices derived from conventional well logs can map fracture trends in some cases. Pirson (1969) indicates that fracture porosity is possibly the factor which better measures quantitatively the intensity of deformational shattering of brittle rocks. According to this rationale, the value of a fracture intensity index (FII) increases as a fault is approached. With the value of FII, it is possible to estimate lateral distance to a fault.

The way in which this fracture intensity index or fracture porosity can be quantified is illustrated in Chapter 3. Once these fracture porosities are obtained it is possible to prepare maps of ISO-fracture intensity index or ISO-fracture porosity.

An application of this method has been presented by Trunz (1966) in a thesis at the University of Texas. He calculated fractured indexes from fifty one Austin Chalk Wells in the Luling area Branyon reservoir of Caldwell County, Texas, using Equation 3–29. From these calculations, the contour map shown on Figure 1–32 was prepared. Note that the fracture index is smaller at the periphery and increases steadily as the faults (heavy lines) are

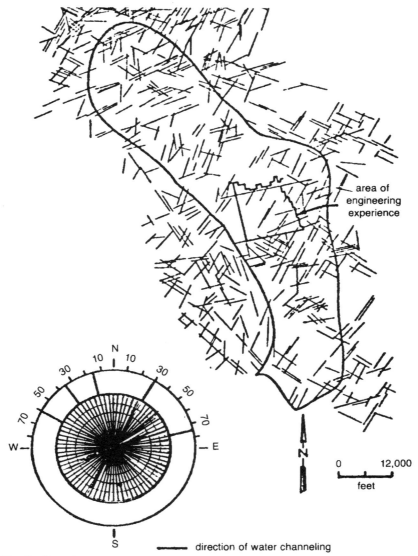

FIGURE 1–29 Surface fracture traces in Field 4 (after Alpay, 1969)

approached. Trunz found that the maximum degree of fracturing occurred at 500 to 750 ft from the fault line. The same type of mapping can be carried out with more modern fracture identification logs.

Remote Sensing Imagery.   This appears to be a promising tool for delineating fractured trends. Rabshevsky (1976) indicates that the most important step in using remote sensing imagery is interpreting and analyzing data. The interpreter must become familiar with image processing, especially when using LANDSAT imagery. In the image processing and interpretation stages, he can use his ingenuity. Other steps, such as mission planning and image acquisition, are beyond the interpreter's control.

Onyedim and Norman (1986) have described lineaments at least 50 km in length seen on LANDSAT images and have used a finite element computer model to demonstrate that stress and strain effects on a deep vertical fault lead to the development of two different types of faults, which are sometimes observed as lineaments at the Earth's surface.

Geochemical and Structural Techniques.   They have proven useful to decipher the origin, history, and extent of fracturing and migration events in the northern Appalachian basin (Tillman, 1983). Key in this method are chemical and textural data obtained from

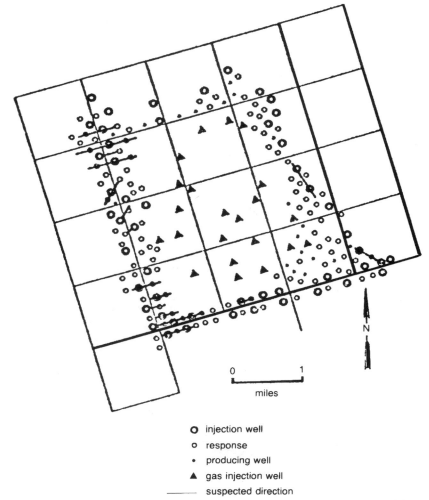

O injection well
o response
• producing well
▲ gas injection well
——— suspected direction

FIGURE 1–30 Waterflood experience to date in Field 4. San Andres Reservoir (after Alpay, 1969)

vein and matrix mineral assemblages. The geochemical records of these minerals can give details of all episodes of fluid migration, can help to differentiate fracture systems, and can help to develop realistic geologic models where areas with a high degree of permeability enhanced fractures are pinpointed.

Radius of Curvature. This concept has proved to be very useful in some naturally fractured reservoirs to identify the areas with the largest degree of fracturing. In a classical example Murray (1968) was able to show that there was a good correlation between the best wells in the Sanish pool, McKenzie County, North Dakota, and the minimum radius of curvature. The radius of curvature development presented below follows very closely the original work of Murray.
Basic assumptions:

1. The schematic of fracture geometry is presented in Figure 1–33. T is the thickness of the brittle bed, R is the radius of curvature, $\Delta\theta$ is the angular increment in radians between adjacent fractures and ΔS is the corresponding increment in the surface of the fracture bed or arc length.
2. The structural configuration is elongated and infinite in the axial directions.
3. The base of the bed is a neutral surface, i.e., there are no changes in length and no strain at the base.

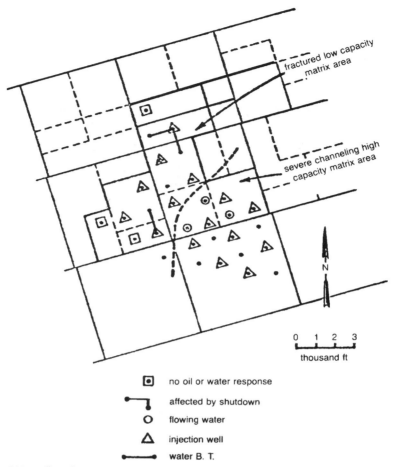

- ▣ no oil or water response
- └┐ affected by shutdown
- ○ flowing water
- △ injection well
- •——• water B. T.

FIGURE 1-31 Waterflood experience in Field 4, Upper Clearfork Reservoir (after Alpay, 1969)

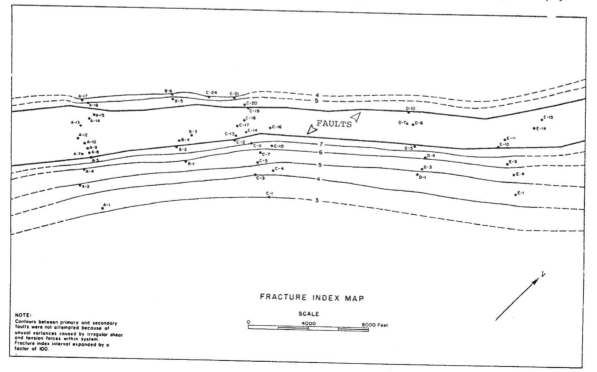

FRACTURE INDEX MAP

SCALE

0   4000   8000 Feet

NOTE:
Contours between primary and secondary
faults were not attempted because of
unusual variances caused by irregular shear
and tension forces within system.
Fracture index interval expanded by a
factor of 100.

FIGURE 1-32 Fracture index map, Austin Chalk, Luling Area, Branyon reservoir, Texas (after Trunz, 1966)

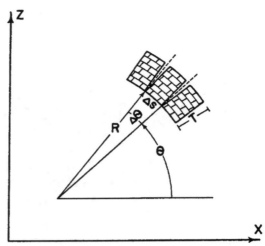

FIGURE 1–33 Geometry of fracture system used in deriving expressions for fracture porosity and fracture permeability (after Murray, 1968)

4. Folding is sharp enough to allow generation of stresses greater than the ultimate tensile strength of the brittle bed. This results in tension fractures.
5. The Y axis, perpendicular to the page of the book, coincides with the direction of the structure. The X axis is perpendicular to the structure.

Fracture porosity attached to bulk properties of the composite system is given by:

$$\phi_2 = \frac{Fracture\ Void\ Space}{Total\ Volume} = \frac{Fracture\ Void\ Area}{Total\ Area} \tag{1–41}$$

Area of a total sector including fractures, A_s, for a circle of radius R + T is given by:

$$A_s = \frac{1}{2}(R + T)^2 \Delta\theta \tag{1–42}$$

Area of a sector without fractures, A_b, below the neutral surface is calculated from:

$$A_b = \frac{1}{2}R^2 \Delta\theta \tag{1–43}$$

Taking the difference between A_s and A_b leads to the area of the brittle bed including fractures, A_t:

$$A_t = A_s - A_b = \frac{1}{2}(2RT + T^2)\Delta\theta \tag{1–44}$$

Area of the brittle bed without fractures, A_m, is given by:

$$A_m = T\Delta S \tag{1–45}$$

Area of only fractures, A_f, is given by:

$$A_t - A_m, \text{ or}$$

$$A_f = \frac{1}{2}(2RT + T^2)\Delta\theta - T\Delta S \tag{1–46}$$

Fracture porosity, ϕ_2, from Equation 1–41 is given by fracture area, A_f, divided by total area, A_t, or

$$\phi_2 = \frac{\frac{1}{2}(2RT + T^2)\Delta\theta - T\Delta S}{\frac{1}{2}(2RT + T^2)\Delta\theta} \tag{1–47}$$

Equation 1–47 can be rewritten as:

$$\phi_2 = \frac{\dfrac{1}{2}(2RT\Delta\theta + T^2\Delta\theta) - T\Delta S}{\dfrac{1}{2}(2RT\Delta\theta + T^2\Delta\theta)} \tag{1–48}$$

Since the arc length, ΔS, is equal to $R\Delta\theta$:

$$\phi_2 = \frac{(2T\Delta S + T^2\Delta\theta) - 2T\Delta S}{(2RT\Delta\theta + T^2\Delta\theta)} \tag{1–49}$$

Which can be simplified to:

$$\phi_2 = \frac{T}{2R + T} \tag{1–50}$$

If the assumption is made that T is very small compared with R Equation 1–50 reduces to:

$$\phi_2 = \frac{T}{2R} \tag{1–51A}$$

In most structural situations, dz/dx is much smaller than unity. Consequently, R is the inverse of the second derivative of z with respect to x, or:

$$R = 1/(d^2z/dx^2) \tag{1–51B}$$

Equation 1–51A can then be written as:

$$\phi_2 = \frac{T}{2}\frac{d^2z}{dx^2} \tag{1–52}$$

The tensile stress, F, in the upper surface of the bed is:

$$F = ET\frac{d^2z}{dx^2} \tag{1–53}$$

where E is Young's modulus for the bed. If F exceeds the ultimate tensile strength of the brittle bed, fractures will develop.

A minimum value of $T(d^2z/dx^2)$ must be exceeded before fracturing is developed. The critical value for limestones seems to approach 1.2×10^{-4}/ft for limestones according to Murray (1968). For the Sanish pool the critical value is approximately 2×10^{-5}/ft. Murray found that for the Sanish pool the curvature was more than 4×10^{-5}/ft in the best wells, $2-4 \times 10^{-5}$/ft in average wells and less than 2×10^{-5}/ft in poor and dry holes.

Figure 1–34 shows a structural contour map of the Sanish pool on top of the Mississippian Bakken formation. Wells are indicated by symbols representing good, mediocre, poor and very poor wells. Notice that the best wells are located where the curvature is more than 4×10^{-5}/ft.

A cross section through A–B is shown on the upper part of Figure 1–35. The middle profile on Figure 1–35 represents the dip magnitude, or dz/dx, using a scale of +5 to -15×10^{-2}/ft/ft. The west dip was chosen arbitrarily as positive and the east dip as negative. The profile at the bottom of Figure 1–35 represents the second derivative, or curvature d^2z/dx^2, in a scale of 10 to -5×10^{-5}/ft. The dashed areas represent the portions of largest fracturing. This is followed by the dotted areas. Empirically, this has been found to be correct as the best wells are located in dashed areas. Mediocre wells are located in the dotted areas and subcommercial wells in areas where the curvature is smaller than 2×10^{-5}/ft.

The schematic of Figure 1–33 has been extended to include subdivisions of the brittle bed as shown on Figure 1–36. The same nomenclature of Figure 1–33 applies in this case but two new parameters are included, h is the distance from the top of the brittle bed to the top of the section for which we want an estimate of porosity, h' is the thickness of the interval for which we want an estimate of fracture porosity. The equation handling this situation can be written as follows (Aguilera, 1983):

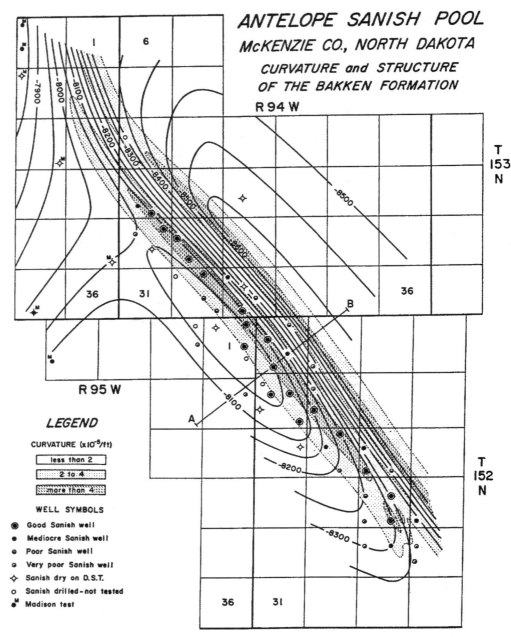

FIGURE 1–34 Structural contour map of Antelope Sanish pool, McKenzie County, North Dakota. Structural datum, top of Mississippian Bakken Formation. As noted on legend, well spots are keyed to their productivity and values of structural curvature are mapped by patterned areas. Contour interval, 50 ft. Contour datum, sea level. Section A–B is location of Figure 1–35 (after Murray, 1968)

$$\phi_2 = \frac{2(T-h) - h'}{2R + 2(T-h) - h'} \tag{1-54}$$

By assuming that T, h and h' are small compared with R, and 1/R is equal to d^2z/dx^2, the above equation can be written as:

$$\phi_2 = \frac{2(T-h) - h'}{2} \frac{d^2z}{dx^2} \tag{1-55}$$

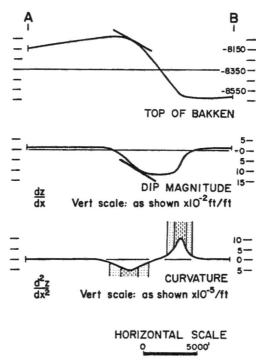

FIGURE 1–35 Comparison profiles of Bakken structure, dip magnitude, and structural curvature along line of section A–B (after Murray, 1968)

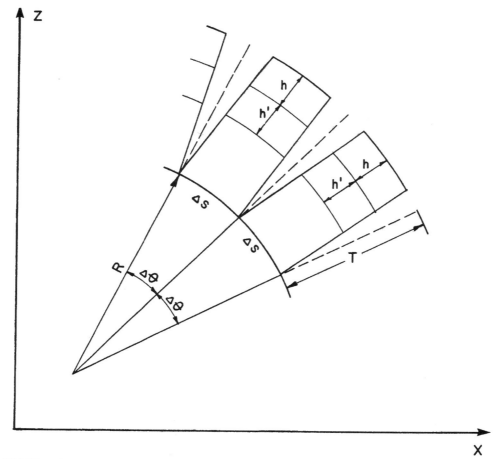

FIGURE 1–36 Schematic of fracture geometry used in deriving equations for calculating fracture porosity at any depth within a competent reservoir (after Aguilera, 1982)

As an example, if we assume a thickness of 40 ft for a portion of the Sanish pool and a curvature of 5×10^{-5}/ft, we calculate an average fracture porosity, $\phi_2 = 0.001$ from equation 1–52. If we want only the fracture porosity of the upper 1 ft, we calculate 0.0198 or 1.98% from Equation 1–55. This example shows that fracture porosity is strongly scale-dependent.

Fracture porosities as calculated from the above equations must be considered only as orders of magnitude. They can become very powerful, however, when we prepare maps of ISO-ϕ_2 for locating new wells in areas with the best fracture porosities.

A computer program for calculating the second vertical derivative and an application to the Pearsal anticline located 60 miles southwest of San Antonio, Texas has been presented by Munoz-Espinosa (1968) in a University of Texas Thesis.

FRACTURED RESERVOIRS

Naturally fractured reservoirs are found in essentially all types of lithologies including sandstones, carbonates, shales, cherts, siltstones, basement rocks and coals. The percentage of the total porosity made from fractures ranges from very small to 100%. This section describes some geologic aspects of fractured reservoirs found in these Lithologies. Production aspects of various naturally fractured reservoirs are found in Chapter 5 dealing with case histories.

Fractured Sandstones

The Spraberry field of West Texas is an example of an underpressured naturally fractured sandstone reservoir. The main producing structure in the Spraberry is a fracture permeability trap on a homoclinal fold which is made of alternate layers of sands, siltstones, shales and limestones. These layers were deposited in the deep Midland basin. The fractures are regional.

Figure 1–37 shows a generalized stratigraphic section of the area. The oil is stored primarily in the sandstone matrix, and paper-thin fractures provide the channels to conduct the oil to the well bore (Type A). The production history of the Spraberry field is presented in Chapter 5 (Case Histories).

Another naturally fractured sandstone reservoir is found in the overpressured Altamont trend, Uinta basin, Utah. The natural fractures are also regional. Production from this reservoir comes from fine-grained sandstones of Tertiary age and fractured rocks consisting of sandstone, carbonate, and highly calcareous shale. Initial production rates of over 1000 bo/d are not uncommon in this Type A reservoir of low porosity ($\phi \cong 3 - 7\%$) and low permeability ($k < 0.01$ md). Figure 1–38 shows a stratigraphic section of the area. Practical experience and detailed economic evaluations indicated that 640 acres was the initial optimum spacing for the Altamont trend. Closer spacing proved uneconomical (See also Chapter 5).

Finn (1949) studied the presence of fractures in the Oriskany sandstone in Pennsylvania and New York. These fractures have been responsible for larger flow rates than expected in the area.

The Palm Valley gas field in Australia produces from Type A Ordovician naturally fractured sandstones and siltstones of very low matrix porosities ($\cong 4$–5%) and very low matrix permeabilities ($\cong 0.01$ md). Natural fractures of tectonic origin provide the necessary permeability that led to a gas rate of over 70 MMSCFD from well PV2 when it had penetrated only about 1 foot of the Pacoota P1 formation. Natural fractures also led to a gas production of about 136 MMSCFD from deviated well PV6B when the vertical well PV6 (same surface borehole) which was air-drilled had not shown any commercial production from the same formation (Milne and Barr, 1990; Aguilera, Au, and Franks, 1993). The production history of Palm Valley Gas field is presented in Chapter 5 (Case Histories).

Fractured Carbonates

Carbonates are limestones, dolomites, and the intermediate rocks between the two.

Daniel (1954) provides an excellent description of three fractured reservoirs of the Middle east: Ain Zalah and Kirkuk in Iraq, and Dukham in Qatar.

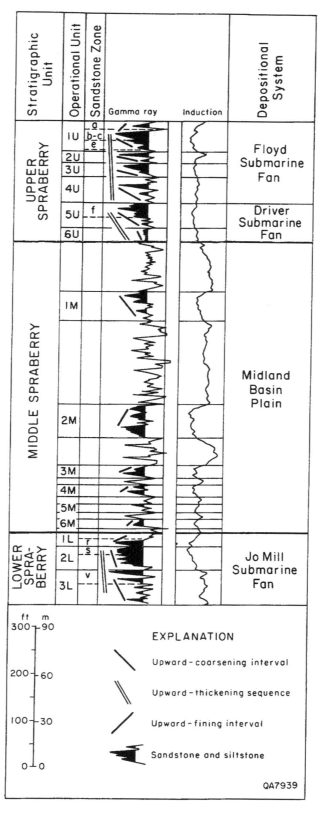

FIGURE 1–37 Stratigraphic subdivisions of the Spraberry Formation, Driver unit (after Guevara, 1988)

Altamont trend

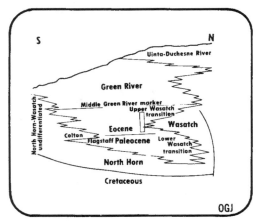

FIGURE 1–38 Abnormally pressured Altamont trend productive section lies in the transition between Green River source rocks and Wasatch red bed facies (after Baker & Lucas, 1972)

Ain Zalah field

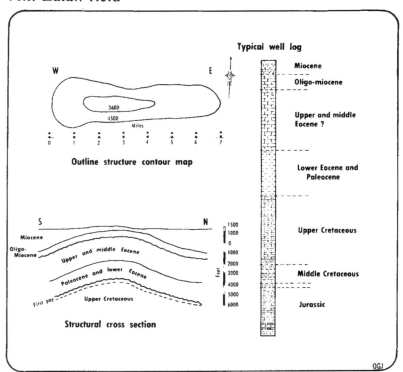

FIGURE 1–39 Ain Zalah oil field. Diagrammatic structure-contour outline and stratigraphic section. Length of field about 7 miles; thickness of stratigraphic column about 8,330 ft (after Daniel, 1954)

The Ain Zalah reservoir is extremely tight and has very low porosity. However, due to the presence of fractures, it can produce at high rates during limited periods. Figure 1–39 shows a generalized stratigraphic section of the field, a structure-contour map, and a structural cross section.

Daniel found that the oil possibly entered the "present reservoir by upward migration along fractures from some deeper zone, perhaps the middle Cretaceous or the Jurassic." Due to the high degree of fracturing of the formation, he indicated that drainage from the first and second pay at Ain Zalah could probably be achieved with two or three wells.

Comparison of the structural map shown on Figure 1–39 with the map shown on Figure 5–47 indicates that the reservoir presented in both maps is the same one. Wells in the crest and toward the crest produced between 1000 and 20,000 bo/d from the upper pay. Fracture permeability was essential to achieve these rates. Wells down structure did not prove any oil. Thus this is the opposite of the Sanish pool (Murray, 1968) where the best wells were in the flanks and the worst wells were in the crest.

Freeman and Natanson (1963) calculated that approximately 40% of the total pore volume was made from fractures in the upper pay (Type B-1 reservoir). The lower or second pay was better that the upper one. Furthermore it was communicated to an aquifer. Additional geologic and engineering aspects of this reservoir are presented in Chapter 5 under the section "Planning and Cooperation in Early Stages of Development Pays Off."

The Kirkuk field limestone reservoir has a higher average porosity and varied permeability which depends on stratigraphy. Figure 1–40 shows structural maps of the Khurmala Avanah and Baba domes.

According to Daniel (1954), the fractures at this 61-mile structure are so closed that only a few wells located at the base of the highest dome (Baba) would be enough to drain the entire reservoir and a 2-mile spacing could give adequate drainage (spacing of approximately of 1280 acres).

Production is from the Eocene and Oligocene fractured limestones. Lithologic rock types, average porosities, and permeabilities vary dramatically throughout the reservoir as indicated in the following table (Daniel, 1954):

	Average Porosity (Percentage)
Transition Zone limestones (extremely variable in all respects)	0–30
Backreef and reef limestones where porcelaineous and unaltered	0–4
Backreef and reef limestones where partly altered by diagenesis	4–10
Forereef limestones where partly recrystallized (which is their normal conditions)	18–36
Globigerina limestones	8–18
Globigerina limestones with variable amounts of course-grade fossil detritus	4–20

	Permeability (Millidarcys)
Backreef and reef, porcelaineous	0–5
Forereef, where altered	50–1000
Globigerina	0–10

According to Nelson (1985) this is a classical example of a reservoir where the fracture permeability assists an already producible reservoir. Without the fractures, however, the rates would have been much smaller than the 20,000 – 30,000 bo/d tested by some wells with a bottomhole pressure drawdown varying from 100 to less than 10 psi.

The Dukhan field (Type A reservoir) has limestones which are moderate to highly porous and permeable. The degree of fracturing is lower than at Ain Zalah and Kirkuk; consequently, the appropriate drainage of the reservoir requires a closer spacing.

Figure 1–41 shows a structural map, structural cross section, and typical lithological log of the Dukhan oil field. The length of the anticline is about 31 miles and maximum well depth is 6600 ft.

The main pays—the No. 3 and No. 4 Upper Jurassic limestones have been cored with an excellent 95% or better core recovery, as opposed to Kirkuk where recovery was in the order of 30%. It is not unusual to have very low core recoveries from fractured intervals. In fact, it has been found in some naturally fractured reservoirs that there is a good correlation between poor core recovery and good producing intervals.

The No. 3 limestone has an average porosity of 16% and average permeabilities of 30 and 15 md parallel and across the bedding, respectively.

Kirkuk structure

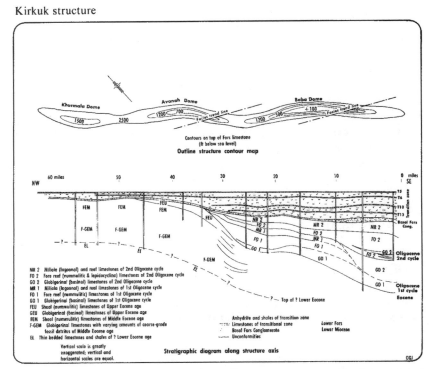

FIGURE 1–40 Kirkuk oil field. Stratigraphic diagram along the axis of the Kirkuk structure; structure-contour map (after Daniel, 1954)

Dukhan field

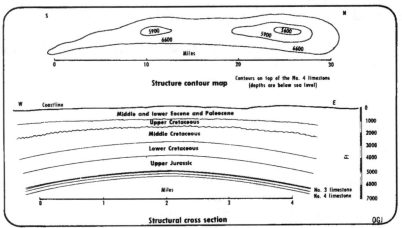

FIGURE 1–41 Dukhan oil field. Diagrammatic structure-contour outline and cross section (after Daniel, 1954)

Fractures are present particularly where the rock is fine textured, they tend to be at right angles to the bedding planes, and subparallel with the strike. The vertical extend is as large as 3 ft and sometimes they end in a strong stylolitic fracture.

The No. 4 limestone has an average porosity of 21% and average permeabilities of 75 and 40 md, parallel and across the bedding, respectively. Fractures are similar but probably more abundant than in the No. 3 limestone.

Production as of 1980 was in the order of 230,000 bo/d from the No. 3 limestone (22 wells) and No. 4 limestone (64 wells) combined. Both reservoirs were under water injection.

The Savanna Creek gas field in the Canadian Rocky Mountains has a 900 ft carbonate (Mississippian Livingstone formation) which is essentially impermeable and produces

Mara-La Paz fields

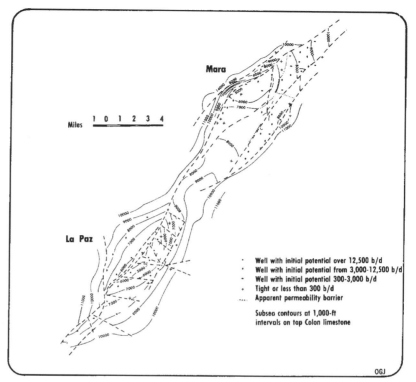

FIGURE 1–42 Structure contours on Mara-La Paz Fields in Venezuela (after Smith, 1951)

through fractures. Wells have produced from zero to 57 MMcfd. Matrix porosity is very low—about 4%. However, it is increased considerably by open fractures and brecciation (Scott et al, 1958). This is probably a reservoir of Type B-II.

Figure 1–42 is a structural map of La Paz-Mara fields in Venezuela. The intergranular porosity of the Cretaceous limestone reservoir (Colon-Cogollo formations) is 2–3%. Permeability is smaller than 0.1 md. However, these fields produced at 250,000 bo/d during 1951 when they were described by Smith. This production was clearly related to fracture porosity and fracture permeability (Type C reservoir).

Lowenstam (1942) found that the Marine reservoir of Illinois was characterized by a matrix limestone porosity, enlarged by a system of fractures. Lowenstam believed that fractures connected the discontinuous producing streaks with each other and with the main reef core.

Boyd (informal communication with Landes, 1959) indicated that fractures contributed significantly to the storage capacity of the Silurian dolomite in the Howell gas field in Michigan. In fact, he noted gas reserves exceeding those calculated by conventional methods and concluded that the excess gas was stored in fracture networks (Type B reservoir).

Braunstein (1953) reported oil production from the Selma fractured chalk (Navarro and Taylor age) in Gilbertown field, Alabama. He indicated that no matrix porosity was found in the chalk, and that fracture porosity was associated with the zone of fault. The fracture porosity provided a secondary trap for oil which migrated from the lower Eutaw sands.

Production from the Selma chalk comes only from wells located in the down-thrown side near the fault. This opposes other fault-related naturally fractured reservoirs where the best fracturing is found in the up-throw block.

Figure 1–43 shows a schematic cross section of Gilbertown field. The dry well encountered a complete section of the Selma chalk, but was far away from the fault and, consequently, did not penetrate any fractures. The middle well near the fault produced oil from the fracture Selma chalk. Finally, the well to the right of the fault produced oil from the Eutaw sand, but not from the Selma Chalk.

Gilbertown field

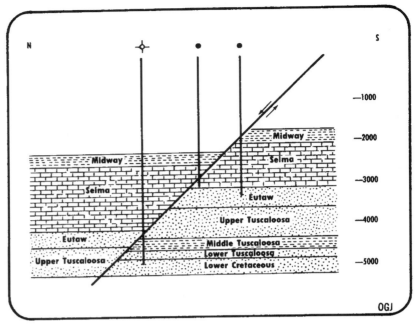

FIGURE 1–43 Diagrammatic north-south section across Gilbertown field (after Braunstein, 1953)

Hanna (1953) and Pirson (1967) in separate studies have also reported fracture porosity associated with faults in the Taylor marl and Austin chalk.

The Asmari limestone of Iran is characterized by an extensive fracture system. Lane (1949) estimated that 80% of the recoverable oil was stored in the matrix and 20% in the fractures (Type B-I reservoir).

The Tamaulipas limestone in Mexico has produced oil of 12.5°API at rates as high as 30,000 bo/d from a single well. The fracture permeability varies so much that wells only 200 ft apart from a producing well are dry.

Excellent oil rates have been obtained from the Reform (dolomitic limestone) giant Cretaceous reservoirs in Mexico. Delgado (1975) describes the reservoir as characterized by dolomitic limestone of low primary porosity and very good secondary permeability, due to the presence of natural fractures and caverns.

Very prolific gas rates have been obtained from Triassic naturally fractured carbonates in the Monkman area of British Columbia (Barss and Montandon, 1982; Cooper, 1991). Matrix porosities are very small ($\cong 5\%$) and matrix permeabilities are a fraction of a millidarcy. Thus the excellent gas rates are associated with natural fractures. The extensive tectonic fracture system has been attributed to the Laramide orogeny which created a series of compressional type structures. Absolute open flow tests as high as 223 MMscfd have been calculated in some wells. The gas has high H_2S and CO_2 content. Figure 1–44 shows a cross-section which highlights the structural style of the Triassic carbonates and Well Triad BP a–43–B 93–P–5, the discovery well of the Sukunka structure. Notice frontal imbrication at the leading edge of the thrust plate. Detachment of the Nordegg, Pardonet and Baldonnel carbonates from the underlying Charlie Lake anhydrite and folding progressed until the rupture threshold was reached and slippage occurred along fault planes (Barss and Montandon, 1982).

Read and Richmond (1993) have described the geology and reservoir characteristics of the Arbuckle Brown zone in the Cottonwood Creek field, Carter County, Oklahoma. The Brown zone is a highly fractured, vuggy to cavernous dolomite that yielded more than 4000 bo/d and 3 MMscfd when well 32–1 discovered the field in November 1987. Much of the fracturing is attributed to Pennsylvanian tectonism, although Lynch and Al-Shaieb

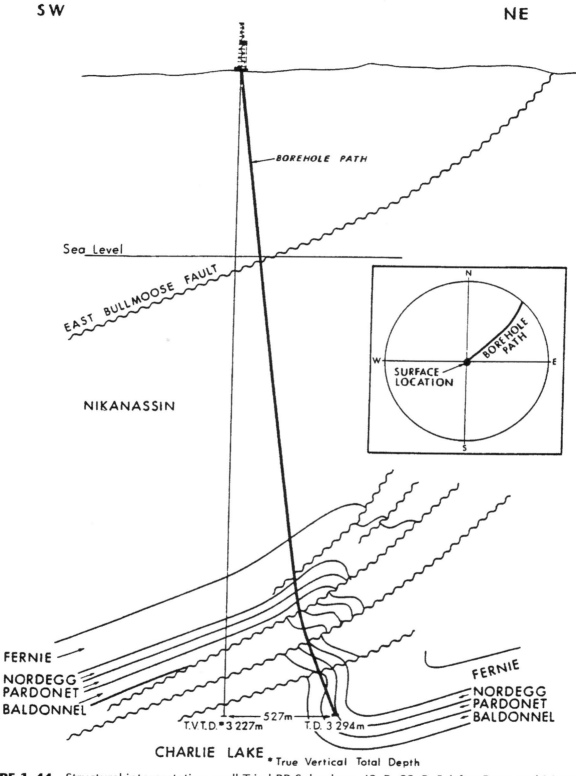

SW

NE

BOREHOLE PATH

Sea Level

EAST BULLMOOSE FAULT

NIKANASSIN

N

W — SURFACE LOCATION — E

BOREHOLE PATH

S

FERNIE

NORDEGG
PARDONET
BALDONNEL

FERNIE

NORDEGG
PARDONET
BALDONNEL

527m

T.V.T.D.*3 227m

T.D. 3 294m

CHARLIE LAKE *True Vertical Total Depth

FIGURE 1–44 Structural interpretation, well Triad BP Sukunka a–43–B 93–P–5 (after Barss and Montandon, 1982)

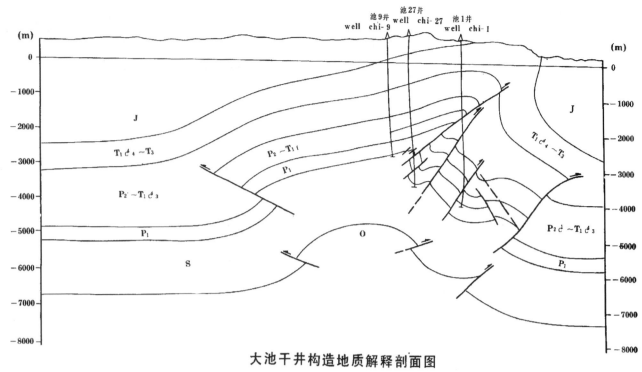

大池干井构造地质解释剖面图

FIGURE 1–45   Geological interpretation profile of Dachiganjing structure (after Guangcan, 1990)

(1991) indicate that extensive brecciation might be due to paleokarstification and solution collapse.

About 100 oil and gas fields have been discovered in naturally fractured carbonates in the Province of Sichuan, China (SPA, 1990). The gas pay zones have very low porosities averaging between 3% and 6%. The main Carboniferous reservoir space is attributed to solution pores and cavities following diagenesis. The permeability is provided by fractures generated as a result of tectonic forces (Huaibo et al, 1991). Figure 1–45 shows a cross-section of the Dachiganjing structure published by Guangcan (1990). Included in the cross-section are wells Chi-9, Chi-27 and Chi-1. Notice the similarity between this Chinese cross-section and the one of the Monkman area in British Columbia, Canada, presented in Figure 1–44.

Fractured Shales

Fractured shales have produced gas since the early 1900's along the western margin of the Appalachian Basin. The amount of gas within these shales has been estimated at 460 quadrillion scf. Since shales do not have effective porosity, production is achieved only through fracture networks (Hunter, 1953). Stimulation is usually necessary to obtain commercial production. Fractured shales can be classified as intermediate between Type B and Type C reservoirs.

Devonian shales are present in 26 states of the United States, six provinces and territories of Canada, along the U.S.-Mexican border, South America, Africa, and Europe. Gas production is also obtained from Cambrian shales in the S. Lawrence lowlands of Quebec.

In some cases, oil production has been obtained from shales. Results presented by Peterson (1955)indicate that high grade oil has been produced from the Mancos fracture shale, Rangely field, Colorado. These fractures are associated with the axial bending and arching of the Rangely anticline.

Capuano (1993) has presented evidence of fluid flow in microfractures in geopressured shales. He indicates that fractured shales could support flow similar to that of tight sandstones, and that fracture-fill deposits or collapse, could reduce this flow.

Araki and Kato (1993) have reported on the discovery of the Ayukawa oil and gas field in the Akita prefecture of Japan. Production comes from a Miocene siliceous shale formation with some characteristics similar to the Monterey of California where the source rock is also the reservoir rock. Intrusion of igneous rocks appear to be common.

Another shale, the Mississippian Bakken, is composed of black shales and siltstones. It is distributed over most of the Williston Basin (Montana and North Dakota), and it is both source and reservoir rock. Porosities and permeabilities are low. Natural fractures are attributed to tectonism and overpressuring. Horizontal wells have proved very successful in the Bakken shale with reserves ranging from 200 to 300 Mstbo per well (Fritz et al, 1993).

The northeast Alden Field in Caddo County (Oklahoma) and several other nearby fields produce from the Woodford shale. Natural fracturing is fault-related and resulted from oblique-slip tectonics. Another shale, the New Albani might account for as much as 90% of the known oil reserves in the Illinois Basin (Fritz et al, 1993).

Fractured Cherts

Regan (1953) studied oil production from fractured Monterey shale and chert reservoirs in the Santa Maria Costal District and the San Joaquin Valley in California. The rock is made of fractured chert and siliceous shale of upper Miocene age. Production ranged from 200 to 1000 bo/d.

The following characteristics of commercial fractured chert reservoirs in California have been listed Regan:

1. The reservoirs are associated with sediments considered to be source rock. Productivity of the reservoir seems to be larger as the amount of chert increases.
2. Most commercial reservoirs are covered by a cap of Pliocene sediments. In many cases there is insufficient clay shale, for an effective upward seal. When the Pliocene is strongly unconformable on the Miocene, as in the case of the Santa Maria District, the oil has very low gravity. When the deposition was nearly continuous, higher gravity oils are present.
3. Fractures appear to be fold-related in the San Joaquin Valley where some of the best wells are located on the crest of the structure. Fracturing in the Santa Maria District is more of the regional type.
4. The fractured rocks are hard and difficult to drill. Core recovery is poor. Oil shows are unimpressive, except for a little free oil on fracture planes and drilling fluids. Drilling fluid losses provide a reliable indication of the presence of fractures. However, there have been excellent wells which have been drilled without any mud losses. Intervals exhibiting a very flat spontaneous-potential curve are generally non-productive.

Fractured Basement Reservoirs

In some cases oil production can be obtained from fractured basement rock. Smith (1956) reported that in 1953 a well drilled in a basement of igneous and metamorphic fractured rocks in the La Paz-Mara field of Venezuela produced 3900 bo/d from a depth of 8889 ft. By 1955 the fields were producing 80,000 bo/d from 29 wells drilled in the fractured basement rocks. Since there is essentially no matrix porosity in basement rocks, it can be inferred that all oil produced from a basement reservoir is stored within the fractures (Type C reservoir).

Koning and Darmono (1984) discussed the geology of the Beruk Northeast field, Central Sumatra, where oil production was obtained from pre-Tertiary Basement rocks. Well No. 1 produced in excess of 1 MMstbo. Other wells were less productive due to separate water oil contacts and possible unrecognized fracture systems.

Walters (1953) researched oil production from fractured pre-Cambrian basement rocks in central Kansas. Sixteen wells in the Orth field, Rice County, Kansas, had produced over one million barrels of oil from fracture pre-Cambrian basement rock by 1953. Oil production had also been obtained from four wells in the Kraft-Prusa field and from one well each in the Beaver, Bloomer, Eveleigh, and Trapp fields in Barton County (Type C reservoirs).

Rice County, Kansas fields

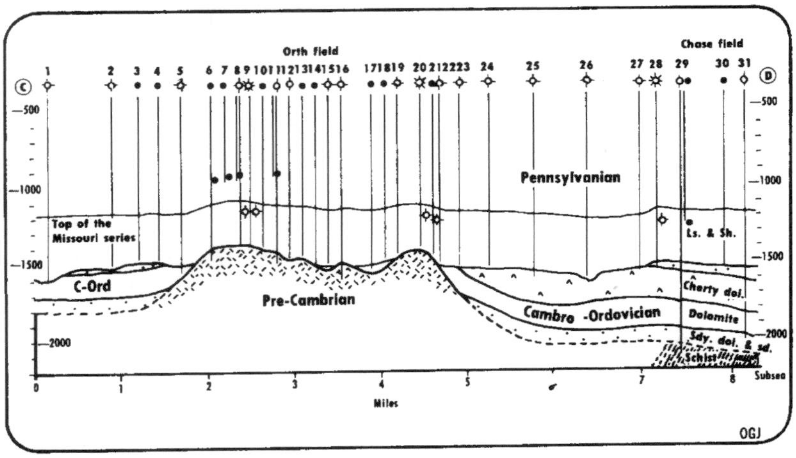

FIGURE 1–46 Orth and Chase fields (after Walters, 1953)

In these reservoirs, no oil was found when the basement rocks were structurally lower. Only when the Pennsylvanian beds were resting directly on the pre-Cambrian was oil found in the basement rocks.

Figure 1–46 shows a NW-SE cross section through the Orth and Chase fields. Wells 6, 10, 11-B, 12, 13, 16, 20, and 21 produce oil from fractured pre-Cambrian quartzite.

Eggleston (1948) did a comprehensive summary of oil production from fractured basement rocks in California. Fifteen thousand bo/d were being produced from fractured basement rocks by 1948. This represented about 1.5% of the total California production at the time (918,000 bo/d). Production from all kinds of fractured reservoirs was 55,000 bo/d, or about 6% of the total state production.

McNaughton (1953) used the theory of dilatancy to explain migration and accumulation of hydrocarbons in various California basement metamorphic rocks, including Placerita Canyon, near Newhall, Ca., playa del Rey field at Venice, California, El Segundo field, Santa Maria Valley field, Edison field southeast of Bakersfield and the Wilmington field. Various of these basement reservoirs were discovered accidentally.

The importance of this information lies in the fact that oil production from basement rocks (Type C reservoirs) had been considered nearly impossible up to that time. Interesting enough there are still many oil companies that stop drilling operations the minute basement is found. It is my strong recommendation that, on the contrary, drilling should be continued into basement for at least 300 m, preferentially with a slanted hole, specially if a source rock rests on top of the basement. I anticipate significant discoveries in basement rock if this procedure is followed.

Coalbed Methane

Estimates of coalbed methane in the world range from 3000 to 9000 trillion cubic feet (Law and Rice, 1993). Proponents of this type of production suggest that coalbed methane is one of the key elements in the future of the energy industry.

Fractures in coalbeds are approximately vertical and perpendicular to bedding.

The coalbed methane industry follows the tradition of British mining engineers and calls the microfractures in coal seams cleats. The larger dominant microfractures are called face cleats. The smaller microfractures that go perpendicular to face cleats are called butt cleats. They are very short and usually are interrupted at the face cleats.

Coalbed seams can be represented by dual-porosity systems composed of matrix and fractures. The matrix blocks contain micropores that are too small to contain water or any other liquids, but store methane molecules absorbed onto the surface of the coal matrix.

Because of the influence of primary controls (coal rank and composition, and depth temperature) and secondary processes (mixing and oxidation), it is suggested that coalbed gases can be better understood if studied individually. Law and Rice (1993, p.171) have reported on characteristics of coalbed gases from selected areas, including Western Germany, Bowen and Sidney Basins (Australia), Eastern China, Lower Silesian Basin (Poland), San Juan Basin (New Mexico and Colorado), Piceance Basin (Colorado), Powder River Basin (Montana and Wyoming), and Black Warrior Basin (Alabama and Mississippi).

EXPLORING FOR NATURALLY FRACTURED RESERVOIRS

Most naturally fractured reservoirs have been discovered by accident. Possible exceptions are some fields of Canada, the USA, Venezuela, Iran, and Iraq, where these types of reservoirs have been sought.

McNaughton and Garb (1975) related exploration methods to the concept of dilatancy illustrated in Figure 1–19. They recommended to focus attention in the search for large bodies of brittle rock which are either underlain or overlain by petroleum source rocks. Preferentially the basin should have structural complexities and stratigraphic conditions should be favorable. Modern seismology may provide a valuable tool for detecting fractured reservoirs. In fact, acoustic velocity should decrease in open, unhealed fractured rocks. As an example McNaughton and Garb (1975) indicated that a decrease in acoustic velocity associated with brittle fracture rocks was established in the Amadeus basin of Central Australia.

Bloxsom Lee (1986) has reported on the use of seismic data to predict reservoir anisotropy. The principle behind this approach is that the direction of the fractures or the direction of the principal stress axes determines how fast and in what manner the shear wave of a given polarization will travel.

Mapping fractured trends facilities searching for areas of more intense shattering. The combination of seismic data, photogeology (Figure 1–29), subsurface information, geochemical techniques, outcrop studies, radius of curvature based on seismic data, and orientation of residual stresses are significant tools for discovering naturally fractured reservoirs.

Stresses in a body cannot be measured directly. However, there are various techniques which allow deducing the stress in tunnels, boreholes, open cuts and other underground openings if the external forces are known. These devices are located in the borehole and measure the strain, from which stresses are calculated (Billings, 1972). One of the methods utilizes photoelastic techniques (Voight, 1967). Other methods are based on signals extracted from mechanical, electrical or hydraulic components (Terzaghi, 1962; Hast, 1958).

The main disadvantage of surface methods is the uncertainty that exists in the extrapolation of these fractures down thousands of feet to deeply buried reservoir rocks. Analysis of "breakout" data corroborates that stress direction changes with depth in many instances.

Figure 1–47 shows surface fracture orientation and residual stress orientation of the fractured Palm Valley field in Australia (Magellan Petroleum Corporation). Fracture rosettes based on photogeology indicate that northwesterly trends dominate in the south, southeast, and northeast of the field. These trends are corroborated by residual stress analysis in test holes. Note that fractured trends in the reservoir tend to be parallel to the principal horizontal stress measured in surface outcrops. Thus this is an instance in which stress orientation does not seem to change with depth.

Direction of principal horizontal stresses in continental United States based on the orientation of vertical hydrofractures has been presented by Haimson (1978) as shown in Figure 1–48A. It is interesting to note that out of 25 principal horizontal stresses shown on the figure only two (one in Texas and one in Colorado) have a NW-SE direction. The other 24 principal horizontal stresses have a general SW-NE direction suggesting that probably most of the natural fractures in the United States have that preferential orientation.

Bell and Babcock (1986) have presented stress trajectories inferred from "breakout data" for the Western Canadian Basin (Figure 1–48B). The upper part of the Figure shows the normal breakout pattern and four possible causes of anomalous breakout patterns. Breakout

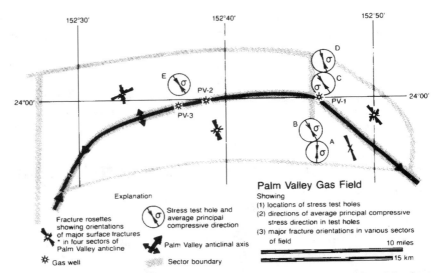

FIGURE 1–47 Palm Valley gas field—surface fracture orientations and residual stress orientations (after McNaughton & Garb, 1975)

FIGURE 1–48A Maximum horizontal principal stress directions in continental United States based on the orientation of vertical hydrofractures (after Haimson, 1978)

refers to the process of borehole wells spalling to relieve in-situ stress. This results in an elliptical borehole. The main breakout axis is parallel to the minimum horizontal stress and perpendicular to the maximum horizontal stress. The lower half of Figure 1-48B shows the possible stress trajectory inferred from "breakout" data.

Most natural fractures below 2500 ft (762 m) resulting from normal faulting or deep erosion of the overburden tend to be vertical or of high inclination. Thrust faulting might

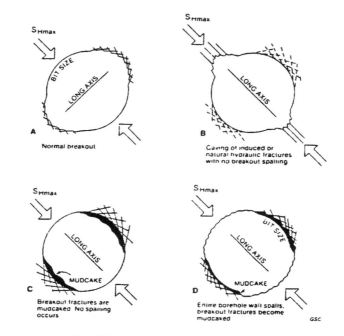

Possible causes of anomalous breakout orientations.

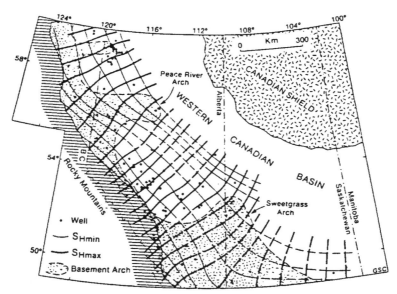

FIGURE 1–48B Postulated stress trajectories inferred from breakout data, Western Canadian Basin (after Bell and Babcock, 1986)

generate low angle and horizontal fractures. The problem with horizontal fractures is that they probably tend to close owing to the effect of overburden. Vuggy fractures and partially mineralized fractures, however, probably would tend to remain open.

Exploring for naturally fractured reservoirs of the A Type probably does not represent too much of a problem, as conventional methods can be used in this case. As an example some of the North Sea reservoirs have porosities in the order of 30% and are naturally fractured. In this case basically all the hydrocarbon storage capacity is in the matrix.

Exploration for Type B-I reservoirs can also be carried out in a conventional form, as the matrix system, in spite of low porosity, has the capability of releasing oil.

When the explorationist starts running into serious problems is when he is looking for Type B-II and Type C reservoirs.

Figure 1–49 shows schematic sketches of three wells drilled in a naturally fractured reservoir. The cross-section extends from the west towards the east. Well 1 was a discovery well as it intercepted the high angle and vertical fractures.

Well 2 did not intercept any fractures. The logs showed, in general, similar responses to the ones of the discovery well. This allowed to establish a good correlation. However, it must be remembered that in many cases, there is no production correlation between wells in naturally fractured reservoirs, in spite of a good log correlation.

Well 2 was drill stem tested at the equivalent producing level of Well 1 and it did not flow. The test further resulted in a very low mud recovery and a small increase in pressure which suggested a very low permeability system. Because of the good correlation with Well 1, Well 2 was completed, perforated, and stimulated with a massive hydraulic fracturing job. The production results were very poor and the well was abandoned.

This author has run into the above experience several times in different places around the world.

What are the probabilities of intercepting a vertical fracture with a vertical hole? Most likely those probabilities are very slim. Therefore, when looking for naturally fractured reservoirs, the chances for success would be better if, instead of drilling vertical holes, directional or horizontal wells perpendicular to the orientation of the fractures were drilled (Aguilera, 1982, 1983). This has been noticed, for instance, in the Mississippian lime of Oklahoma, the Austin Chalk, certain offshore plays where directional holes are necessary and the Monterey shale of California (Figure 1–50).

Well No. 3 on Figure 1–49 was drilled according to this reasoning and was completed as a production well after a mild acid job. In some cases the network of fractures might be regional or contractional in nature. For this situation directional holes would also stand better probabilities of success.

Since many naturally fractured reservoirs have important thicknesses, deviated holes (not necessarily horizontal) appear very sound. The larger the spacing between fractures the more important the idea of deviated holes. As shown on Figure 1–8 thinner beds contain closer spaced fractures. For these situations the directional hole probably would not be necessary. As the bed thickness increases, however, the distance between fractures also increases and a directional or horizontal hole increases the probability of success. A survey conducted by ARCO indicates that 65% of the naturally fractured reservoirs in the U.S.A. have a thickness of 400 ft or more (Dech et al, 1986).

In the schematic of Figure 1–49, Well 1 was luckily a discovery well. But what would have happened if Well 2 had been drilled first? Probably this potentially good reservoir would have been abandoned. This author believes that many good reservoirs have been left behind because the wells have not intercepted the natural fractures.

This is more dramatic where drilling investments are very large; for example in offshore wells. If the explorationist thinks in a conventional way the first well will be drilled vertically and right on the crestal area of the anticline. This might be a double pitfall. First, the vertical hole might not intercept the vertical or high inclination fractures. Second, contrary to the popular belief the crestal area of an asymmetrical anticline might contain little fracture porosity (Martin, 1963).

The best wells of Kirkuk field in Iraq, La Paz field in Venezuela, Sanish pool in North Dakota and Raman field in Turkey are located over the gentler flanks. Crestal wells in these reservoirs are rare or less prolific than the ones in the plunging ends.

Martin (1963, 1992) shows how petrofabric studies may help find fracture porosity reservoirs. Figure 1–51 is a schematic top view of an asymmetrical anticline. Open tension fractures occur in the plunging ends and on the gentle flank. The steep flank is a zone of compressional movements where fractures are often brecciated and cemented by secondary mineralization.

Important consideration in this model is given to faulting. The best possibility of finding open fractures is in the upthrow block because of tension while fractures in the downthrow block may be closed due to compression. An example of this type of situation is provided by the Austin chalk in some areas.

Reservoirs of Types B-II and C do not necessarily need to have structural closure as shown in Figure 1–52. In this sense, they might be regarded as stratigraphic traps.

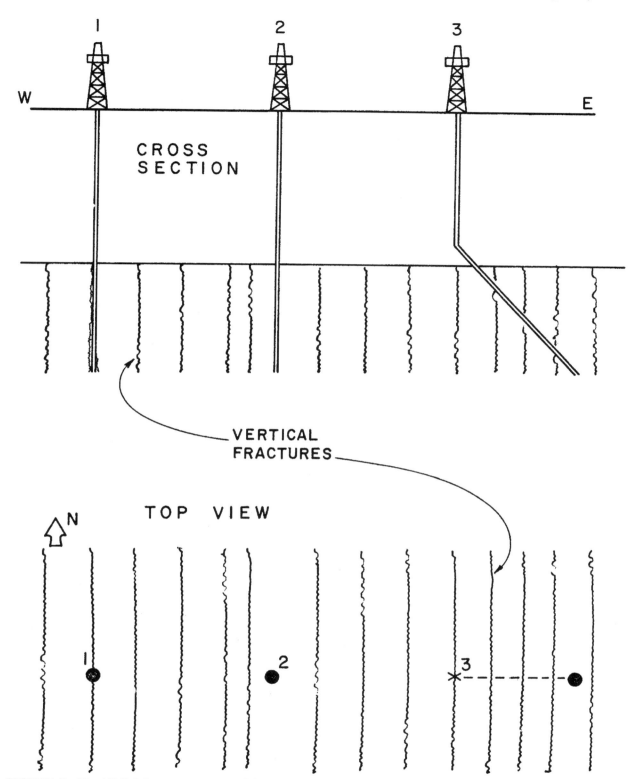

FIGURE 1–49 Well 1 intercepts vertical fractures. It is a discovery well. Well 2 does not intercept any fractures. It is a dry hole. Well 3 intercepts several vertical fractures. It is a directional production well (after Aguilera, 1982)

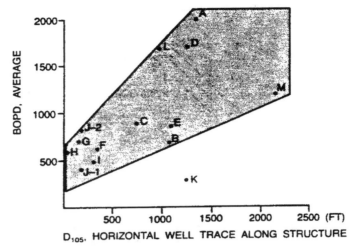

FIGURE 1–50 Length of projected traces of well tracks in 105° direction versus production in barrels of oil per day. The trend indicates that higher production is found for deviated wells drilled perpendicular to the predominant fracture set in South Ellwood Field, Santa Barbara Channel, California. Letters designate individual wells (after Belfield et al, 1983)

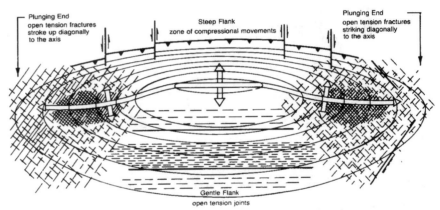

FIGURE 1–51 Open fractures and fracture pattern in asymmetrical anticline (after Martin, 1963).

Figure 1–52A shows a schematic sketch of how deep erosion of the overburden permitted uplift of the rock, generating fractures in the uplifted hill. In Figure 1–52B the hill became buried with continuous subsidence and sedimentation. In Figure 1–52C the whole basin was tilted. In Figure 1–52D a naturally fractured reservoir of Type B-II or C, without structural closure was discovered by keeping track of thickness reductions of the formation above the fractured reservoir. Note that thickness, h, above the uplifted reservoir is smaller than thickness, H, above the section which was not uplifted.

Following the study of several North American folded carbonate sections, Nelson (1994) has indicated that although many hinge zones have a well-developed fracture intensity, flank positions can contain layers with fracture intensities that might be as good and better than intensities in the hinge and forelimb zones. The conclusion follows that directional or horizontal wells in flank positions (backlimb or forelimb) can produce at better oil and gas rates in the flank than in the hinge.

Figure 1–53 shows a ranking of optimum deviated paths for an asymmetric carbonate fold as presented by Nelson (1994). Path A oblique across hinge in dip direction would intercept Type II fractures (refer to section on fold-related fractures) and could be anticipated to produce at the highest rates. Well A could be deepened to develop the forelimb following path B to intercept Type II fractures or path B' to intercept both Type I and Type II fractures. Wells B and B' would probably have the second best production. Type I fractures in the crest would

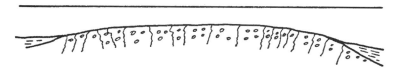

(A) Uplift and erosion forming "hill" capped
    by weathered rock debris.

(B) Subsidence, sedimentation, petroleum
    migration, and accumulation in fractured system.

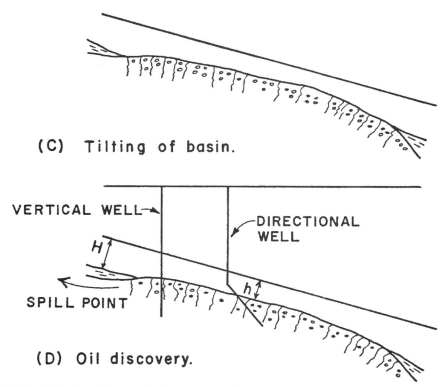

(C) Tilting of basin.

(D) Oil discovery.

FIGURE 1-52 Possible geologic evolution leading to petroleum accumulations in B-II and C reservoirs without structural closure. Stages A and B have been adapted from McNaughton and Garb (after Aguilera, 1982)

have been developed as a result of folding. Type II fractures in the forelimb would have been developed as a result of a "migrating hinge" during fold development.

Well C would penetrate Type I fractures in the back limb and would probably have the next best production. Type I fractures in the back limb are not as abundant but they might be larger laterally and vertically communicating thus with larger volumes of hydrocarbons

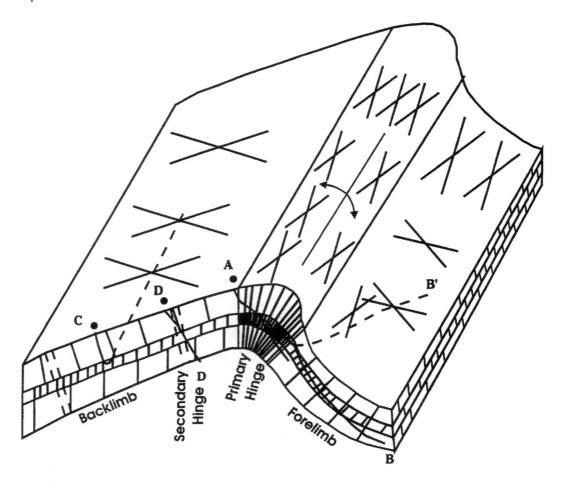

A - Oblique Across Hinge In Dip Direction
B - As With "A" But In Forelimb
B' - Alternate to "B" Oblique To Strike & Dip
C - Parallel To Strike In Backlimb In Most Fractured Layer (s)
D - Oblique To Both Secondary Hinge and Layering

FIGURE 1–53 Schematic diagram of a typical fractured asymmetric fold with highgraded optimum well paths (A is best) for a brittle carbonate. Well path rankings will vary with rock type (after Nelson, 1994)

(and water). Well D would be oblique to both secondary hinge and layering. This wellbore path would communicate small swarms of Type II fractures. Production probably would not be as good as in the previous cases.

SUBSURFACE FRACTURE SPACING

Many of the calculations carried out in reservoir engineering practice require knowledge of the distance between fractures or fracture spacing. This distance is also known as size of the matrix blocks (h_m). In some instances this spacing can be determined from well logs (Chapter 3) and/or well testing data. In the case of well testing data the assumption is made that the fractures are uniformly distributed throughout the area investigated during the test.

A method has been developed for determining fracture spacing based on core data. The geological model has been described by Narr and Lerche (1984). Figure 1–54 shows a schematic block diagram of a wellbore through fractured beds of two different thickness. The fractures are perpendicular to bedding, parallel to each other and systematically spaced. Cores cut in the two central beds are unfractured, but cores from the upper and lower beds

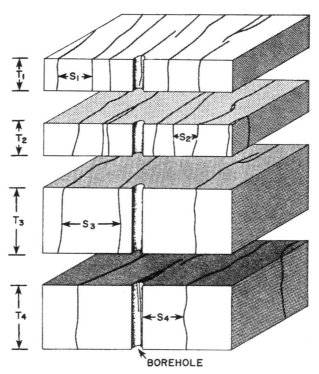

FIGURE 1–54 Schematic block diagram of a wellbore through fractured strata of 2 different thicknesses. Cores cut in upper and lower beds would intersect fractures, but core from 2 central beds is unfractured. Note closer fracture spacing in the 2 upper, thinner beds. Compare this diagram with joint spacing shown in outcrop photograph of Figure 1–55.

would intercept the fractures. Note that fracture spacing is closer in the two upper thinner beds. Compare this diagram with the outcrop photograph shown on Figure 1–55.

Outcrop 1 (Figure 1–55) is located 10 km (6.2 miles) north of Water Street, Pennsylvania along Truck Route 45. The photograph shows a sequence of fine-grained limestone beds of the Ordovician Axeman Formation. Thickness of beds A through E and fracture spacings were measured on site. Fracture indexes (I_f) were calculated from the equation:

$$I_f = \frac{T_i}{S_i} \tag{1–56}$$

where T_i =   thickness of the ith bed, cm
            S_i =   distance between fractures or fracture spacing, cm

Parallel white lines on Figure 1–55 separate 20 hypothetical "cores", each one of 10 cm diameter cut perpendicular to bedding. This large number of "cores" provided a meaningful statistical base. The number of fractures penetrated by each "core" were counted and these data served to prepare the histogram shown on Figure 1–56. The mean number of fractured beds intercepted by a "core" was 2.45.

Table 1–2 shows median fracture spacing, median bed thickness, number of fracture spacings measured and fracture index. The outcrop average fracture index is 0.98. The standard deviation is 0.34.

For a given average bed thickness (T_{avg}) and core diameter (D), it is possible to calculate the probability of any combination of fractured and unfractured beds from the binomial theorem:

$$(Q + P)^N = Q^N + NQ^{N-1}P + \frac{N(N-1)}{1 \times 2} Q^{N-2}P^2 + \ldots$$

$$+ \frac{N(N-1)(N-2) \ldots (N-r+2)}{1 \times 2 \times \ldots \times (r-1)} Q^{N-r+1}P^{r-1} + P^N \tag{1–57}$$

FIGURE 1–55 Joints in Axemann Formation in central Pennsylvania at outcrop 1. Notice closer spacing of joints in thinner beds. Meter stick for scale. White lines delimit 20 hypothetical 10-cm diameter cores used to test subsurface fracture index solution. Measured beds are labelled A through E (after Narr and Lerche, 1984)

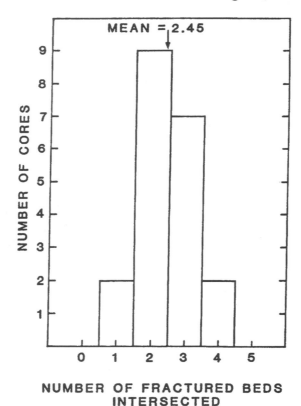

FIGURE 1–56 Number of fractured beds versus number of cores for outcrop 1 (Fig. 1–55) with core axis 90° to bedding. Mean number of fractured beds intersected by a core is 2.45 (after Narr and Lerche, 1984)

TABLE 1–2 Fracture Index Data for Outcrop 1* (after Narr and Lerche, 1984)

Bed	Median Joint Spacing (cm)	Median Bed Thickness (cm)	Number of Joint Spacings Measured	Fracture Index
A	50.0	46.5	5	0.93
B	9.0	14.0	16	1.55
C	7.5	7.5	16	1.00
D	27.5	18.5	10	0.67
E	40.0	30.5	9	0.76

*Outcrop average index I = 0.98
Standard deviation σ = 0.34.

TABLE 1–3 Probability of Intercepting a Fracture in a Bed at Fracture Index (I) of One, Outcrop 1, Case 1* (after Aguilera, 1988)

Combination of Events		Probability of Combination (%)	Probability of Success** (%)
Unfractured	Fractured		
5	0	6.16	
4	1	22.97	93.84
3	2	34.29	70.87
2	3	25.60	36.58
1	4	9.55	10.98
0	5	1.43	1.43

*Probability of not intercepting a fracture in a bed = 0.5726, probability of intercepting a fracture in a bed = 0.4274, total number of intercepted beds = 5. Location from Narr and Lerche (1984).
**Probability of success means probability of intercepting at least the number of fractured beds indicated under combination of events column.

where:

Q = probability of intercepting unfractured beds = $1 - P$
P = probability of intercepting fractured beds = $1 - Q$
r = successive number of beds (all beds contain vertical fractures which may or may not be intersected by the core)
N = total number of beds intercepted

The probability, P, of intercepting a fractured bed is given by:

$$P = \frac{D}{S} = \frac{DI_f}{T_{avg}}$$

(1–58)

where S is the distance between fractures, D is core diameter, and I_f is a fracture index given by Equation 1–56.

Example 1–5. Calculate fracture index and fracture spacing for the outcrop shown on Figure 1–55, assuming that we cut 20 cores of 10 cm-diameter each one, perpendicular to bedding as shown by the white lines. Thickness in cm for beds A – E are shown on Table 1–2. The core axis are parallel to the fractures. The fractures are parallel to each other and perpendicular to bedding.

Solution. Average thickness is 23.4 cm. The probability of intercepting a fractured bed is $P = 10\,I_f/23.4 = 0.4274\,I_f$. A probability evaluation for the case of $I_f = 1$, is shown on Table 1–3. These data are plotted on Figure 1–57. The number of fractured beds at 50% probability is 2.6. This is plotted vs. a fracture index of 1.0 on Figure 1–58 and a straight line is drawn between this point and the origin.

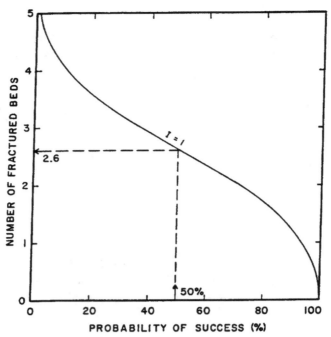

FIGURE 1–57 Probability of intercepting a certain number of fractured beds at fracture index (I) of one, outcrop 1, case 1. Number of fractured beds at 50% probability is 2.6 (after Aguilera, 1988)

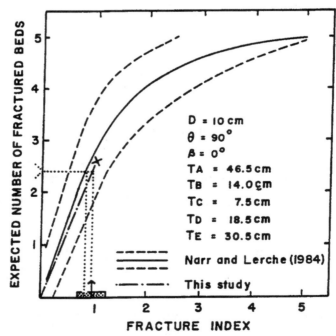

FIGURE 1–58 Expected number of fractured beds vs. fracture index (and its uncertainty) for beds A–E, outcrop 1, case 1. Mean number of fractured beds intersected by hypothetical core plotted at 2.45. Observed fracture index measured at outcrop indicated by arrow at I = 0.98. Stippled bar (lower left corner) indicates one standard deviation (adapted from Narr and Lerche, 1984)

The number of fractured beds intercepted by the core (2.45) as reported by Narr and Lerche (1984) is entered on Figure 1–58 and a fracture index of 0.94 is read from the Figure. This number compares with 0.98 measured at the outcrop (Table 1–2) and 0.83 calculated by Narr and Lerche (1984).

Example 1–6. Calculate fracture index and fracture spacing for the schematic shown on Figure 1–59 assuming an angle between core axis and bedding, θ, equal to 60°, and an angle between core axis and fractures, β, equal to 30°. Apparent fracture thickness in cm of penetrated beds was as follows: $T_A = 53.7$, $T_B = 16.2$, $T_C = 8.7$, $T_D = 21.4$, $T_E = 35.2$ cm. The mean number of fractured beds penetrated was found to be 3.70 from "core" data as shown on Figure 1–60.

Solution. The average apparent thickness T_{eavg} is 27.04 cm. The probability of intercepting a fracture bed is $P = 10\,I_e/27.04 = 0.3698\,I_e$ from equation 1–58 where I_e is an apparent rather than a true fracture index. Tables 1–4 and 1–5 present probability evaluations for $I_e = 1$ and $I_e = 2$ from which Figure 1–61 has been prepared. The number of fractured beds at 50% probability for a fracture index equal to 1 is 2.31 and for a fracture index equal to 2 is 4.27.

On the basis of the angles θ and β a true fracture index is calculated from the equation:

$$I = \frac{I_e \sin\theta \cos\beta}{\sin\theta + I_e \sin\beta \cos\beta} \tag{1–59}$$

Equation 1–59 indicates that $I = 0.577$ when $I_e = 1$, and $I = 0.866$ when $I_e = 2$. The true fracture indexes (0.577 and 0.866) are crossplotted vs. number of fractured beds intercepted (2.31 and 4.27) as shown on Figure 1–62. The number of fractured beds intercepted by the core (3.70 from Figure 1–60) allows to read a fracture index of 0.79 from Figure 1–62. This number compares with a 0.98 measured fracture intensity index and 0.81 calculated by Narr and Lerche (1984).

Example 1–7. Use the binomial theorem to calculate probability of occurrence of a given combination of fractured and unfractured beds (PC), and the probability of intercepting at least the number of fractured beds indicated in each combination (PS). The following data are available:

Probability of intercepting unfractured beds = 0.4545
Probability of intercepting fractured beds = 0.5455

Note: The same procedure presented in this example was used to generate Tables 1–3 through 1–5.

6 Unfractured beds:

$$PC = (0.4545)^6 = 0.0088$$

5 Unfractured beds + 1 fractured bed combination

$$PC = (6)\,(.5455)\,(0.4545)^{6-1} = 0.0635$$
$$PS = 1.0 - 0.0088 = 0.9912$$

4 Unfractured beds + 2 fractured beds combination

$$PC = \frac{6(6-1)(0.4545)^{6-2}(0.5455)^2}{1 \times 2} = 0.1905$$
$$PS = 1.00 - 0.0635 - 0.0088 = 0.9277$$

3 Unfractured beds + 3 fractured beds combination

$$PC = \frac{6(6-1)(6-2)(0.4545)^{6-3}(0.5455)^3}{1 \times 2 \times 3} = 0.3048$$
$$PS = 1 - 0.1905 - 0.0635 - 0.0088 = 0.7372$$

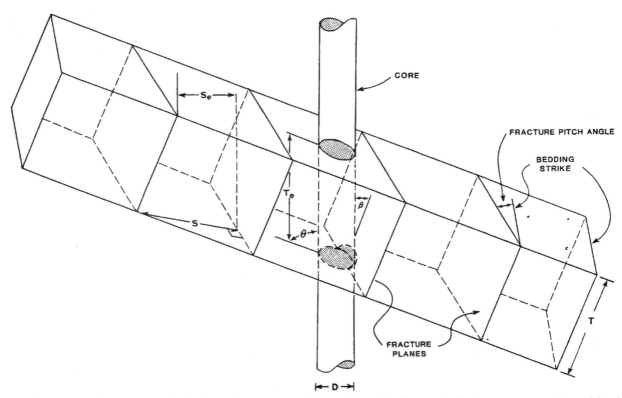

FIGURE 1–59 Diagrammatic fractured stratum and core sample. β = angle between core axis and bedding, S = spacing of fractures, S$_e$ = effective fracture spacing (relative to borehole incidence), T = bed thickness, T$_e$ = effective bed thickness (parallel with core axis) (after Narr and Lerche, 1984)

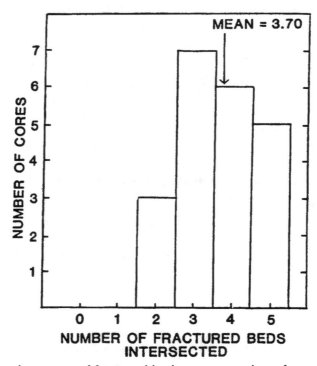

FIGURE 1–60 Number space of fractured beds versus number of cores for outcrop 1 (Fig 1–55) with core axis 60° to bedding. Mean number of fracture beds intersected by a core is 3.70 (after Narr and Lerche, 1984)

TABLE 1–4 Probability of intercepting a fracture in a Bed at an Apparent Fracture Index (I_e) of One, Outcrop 1, Case 2* (after Aguilera, 1988)

Combination of Events		Probability of Combination (%)	Probability of Success** (%)
Unfractured	Fractured		
5	0	9.94	—
4	1	29.16	90.06
3	2	34.23	60.90
2	3	20.08	26.67
1	4	5.89	6.58
0	5	0.69	0.69

*Probability of not intercepting a fracture in a bed = 0.6302, probability of intercepting a fracture in a bed = 0.3698, total number of intercepted beds = 5. Location from Narr and Lerche (1984).
**Probability of success means probability of intercepting at least the number of fractured beds indicated under combination of events column.

TABLE 1–5 Probability of intercepting a fracture in a Bed at an Apparent Fracture Index (I_e) of Two, Outcrop 1, Case 2** (after Aguilera, 1988)

Combination of Events		Probability of Combination (%)	Probability of Success** (%)
Unfractured	Fractured		
5	0	0.12	—
4	1	1.70	99.88
3	2	9.66	98.18
2	3	27.43	88.52
1	4	38.96	61.09
0	5	22.13	22.13

*Probability of not intercepting a fracture in a bed = 0.2604, probability of intercepting a fracture in a bed = 0.7396, total number of intercepted beds = 5. Location from Narr and Lerche (1984).
**Probability of success means probability of intercepting at least the number of fractured beds indicated under combination of events column.

2 Unfractured beds + 4 fractured beds combination

$$PC = \frac{6(6-1)(6-2)(6-3)(0.4545)^{6-4}(0.5455)^4}{1 \times 2 \times 3 \times 4} = 0.2744$$

$$PS = 1 - 0.3048 - 0.1905 - 0.0635 - 0.0088 = 0.4324$$

1 Unfractured bed + 5 fractured beds combination

$$PC = \frac{6(6-1)(6-2)(6-3)(6-4)(0.4545)^{6-5}(0.5455)^5}{1 \times 2 \times 3 \times 4 \times 5} = 0.1317$$

$$PS = 1 - 0.2744 - 0.3048 - 0.1905 - 0.0635 - 0.0088 = 0.158$$

6 Fractured beds

$$PC = (0.5455)^6 = 0.0263$$

$$PS = 1 - 0.1317 - 0.2744 - 0.3048 - 0.1905 - 0.0635 - 0.0088 = 0.0263$$

An application of a different statistical method for determining the size of matrix blocks in the naturally fractured Eschau Field of France has been presented by Janot (1973).

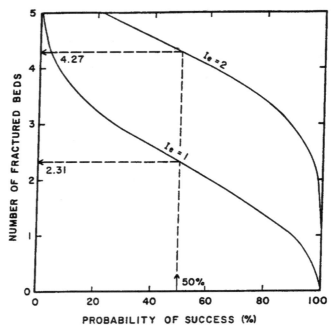

FIGURE 1–61 Probability of intercepting certain number of fractured beds at apparent fracture indexes (I_e) of one and two, outcrop 1, case 2. Number of fractured beds at 50% probability are 2.31 and 4.27, respectively (after Aguilera, 1988)

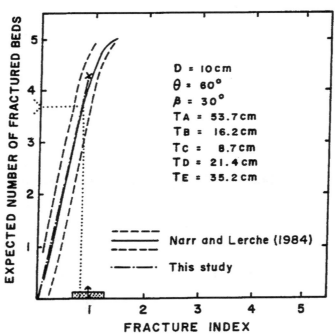

FIGURE 1–62 Expected number of fractured beds vs. fracture index (and its uncertainty) computed for beds A–E, outcrop 1. Mean number of fractured beds intersected by hypothetical core plotted at 3.70. Observed fracture index measured at outcrop indicated by arrow at I = 0.98. Bar indicates one standard deviation (adapted from Narr and Lerche, 1984)

IMPACT FRACTURES

Figure 1–13 illustrated strain rates (ė) associated with various events including nuclear explosions and meteorite impact. The logarithm of strain rate for these events was shown to range between approximately 6 and 3.

Impact fractures are generated as a result of a sudden shock as in the case of an explosion or a meteorite impact. They tend to be circular or ellipsoid, with diameter ranging from less than a mile to 50 miles and more.

Figure 1–63 shows the evolution of a meteor crater. Under certain conditions the breccia in the surroundings of the crater could become hydrocarbon reservoirs. This would require to have a source bed next to the impact generated fractures.

One example of hydrocarbon reservoirs extensively fractured due to meteorite impact is provided by the "Ames Crater" on the northern shelf of the Anadarko basin (Kuykendall, 1994). Structural features and petrographic evidence from cores and cuttings support the "meteorite impact" origin of the structure. Evaluation is difficult due to unforseen changes in lithology and mineralogy, pore geometry and pore size, water salinities and water saturations.

Butler (1989) has discussed a controversial theory by which all geologic phenomena are created immediately when a meteorite impacts the earth. Based on this reasoning the center of a basin is created at the point of impact. Mountains are built and continents are uplifted as a result of the impact's force. He indicates that continents can be reduced to sea level in about 60 million years by erosion. Thus there must be a continual renewal of energy to sustain the existence of continents which is probably the result of meteor impact. Butler's theory opposes traditional geologic concepts that indicate that basins evolve and grow slowly over time with deposition of sediments.

EFFECT OF FRACTURES ON FLOW BEHAVIOR

Publications on Burbank (Hagen, 1972), Norman Wells (Kempthorne and Irish, 1981), Gach Saran (McQuillan 1985) and many other reservoirs suggest that natural fractures control fluid behavior to a large extent. Park Dickey (1984) suggests that a realistic model to many naturally fractured reservoirs is as presented in Figure 1–64. One set of fractures is indefinitely long, while a conjugate set runs from one principal fracture to the next. In his model there are no horizontal fractures, rather impermeable layers that interrupt vertical communication. This could be handled with the match sticks model presented in Figure 1–2B. The lack of symmetry in Figure 1–64 can be handled by introducing a tortuosity greater than 1.0 in Figure 1–2B.

UNDISCOVERED NATURALLY FRACTURED RESERVOIRS, WHY AND HOW?

Many hydrocarbon naturally fractured reservoirs around the world have not become profitable discoveries and have been abandoned because of (1) incorrect pressure extrapolations, (2) poor completions, and/or (3) failure to intersect the natural fractures (Aguilera, 1992).

Incorrect pressure extrapolations might occur when the infinite acting radial flow period has not been reached during a pressure transient test. This can lead to the erroneous conclusion that the reservoir is depleting.

Conventional completions are typically performed in intervals that meet certain porosity, permeability, and water saturation cutoff criteria. This might be dangerous practice in some naturally fractured reservoirs where the largest degree of natural fracturing could be associated with the lowest porosities. Furthermore, there are instances where the largest fracturing is found in the thinner beds.

Vertical wells might not intercept vertical natural fractures. In these cases, a conventional test might yield negative results, even if the matrix blocks are hydrocarbon saturated. These poor results are due to the usually low matrix permeability of naturally fractured reservoirs. Commercial production of hydrocarbons is not possible from the tight

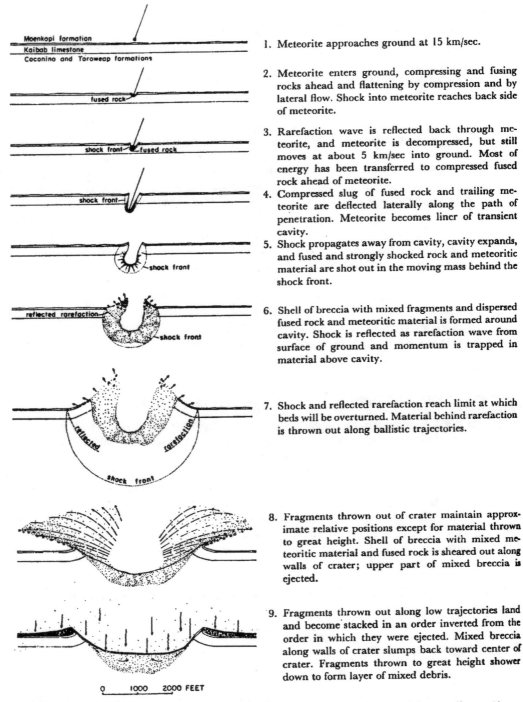

1. Meteorite approaches ground at 15 km/sec.

2. Meteorite enters ground, compressing and fusing rocks ahead and flattening by compression and by lateral flow. Shock into meteorite reaches back side of meteorite.

3. Rarefaction wave is reflected back through meteorite, and meteorite is decompressed, but still moves at about 5 km/sec into ground. Most of energy has been transferred to compressed fused rock ahead of meteorite.

4. Compressed slug of fused rock and trailing meteorite are deflected laterally along the path of penetration. Meteorite becomes liner of transient cavity.

5. Shock propagates away from cavity, cavity expands, and fused and strongly shocked rock and meteoritic material are shot out in the moving mass behind the shock front.

6. Shell of breccia with mixed fragments and dispersed fused rock and meteoritic material is formed around cavity. Shock is reflected as rarefaction wave from surface of ground and momentum is trapped in material above cavity.

7. Shock and reflected rarefaction reach limit at which beds will be overturned. Material behind rarefaction is thrown out along ballistic trajectories.

8. Fragments thrown out of crater maintain approximate relative positions except for material thrown to great height. Shell of breccia with mixed meteoritic material and fused rock is sheared out along walls of crater; upper part of mixed breccia is ejected.

9. Fragments thrown out along low trajectories land and become stacked in an order inverted from the order in which they were ejected. Mixed breccia along walls of crater slumps back toward center of crater. Fragments thrown to great height shower down to form layer of mixed debris.

FIGURE 1–63 Evolution of a meteor crater. Based on Meteor Crater, Arizona (from Shoemaker, 1960)

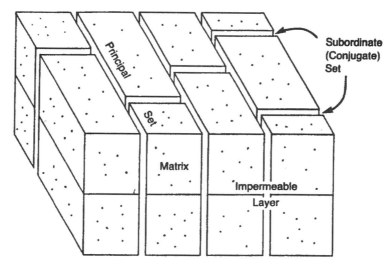

FIGURE 1–64   Model of fractures as they usually occur in the earth as visualized by Dickey. There is one principal set of fractures whose orientation is the same over large areas. There is a subordinate set at right angles, which run from one principal fracture to the next. The spacing depends on the thickness of the hard stratum. There are no open horizontal fractures, but horizontal impermeable layers are common (after Dickey, 1986)

matrix into the wellbore. However, hydrocarbons can flow very efficiently from the tight matrix into the natural fractures. The key to success is to ensure the vertical fractures are intersected via directional or horizontal wells.

Incorrect Pressure Extrapolations

The typical signature of a properly designed buildup test in a naturally fractured carbonate of the Monkman area (British Columbia) is presented in Figure 1–65. Segment A corresponds to a flow period dominated by the natural fractures. Segment B corresponds to interflow from the matrix into the fracture system. Segment C corresponds to a flow period dominated by the composite system of matrix and natural fractures.

The key to a successful test in naturally fractured reservoirs is reaching segment C in such a way that a correct pressure extrapolation may be made.

However, I have witnessed cases in which a buildup test has been terminated at some time during segment B. In this situation an incorrect pressure extrapolation would be made as shown in Figure 1–66 (using the same data of Figure 1–65). This incorrect extrapolation could lead to the erroneous conclusion that the reservoir is small and in depletion. Under these circumstances a good, potentially commercial fracture reservoir could be abandoned.

Poor Completions

The conventional wisdom is that a good well completion should be carried out in intervals that meet certain porosity, permeability, and water saturation cutoff criteria.

Porosity.   The standard practice, in conventional reservoirs is to select intervals with porosities greater than a certain minimum cutoff. The approach has served well in many "conventional unfractured reservoirs."

In naturally fractured reservoirs, however, it has been found through laboratory work, examination of outcrops, cores, and production logs that, other things being equal and for the same physical environment, the degree of natural fracturing and consequently the level of fracture permeability is more intense in those intervals with the lowest porosities.

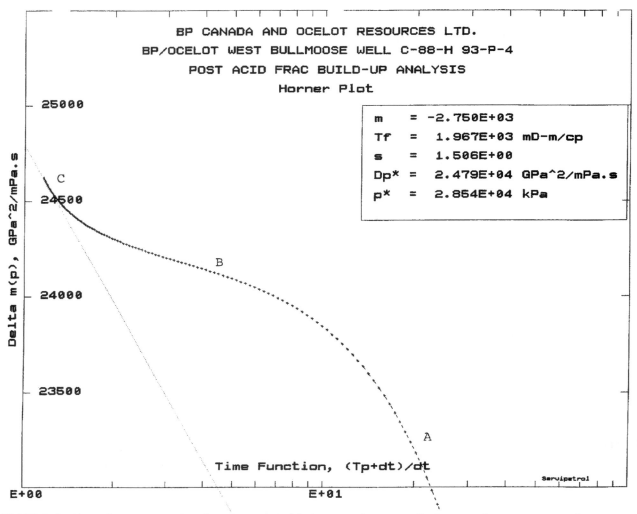

FIGURE 1–65 Horner crossplot showing typical behavior of a naturally fractured reservoir, Monkman area, British Columbia, Canada (after Aguilera, 1992)

Consequently, if we ignore the intervals with the lowest porosities during testing and completion, we might be bypassing the intervals with the best fracture development, i.e., the intervals that will provide the necessary rates to establish commercial production.

Figure 1–67 shows perforated intervals in a well across the naturally fractured Trenton carbonate in Ontario, Canada. A production log indicated that in the upper perforated interval, zone 775.5–780.5m did not contribute any production while zone 780.5–785 m contributed 1009 bo/d. Note that the porosity of the non-productive zone 775.5–780.5 m is higher than the porosity of the oil-producing zone 775–5–780 m. This does not mean that the non-productive zone does not contain any oil. It simply means that the non-productive zone is not fractured. Thus any oil in this zone cannot flow to the wellbore but it can flow very efficiently to the fracture interval at 780.5–785 m and then to the wellbore via the fractures.

The same holds true for the lower two perforated intervals. A production log indicated that zone 799–800.5 m produced 1286 bo/d while zone 802–807 m did not contribute any production. Note once again that the porosity of the oil-producing zone 799–800.5 m is lower than the porosity of unproductive zone 802–807 m. Furthermore notice that the productive interval is thinner than the non-productive interval.

The point that I am trying to emphasize is not that the intervals with the highest porosities should not be tested. Rather that the intervals with the lowest porosities in naturally fractured reservoirs should not be overlooked.

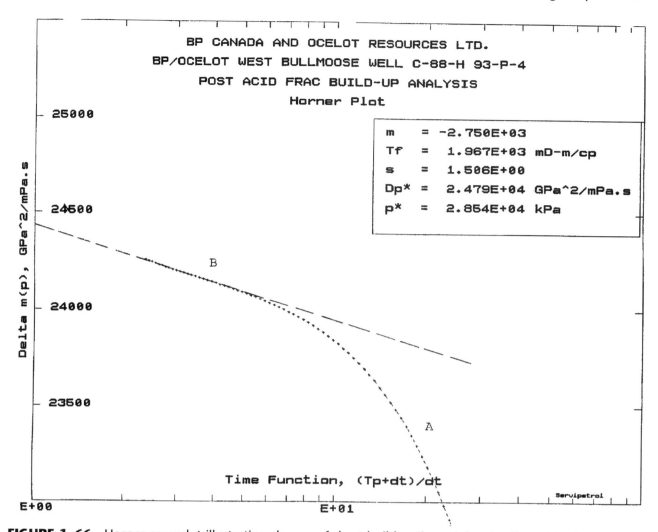

FIGURE 1–66 Horner crossplot illustrating danger of short buildup times (after Aguilera, 1992)

Permeability. The use of permeability cutoff is a well established means of determining net pay in conventional non-fractured reservoirs. In my opinion, however, this is a poor practice in naturally fractured reservoirs.

A matrix permeability of 0.01 md is not capable, in general, of contributing commercial production into a wellbore because of the small surface area of the matrix exposed to the well-bore. However, the same 0.01 md matrix permeability can allow very efficient flow of hydro-carbons from the matrix into a well developed system of natural fractures. This occurs because in this instance there is a large surface area of the matrix exposed to the natural fractures.

The flow of hydrocarbons from a tight matrix into natural fractures has been well demonstrated through microsimulation of naturally fractured cores (Au and Aguilera, 1994), well testing and numerical simulation. Analysis of capillary pressures might lead in some instances to the same conclusion (Aguilera et al, 1992).

Water Saturation. The use of water saturation as a cutoff criteria in naturally fractured reservoirs must be handled carefully. There are instances in which the water saturation of the tight matrix might approach 100% and the hydrocarbon saturation of the natural frac-tures might approach 100%.

If we assume the above saturations for an interval with a fracture porosity of 0.5% and a matrix porosity of 4%, then the average water saturation (S_w) of the composite system seen by the well logs would be equal to:

$$S_w = S_{wm} (1 - v) + S_{wf} v$$

where v is the partitioning coefficient (v = 0.5/4.5 = 0.111), and the subscripts m and f refer to the matrix and the fractures respectively.

For the above data the average water saturation of the composite system seen by the well logs would be in the order of 88.89%. Thus based on a conventional well log interpretation, this interval would be discarded from the testing procedure, even if in reality it would flow hydrocarbons with a water cut very close to zero.

A sound well log interpretation using the appropriate values of the exponents m and n in Archie's equations (or similar type of equations) should help to alleviate this problem For techniques associated with this type of analysis please see Chapter 3.

Bed Thickness.   Sometimes a thin bed might not be considered important from an economic point of view, especially in an exploration well. It must be kept in mind, however, that other things being equal and for the same physical environment, the largest degree of natural fracturing occurs in the thinner beds. This was demonstrated in Figure 1–67 where thin interval 799–800.5 m (1.5 m) produced 1286 bo/d as opposed to thicker interval 802–807 m (5 m) which was non-productive.

Thin high permeability beds might not be important from a hydrocarbon-storage point of view but they might communicate with other parts of the reservoir where substantial amounts of hydrocarbons might be located. As an example, there is an oil well in Alberta producing from a thin fractured interval which has accumulated large amounts of oil. This well is surrounded by at least six dry holes.

DSTs and RFTs.   Although these are powerful techniques, care must be exercised in their interpretation because they are not fully diagnostic in naturally fractured reservoirs. For example, if only the matrix is tested these tools will indicate correctly very low permeability and no flow capabilities. Even if the fracture is tested sometimes the recovery might be only mud that has been lost via the natural fractures during drilling operations.

Failure to Intersect Natural Fractures

Most natural fractures of commercial importance are vertical or subvertical. Under these circumstances vertical wells do not stand the same probability of success as directional or horizontal wells in naturally fractured reservoirs. These type of failures with vertical wells have been a common occurrence.

For example, many years ago three vertical wells were drilled and fraced in the Dilly field's Austin chalk unsuccessfully. As a result the corresponding Lease was dropped back in 1979.

More recently two horizontal wells were drilled in the same general area. The results have been discussed by McNaughton (1992):

> "Both wells had roughly 3000 foot horizontal transects in the chalk. Well "A" was drilled at an acute angle (25°) and well "B" was drilled at a high angle (90°) to my estimated trend (NE-SW) of open fractures in the Dilly area. Well "A" and well "B" have produced about the same gross revenues but well "A" has been producing for about two months longer than "B". Thus, from this very limited production history, well "B" appears to be the better well.
>
> Both wells encountered numerous fracture cells separated by barren septa of chalk lacking fluorescence in cuttings and lacking significant increases in gas production rates during drilling. Reservoir pressures in fracture cells varied greatly. Some cells had very low reservoir pressures while others had sufficiently high pressures to produce semi-controlled blowouts during drilling.
>
> All geologists and engineers attributed low pressure cells to prior drainage in this partly depleted oil field. They also attributed high pressure cells to lack of prior drainage.
>
> Acceptance of these explanations leads to an obvious conclusion, i.e., geologic barriers to cross fracture flow must exist around both high pressure and low pressure cells. The field had been shut in for six or seven years so during that period of time, there appears to have been little recharge or leakage involved in so far as these two different types of fracture cells are concerned."

Another example where a vertical well has failed to yield commercial production is provided by well PV6 in the Palm Valley gas field of Australia (Figure 1–68). The well was air drilled but it did not intercept any significant vertical fractures. The main producing bed

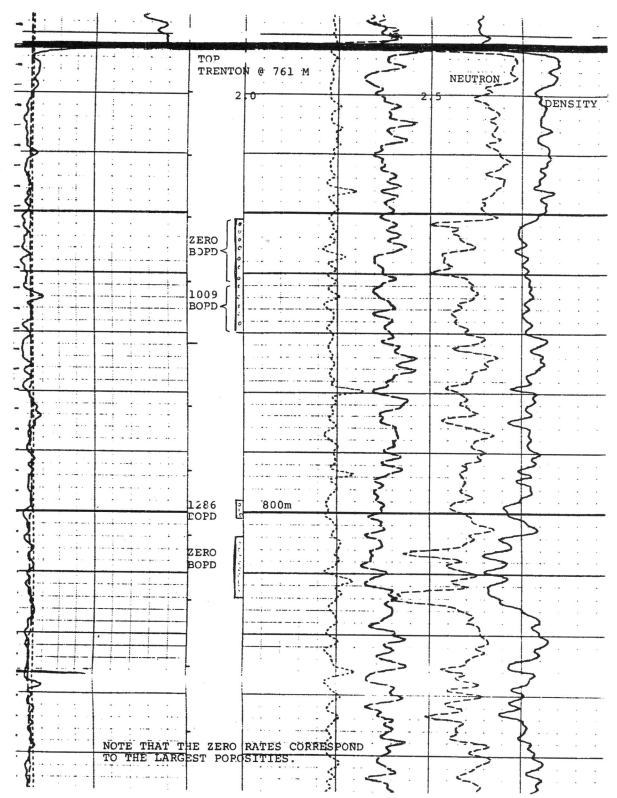

FIGURE 1–67 Well logs of naturally fractured Trenton carbonate, Ontario, Canada. Notice that the best producing intervals have the lowest porosities (after Aguilera, 1992)

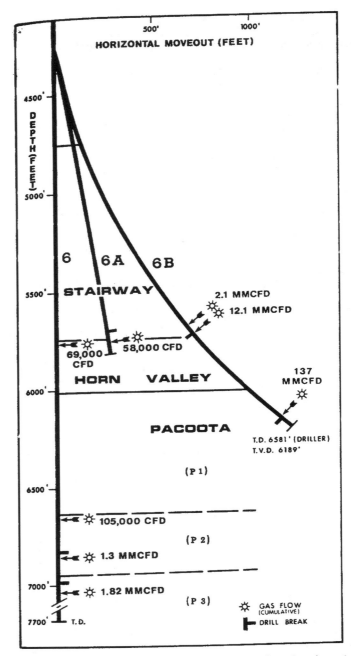

FIGURE 1–68 Well deviation plot - Palm Valley 6, 6A, 6B—showing location of drill breaks and cumulative gas flows.

(Pacoota P1) in other wells was non-productive in well PV6. Maximum gas rate in a deeper bed was 1.82 MMscfd, which was not economic for this area. The well was subsequently deviated as well 6–A but it never reached the desired deviation due to mechanical problems.

A new attempt from the same borehole was provided by well 6B, which intercepted a high inclination natural fracture. As shown in the schematic of Figure 1–68 the Pacoota P1 formation yielded 137 MMscfd from this directional hole while the vertical well did not produce any gas from the same formation.

It is important to visualize in this case that the tight Pacoota sandstone matrix is gas saturated. The very low matrix permeability did not allow a commercial flow rate from the matrix into the vertical wellbore. The same tight matrix, however, can flow very efficiently into the natural fractures.

Conclusion

Based on my experience to date I am convinced that many hydrocarbon naturally fractured reservoirs around the world have not become profitable discoveries and have been abandoned because of (1) incorrect pressure extrapolations, (2) poor completions, and/or (3) failure to intercept the natural fractures.

Proper geological and engineering techniques developed specifically for evaluating and exploiting naturally fractured reservoirs should help to avoid these potential problems.

PRACTICAL APPLICATIONS

Estimating Type of Sub-surface Natural Fractures to be Found at Various Depths

The state of stress can provide an insight into the type of fractures to be found at a given depth. This is better illustrated with an example (1–8).

Example 1–8. The following information is available for a given area:

Depth, h = 1000 ft
Overburden grain density, ρ = 2.71 gr/cc
Poisson's ratio, μ = 0.3
Tectonic stress in horizontal direction xx = 1700 psi
Tectonic stress in horizontal direction yy = 1100 psi

Estimate the probable depths at which vertical, horizontal and strike slip fractures might be found.

Solution. The vertical stress, σ_{zz}, is given by:

$$\sigma_{zz} = \rho gh$$

$$= 2.71 \text{ gr/cc} \times 981 \text{ cm/sec}^2 \times (1000 \text{ ft} \times 100/3.28)$$

$$= 81.05 \times 10^6 \text{ dynes/cm}^2 = 1{,}173 \text{ psi @ 1,000 ft} \qquad (1\text{–}60)$$

Note also, σ_{zz} = 2.71 gr/cc × 0.433 psi/ft × 1,000 ft = 1,173 psi
The horizontal stress, σ_{xx}, is given by:

$$\sigma_{xx} = \left(\frac{\mu}{1-\mu}\right)\sigma_{zz} + (tectonic\ stress)_{xx} \qquad (1\text{–}61)$$

$$= \left(\frac{0.3}{1-0.3}\right)1{,}173 + 1{,}700 = 2{,}202\ psi$$

The horizontal stress, σ_{yy}, is given by:

$$\sigma_{yy} = \left(\frac{\mu}{1-\mu}\right)\sigma_{zz} + (tectonic\ stress)_{yy} \qquad (1\text{–}62)$$

$$= \left(\frac{0.3}{1-0.3}\right)1{,}173 + 1{,}100 = 1{,}602\ psi$$

The same approach is used to calculate stresses at other depths of interest. A crossplot of depth vs. stress following this procedure is presented in Figure 1–69.

At formations shallower than about 1,650 ft, $\sigma_{xx} > \sigma_{yy} > \sigma_{zz}$, which suggests that in all likelihood horizontal fractures will be generated. Wrench or strike slip faulting and fracturing will be generated between approximately 1650 and 2550 ft where $\sigma_{xx} > \sigma_{zz} > \sigma_{yy}$. Vertical fractures will most likely be generated below 2550 ft where $\sigma_{zz} > \sigma_{xx} > \sigma_{yy}$. These results should be considered only as estimates, since different beds might have variations in environmental and intrinsic properties. Furthermore, a formation might have been subjected to many different stress states over time. Finally keep in mind that tectonic stresses are non-linear at shallow depths and get closer to zero as they approach the surface.

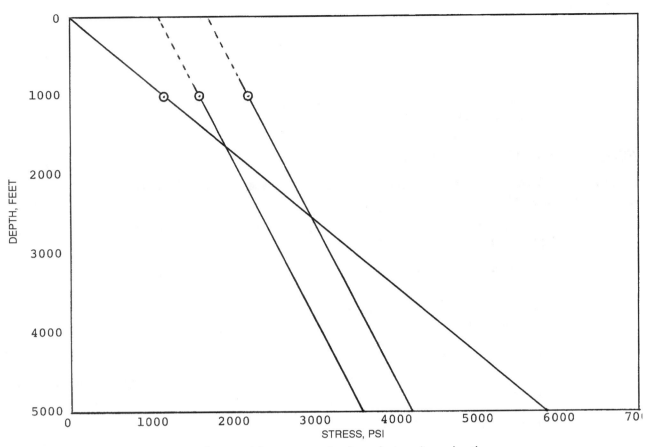

FIGURE 1–69 Estimating type of natural fractures to be found at various depths.

Estimating Reduction in Fracture Permeability as a Function of In-Situ Normal and Shear Stresses

Studies conducted by Stearns (1994) indicate that both in-situ normal (σ_n) and shear (τ) stresses and the ratio τ/σ_n have a pronounced effect on fracture permeabilities. Historically most of the reservoir engineering work and simulation studies of naturally fractured reservoirs have been conducted using reductions in fracture permeabilities determined from normal stress laboratory studies without paying much attention to the shear stresses. Not including the effect of shear stresses might lead to erroneous estimates of ultimate recoveries. The approach suggested by Stearns (1994) to handle this problem is better illustrated with a couple of examples (1–9 and 1–10). This presentation follows very closely Stearns original work.

Example 1–9. The following data are available for a hypothetical naturally fractured reservoir in an anticline:

Depth = 18,000 ft
Lithostatic pressure gradient = 1 psi/ft
Pore pressure gradient 0.4 psi/ft

Reductions of normalized fracture permeabilities as a function of normal (σ_n) and shear stresses (τ) are presented in Figure 1–70. A contoured surface of pseudo fracture permeability above the normal stress-shear stress plane is presented in Figure 1–71. The data are normalized to pseudo fracture permeability at $\sigma_n = 35$ psi. These data have been acquired using specialized experimental equipment that allows τ and σ_n be increased together.

The in-situ stress is non-hydrostatic. The vertical component of stress exceeds the horizontal component of stress by a factor of 3. The assumption is made that only fold-

A

B

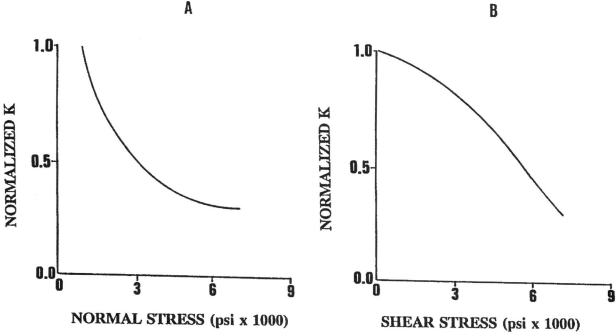

FIGURE 1–70 (A) Normalized pseudo fracture permeability versus normal stress across an artificial fracture in Jackfork Sandstone. Tests run under different hydrostatic stress states so that shear fracture stress remains zero during all the tests. (B) Normalized pseudo fracture permeability versus shear stress across artificial fracture in Jackfork Sandstone (average of two tests). Data normalized to pseudo fracture permeability at normal stress for first application of shear stress (after Stearns, 1994)

related natural fractures of Type 1 and Type 2 are present in this anticline (Refer to Figure 1–15). Fracture dips for Type 1 and Type 2 fracture assemblages vs bed dip assuming an acute angle of 60° between Type 1 shear fractures (T_{1s}) and an acute angle of 40° between Type 2 shear fractures (T_{2s}) are presented in Figure 1–72. The fracture morphology is such that surface roughness (asperity size) and fracture spacing do not change throughout the fold trap. The fluid flow direction along a fracture is parallel to the strike of the fracture.

Calculate shear stress and normal stress acting across shear fractures of Type 1 and Type 2, and extension fractures of Type 2. Run the calculations for various fracture dips.

Solution. The gross overburden is 18,000 ft $\times$ 1 psi/ft = 18,000 psi. The pore pressure is 18,000 ft $\times$ 0.4 psi/ft = 7,200 psi. The vertical stress, σ_1, is 18,000 – 7,200 = 10,800 psi. The horizontal stress ($\sigma_H = \sigma_2 = \sigma_3$) is 10,800/3 = 3,600 psi.

The shearing (τ) and normal (σ) stress are given by:

$$\tau_\alpha = \frac{\sigma_1 - \sigma_3}{2} \sin 2 \, (90 - \alpha) \tag{1–63}$$

and,

$$\sigma_\alpha = \frac{\sigma_1 + \sigma_3}{2} - \frac{\sigma_1 - \sigma_3}{2} \cos 2 \, (90 - \alpha) \tag{1–64}$$

where α is the fracture dip and other nomenclature has been defined previously. Including the previously estimated values of σ_1 and σ_3 leads to:

$$\tau_\alpha = 3,600 \sin 2 \, (90 - \alpha) \text{ psi}$$

and

$$\sigma_\alpha = 3,600 \, [\, 2 - \cos 2 \, (90 - \alpha) \,] \text{ psi}$$

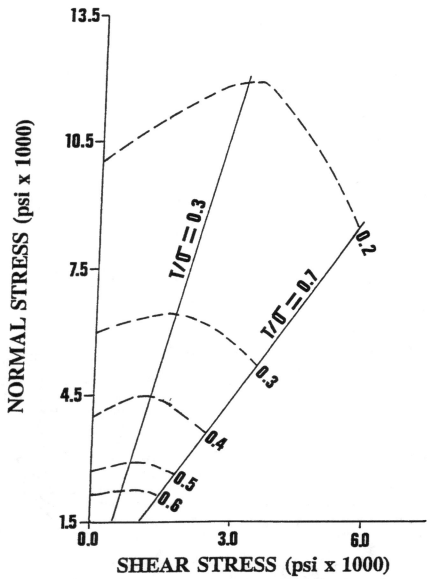

FIGURE 1–71 Contoured surface of pseudo fracture permeability above the normal stress-shear stress plane. Data normalized to pseudo fracture permeability at σ_n = 35 psi (after Stearns, 1994)

Thus for any fracture dip the shearing and normal stresses can be calculated as shown on Table 1–6, where the first column is bed dip, the second column is fracture dip read from Figure 1–72, the third column is shear stress and the last column is normal stress.

Notice that the results presented in Table 1–6 have been calculated for Type 1 shear fractures (T_{1s}), Type 2 shear fractures (T_{2s}), and Type 2 extension fractures (T_{2e}). The Type 1 extension fractures (T_{1e}) always remain vertical independent of bed dip and consequently T_{1e} fractures always have the same normal stress (σ_n) and the same shear stress (τ) acting across them independent of the bed dip. A common observation is that extension fractures tend to be more mineralized than shear fractures.

Example 1–10. Estimate reductions in fracture permeability as a function of bed dip for shear fractures and extension fractures of Type 1 and Type 2. Use data and results from the previous example.

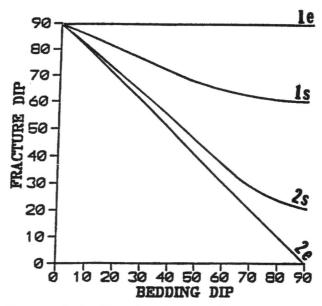

FIGURE 1–72 Fracture dip for Type 1 and Type 2 fracture assemblages versus bed dip assuming an acute angle of 60° between Type 1 shear fractures and an acute angle of 40° between Type 2 shear fractures (after Stearns, 1994)

TABLE 1–6 Shear and Normal Stresses for Various Bed and Fracture Dips (after Stearns, 1994)

Bed Dip	Fracture Dip	τ (psi)	σ (psi)
Type 1 Shear Fractures			
0°	90°	0	3600
15°	83°	870	3700
30°	78°	1460	3910
45°	69°	2400	4525
60°	64°	2835	4980
75°	61°	3050	5290
90°	60°	3115	5400
Type 2 Shear Fractures			
0°	90°	0	3600
15°	76°	1690	4020
30°	64°	2835	4980
45°	50°	3545	6575
60°	37°	3460	8190
75°	28°	2985	9215
90°	20°	2314	9960
Type 2 Extension Fractures			
0°	90°	0	3600
15°	75°	1800	4080
30°	60°	3120	5400
45°	45°	3600	7200
60°	30°	3120	9000
75°	15°	1800	10320
90°	0°	0	10800

The values of the shear (τ) and normal (σ) stress components across Type 1 shear, and Type 2 shear and extension, fractures for selected bed dips. σ_1 and σ_3 values used in these calculations are taken from the example case in the text.

TABLE 1–7 Normalized Pseudo Fracture Permeabilities for Various Bed Dips (after Stearns, 1994)

Bed Dip	K_{FS1}/k_{FS1V}	K_{FS2}/K_{FS2V}	K_{FE2}/K_{FE2V}	K_{FE1}/K_{FE1V}
0°	1.00	1.00	1.00	1.00
30°	1.00	0.79	0.71	1.00
60°	0.75	0.60	0.57	1.00
90°	0.71	0.55	0.41	1.00

Normalized pseudo fracture permeabilities for Type 1 shear fractures (K_{FS1}/K_{FE1V}), Type 2 shear fractures (K_{FS2}/K_{FS2V}), Type 2 extension fractures (K_{FE2}/K_{FE2V}) and Type 1 extension fractures (K_{FE1}/K_{FE1V}) as a function of bed dip (normalized to permeabilities for fracture when vertical, i.e., at minimum normal stress and no shear stress). Permeabilities are read from Figure 1-71 according to τ and σ values taken from Table 1-6 for each fracture orientation.

Solution. The normalized permeabilities for all fractures when bed dip is equal to 0°, i.e., when all fractures are vertical is equal to 1.0. The permeabilities of non-vertical fractures are presented as a fraction of the maximum permeability of the fracture when it is vertical in order to compare the changes as a function of dip.

If for example the bed dip is 60° then for Type 2 shear fractures (T_{2s}) the fracture dip is 37° from Figure 1–72 and Table 1–6. For this case the shear stress (τ) is 3460 psi and the normal stress (σ) is 8190 psi from Table 1–6. Input these two stresses on Figure 1–71 and read a normalized pseudo fracture permeability (k_{Fs2}) of about 0.264. This pseudo fracture permeability is normalized to permeabilities for fractures when vertical, i.e., at minimum normal stress (3,600 psi) and zero shear stress. From Figure 1–71 the pseudo fracture vertical permeability for shear fractures of Type 2 (k_{Fs2v}) is approximately 0.44. Thus the ratio $k_{Fs2}/k_{fs2v} = 0.264/0.44 = 0.60$

This indicates that a Type 2 shear fracture (T_{2s}) with a 37° dip (bed dip is 60°) will have a permeability that is 60% of a vertical (90° dip) Type 2 shear fracture. In other words the T_{2s} fracture will lose 40% of its original permeability by decreasing the fracture dip from 90° to 37°.

Results of calculations for other bed dips are presented in Table 1–7, where k stands for permeability, and the subscripts, F = fracture, S = shear, E = extension, V = vertical, 1 = Type 1 fracture, and 2 = Type 2 fracture.

Figure 1–71 illustrates the importance of the τ/σ_n ratio on fracture permeability reduction. For a given normal stress, fractures subjected to a moderate τ/σ_n ratio ($\tau/\sigma_n = 0.3$) have higher permeabilities than do fractures subjected either to no shear stress or to a higher τ/σ_n ratio of for example 0.7.

Estimating Optimum Horizontal Drill Direction

As discussed previously in this chapter slanted and horizontal boreholes provide the most sound approach for optimum exploitation of naturally fractured reservoirs. The idea is to drill the directional hole in such a way that it goes perpendicular to the orientation of the open fractures. If there is more than one set of open fractures the best rates will come from wellbores that intercept the most permeable fractures. Care must be exercised, however, when water is present in the reservoir.

An approach for estimating the optimum horizontal drill direction in these situations has been presented by Nolen-Hoeksema and Howard (1987). For the case in which the open fractures are vertical or close to vertical the optimum drilling direction is given by the equation:

$$K_f = \arctan \left[\frac{\dfrac{(e_1)^3}{(S_1)^*} \sin\theta_1 + \dfrac{(e_2)^3}{(S_2)^*} \sin\theta_2}{\dfrac{(e_1)^3}{(S_1)^*} \cos\theta_1 + \dfrac{(e_2)^3}{(S_2)^*} \cos\theta_2} \right] \tag{1–65}$$

where:

$$K_f = \text{Optimum drilling direction}$$
$$e_n = \text{Width (aperture) of fracture set n.}$$
$$\theta_n = \text{Direction perpendicular to trend of fracture set n}$$
$$S^*_n = \text{Fracture spacing (distance between fractures) of set n.}$$

The use of equation 1–65 is better illustrated with a couple of examples (1–11 and 1–12).

Example 1–11. A naturally fractured reservoir has two fracture sets, each normal to bedding and forming traces on a horizontal surface. Set 1 trends 300° (N60°W) and set 2 trends 0° (North). Fracture spacing for sets 1 and 2 are 2 ft and 1 ft, respectively. Fracture widths are 0.002 in. for each set. Determine the optimum drilling direction.

Solution.

$\theta_1 = 30°$		$\theta_2 = 90°$
$(S_1)^* = 24$ in.		$(S_2)^* = 12$ in.
$e_1 = 0.002$ in.		$e_2 = 0.002$ in.

Ideally, the well should go perpendicular to the natural fractures. Thus, angle θ_1 is 30° and angle θ_2 is 90°. Inserting these data into eq. 1–65 leads to an optimum drilling direction (K_f) of 70.9°.

Example 1–12. Determine the optimum drilling direction in example 1–11 if fracture with e_2 is 0.001 in.

Solution. If e_2 is 0.001 inches the optimum drilling direction is 40.9° from equation 1–65. Notice that changing e_2 from 0.002 in. to 0.001 in. leads to a change in optimum drilling direction from 70.9° to 40.9°, a shift of 30° from the calculation presented in Example 1–11. The drastic variation emphasizes the need for accurate input data when performing all these calculations.

REFERENCES

Aguilera, Roberto and H.K. van Poollen. "Geologic Aspects of Naturally Fractured Reservoirs Explained." *Oil and Gas Journal* (Dec. 18, 1978), 47–51.

Aguilera, Roberto. "Log Analysis of Gas-Bearing Fractured Shales in the Saint Lawrence Lowlands of Quebec." SPE 7445 Presented at the Annual Meeting of SPE of AIME, Houston (October 1–4, 1978).

Aguilera, Roberto. "Exploring for Naturally Fractured Reservoirs-A Petroleum Engineer's Point of View." Abstract in *Bulletin. AAPG* (Feb. 1982) v.66, p. 242.

Aguilera, Roberto. "Relative Permeability Concepts for Predicting the Performance of Naturally Fractured Reservoirs", *Journal of Canadian Petroleum Technology* (Sept.—Oct. 1982), 41–48.

Aguilera, Roberto. "Exploring for Naturally Fractured Reservoirs." SPWLA Twenty-Fourth Annual Logging Symposium, paper C (June 27–30, 1983).

Aguilera, Roberto. "Determination of Subsurface Distance Between Vertical Parallel Natural Fractures Based on Core Data." *Bulletin. AAPG* (July 1988), 845–851.

Aguilera, Roberto. "Undiscovered Naturally Fractured Reservoirs, Why and How?" *The Petroleum Explorer* (December 1992), 8–11.

Aguilera, R., J.S. Artindale, G.M. Cordell, M.C. Ng, G.W. Nicholl, and G.A. Runions. *Horizontal Wells, Formation Evaluation, Drilling and Production, Including Heavy Oil Recovery.* Contributions in Petroleum Geology & Engineering No. 9, Gulf Publishing Company, Houston, Texas (1991), 401 p.

Aguilera, R., A.D. Au, and L.N Franks, "Isochronal Testing of a Naturally Fractured Gas Reservoir—A Case History" *The Journal of Canadian Petroleum Technology* (October 1993), 25–30.

Aguilera, Roberto. "Advances in the Study of Naturally Fractured Reservoirs." *The Journal of Canadian Petroleum Technology* (May 1993), 24–26.

Aguilera, Roberto. *AAPG Fractured Reservoir School Notes*, Great Falls, Montana (1982–1994).

Aguilera, R, Au, A.D. and Franks, L.N. "Well Test Analysis of a Naturally Fractured Gas Reservoir—A Case History," *Journal of Canadian Petroleum Technology* (April 1992), 46–51.

Araki, N. and K. Susumu. "A Discovery of the Ayukawa Oil and Gas Field, Akita Prefecture." *Journal of the Japanese Association for Petroleum Technology* (March 1993), 119–127.

Au, A.D. and R. Aguilera. "Micro-simulation of Naturally Fractured Cores." Petroleum Society of CIM paper 94–79 presented at the Annual Technical Conference in Calgary, Canada (June 12–15, 1994).

Alpay, O.A. "Application of Aerial Photographic Interpretation to the Study of Reservoir Natural Fracture Systems." SPE 2567 presented at the 44th Annual Meeting of the SPE of AIME, Denver (September 28—October 1, 1969).

Amaefule, J.O. et al. "Laboratory Determination of Effective Liquid Permeability in Low Quality Reservoir Rocks by the Pulse Decay Technique", SPE paper 15149 presented at the 56th Californian regional meeting held in Oakland, California (April 2–4, 1986).

American Petroleum Institute: *API Recommended Practice for Core-Analysis Procedure*, American Petroleum Institute, Dallas (1960) API RP 40, pp. 18, 20, 55.

Amyx, J.W., D.M. Bass, and R.L. Whiting. *Petroleum Reservoir Engineering—Physical Properties*, McGraw-Hill Book Company (1960), 538.

Anderson, T.O., and E.J. Stahl. "A Study of Induced Fracturing Using and Instrumental Approach." *Journal of Petroleum Technology* (February 1967).

Atkinson, B., and D. Johnston. "Core Analysis of Fractured Dolomite in the Permian Basin." *Transactions*. AIME (1949), 128-132.

Bagnel, W.D. and W.M. Ryan. "The Geology, Reserves and Production Characteristics of the Devonian Shales in Southwestern West Virginia." Proceedings of the Seventh Appalachian Petroleum Geology Symposium, Morgantown, West Virginia. (March 1–4, 1976), 41.

Baker, D.A., and P.T. Lucas.: "Strat Trap Production May Cover 280 Plus Square Miles." *World Oil* (April 1972), 65–68.

Barss, D.L. and F.A. Montandon. "Sukunka-Bullmoose Gas Fields: Models for a Developing Trend in the Southern Foothills of Northeast British Colombia." *Rocky Mountain Association of Geologists* (1982), 549–574.

Beach, J.H. "Geology of Edison Oil Field, Kern County, California." Structure of Typical American Oil Fields. *Bulletin*. AAPG (1948), 58–85.

Beiers, R.J. "Quebec Lowlands: Overview and Hydrocarbon Potential." Proceeding of the Seventh Appalachian Petroleum Geology Symposium, Morgantown, West Virginia (March 1–4, 1976), 142.

Belfield, W.C. et al. "South Ellwood Oil Field, Santa Barbara Channel, California, A Monterey Formation Fractured Reservoir." *Petroleum Generation and Occurrence in the Miocene Monterey Formation, California*, edited by Caroline Isaacs and Robert Garrison, Society of Economic Paleontologists and Mineralogists, Los Angeles, California (May 20–22, 1983).

Belfield, W.C. and J.P. Sovich. "Fracture Statistics from Horizontal Wells." The Canadian SPE/CIM/CANMET International Conference on Recent Advances in Horizontal Well Applications, Calgary, Canada (March 20–23, 1994)

Bell, J.S. and E.A. Babcock. "The Stress Regime of the Western Canadian Basin and Implications for Hydrocarbon Production." *Bulletin of Canadian Petroleum Geology*, vol. 34, No. 3 (September 1986) 364–378.

Bergosh, J.L. and G.D. Lord. "New Developments in the Analysis of Cores from Naturally Fractured Reservoirs," paper SPE 16805 presented at the 1987 SPE Annual Technical Conference and Exhibition, Dallas (Sept. 27–30, 1987).

Bergosh, J.L., T.R. Marks, and A.F. Mitkus. "New Core Analysis Techniques for Naturally Fractured Reservoirs." SPE paper 13653 presented at California Regional Meeting held in Bakersfield, California (March 27–29, 1985).

Billings, M.P. *Structural Geology*. Prentice Hall Inc., Englewood Cliffs, New Jersey (1972), 364 p.

Bloxsom Lynn, H. "Seismic Detection of Oriented Fractures", *Oil and Gas Journal* (August 4, 1986) 54–55.

Bock, H. "Ueber die Abhaengigkeit von Kluftabstaenden und Schichtmaechtigkeiten". *Neues Jahrbuch fur Geologie and Palaeontologie Monatshefte*, (1971), n. 9. p. 517–531.

Braunstein, J. "Fracture-Controlled Production in Gilbertown Field, Alabama." *Bulletin*. AAPG (February 1953), 245–249.

Butler, M.D. "Impact Craters: The Key Ingredient?" *AAPG Explorer* (April 1989), 18.

Capuano, R.M. "Evidence of Fluid Flow in Microfractures in Geopressured Shales." *Bulletin*. AAPG (August 1993), 1303–1314.

Cooper, M. "The Analysis of Fracture Systems in Subsurface Thrust Structures from the Foothills of the Canadian Rockies." *Thrust Tectonics*, (ed.) K.R. McClay, Unwin Hyman (1991), 18p.

Craft, B.C., and M.F. Hawkins. *Applied Petroleum Reservoir Engineering*. Prentice-Hall Inc. (1962).

Daniel, E.J. "Fractured Reservoirs of Middle East." *Bulletin*. AAPG (May 1954), 774–815

Dech, J.A. et al. "New Tools Allow Medium-Radius Horizontal Drilling." *Oil and Gas Journal* (July 14, 1986) 95–99.

Delgado, O.R., and E.G. Loreto. "Reforma's Cretaceous Reservoirs: An Engineering Challenge." *Petroleum Engineer* (December 1975), 56–66.

Dempsey, J.C., and J.R. Hickey. "Use of Borehole Camera for Visual Inspection of Hydraulically Induced Fractures." *Producers Monthly* (April 1958), 18–21.

Dickey, P. *Petroleum Development Geology*, PennWell Books, Second Edition (1986)

Eggleston, W.S. "Summary of Oil Production from Fractured Rock Reservoirs in California." *Bulletin*. AAPG (July 1948), 1352–1355.

Elkins, L.F. "Reservoir Performance and Well Spacing, Spraberry Trend Area Field of West Texas." *Transactions*. AIME (1953) 301–304.

Fatt, I. "The Network Model of Porous Media, II Dynamic properties of a Single Size Tube Network", *Transactions*, AIME (1956) 207, 160–163.

Finn, F.H. "Geology and Occurrence of Natural Gas in Oriskany Sandstone in Pennsylvania and New York." *Bulletin*. AAPG (March 1949), 303–335.

Focardi, P., S. Gandolfi, and M. Mirto. "Frequency of Joints in Turbidite Sandstone." Second Congress of International Society of Rock Mechanics Proceedings, (1970), v. 1, p.97–101.

Fons, L.C. "Downhole Camera Helps Solve Production Problems." *World Oil* (1960), 150–152.

Fraser, C.D., and C.D. Pettit. "Results of a Field Test to Determine the Type and Orientation of a Hydraulically Induced Formation Fracture." *Journal of Petroleum Technology* (1961), 463–466.

Fraser, H.J., and L.C. Graton. "Systematic Packing of Spheres--With Particular Relation to Porosity and Permeability." *Journal of Geology* (November–December 1935), 785–909.

Freeman, H.A., and S.G. Natanson. "A Reservoir Begins Production: A Study on Planning and Cooperation", *Sixth World Petroleum Congress*, Section II (June 1963), 407–414.

Friedman, M. and D.E. McKiernan, "Extrapolation of Fracture data from Outcrops of the Austin Chalk to Corresponding Petroleum Reservoirs at Depth." The Canadian SPE/CIM/CANMET International Conference on Recent Advances in Horizontal Well Applications, Calgary, Canada (March 20–23, 1994)

Friedman, Melvin. *AAPG Fractured Reservoir School Notes*, Great Falls, Montana (1982–1994).

Fritz, R.D. et al. *Geological Aspect of Horizontal Drilling*, AAPG Continuing Education Course Note Series #33 (1991).

Georgi, D.T. and S.C. Jones, "Application of Pressure-Decay Profile Permeametry to Reservoir Description." SPENC 9212 presented at the 16th Annual SPE (Nigeria Council) International Conference and Exhibition held in Lagos, Nigeria (August 26–27, 1992).

Griggs, D.T. and J.W. Handin. "Observations of Fracture and a Hypothesis of Earthquakes." *Geological Society of America*, Mem 79 (1960), 347–364.

Griggs D.T. "Deformation of Yule Marble: Part IV, Effects at 150°." *Geological Society of America Bulletin* (1951), vol. 62, 1385–1406.

Griggs D.T. "Creep of Rocks." *Journal of Geology* (1939), vol. 47, 225–251.

Gronseth, J.M. and P.R. Kry, "In Situ Stresses and the Norman Wells Expansion Project." Petroleum Society of CIM paper 87–38–57 presented ast the 38th Annual Technical Meeting held in Calgary, Canada (June 7–10, 1987).

Guangcan, Hu. "A Brief Introduction to Oil and Gas Geology in Sichuan Basin," SPG/SEG Symposium (1990).

Hagen, Kurt. "Mapping of Surface Joints on Air Photos Can Help Understand Waterflood Performance at North Burbank Unit, Osage and Kay Counties, Oklahoma." M.S. Thesis, University of Tulsa, (1972).

Haimson, B.C. "The Hydrofracturing Stress Measuring Method and Recent Field Results", *International Journal of Rock Mechanics, Mining Sciences and Geomechanical Abstracts* (1978), v. 15, p. 167–178.

Handin, J.W. "Experimental Deformation of Sedimentary Rocks Under Confining Pressure: Pore Pressure Test." *Bulletin*. AAPG (1963), 717–755.

Hanna, M.A. "Fracture Porosity in Gulf Coast," *Bulletin*. AAPG (February 1953), 266–281.

Harp, L.D. "Do not Overlook Fractured Zones." *World Oil* (April 1966), 119–123.

Hast N. "The Measurement of Rock Pressure in Mines", *Syveriges Geologisko Undersökning*, Avhanlingar Och Uppsator, ser. C, 560, Arksbok, (1958) vol. 52, pt. 3, 183 pp.

Hensel, W.M. "A Perspective Look at Fracture Porosity." *SPE Formation Evaluation* (December 1989), 531–534.

Hilchie, D.E., and S.J. Pirson. "Water Cut Determination from Well Logs in Fractured and Vuggy Formations", *Transactions*, SPWLA (Dallas, 1961).

Hohlt, R.B. "The Nature of Origin of Limestone Porosity." *Colorado School of Mines Quarterly* (1948) No. 4.

Howard, J.H. and R.C. Nolen-Hoeksema, "Description of Natural Fracture Systems for Quantitative Use in Petroleum Geology." *Bulletin*. AAPG (February 1990), 151–162.

Huaibu, L. et al. "Upper Permian Carbonate Buildups and Associated Lithofacies, Western Hubei-Eastern Sichuan Provinces, China." *Bulletin*. AAPG (September 1991), 1447–1467.

Hunt, W.C., L.G. Collings, and E.A. Skobelin. *Expanding Geospheres, Energy and Mass Transfers From Earth's Interior*, Polar Publishing, Calgary, Canada (1992), 421 p.

Hunter, C.D., and D.M. Young. "Relationship of Natural Gas Occurrence and Production in Eastern Kentucky (Big Sandy Gas Field) to Joints and Fractures in Devonian Bituminous Shale." *Bulletin*. AAPG, No. 2. (February 1953), 282–299.

Janot, P. "Determining the Elementary Matrix Block in a Fissured Reservoir—Eschau Field, France." *Journal of Petroleum Technology* (May 1973) 523–530.

Jones, S.C. "The Profile Permeameter—A New, Fast, Accurate Minipermeameter." SPE 24757 presented at the 67th Annual SPE Conference and Exhibition held in Washington D.C. (October 1992).

Kahle, C.F. and J.C. Floyd. "Stratigraphic and Environmental Significance of Sedimentary Structures in Cayugan (Silurian) Tidal Flat Carbonates, Northwestern Ohio." *Bulletin*. AAPG, (1971), vol 82, 2071–2098.

Kamath, J., R.E. Boyer, and F.M. Nakagawa. "Characterization of Core-Scale Heterogeneities Using Laboratory Pressure Transients." *SPE Formation Evaluation* (September 1992) 219–227.

Kanamori, Kiroo: "Earthquake Prediction." California Institute of Technology (1974).

Kelton, F.C. "Analysis of Fractured Limestone Cores," *Transactions*. AIME, 189 (1950) 225–234.

Kempthorne, R.H., and J.P.R. Irish, "Norman Wells—a New Look at One of Canada's Largest Fields." *Journal of Petroleum Technology* (June, 1981) 985–991.

Koning, T. and F.X. Darmono, "The Geology of the Beruk Field, Central Sumatra, Oil Production from Pre-Terciary Basement Rocks," Indonesian Petroleum Association, 13th Annual Convention, Jakarta, Indonesia (May 29–30, 1984).

Kuykendall, M.D. et al. "Petrophysical Evaluation of Reservoir Lithofacies of a Complex Impact Crater: Ames Crater, Northern Shelf, Anadarko Basin," Abstract in *The Log Analyst* (March–April 1994), 99.

Ladeira, F.L. and N.J. Price. "Relationship Between Fracture Spacing and Bed Thickness." *Journal of Structural Geology*, (1981), v.3, p.179–183.

Landes, K.K. *Petroleum Geology.* 2 ed. John Wiley and Sons Inc. (1959).

Lane, H.W. "Oil Production in Iran." *Oil and Gas Journal* (August 18, 1949), 128.

Law, B.E. and D.D. Rice. (editors). *Hydrocarbons from Coal*, AAPG Studies in Geology #38 (1993), 400p.

Levorsen, A.I. *Geology of Petroleum.* 2 ed. W.H. Freeman and Company (1967).

Lin, P. and T.G. Ray. "A New Method for Direct Measurement of In-Situ Stress Directions and Formation Rock Properties." *Journal of Petroleum Technology* (March 1994), 249–254.

Locke, L.C., and J.E. Bliss. "Core Analysis Technique for Limestone and Dolomite." *World Oil* (September 1950), 204.

Lowenstan, H.A. "Marine Pool, Madison County, Illinois, Silurian Reef Producer." Illinois Geological Survey 131 (1942).

McLellan, P. "In Situ Stress Prediction and Measurement by Hydraulic Fracturing, Wapiti, Alberta." *Journal of Canadian Petroleum Technology* (March–April 1988), 85–95.

McNaughton, D.A. "Dilatancy in Migration and Accumulation of Oil in Metamorphic Rocks." *Bulletin.* AAPG No.2 (February 1953) 217–231.

McNaughton, D.A., and F.A. Garb. "Finding and Evaluating Petroleum Accumulations in Fractured Reservoir Rock." *Exploration and Economics of the Petroleum Industry*, Vol. 13 Matthew Bender & Company Inc. (1975).

McNaughton, D.A. Personal Communication with Roberto Aguilera (April 20, 1992).

McQuillan, H. "Small-Scale Fracture Density in Asmari Formation of Southwest Iran and its Relations to Bed Thickness and Structural Setting." *Bulletin.* AAPG (1973), vol. 4.7, No. 12, 2367–2385.

McQuillan, H. "Fracture-Controlled Production from the Oligo-Miocene Asmari Formation in Gach Saran and Bibi Hakimeh Fields, Southwest Iran." P.O. Roehl and P.W. Choquette, eds, *Carbonate Petroleum Reservoirs*, (1985), pp. 511–524.

Magara, K. "Mechanisms of Natural Fracturing in a Sedimentary Basin." *Bulletin.* AAPG (April 1987), 357–367.

Martin, G.H. "Kluftporosität und Permeabilität in Erdöl-Erdas Laggerstätten." *Z. angew. Geowiss*, Heft 11, 5–5–19 (1992).

Martin, G.H. "Petrofabric Studies May Find Fracture-Porosity Reservoirs." *World Oil* (February 1, 1963), 52–54.

Mead, W.J. "The Geologic Role of Dilatancy." *Journal of Geology* (1925), 685–698.

Milne, N.A. and D.C. Barr. "Subsurface Fracture Analysis, Palm Valley Gas Field." *The APEA Journal* (1990), 321–341.

Muir, J.M. "Limestone Reservoir Rocks in the Mexican Oil Fields." Problems of Petroleum Geology. *Bulletin.* AAPG (1934), 382.

Muñoz Espinoza, R.E. "Fractured Finding by Structural Curvature Mapping." M. Sc. Thesis, The University of Texas at Austin (January 1968).

Murray, G.H. "Quantitative Fracture Study—Sanish Pool, McKenzie County, North Dakota." *Bulletin.* AAPG (September 1984), Vol. 52, No. 1,57–65.

Nardon, S. et al. "Fractured Carbonate Reservoir Characterization and Modelling: A multidisciplinary Case Study for the Cavone Oil Field, Italy." *First Break* (December 1991), 553–565.

Narr W. "The Origin of Fractures in Tertiary Strata of the Altamont Field, Uinta Basin, Utah." Master's Thesis, University of Toronto, Toronto, Ontario, (1977).

Narr, W. and J.B. Currie. "Origin of Fracture Porosity—Example from Altamont Field, Utah." *Bulletin.* AAPG (1982), v. 66, 1231–1247.

Narr, W. and R.C. Burrus. "Origin of Reservoir Fractures in Little Knife Field, North Dakota." *Bulletin.* AAPG (September 1984), vol. 68, No. 9, 1087–1100.

Narr W., and I. Lerche. "A Method for Estimating Subsurface Fracture Density in Core", *Bulletin.* AAPG (May 1984) V. 68, 637–648.

Nelson, R.A. "Natural Fracture Systems: Description and Classification." *Bulletin.* AAPG (1979), vol. 63 No. 12, 2214–2221.

Nelson, R.A. "Significance of Fracture Sets Associated with Stylolite Zones." *Bulletin.* AAPG (1981), vol. 65, No. 11.

Nelson, R.A. "An Approach to Evaluating Fractured Reservoirs." *Journal of Petroleum Technology* (September 1982), 2167–2170.

Nelson, R.A. *Geologic Analysis of Naturally Fractured Reservoirs.* Gulf Publishing company, Houston, Texas (1985).

Nelson, R.A. et al. "Oriented Core: Its Use, Error and Uncertainty," *Bulletin.* AAPG (April 1987), 357–367.

Nelson, R.A. and S. Serra. "Vertical and Lateral Variations in Fracture Spacing in Folded Carbonate Sections and its Relation to Locating Horizontal Wells." The Canadian SPC/CIM/CANMET International Conference on Recent Advances in Horizontal Well Applications, Calgary, Canada (March 20–23, 1994)

Nelson, R.A. *AAPG Fractured Reservoir School Notes*, Great Falls, Montana (1982–1994).

Netoff, D.I. "Polygonal Jointing in Sandstone Near Boulder, Colorado." *Mountain Geologist* (1971), vol. 8, 17–24.

Norris, D.K. "Structural Conditions in Canadian Coal Mines." *Geological Survey of Canada Bulletin*, (1958) 44, 54 p.

Nolen-Hoeksema, R.C. and J.H. Howard. "Estimating Drilling Direction for Optimum Production in a Fractured Reservoir." *Bulletin.* AAPG (August 1987), 958–966.

Onyedim, G.C. and J.W. Norman. "Some Appearances and Cases of Lineaments Seen on Landsat Images." *Journal of Petroleum Geology* (1986), 179–194.

Overbey, W.K. et al. "Analysis of Natural Fractures Observed by Borehole Video Camera in a Horizontal Well." SPE paper 17760 presented at the SPE Joint Rocky Mountain Regional/Low Permeability Reservoir Symposium and Exhibition, Denver, Colorado (March 6–8, 1989).

Peterson, V.E. "Fracture Production from Mancos Shale, Rangely Field, Rio Blanco County, Colorado." *Bulletin.* AAPG 39 (April 1955), 532.

Piccard, M.D. "Oriented Linear Shrinkage Cracks in Green River Formation (Eocene), Raven Ridge Area, Uinta Basin, Utah." *Jour. Sed. Petrology* (1966), vol. 36, No. 4, 1050–1057.

Pickett, G.R., and E.B. Reynolds. "Evaluation of Fractured Reservoirs." *Society of Petroleum Engineers Journal* (March 1969), 28.

Pirson, S.J. "How to Map Fracture Development From Well Logs." *World Oil* (March 1967), 106–114.

Pirson, S.J. "Petrophysical Interpretation of formation Tester Pressure Buildup Records." *Transactions.* SPWLA (May 17–18, 1962).

Porter, L.E. "El Segundo Oil Field, California." *Transactions.* AIME (1943), 451.

Pittman, E.D.: "Porosity, Diagenesis and Productive Capability of Sandstones Reservoirs", *SEPM Special Publication No. 26*, Soc. of Economic Paleontologists and Mineralogists, Tulsa, OK (March 1979) 159–73.

Price, N.J. *Fault and Joint Development in Brittle and Semibrittle Rock*, London, Pergamon Press (1966), 176 p.

Provo, L.J. "Upper Devonian Black Shale--Worldwide Distribution and What It Means." Proceeding of the Seventh Appalachian Petroleum Geology Symposium, Morgantown, West Virginia (March 1–4, 1976).

Rabshevsky, G.A. "Optical Processing of Remote Sensing Imagery." Proceeding of the Seventh Appalachian Petroleum Geology Symposium, Morgantown, West Virginia (March 1–4, 1976), 100.

Read, D.L. and Richmond, G.L. "Geology and Reservoir Characteristics of the Arbuckle Brown Zone in the Cottonwood Creek Field, Carter County, Oklahoma," in Johnson, K.S. and Campbell J.A. (eds.), Petroleum Reservoir Geology in the Southern Midcontinent, 1991 Symposium: Oklahoma Geological Survey Circular 95, p.113–125.

"Recommended Practice for Determining Permeability of Porous Media." American Petroleum Institute (September 1951).

Redwine L,: "Hypothesis Combining Dilation, Natural Hydraulic Fracturing, and Dolomitization to Explain Petroleum Reservoirs in Monterey Shale, Santa Maria Area, California, *The Monterey Formation and Related Siliceous Rocks of California*, Society of Economic Paleontologists and Mineralogists, Pac. Sec. (1981) 221–48.

Regan L.J. "Fractured Shale Reservoirs of California." *Bulletin 37.* AAPG (February 1953), 201–216.

Reiss, L.H. *The Reservoir Engineering Aspects of Fractured Formations*, Gulf Publishing company, Houston, Texas (1980).

Robertson, E.C. "Experimental Study of the Strength of Rocks." *Geological Society of America Bulletin* (1955), vol. 66, 1294–1314.

Rowley, D.S. et al. "Oriented Cores." Christensen Publication, 9th Printing (June 1971).

Sahuquet, B.C. and J.J. Ferrier. "Steam-Drive Pilot in a Fractured Carbonate Reservoir: Lacq Superieur Field." *Journal of Petroleum Technology* (April 1982), 873–880.

Sangree, J.B. "What You Should Know to Analyze Core Fractures." *World Oil* (April 1969), 69–72.

Scott, J.C., W.J. Hennessey, and R.S. Lamon. "Savanna Creek Gas Field, Alberta." *Bulletin of Canadian Mineralogists and Metallurgists*, 533 (1958). 270–278.

Shoemaker, E.M. *Penetration Mechanics of High Velocity Meteorites, Illustrated by Meteor Crater, Arizona.* Report 21st Int. Geol. Congr., Session Norden, Part XVIII (1960) 418–434.

Sinclair, S.W. "Analysis of Macroscopic Fractures on Teton Anticline, Northwestern Montana." M.Sc. Thesis, Department of Geology, Texas A & M University (1980).

Skopec, R.A. "Proper Coring and Wellsite Core Handling Procedures: The First Step Towards Reliable Core Analysis." *Journal of Petroleum Technology* (April 1994), 280.

Smith, J.E. "The Cretaceous Limestone-Producing Areas of the Mara and Maracaibo District—Venezuela." Proceedings of the Third World Petroleum Congress (1951), 56–71.

Smith, J.E. "Basement Reservoir of La Paz-Mara Oil Fields, Western Venezuela." *Bulletin 40*, AAPG (February 1956), 380–385.

Snider, L.C. "Current Ideas Regarding Source Beds for Petroleum." Problems of Petroleum Geology. *Bulletin.* AAPG (1934), 51–66.

Snow, D.T.: "Rock Fracture Spacing Openings and Porosities," American Soc. of Civil Engineers, Soil Mechanics and Foundation Div. (1968) 94, 73–91.

Snyder, R.H. and M. Craft. "Evaluation of Austin and Buda Formations from Core and Fracture Analysis." *Transactions.* Coast Association of Geological Societies (1977), vol. XXVII, 376–382.

SPA. *Petroleum Geology of China Sichuan Basin.* Editor-in-Chief Wang Fujun, Sichuan Petroleum Administration (Circa 1990).

Sprunt, E.S. and N.V. Humphreys. "Obtaining Reliable Laboratory Measurements." *Journal of Petroleum Technology* (April 1994), 281.

Stearns, D.W. "Certain Aspects of Fracture in Naturally Deformed Rocks." NSF Advanced Science Seminar in Rock Mechanics: Bedford, Massachusetts. R.E. Ricker, ed. (1967), 97–118.

Stearns, D.W., and Linscott, J.P. "The Influence of In-Situ Stress Conditions on Fold-Related Fracture Permeability." The Canadian SPE/CIM/CANMET International Conference on Recent Advances in Horizontal Well Applications, Calgary, Canada, (March 20–23, 1994).

Stearns, D.W., and M. Friedman. "Reservoirs in Fractured Rock." AAPG Memoir (1972), 82–106.

Stearns, D.W. *AAPG Fractured Reservoir School Notes*, Great Falls, Montana (1982–1994).

Stone, D. "Sub-Surface Fracture Maps Predicted from Borehole Data: An Example from the Eye-Dashwa Pluton, Atikokan, Canada." *Int. J. Rock Mech. Sci. & Geomech.* (1984) 183–194.

Terzaghi, K. "Measurements of Stress in Rock", *Geotechnique* (1962) 12: 105–24.

Thirumalai, K. "Process of Thermal Spalling Behavior in Rocks—An Exploratory Study." 11th Symposium on Rock Mechanics, Berkeley, California, Proc. (1969), 705–727.

Thomas, G.E. "Effects of Differential Compaction Fracturing Shown in Four Reservoirs." *Oil and Gas Journal* (February 3, 1992), 54–57.

Thomas, E.P. "Understanding Fractured Oil Reservoirs." *Oil & Gas Journal* (July 7, 1986), 75–79.

Tillman, J.E. "Exploration for Reservoirs With Fractured-Enhanced Permeability." *Oil and Gas Journal* (February 21, 1983), 165–179.

Trunz, J.P. "Some New Interpretations of Reservoir Properties by Well Log Analysis." M.Sc. Thesis, U. of Texas at Austin (August 1966).

Voight, Barry. "On Photoelastic Techniques, In Situ Stress and Strain Measurement, and the Field Geologist", *Journal of Geology* (1967) 75: 46–58.

Walters, R.F. "Oil Production from Fractured Pre-Cambrian Basement Rocks in Central Kansas." *Bulletin* 37. AAPG (February 1953), 300–313.

Weber, K.J. and Bakker, M.: "Fracture and Vuggy Porosity," paper SPE 10332 presented at the 1981 SPE Annual Technical Conference and Exhibition, San Antonio, Oct. 5–7.

Wilkinson, W.M. "Fracturing in Spraberry Reservoir, West Texas." Bulletin 37. *Bulletin*. AAPG (February 1953), 250–265.

Woodland, D.C. and J.S. Bell. "In Situ Stress Magnitudes from Mini-Frac Records in Western Canada," *Journal of Petroleum Technology* (September–October 1989), 22–31.

CHAPTER 2

Drilling and Completion Methods

Drilling directional or horizontal wells perpendicular to the preferential orientation of natural fractures increases the probability of success in these types of reservoirs. The rationale behind this idea is that many fractures below 3000 ft are nearly vertical or of high inclination. Consequently, a deviated or horizontal hole will have more chances of intercepting a vertical fracture than a vertical hole as shown on Figure 2–1. Although a vertical hole may intercept some fractures, experience indicates that directional holes represent a more sound approach for the development of naturally fractured reservoirs. This has been observed in practice in many places including the Mississippian lime of Oklahoma, certain offshore plays where directional holes are necessary, the Monterey shale of California, the Austin chalk of Texas, and the Palm Valley gas field of Central Australia as discussed in Chapter 1.

DRILLING FLUIDS

Naturally fractured reservoirs are found in hard rocks which can usually be drilled with inexpensive muds. Soft rocks, on the other hand, must be drilled with muds of very precise properties which usually make the drilling muds very expensive.

In some cases it is possible to drill hard rock using air or gas as a drilling fluid with excellent results. However, high reservoir pressures and the presence of water in the subsurface put drastic limitations on the use of air and gas. It is impossible to control subsurface pressures when using air or gas as a drilling fluid, and the presence of water generates stickiness in the hole which may result in a stuck drillpipe.

DRILLING MUD

Mud properties must be controlled continuously for an efficient drilling job. Mud testing procedures are covered in detail in API Recommended Practices No. 29. Table 2–1 summarizes the most important mud tests, apparatus, and objectives of the tests. μ_p in Table 2–1 stands for plastic viscosity (cp), μ_{aF} is apparent viscosity (cp), γ_b is Bingham yield point (lb/100 ft^2) and γ_t is true yield point (lb/100 ft^2).

According to Dresser Magcobar's Mud Engineering these are some of the functions of drilling mud:

1. To remove the cuttings from the bottom of the hole and transport them to the surface
2. To cool and lubricate the bit and drill string
3. To wall the hole with an impermeable filter cake
4. To control encountered subsurface pressures
5. To suspend cuttings and weight materials when circulation is stopped
6. To release sand and cuttings at the surface
7. To support part of the weight of drillpipe and casing
8. To reduce adverse effects upon the potential pay sections
9. To transmit hydraulic horsepower to the bit

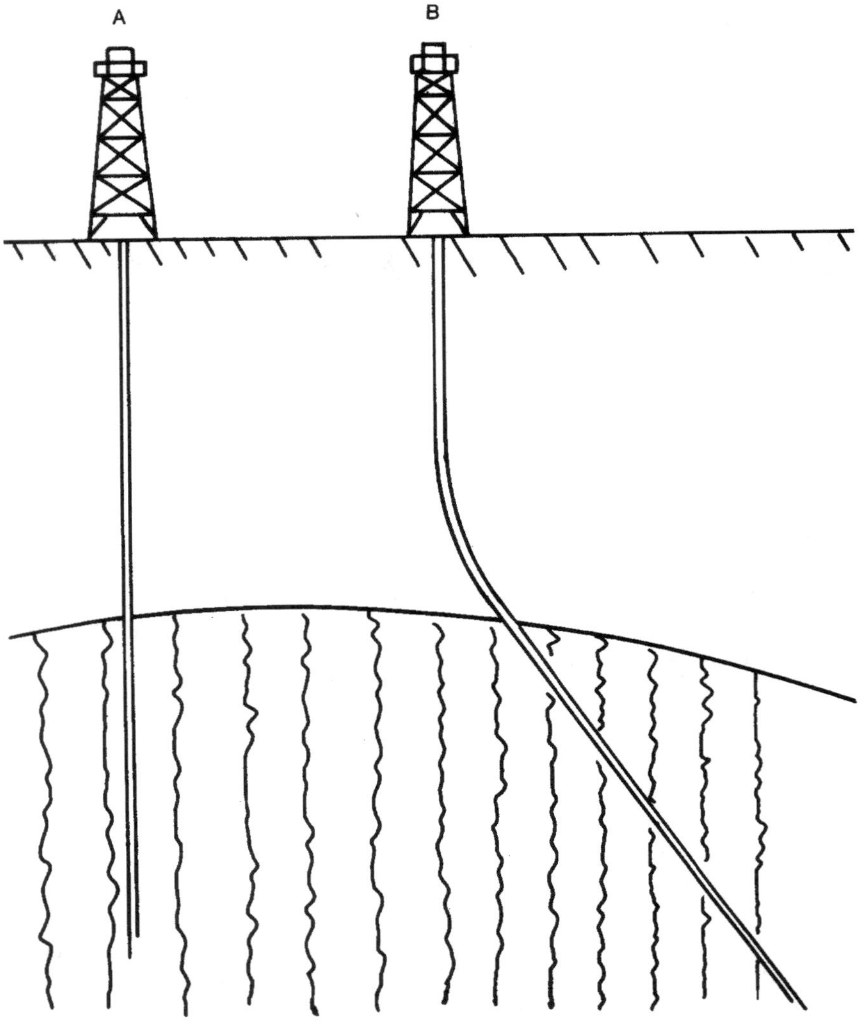

FIGURE 2–1   Vertical and deviated holes in a naturally fractured reservoir with vertical fractures. Well A (vertical) is dry; well B (deviated) is a discovery well.

TABLE 2–1   Mud testing summary

Test	Apparatus	Purpose
a. Mud density	Mud balance	Measures density of drilling fluid
b. Mud viscosity	Marsh funnel	Measures time to discharge 946 cc out of the funnel which contains 1500 cc
	Stormer viscosimeter	The driving force in grams is used with a calibration chart to obtain mud viscosity
	Fann V-G viscosimeter	Calculates μ_p, μ_{aF}, γ_b, and γ_t
c. Gel strength	Fann V-G viscosimeter	Measure of shearing stress necessary to initiate a finite rate of shear ($\simeq \gamma_t$)
d. Filtration	Filter press	Measure of water loss and cake thickness
e. pH	pH papers or pH meters	Measure of acidity or alkalinity of the mud
f. Sand content	Centrifuge, screens, measuring tube	Determines sand content of the mud

Removal of Cuttings

Removal depends mainly on drilling fluid velocity. Experience indicates that annular velocities between 100 and 200 ft/min are sufficient to transport the cuttings to the surface. The annular velocity can be calculated from the relationship:

$$\text{annular velocity} = \frac{\text{pump output (b/min)}}{\text{annular volume (bbl/100 ft)}} \qquad (2\text{--}1)$$

The lower value of velocity could apply to drilling hard rock (fracture) reservoirs. The upper values apply to drilling soft formations.

Other factors affecting the removal and transport of cuttings include the density and viscosity of the mud. The density is usually measured with a mud balance and reported in pounds per gallon (lb/gal). The balance is easily calibrated with water of known density (for instance 8.33 lb/gal). The higher the density of the mud, the higher its transportation capacity.

Mud viscosity is measured sometimes in seconds with the old marsh funnel. In this technique, the funnel is filled with 1500 cc mud. The time required to discharge 946 cc is recorded and reported as mud viscosity. Water has a funnel viscosity of about 26 sec.

Better measurements of viscosity are obtained with the Fann V-G (viscosity-gel) meter which calculates plastic viscosity (cp), apparent viscosity (cp), Bingham yield point (lb/100 ft^2), and true yield point (lb/100 ft^2).

Most muds are colloids and/or emulsions which behave as non-newtonian or plastic fluids, as opposed to oil and water which are newtonian fluids, i.e., fluids whose viscosity does not change with the rate of shear.

Figure 2–2 shows the behavior of newtonian and non-newtonian (plastic) fluids in a cross plot of shear stress vs. rate of shear or velocity. For plastic fluids, a certain shear stress is necessary to initiate the rate of shearing. This is followed by transition from plug to viscous or laminar flow.

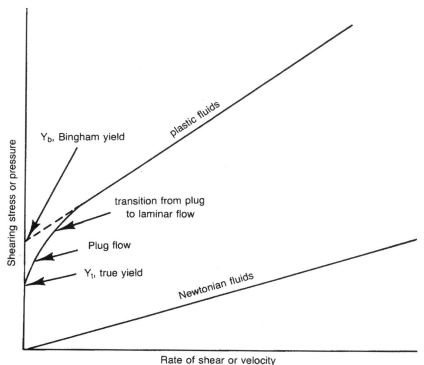

FIGURE 2–2 Flow behavior of plastic and newtonian fluids (after Gatlin, 1960)

For the newtonian fluid, the viscosity is constant and is represented by the slope of the straight line, or:

$$\mu = \frac{\text{shearing stress}}{\text{rate of shear}} \qquad (2\text{–}2)$$

Increasing the viscosity of the mud increases its transportation capacity.

Cooling and Lubricating Bit and Drillstring

The heat generated by friction between the bit and drillstring in contact with the formation is removed by the circulating fluid. Friction is more severe when drilling directional holes. The heat, which has been transmitted from friction points to the mud, is disposed as the mud reaches the surface.

Lubrication is carried out by mud, especially when coupled with various emulsifying agents. This increases bit life, reduces pump pressure, and decreases torque. Under some extreme load conditions, it is advisable to use extreme pressure (E.P.) lubricants.

Wall Building

This depends mostly on filtration properties of the mud. A good mud should deposit an impermeable filter cake rather quickly to retard the passage of fluids into the formation and thus prevent possible damage. The addition of fluid loss control additives helps obtain good wall building. This, however, substantially increases the price of the mud.

Lost Circulation

In some cases, there may be lost circulation through fractures and other large pores. In Figure 2–3, section A represents permeable unconsolidated formations; section B represents angular and cavernous formations; section C represents natural fractures; and section D represents induced fractures.

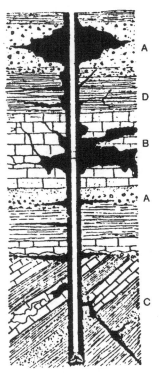

FIGURE 2–3 Lost circulation sections (after Dresser Magcobar)

Lost circulation must be prevented for many reasons: 1) the possibility of blowout due to drop of mud level on the annulus; 2) excess drilling costs when there are no returns; 3) considerable damage of the pay zone if it happens to be the zone of circulation loss; 4) loss of drilling time; and 5) possibility of drillpipe sticking.

In general, an unconsolidated formation should have at least 14 darcys of permeability to admit the entrance of mud. However, the situation depends mostly on the ratio between pore openings and sizes of the particles constituting the mud. Losses in vugular and cavernous formations can usually be predicted by correlation with nearby wells.

When there is no mud loss correlation between wells, the presence of fractures can be inferred. These fractures can be induced or natural. When these fractures are natural, they may be opened by a certain pressure level, permitting the entrance of the mud. It is better to prevent lost circulation than allow it to occur.

This is illustrated in Chapter 3 (Figure 3–9) with the use of induction logs run in a Gulf Coast well drilled with an inverted oil emulsion mud. The middle induction log (B) shows an increase in resistivity as compared with log A. This is due to an increase in mud density from 17.1 to 17.5 lb/gal which allowed the mud to flow in the formation probably through a natural fracture which was open by the increase in mud density or through a hydraulically induced fracture. When density was decreased to 17.3 lb/gal the mud flowed back into the wellbore. This is corroborated by induction log C on Figure 3–9.

Mechanically induced fractures resulting from very high pressures imposed on the formation also result in strong loss circulations. Thus, it is important to locate the zones of lost circulation associated with natural or induced fractures. Fortunately, many methods do this, such as the spinner survey, temperature survey, radioactive tracer survey, hot wire survey, and pressure transducer survey.

The Spinner Survey. This detects horizontal entrance of the mud into the formation by a rotor whose velocity increases in the zone of lost circulation. The main disadvantage of this method is that it requires deliberate loss of considerable volumes of mud.

Temperature Survey. This method indicates the zone of loss by comparing two temperature surveys. One is run to determine the temperature gradient at stabilized conditions; the other is run after adding cool mud to the well. A great difference occurs between the two surveys at the point of loss (see also Temperature log in Chapter 3).

Radioactive Tracer Survey. This consists of two gamma ray surveys. One is run before and the other after the introduction of radioactive material. A vast difference occurs between the two surveys at the point of loss, due to the deposition of radioactive material at such point.

Hot Wire Survey. This survey is run with a resistance wire which is sensitive to temperature changes. The wire is run to a certain point in the well and fresh mud is pumped. If there is no change in the resistance wire, then the point of loss is located above the tool. Conversely, if the resistance changes, then the point of loss can be inferred to exist below the tool. This is, then, a trial and error procedure which requires large amounts of mud to perform the test.

Pressure Transducer Survey. This estimates the flow rate of the mud and where the mud becomes static. This method also requires large amounts of mud flow.

Lost circulation can be combatted by reducing mud density, spotting plugs, or using fibrous, micaceous, or granular bridging materials. Mud density reduction is used only when there is no danger of fluids coming from the formation which could cause a blowout. Spotting plugs are made of cement, mud, or lost circulation material. The plug is placed at the desired depth in the hole, which requires knowing the point of loss. Fibrous materials are bark, hay, cork, and cottonseed hulls which plug intergranular porous media of high permeability. Micaceous materials are very effective in intergranular porous media of relatively small voids.

Material	Type	Description	Concen-tration LBS/BBL	Largest Fracture Sealed IN.
Nut shell	Granular	50%-3/16+10 mesh 50%-10+100 mesh	20	
Plastic	"	"	20	
Limestone	"	"	40	
Sulphur	"	"	120	
Nut shell	"	50%-10+16 mesh 50%-30+100 mesh	20	
Expanded perlite	"	50%-3/16+10 mesh 50%-10+100 mesh	60	
Cellophane	Lamellated	3/4 in. flakes	8	
Sawdust	Fibrous	1/4 in. particles	10	
Prairie hay	"	1/2 in. fibers	10	
Bark	"	3/8 in. fibers	10	
Cottonseed hulls	Granular	fine	10	
Prairie hay	Fibrous	3/8 in. particles	12	
Cellophane	Lamellated	1/2 in. flakes	8	
Shredded wood	Fibrous	1/4 in. fibers	8	
Sawdust	"	1/16 in. particles	20	

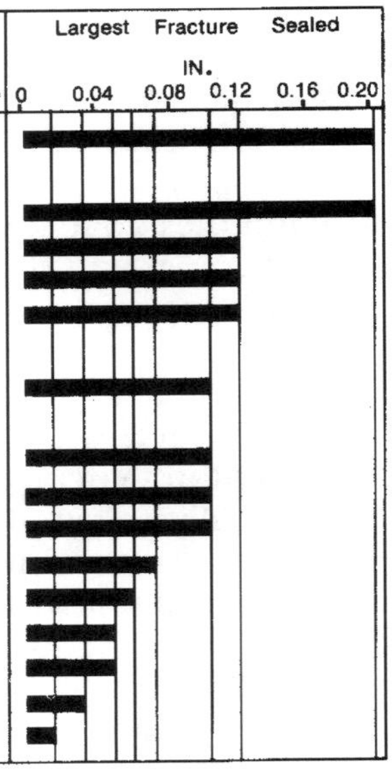

FIGURE 2–4 Summary of material evaluation tests (after Howard and Scott, 1951)

Of all these methods, granular bridging materials are especially effective in fracture reservoirs. Studies conducted by Howard and Scott (1951) indicate that granular bridging materials work well in fractured media, and that concentrations of 20 lb/bbl are enough for sound plugging in many instances. Figure 2–4 summarizes the experiments of Howard and Scott (1951) regarding lost circulation for various fracture widths.

The bridging can be made initially with spherical grains of one half the fracture thickness. In addition, the grains must have sufficient strength to support the pressure differential across the bridge.

Control of Subsurface Pressures

This depends on the mud density. The static pressure that a column of mud exerts upon any point in the hole can be calculated from:

$$p_m = \frac{\rho_m}{8.33} \times 0.433D = 0.052\rho_m D \qquad (2\text{–}3)$$

where:
p_m = hydrostatic pressure exerted by column of mud at depth D, psi
ρ_m = mud density, lb/gal
D = depth, ft

Usually the weight of water plus solids and clays encountered during drilling are sufficient to control formation pressures. Sometimes, however, it is necessary to increase the mud density by adding weighting materials such as barite.

The presence of natural fractures can be detected when barite is deposited in the fractures using P_e curve of the lithodensity log as illustrated on Figure 3–37.

Cuttings and Weight Material Suspension

Suspension depends mainly on gel strength, density, and viscosity. A good mud, because of its thixotropic characteristics, should hold cuttings and weight material in suspension after perforation is stopped. When circulation is reinitiated, the mud should revert to its previous fluid condition.

Cuttings suspension is important because it prevents drillpipe sticking when drilling operations are stopped.

Under certain conditions drill cuttings may detect natural tectonic fractures which tend to be pervasive down to the grain scale. However, it is possible that regional and contractional natural fractures may not be preserved in cuttings due to breakage along the fractures.

Releasing Sand and Cuttings at the Surface

Release of sand at the surface is very important because sand is abrasive. If it recirculates through the system, it will damage pumps and fittings. Generally the sand content should never be allowed to be larger than 2%. It is possible to determine if the sand is being properly released at the surface by comparing sand content of samples taken at the flow line and at the suction.

Supporting of Weight of Drillpipe and Casing

Due to buoyancy effects equivalent to the weight of mud displaced by drillpipe and casing, there is a reduction in the total weight that the surface equipment must support. The higher the mud density, the lower the weight that surface equipment must support.

Protecting Potential Pay Sections

Protection can be achieved by optimizing the properties of the drilling fluids. Proper control of fluid loss helps avoid damage to the formation. In some cases, optimum properties must be sacrificed to insure maximum hole information. For example, oil might improve the characteristics of a mud, but if it interferes with the fluorescence cutting analysis of the geologists, it may be forbidden.

In general, it has been found that there is a tendency to an increase in damage as we move from A towards C in the naturally fractured reservoir classification shown on Figure 1–6.

AIR AND GAS AS DRILLING FLUIDS

Air and gas have been used successfully in some naturally fractured reservoirs with excellent penetration rates and savings in drilling costs and damage.

Air is provided by compressors. When gas is used, it usually comes from nearby gas wells. After the gas is circulated through the well, it is burned out.

Some of the main functions of drilling mud are also carried out by air and gas. For example, air and gas are very efficient for removing cuttings from the bottom of the hole, transporting them to the surface, and cooling and lubricating the drillstring.

Other tasks, however, cannot be handled efficiently by air and gas. For example, air and gas cannot wall the hole with an impermeable filter cake, control encountered subsurface pressures, or suspend cuttings and weight material when air circulation is stopped.

Water zones represent a paramount problem for efficient air and gas drilling. Large amounts of water production require prohibitive circulation air rates for efficient water removal. Small amounts of water production make the dust-like cutting particles stick to each other, forming balls which may adhere to the drillpipe and result in expensive fishing jobs. Consequently, the technique is not applicable for drilling in water zones.

In some cases, it has been possible to air-drill water zones by adding foaming agents. Reed (1958) reported that this technique was used to air-drill wells producing at 50 b/hr of water in the Texas Panhandle.

Excellent penetration rates are usually obtained when drilling with air in the absence of the problems discussed above. For example, gas-bearing fractured shales are excellent candidates for air drilling. Nicholson (1954) provides an example of the successful application of air and gas drilling in the San Juan Basin of New Mexico (Table 2–2). In addition to the economic advantage resulting from excellent penetration rates, the airdrilling technique produced better-productivity wells without formation damage. However, extreme care must be exercised when drilling with gas and air because the danger of fires and explosions always exists.

Foam as a Drilling Fluid

Preformed stable foam has been used successfully for workovers and drilling in some naturally fractured reservoirs. The AEC (Atomic Energy Commission) stable foam, usually

TABLE 2–2 Comparison of drilling methods for the Mesa Verde section, San Juan Basin (after Nicholson, 1954)

Drilling method	Time required	Bits used	Remarks
Rotary, conventional mud	10–20 days	8–10	Severe formation damage
Cable tool	5–8 weeks	—	Continuous fishing jobs and waste of gas
Rotary, gas drilling	4½ days	3–4	Higher well productivity

referred to as "stiff" foam, is made of a gel-based mud containing compatible foaming surfactants. Pre-formed stable foam is made by injecting a solution of fresh or salt water and a foaming surfactant into a gaseous phase in a foam generator. This type of foam tolerates reasonably well contamination from calcium, solids, salt waters and crude oil. Pre-formed stable foam has good hole cleaning ability at low annular velocity as well as good thermal properties for high temperature and permafrost operations (Hutchinson and Anderson, 1974).

Lincicome (1984) has described the use of foam to successfully drill underpressured zones ($p \simeq 2700$ psig) at a depth greater than 22,000 ft. The formation of interest is the Ellenburger (West Texas), a brittle dolomite with porosities ranging between 3% and 8%, and production attributed mainly to natural fractures. In this case, air and mist were eliminated from consideration due to excessive corrosion at depths and temperatures (240°F) encountered and the potential for downhole explosions. Natural gas or nitrogen were eliminated because of the large volumes required and potential dangers in the surface.

Bentsen and Veny (1976) have described the successful application of pre-formed stable foam to drill shallow gas wells ($\simeq 1200$ ft) in the Calling Lake-Algar area of north eastern Alberta. The producing formation is a vuggy and fractured reef that led to large volumes of fluid losses while drilling the original wells with conventional muds. Some of the advantages of using foam included reduction in mud costs, elimination of conventional DST's by using a system that provided continuous formation evaluation data during drilling, and elimination of formation damage.

The following table compares typical mud densities in pounds per gallon of various well fluids available:

	Density (lb/gal)
Gas, air and mist	0.01–0.3
Stable foam	0.30–7.0
Oil	7.00–8.3
Fresh water and cut brine	8.30–10.0
Brine water	10.00–11.1
Weighted muds	11.10– > 20.0

DIRECTIONAL DRILLING

Directional drilling is beneficial when exploring for naturally fractured reservoirs. Figure 2–1 shows that the chances for successful drilling are better if the hole is deviated rather than vertical.

Directional drilling has many applications in the oil industry. It has been used in offshore drilling where several holes may be drilled from each platform (Cook, 1957); in undersea drilling from the shore; to drill under a city from a nearby location; to drill under a salt dome; to extinguish an oil fire by drilling from a site away from the original site and pumping mud through the new well; and to reach under a river or lake from the shore. More recently, directional and horizontal drilling has been used successfully for exploiting naturally fractured reservoirs.

Direction of the drilling string to any desired location can be carried out by means of various deflection tools. The main problem with this approach is the amount of time involved. There are two general types of tools used for directional drilling: whipstocks and knuckle joints.

Many surveying instruments indicate both vertical deviation and horizontal direction. Modern portable minicomputers can provide dogleg severity data and complete survey computations using trapezoidal, average angle, or minimum radius of curvature techniques.

Directional tendencies of the bit are affected mainly by the subsurface geology of each region. In addition, weight, rotary speed, and circulation rate have a pronounced effect on directional tendencies.

Basic Hole Patterns

There are three basic hole patterns in most deflection drilled holes (Figure 2–5).

Type 1 pattern produces the initial deflection angle near the surface. From this point, the angle is maintained as a straight line until the depth objective is reached. This slanted hole can intercept many vertical and subvertical fractures.

Type 2 produces the initial deflection near the surface. From this point, the angle is maintained until reaching the desired depth. Finally, the hole is returned to a vertical position and is drilled until the objective is reached.

Type 3 produces the desired deflection well below the surface. From there, the established angle is maintained until the target is reached. This type of pattern is useful for fault and fracture drilling.

Hole Deviation Problems

Graphical solutions to hole deviation problems in directional drilling as presented by Woods and Lubinski (1954) consider three elements.

First are established data, which include the outside diameter of collars (in), weight on bit (lb), hole size (in), hole inclination (degrees), and formation dip (degrees). Second are problem data, which include the values for all but one of the quantities listed under established

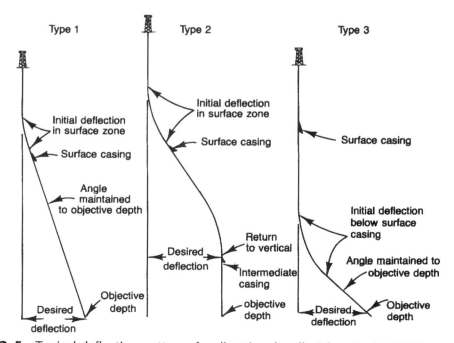

FIGURE 2–5 Typical deflection patterns for directional wells (after Cook, 1957)

data, i.e., one of the quantities is different from established data. Third is unknown, or the value of the quantity not given in the problem data.

The reader is referred to the Woods and Lubinski papers (1954) for detailed examples using their graphical solutions.

Calculation of Directional Drilling Plan

Once the directional drilling program has been determined, it must be put into plan form. Calculation from directional well surveys require three-dimensional locations of the bottom hole position at each survey (Figure 2–6). Calculations can be carried out in many different forms. One possibility is to use the equations:

$$Z = MD \cos \alpha$$
$$H = MD \sin a$$
$$Y = H \cos \beta$$
$$X = H \sin \beta \qquad\qquad (2\text{–}4)$$

where:
$$\begin{aligned}
Z &= \text{true vertical depth between survey points 1 and 2} \\
MD &= \text{measured depth (length of drillstring)} \\
H &= \text{horizontal displacement of hole} \\
Y &= \text{latitude, distance north or south of the east-west axis} \\
X &= \text{departure, distance east or west of the north-south axis} \\
\alpha &= \text{vertical deviation angle} \\
\beta &= \text{horizontal angle (compass direction)}
\end{aligned}$$

Table 2–3 shows observed and calculated data from a directional well survey.

Eastman Wipstock has published some useful charts and tables that allow calculations of Types 1, 2, and 3 patterns illustrated in Figure 2–5.

Uniform increase in drift per 100 ft of hole drilled is presented in Table 2–4 and Figures 2–7 to 2–18. These charts are designed to simplify calculations of horizontal and vertical projections of Types 1 and 3 patterns, i.e., the "build and maintain" type of directional holes. Their use is best illustrated with an example.

Example 2–1. This example is extracted from "Introduction to Drilling" by Eastman Whipstock.

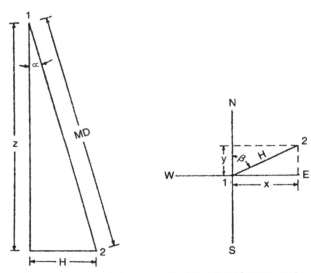

(a) Vertical cross-section between survey points 1 and 2

(b) Plan view of same section

FIGURE 2–6 A) Vertical cross section between survey points 1 and 2. B) Plan view of same section (after Gatlin, 1960)

TABLE 2-3 Observed and calculated data from directional well survey (after Stearns, 1953)

Test no.	Depth	Course length	α, Deviation angle	Cos α	Vertical depth Course	Vertical depth Total	Sin α	β, Course Deviation	Magnetic Bearing β
1	100	100	1°15'	.9998	99.98	99.98	.0218	2.18	N8°00'W
2	200	100	1°45'	.9995	99.95	199.93	.0305	3.05	N30°30'W
3	300	100	1°45'	.9995	99.95	299.88	.0305	3.05	N50°30'W
4	400	100	2°45'	.9988	99.88	399.76	.0480	4.80	N20°00'W
5	500	100	4°30'	.9969	99.69	499.45	.0785	7.85	N15°00'E
6	600	100	7°30'	.9914	99.14	598.59	.1305	13.05	N9°30'E
7	700	100	9°45'	.9855	98.55	697.14	.1693	16.93	N19°00'E
8	800	100	13°15'	.9734	97.34	794.48	.2292	22.92	N47°00'E
9	900	100	19°15'	.9441	94.41	888.89	.3297	32.97	N53°00'E
10	1000	100	22°00'	.9272	92.72	981.61	.3746	37.46	N75°00'E
11	1100	100	24°30'	.9100	91.00	1072.61	.4147	41.47	N84°00'E
12	1200	100	26°00'	.8988	89.88	1162.49	.4384	43.84	S86°00'E

Test no.	Cos β	Sin β	Course y, latitude N	Course y, latitude S	Course x, departure E	Course x, departure W	Total y, latitude N	Total y, latitude S	Total x, departure E	Total x, departure W
1	.9903	.1392	2.16	—	—	0.30	2.16	—	—	0.30
2	.8616	.5075	2.63	—	—	1.55	4.79	—	—	1.80
3	.6361	.7716	1.94	—	—	2.35	6.73	—	—	4.20
4	.9397	.3420	4.51	—	—	1.64	11.24	—	—	5.84
5	.9659	.2588	7.58	—	2.03	—	18.82	—	—	3.81
6	.9863	.1650	12.87	—	2.15	—	31.69	—	—	1.66
7	.9455	.3256	16.01	—	5.51	—	47.70	—	3.85	—
8	.6820	.7313	15.63	—	16.76	—	63.30	—	20.61	—
9	.6018	.7986	19.84	—	26.33	—	83.17	—	46.94	—
10	.2588	.9659	9.69	—	36.18	—	92.86	—	83.12	—
11	.1045	.9945	4.33	—	41.24	—	97.19	—	124.36	—
12	.0698	.9976	—	3.06	43.73	—	94.13	—	168.09	—

TABLE 2–4 Uniform 1°00' increase in drift per 100' of hole drilled (after Eastman and Whipstock)

Measured Depth	Verticle Depth	Deviation	Drift	Measured Depth	Vertical Depth	Deviation	Drift
100.00	99.99	0.87	1°00'	3100.00	2950.95	818.37	31°00'
200.00	199.96	3.49	2°00'	3200.00	3036.21	870.62	32°00'
300.00	299.86	7.85	3°00'	3300.00	3120.55	924.35	33°00'
400.00	399.68	13.96	4°00'	3400.00	3203.94	979.54	34°00'
500.00	499.37	21.80	5°00'	3500.00	3286.35	1036.18	35°00'
600.00	598.90	31.39	6°00'	3600.00	3367.76	1094.25	36°00'
700.00	698.26	42.71	7°00'	3700.00	3448.15	1153.73	37°00'
800.00	797.40	55.76	8°00'	3800.00	3527.48	1214.61	38°00'
900.00	896.30	70.54	9°00'	3900.00	3605.74	1276.86	39°00'
1000.00	994.93	87.04	10°00'	4000.00	3682.90	1340.47	40°00'
1100.00	1093.26	105.27	11°00'	4100.00	3758.94	1405.41	41°00'
1200.00	1191.25	125.21	12°00'	4200.00	3833.84	1471.67	42°00'
1300.00	1288.87	146.85	13°00'	4300.00	3907.56	1593.23	43°00'
1400.00	1386.11	170.19	14°00'	4400.00	3980.10	1608.06	44°00'
1500.00	1482.92	195.23	15°00'	4500.00	4051.42	1678.16	45°00'
1600.00	1579.29	221.95	16°00'	4600.00	4121.51	1749.48	46°00'
1700.00	1675.17	250.36	17°00'	4700.00	4190.35	1822.02	47°00'
1800.00	1770.54	280.43	18°00'	4800.00	4257.91	1895.74	48°00'
1900.00	1865.37	312.16	19°00'	4900.00	4324.17	1970.64	49°00'
2000.00	1959.63	345.54	20°00'	5000.00	4389.11	2046.68	50°00'
2100.00	2053.30	380.56	21°00'	5100.00	4452.72	2123.84	51°00'
2200.00	2146.34	417.21	22°00'	5200.00	4514.97	2202.10	52°00'
2300.00	2238.72	455.47	23°00'	5300.00	4575.84	2281.43	53°00'
2400.00	2330.43	495.35	24°00'	5400.00	4635.33	2361.82	54°00'
2500.00	2421.42	536.82	25°00'	5500.00	4693.40	2443.23	55°00'
2600.00	2511.68	579.87	26°00'	5600.00	4750.04	2525.64	56°00'
2700.00	2601.17	624.49	27°00'	5700.00	4805.23	2609.03	57°00'
2800.00	2689.87	670.66	28°00'	5800.00	4858.96	2693.36	58°00'
2900.00	2777.75	718.38	29°00'	5900.00	4911.21	2778.63	59°00'
3000.00	2864.79	767.72	30°00'	6000.00	4961.96	2864.79	60°00'

Given:	K.O.P. (Kick Off Point):	2000 feet
	T.V.D. (Total Vertical Depth):	10,000 feet
	Dev. (Horizontal Deviation):	2455 feet
	Buildup:	2° per 100 feet

Select the correct buildup chart (2° increase per 100 feet) on Figure 2–8. Find the usable vertical depth which is equal to the difference between the depth at the K.O.P. and T.V.D. = 10,000–2000 feet = 8,000 feet.

On the graphic part of the chart, find the intersection of the usable depth (vertical scale) and the horizontal deviation (horizontal scale) or, in this case, the intersection of 8000 feet and 2455 feet. This gives the maximum angle of deviation from the slant or degree line which is equal to 18°.

On the chart inset, obtain the Measured Depth, True Vertical Depth, and horizontal Deviation corresponding to the maximum angle at completion of buildup: for 18°, the inset shows a measured depth of 9000 feet, a vertical depth of 885.27 feet and a deviation of 140.21 feet.

Add the given K.O.P. depth to these measurements to find the actual depths at completion of buildup or 2000 + 900 = 2900 feet Measured Depth (M.D.), and 2000 + 885.3 = 2885.3 feet True Vertical Depth.

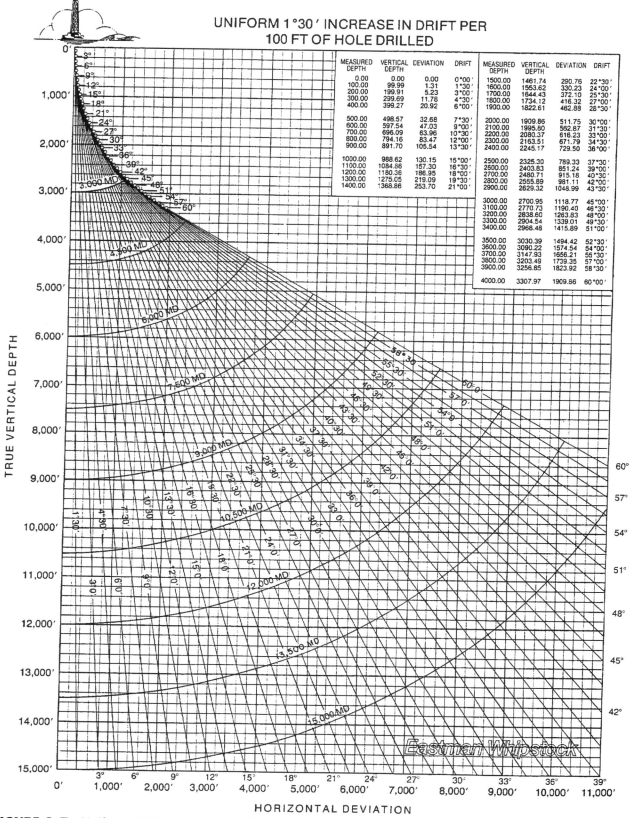

FIGURE 2–7 Uniform 1°30′ increase in drift per 100 ft of hole drilled (after Eastman Whipstock)

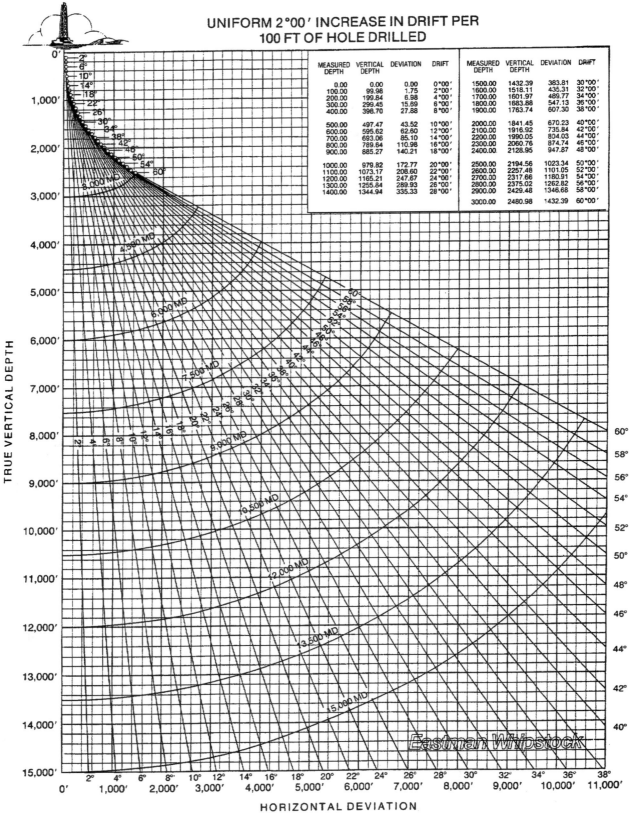

UNIFORM 2°00′ INCREASE IN DRIFT PER 100 FT OF HOLE DRILLED

MEASURED DEPTH	VERTICAL DEPTH	DEVIATION	DRIFT	MEASURED DEPTH	VERTICAL DEPTH	DEVIATION	DRIFT
0.00	0.00	0.00	0°00′	1500.00	1432.39	383.81	30°00′
100.00	99.98	1.75	2°00′	1600.00	1518.11	435.31	32°00′
200.00	199.84	6.98	4°00′	1700.00	1601.97	489.77	34°00′
300.00	299.45	15.69	6°00′	1800.00	1683.88	547.13	36°00′
400.00	398.70	27.88	8°00′	1900.00	1763.74	607.30	38°00′
500.00	497.47	43.52	10°00′	2000.00	1841.45	670.23	40°00′
600.00	595.62	62.60	12°00′	2100.00	1916.92	735.84	42°00′
700.00	693.06	85.10	14°00′	2200.00	1990.05	804.03	44°00′
800.00	789.64	110.98	16°00′	2300.00	2060.76	874.74	46°00′
900.00	885.27	140.21	18°00′	2400.00	2128.95	947.87	48°00′
1000.00	979.82	172.77	20°00′	2500.00	2194.56	1023.34	50°00′
1100.00	1073.17	208.60	22°00′	2600.00	2257.48	1101.05	52°00′
1200.00	1165.21	247.67	24°00′	2700.00	2317.66	1180.91	54°00′
1300.00	1255.84	289.93	26°00′	2800.00	2375.02	1262.82	56°00′
1400.00	1344.94	335.33	28°00′	2900.00	2429.48	1346.68	58°00′
				3000.00	2480.98	1432.39	60°00′

FIGURE 2–8 Uniform 2°00′ increase in drift per 100 ft of hole drilled (after Eastman Whipstock)

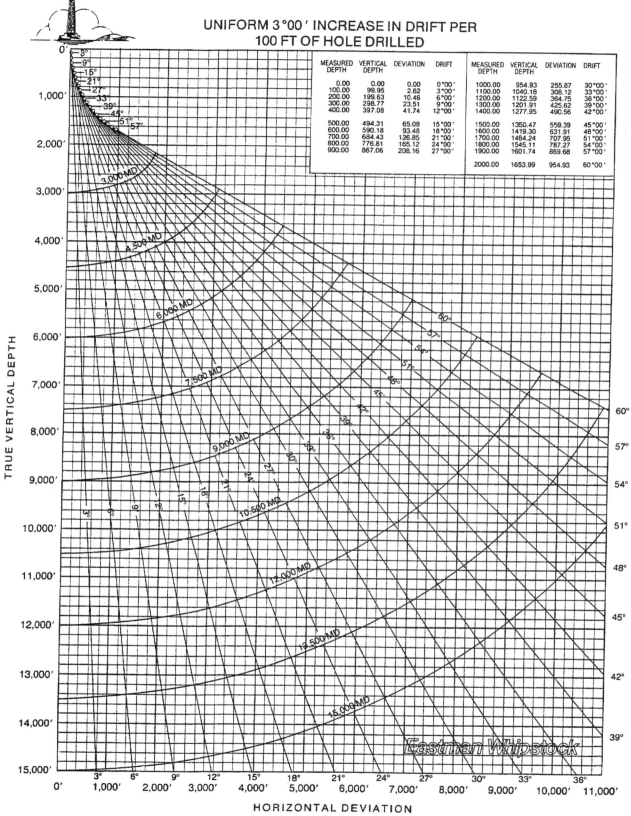

FIGURE 2–9 Uniform 3°00′ increase in drift per 100 ft of hole drilled (after Eastman Whipstock)

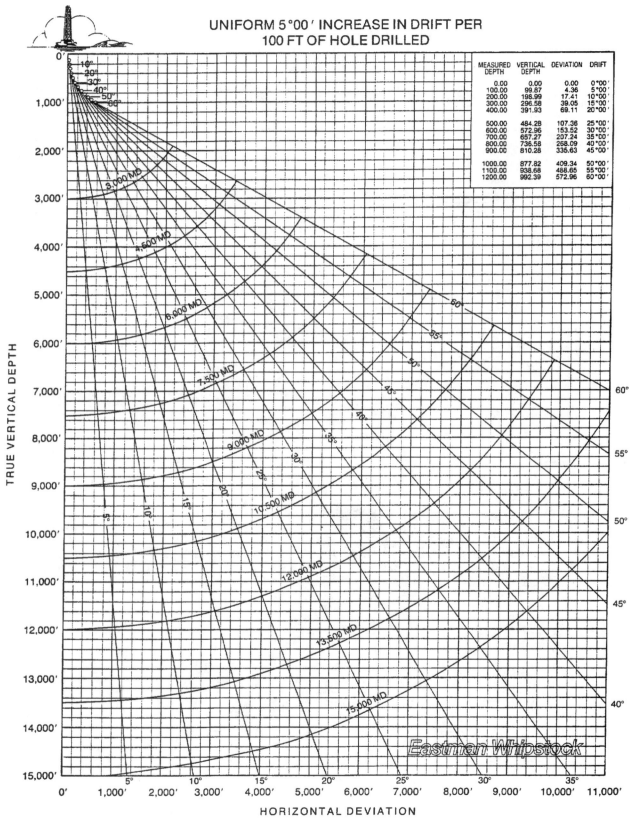

The following data appears in the table within the figure:

MEASURED DEPTH	VERTICAL DEPTH	DEVIATION	DRIFT
0.00	0.00	0.00	0°00′
100.00	99.87	4.36	5°00′
200.00	198.99	17.41	10°00′
300.00	296.58	39.05	15°00′
400.00	391.93	69.11	20°00′
500.00	484.28	107.36	25°00′
600.00	572.96	153.52	30°00′
700.00	657.27	207.24	35°00′
800.00	736.58	268.09	40°00′
900.00	810.28	335.63	45°00′
1000.00	877.82	409.34	50°00′
1100.00	938.68	488.65	55°00′
1200.00	992.39	572.96	60°00′

UNIFORM 5°00′ INCREASE IN DRIFT PER 100 FT OF HOLE DRILLED

FIGURE 2–10 Uniform 5°00′ increase in drift per 100 ft of hole drilled (after Eastman Whipstock)

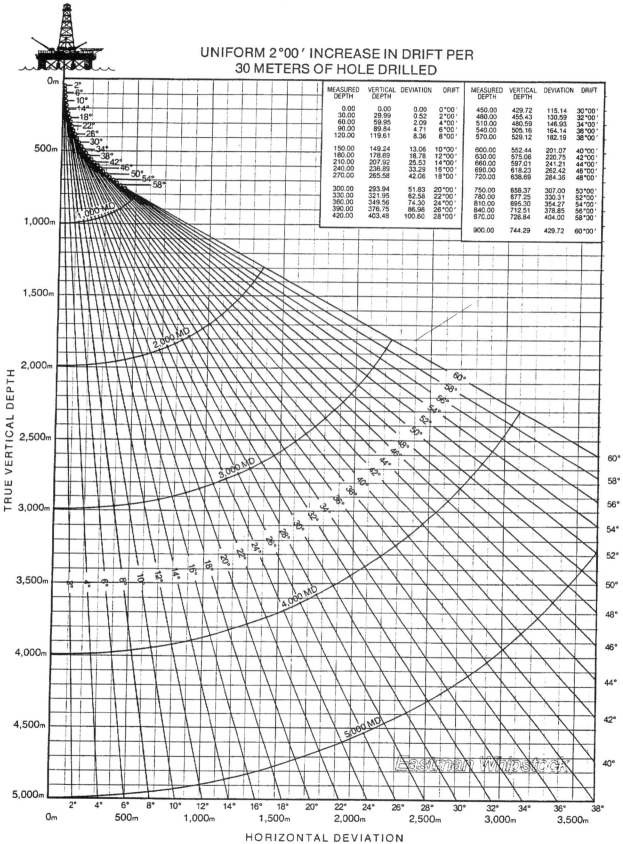

UNIFORM 2°00′ INCREASE IN DRIFT PER 30 METERS OF HOLE DRILLED

MEASURED DEPTH	VERTICAL DEPTH	DEVIATION	DRIFT	MEASURED DEPTH	VERTICAL DEPTH	DEVIATION	DRIFT
0.00	0.00	0.00	0°00′	450.00	429.72	115.14	30°00′
30.00	29.99	0.52	2°00′	480.00	455.43	130.59	32°00′
60.00	59.95	2.09	4°00′	510.00	480.59	146.93	34°00′
90.00	89.84	4.71	6°00′	540.00	505.16	164.14	36°00′
120.00	119.61	8.36	8°00′	570.00	529.12	182.19	38°00′
150.00	149.24	13.06	10°00′	600.00	552.44	201.07	40°00′
180.00	178.69	18.78	12°00′	630.00	575.08	220.75	42°00′
210.00	207.92	25.53	14°00′	660.00	597.01	241.21	44°00′
240.00	236.89	33.29	16°00′	690.00	618.23	262.42	46°00′
270.00	265.58	42.06	18°00′	720.00	638.69	284.36	48°00′
300.00	293.94	51.83	20°00′	750.00	658.37	307.00	50°00′
330.00	321.95	62.58	22°00′	780.00	677.25	330.31	52°00′
360.00	349.56	74.30	24°00′	810.00	695.30	354.27	54°00′
390.00	376.75	86.98	26°00′	840.00	712.51	378.85	56°00′
420.00	403.48	100.60	28°00′	870.00	728.84	404.00	58°00′
				900.00	744.29	429.72	60°00′

FIGURE 2–11 Uniform 2°00′ increase in drift per 30 m of hole drilled (after Eastman Whipstock)

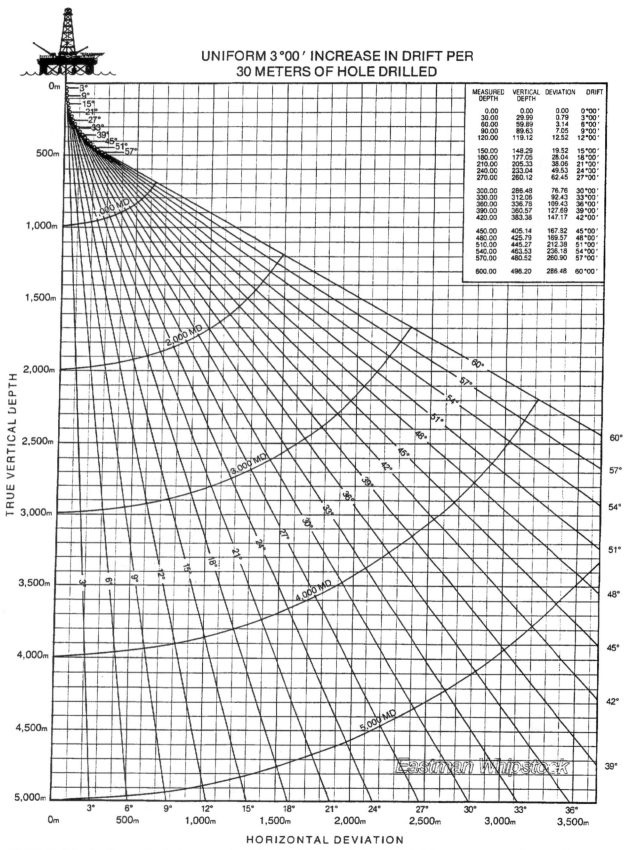

MEASURED DEPTH	VERTICAL DEPTH	DEVIATION	DRIFT
0.00	0.00	0.00	0°00'
30.00	29.99	0.79	3°00'
60.00	59.89	3.14	6°00'
90.00	89.63	7.05	9°00'
120.00	119.12	12.52	12°00'
150.00	148.29	19.52	15°00'
180.00	177.05	28.04	18°00'
210.00	205.33	38.06	21°00'
240.00	233.04	49.53	24°00'
270.00	260.12	62.45	27°00'
300.00	286.48	76.76	30°00'
330.00	312.06	92.43	33°00'
360.00	336.78	109.43	36°00'
390.00	360.57	127.69	39°00'
420.00	383.38	147.17	42°00'
450.00	405.14	167.82	45°00'
480.00	425.79	189.57	48°00'
510.00	445.27	212.38	51°00'
540.00	463.53	236.18	54°00'
570.00	480.52	260.90	57°00'
600.00	496.20	286.48	60°00'

FIGURE 2–12   Uniform 3°00′ increase in drift per 30 m of hole drilled (after Eastman Whipstock)

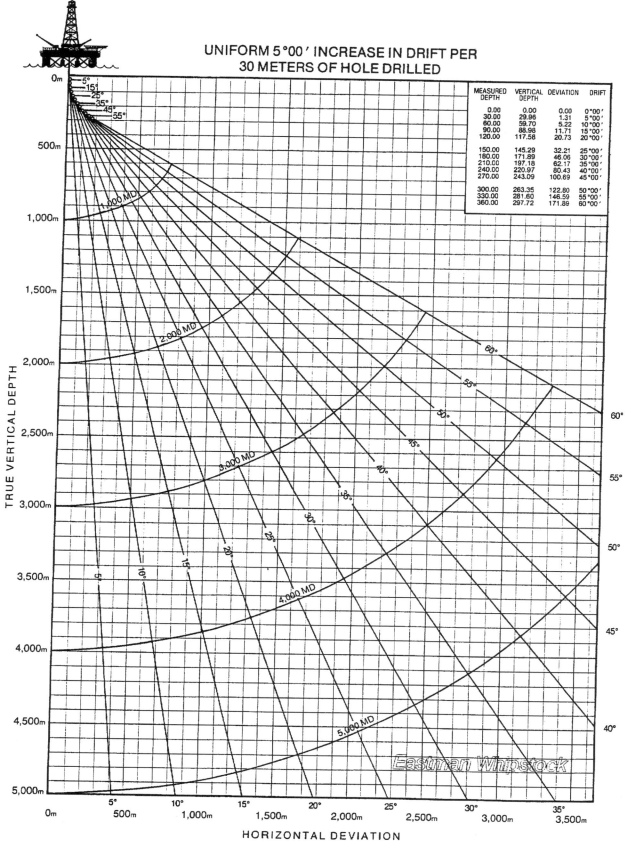

UNIFORM 5°00' INCREASE IN DRIFT PER
30 METERS OF HOLE DRILLED

MEASURED DEPTH	VERTICAL DEPTH	DEVIATION	DRIFT
0.00	0.00	0.00	0°00'
30.00	29.96	1.31	5°00'
60.00	59.70	5.22	10°00'
90.00	88.98	11.71	15°00'
120.00	117.58	20.73	20°00'
150.00	145.29	32.21	25°00'
180.00	171.89	46.06	30°00'
210.00	197.18	62.17	35°00'
240.00	220.97	80.43	40°00'
270.00	243.09	100.69	45°00'
300.00	263.35	122.80	50°00'
330.00	281.60	146.59	55°00'
360.00	297.72	171.89	60°00'

FIGURE 2–13 Uniform 5°00' increase in drift per 30 m of hole drilled (after Eastman Whipstock)

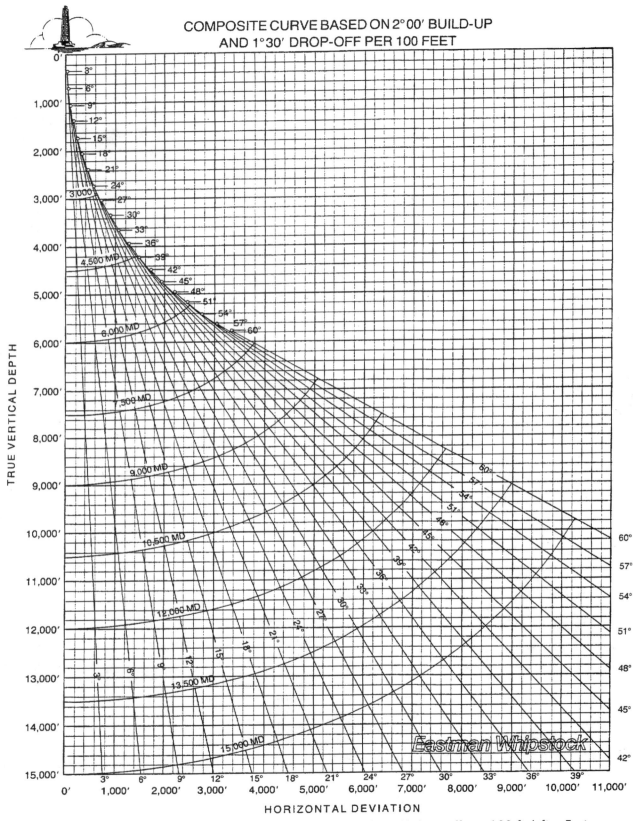

COMPOSITE CURVE BASED ON 2°00' BUILD-UP
AND 1°30' DROP-OFF PER 100 FEET

HORIZONTAL DEVIATION

TRUE VERTICAL DEPTH

FIGURE 2–14 Composite curve based on 2°00' build-up and 1°30' drop-off per 100 ft (after Eastman Whipstock)

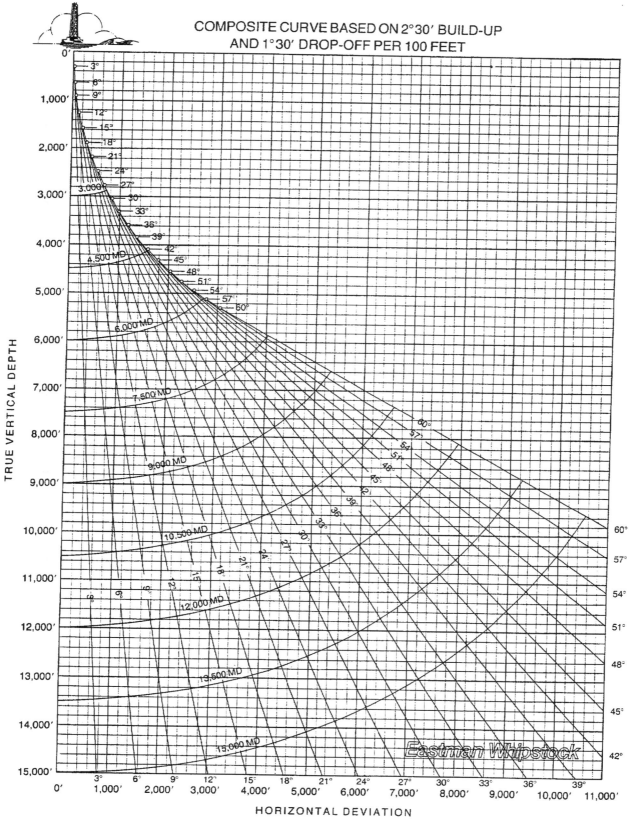

FIGURE 2–15 Composite curve based on 2°30′ build-up and 1°30′ drop-off per 100 ft (after Eastman Whipstock)

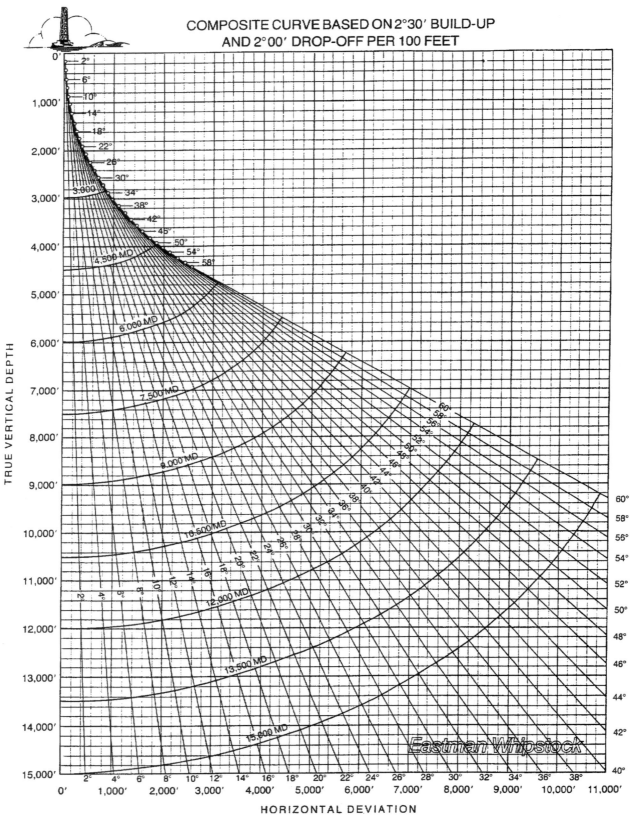

FIGURE 2–16 Composite curve based on 2°30′ build-up and 2°00′ drop-off per 100 ft (after Eastman Whipstock)

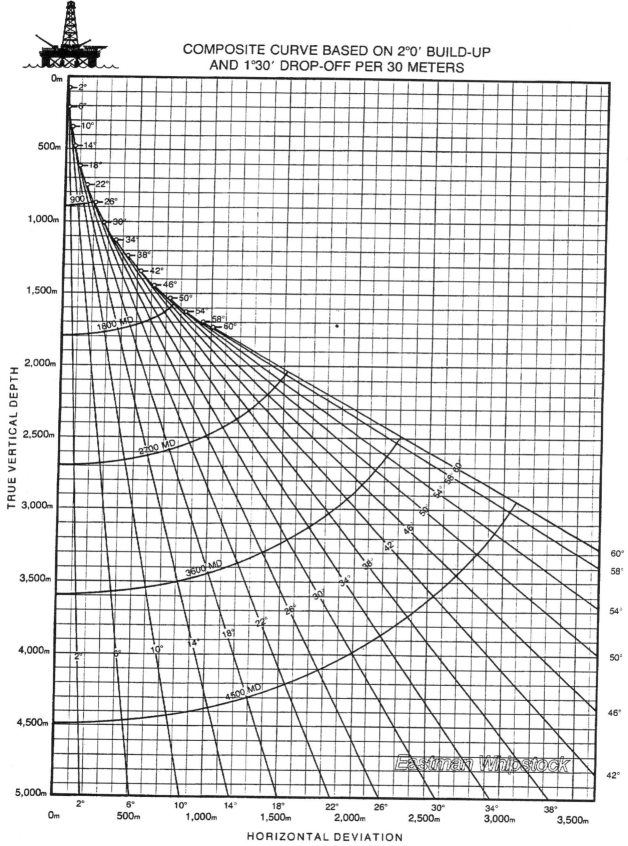

FIGURE 2–17 Composite curve based on 2°00′ build-up and 1°30′ drop-off per 30 m (after Eastman Whipstock)

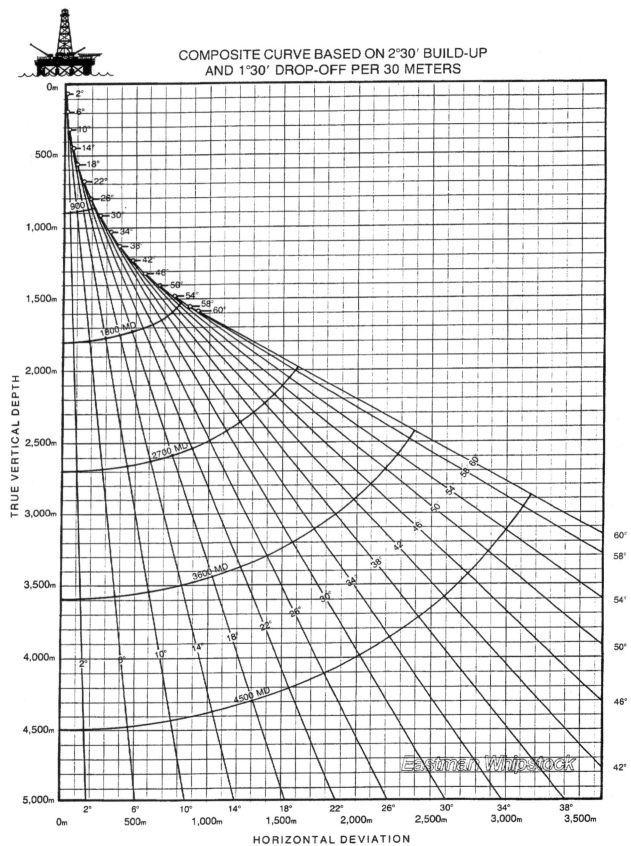

FIGURE 2–18 Composite curve based on 2°30' build-up and 1°30' drop-off per 30 m (after Eastman Whipstock)

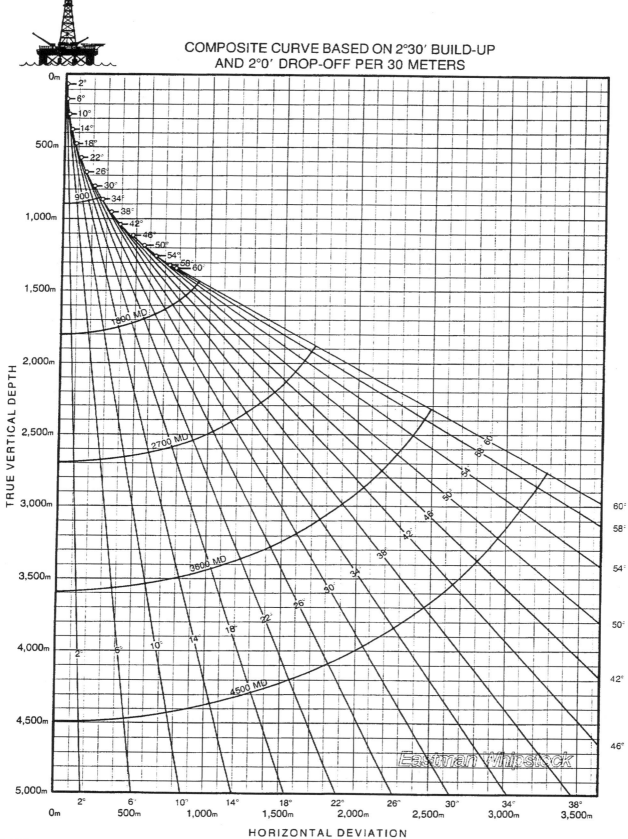

FIGURE 2–19 Composite curve based on 2°30′ build-up and 2°00′ drop-off per 30 m (after Eastman Whipstock)

Find the measurements below the buildup by solving the values of the remaining triangle: the actual Vertical Depth (V.D.) below buildup is equal to the difference between the given T.V.D. and the Vertical Depth at buildup or 10,000 – 2885.3 = 7114.73 feet. The total Measured Depth (TMD) is therefore calculated from:

$$\text{TMD} = \text{M.D. at buildup} + \frac{\text{V.D. below buildup}}{\text{cos. max. angle}} \qquad (2\text{--}5)$$

$$\text{TMD} = 2900 + \frac{7114.73}{\cos 18°} = 10{,}381 \text{ ft}$$

Composite curves for type 2 pattern are presented in Figures 2–14 through 2–19. The calculation procedure varies with respect to the previous presentation as in addition to the inclination buildup, it is necessary to return back to the vertical. Their use is better illustrated with an example problem.

Example 2–2.

Given:			
	K.O.P.	=	2000 feet
	T.V.D.	=	10,000 feet
	deviation	=	2258 feet
	buildup	=	2° per 100 ft
	drop off	=	1° 30′ per 100 ft

The first part of the calculation is the same as described in example 2–1.

- Find the usable depth (8000 ft).
- Find maximum angle at completion of buildup (18°) from Figure 2–14.
- Find Measured Depth (M.D.), Vertical Depth (V.D.), and deviation at completion of buildup (M.D. = 900 ft, V.D. = 885.27 ft, deviation = 140.21 ft) from the inset of Figure 2–8, which is for a uniform 2° increase in drift per 100 ft of hole drilled.

The second part of the calculation is carried out as follows:
Find the Measured Depth, Vertical Depth and Deviation necessary to bring the well back to vertical (from 18° to 0°). They can be obtained from the chart inset for 1° 30′ off to vertical. Here, for an 18° angle the Measured Depth during drop off is 1200 feet, the Vertical Depth is 1180.36 feet, and the deviation 186.95 feet from the inset of Figure 2–7.

Solve the values of the triangle in the middle of the plat. The maximum angle is obtained by solving the equation:

$$\frac{\text{Dev.} - (\text{Dev. for buildup} + \text{Dev. for drop off})}{\text{usable depth} - (\text{V.D. for buildup} + \text{V.D. for drop off})}$$

$$= \text{Tang. of the Max. Angle} \qquad (2\text{--}6)$$

$$\frac{2258 - (140.21 + 186.95)}{8000 - (885.27 + 1180.36)} =$$

$$\text{Tangent of the Max. Angle} = 0.3254$$

$$\therefore \text{Maximum Angle} = 18.02°$$

The measured length of locked-in segment is given by:

$$\frac{\textit{Usable depth} - (\textit{V.D. for buildup} + \textit{V.D. for drop off})}{\textit{cos. maximum angle}}$$

$$= \frac{8000 - (885.27 + 1180.36)}{\cos (18.02)} = 6240.58 \textit{ ft} \qquad (2\text{--}7)$$

The total measured depth (TMD) is given by:

$$TMD = M.D.\ at\ K.O.P. + M.D.\ during\ buildup + M.D.$$
$$of\ lock\text{-}in\ segment + M.D.\ during\ drop\ off \tag{2–8}$$

$$TMD = 2000 + 900 + 6240.58 + 1200 = 10340.58\ ft$$

In practice, there is a level of uncertainty inherent in the calculations of bottom hole position. Walstrom et al (1969) provide two methods for evaluating this uncertainty. The first is analytical and is based on principles of probability and statistics. The second uses Monte Carlo simulations. Results from both methods are comparable, as shown in Figures 2–20 and 2–21.

Errors in directional surveying can result from limitations of instrument accuracy, reading errors resulting from the tool not being positioned concentrically with the hole and other random errors.

Directional drilling has been greatly improved thanks to the Dyna-Drill tool. The power-generating source of the tool is a multistage Moyno pump used in a reverse application as a motor. The motor consist of a rotor and a stator. The rotor is a solid steel spiral shaft of round cross section. The stator is a rubber-like element molded with a spiral passageway having an obround cross section.

When fluid is forced under pressure into the cavities between rotor and stator, rotation is generated. The lower end of the rotor is attached to a connecting rod which in turn is connected to the drive shaft. The connecting rod converts eccentric rotation of the rotor to concentric rotation of the drive shaft. The lower part of the shaft assembly accommodates the drill bit. The upper end of the rotor is unattached and exposed to the flow of drilling fluid.

Without motor rotation, the fluid will not flow through the tool. Consequently, a dump (bypass) valve assembly is placed in the upper end of the tool. This valve permits the drillpipe to fill or drain during trips in and out of the hole.

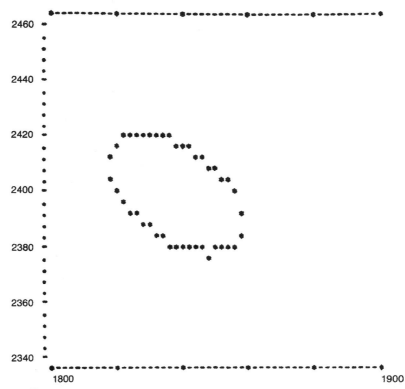

FIGURE 2–20 Ellipse of uncertainty of bottom-hole position, Hole No. 3, probability of 0.99 (after Walstrom et al, 1969)

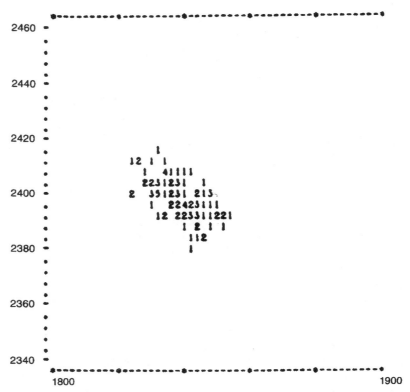

FIGURE 2–21 Bottom-hole coordinates of 100 simulated surveys, Hole No. 3., vertical axis is north (after Walstrom et al, 1969)

HORIZONTAL DRILLING

Horizontal drilling has come of age and it has proved very powerful for development and exploitation of naturally fractured reservoirs (Joshi, 1990; Aguilera et al, 1991; Butler, 1994). It should be useful also in the case of thin layer reservoirs, heavy oil and reservoirs with gas and/or water coning problems. Efforts to drill horizontally were initiated in the United States in the 1940s and the Eastern Bloc in the 1950s (Cranfield, 1984), but without any noticeable followup. More recently many companies including Elf Aquitaine, Arco, Standard Oil (Sohio), Texaco, Esso Resources of Canada, Meridian, and Texas Eastern Corporation have reported successful drilling of horizontal wells.

Horizontal drilling techniques can be classified according to their angle buildup rates as follows (Dech et al, 1986):

- Conventional (or low) curvature method
- Medium curvature method
- High curvature method

The main features of these three drilling methods are presented on Table 2–5 and Figure 2–22. The approximate build radius is 30 ft in the high curvature method, 3000–1200 ft in the conventional curvature method, and 286 ft in the medium curvature method. Horizontal wells with medium and conventional curvatures can be cased and logged; wells with high curvature cannot.

Depth ranges to date are 1600–12000 ft for wells drilled with rigs standing conventionally in a vertical position. Texaco used an ingenious procedure in Canada (Loxam, 1982) and drilled 1095 ft of horizontal hole at a depth of only 413 ft using a slanted rig.

Table 2–6 shows a summary of horizontal drilling characteristics of some of the wells drilled to date including the operating company, true vertical depth, horizontal length, diameter of horizontal section, source of information and curvature method.

TABLE 2–5 Main features of directional drilling methods

> Conventional (or low) curvature
>   Rate of build: 0.03 to 0.05° per ft
>   Build radius: 3000–1200 ft
>   Wells can be cased and logged
> Medium curvature
>   Rate of build: 0.2° per ft
>   Build radius: 286 ft
>   Wells can be cased and logged
> High curvature
>   Build radius: 22–35 ft
>   Wells cannot be cased nor logged

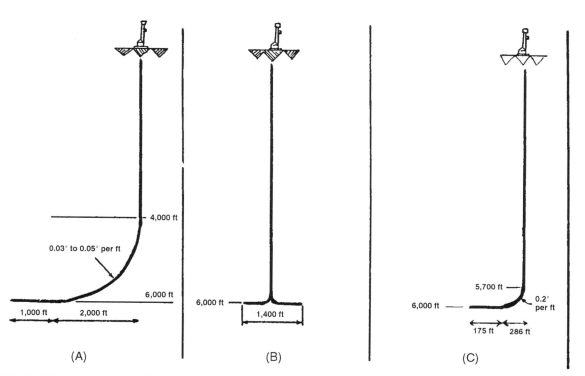

FIGURE 2–22 Horizontal drilling comparison. (A) Conventional or low curvature, (B) high curvature, and (C) medium curvature method

Conventional or Low Curvature Method

In this approach the rotary and motor assemblies, drilling tools, and drilling operations are essentially conventional (Dech et al, 1986). Costs, however, are much higher due to the very long angle-build section. Open logging can be drillpipe conveyed.

The Elf Aquitaine (and Horwell S.A., an Elf Aquitaine subsidiary) experience have included drilling of wells Lacq 90, Lacq 91 in France, Rospo Mare 6D in Italy and Castera Lou in southwest France. Drilling descriptions presented below for wells Lacq 90, Lacq 91, and Rospo Mare 6D follow very closely the original work of Dorel (1983).

Well Lacq 90. It was drilled using 17½, 12¼, and 8½ inch bits in 1028 hours. Casing strings included 13⅜–in., set at 1071 ft (measured depth), 9⅝–in., set at 2529 ft (measured depth), and a 7–in. uncemented slotted liner (Figure 2–23). Total measured depth was 3563 ft. Some 328 ft were penetrated horizontally at angles between 89° and 92°. The hole contacted 896 ft of reservoir.

TABLE 2–6 Characteristics of some horizontal wells

Company field	TVD feet	Length of horizontal feet	Diameter of horizontal section feet	Date of completion	Source	Curvature Method
TEXACO Fort McMurray Canada	413	1095	9⅝"	1981	Loxam (1982)	Low (Slanted Rig)
Esso Resources Cold Lake Canada	1560	1060	9⅝"	1978	Esso (1984)	Low
Elf/IFP Lacq France	2300	1200	8½"	1981	Giannesini (1984)	Low
Elf/IFP Rospo Mare Italy	4500	1542	8½"	1982	Giannesini (1984)	Low
Elf/IFP Castera Lou France	9400	1150	8½"	1983	Giannesini (1984)	Low
Esso Resources Cold Lake Canada	1773	1469	9⅝"	1984	Esso (1984)	Low
Sohio Olmos Sand Texas	10300	1900	8½"	1985	Littleton (1985)	Low
Sohio Prudhoe Bay Alaska	11700	1400	8½"	1986	Littleton (1986)	Low
ARCO Austin Chalk Texas	1380	1340	6"	1986	Dech et al (1986)	Medium
TEXAS EASTERN Grassy Trails Utah	3443	220	4½"	1984	Gorody (1984)	High
ARCO Empire Abo New Mexico	6250	175	4"	1981	Striegler (1982)	High

A fresh water gel mud was used to drill the first 2133 ft. Then 10% diesel oil was added in preparation for running 9⅝–in. casing. Cleaning the hole was accomplished with a mud of 60 sec/qt viscosity (1 qt = 946.35 cc). The rest of the hole was drilled adding diesel in the mud to 20%. Mud weight was maintained at 8.9 ppg and viscosity at 60 sec/qt. Solids were kept below 5% by volume.

Kickoff point was located at 289 ft. A 2½° bent sub was used that yielded a 28° angle at casing depth. Following cementing of the 13⅜–in. casing, the well was continued with a conventional rotary drilling and limber string. Directional drilling was used between 1608 and 1772 ft to build hole angle. The 9⅝–in. casing was set and cemented at 2530 ft where the angle was 70°.

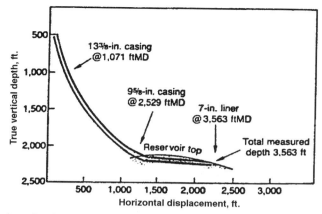

FIGURE 2–23 Relative displacement angles, casing points, and reservoir position for the Lacq 90 well, drilled in France (after Giannesini, 1984)

A downhole motor with a 1½° bent sub was used to drill an 8½–in. hole to a depth of 3202 ft where inclination reached 89°. The bent sub was retrieved at this point. From there on the well was turbo-drilled to its final measured depth of 3563 ft equivalent to 2192 ft (TVD).

Three cores were cut between 3202 ft and TD. A gamma ray log was run through the drillpipe. No problems were found during tripping or connections, and all three casings were run without difficulty. Ninety-five directional surveys were taken, 76 with simple shot and 19 with a steering tool.

Well Lacq 91.   It was drilled by Elf Aquitaine using the same hole sizes, casing programs, and buildup angles of the previous well (Figure 2–24). A significant change was an increase to a 180° angle at the 9⅝–in. casing shoe. The horizontal portion of the well was extended, and the hole contacted 1452 ft of the reservoir.

Well Rospo Mare 6D.   It was drilled by Elf Aquitaine in the Italian Adriatic Sea in 71 days (Figure 2–25). The objective was a vuggy reservoir containing heavy, viscous oil. The well contacted 1978 ft of reservoir at angles between 75° and 96°, which included 1542 ft drilled after reaching 90°.

Problems with shales were alleviated with KCL mud while drilling the 17½–in. and 12¼–in. holes. The 13⅜–in. casing was set at 3251 ft and the 9⅝–in. casing was set at the top of the reservoir at 5251 ft (MD) when the hole angle had reached 80°. Of the 1978 ft drilled in the reservoir 1540 ft were drilled horizontally.

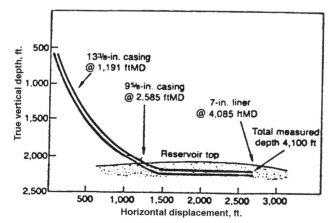

FIGURE 2–24 Angles, casing points, and reservoir position for well Lacq 91 in France (after Giannesini, 1984)

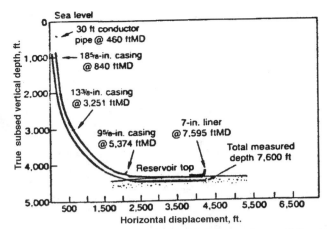

FIGURE 2–25 Angles, casing points, and reservoir position for well Rospo Mare 6D, Italian Adriatic Sea (after Giannesini, 1984)

Production from the Rospo Mare horizontal well started in October, 1980. After four years of history, productivity had been about 15 times that of nearby vertical wells. This is excellent as the 12–15° API oil is found in a thin reservoir underlain by water, and water coning is a tough problem for most vertical wells.

Well Castera Lou 110 (CLU 110). It was spudded on May 28, 1983, by Elf Aquitaine and reached a measured final depth of 12,025 ft (3665 m), on September 23, 1983, with close to 1112 ft (339 m) drilled horizontally through the Garlin dolomite, a reservoir about 200 ft (61 m) thick with natural fractures (Figure 2–26).

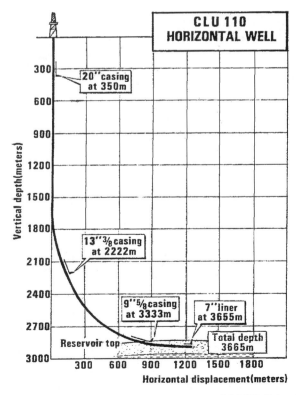

FIGURE 2–26 Well Castera Lou 110 in France penetrated 1150 ft horizontally (after Giannesini, 1984)

The drilling program was as follows (Jordan and Baron, 1984):

1. Drill a 26–in. hole and set 20–in. surface casing about 1148 ft (350 m).
2. Drill a vertical hole of 17½–in. down to approximately 5500 ft (1677 m).
3. Kick off at 5500 ft and build the angle at 1.5°/100 ft. After obtaining a 25° hole angle, run and cement 13⅜–in. casing at 7288 ft (2222 m).
4. Drill out cement and a portion of the new 12¼–in. straight hole.
5. Kick off the 12¼–in. hole, building angle at 1.5°/100 ft to reach 70° at the 9⅝–in. casing point, as close as possible to the reservoir entry (10,932 ft or 3332 m). Run and cement 9⅝–in. casing.
6. Drill out the cement and then the 8½–in. hole, building angle at a rate which will depend upon the exact vertical position of the reservoir. The Maximum rate is 6°/100 ft.
7. On reaching 90°, continue to drill a straight horizontal hole until the target (eastern fault) is located or until problems make further drilling hazardous.
8. Run logging with the system Simphor (Baron and Wittrisch, 1982). See also logging of horizontal wells at the end of this chapter.
9. Run a 7–in. slotted liner.

The main objectives were achieved. The operation, however, created several problems associated with (1) transmitting weight and torque, (2) "locking in" the bent sub and surveying with accuracy the tool face with MWD equipment, and (3) inclination drop-off.

Hole cleaning was satisfactory with a bentonite-treated mud. A total loss of mud circulation was encountered at about 11,960 ft (3646 m). The hole was more or less stabilized after circulating calcium carbonate lost-circulation materials.

Following the coring of a 12 ft interval, mud pit gains and differential sticking were noted. The well was then shut in for a period of stabilization. Two other differential pipe stickings were solved by acidizing.

Following completion of the well, it produced about 20 times better than surrounding wells with a zero water cut.

Braune-Wieding N°1. It was drilled by Sohio in 105 days, 35 days longer than originally estimated. Once the vertical depth was reached, it took about eight days to drill an estimated 1500 ft horizontally. The well was stopped because it reached the Texas Railroad Commission's 467 ft standoff from the edge of the lease.

Existing technology and conventional tools were used to drill the well. A downhole motor was utilized at various points on the well and to drill the entire horizontal section. A top drive system, used throughout the entire operation, provided the following advantages (Littleton, 1985):

1. It delivered high torques smoothly in contrast to roughness of a rotary table.
2. It was more efficient in tripping operations when a tight hole resulted in substantial drag on the drillstring.
3. It provided the ability to simultaneously rotate, circulate, and trip, a necessity in this type of project.

Upon completion of the well, about 1908 ft in excess of 83° had been drilled and 1400 ft was in excess of 90°. The well was logged with drillpipe conveyed tools and a 5½–in. uncemented, slotted liner was run through the formation.

Well JX-2. It was spudded on October 6, 1985, by Sohio in the Prudhoe Bay field, Alaska (Petzed, 1986). Figure 2–27 shows a directional profile of the well. The inset shows depths and angles. Technology acquired by Sohio in Texas was transferred to Prudhoe Bay.

The well was drilled vertically to the surface casing point and set 13⅜–in. casing at 2670 ft. This was followed by a 12¼–in. directional hole to 9269 ft measured depth (MD), or 8613 ft true vertical depth (TVD), at a 49° angle as close to the planned course line as possible to reduce torque and drag. This portion was drilled with conventional rotary and mud motor assemblies. Then, 9⅝–in. intermediate casing was run and cemented conventionally. This is where normal drilling stopped and a new drilling/completion technique began.

Directional profile of Sohio's JX-2 well

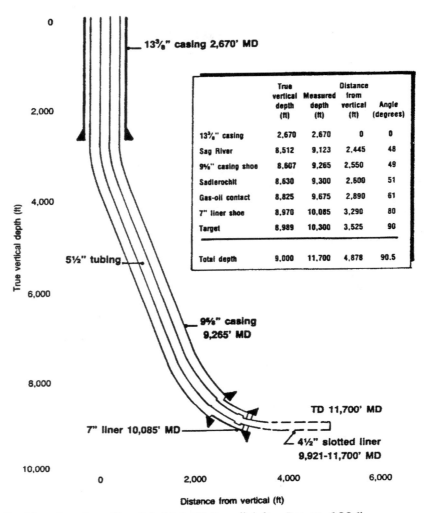

	True vertical depth (ft)	Measured depth (ft)	Distance from vertical (ft)	Angle (degrees)
13⅜" casing	2,670	2,670	0	0
Sag River	8,512	9,123	2,445	48
9⅝" casing shoe	8,607	9,265	2,550	49
Sadlerochit	8,630	9,300	2,600	51
Gas-oil contact	8,825	9,675	2,890	61
7" liner shoe	8,970	10,085	3,290	80
Target	8,989	10,300	3,525	90
Total depth	9,000	11,700	4,878	90.5

FIGURE 2–27   Directional profile of Sohio's JX-2 well (after Petzet, 1986)

The fresh water gel mud system was replaced with a saturated brine/diesel/sized salt system to reduce torque and drag and provide fluid loss control. The first attempt at building angle to 90° in the 8½–in. hole did not succeed. That required plugging back to the 9⅝–in. casing shoe. A later attempt succeeded in landing the horizontal section at 10,300 ft MD, 8989 ft TVD, within 1 ft of proposed target depth.

The horizontal section was drilled to 11,700 ft MD, 9000 ft MD, with the deepest point of the horizontal wellbore at 9000 ft TVD and the shallowest point at 8983 ft TVD. This yielded 1400 ft of horizontal 8½–in. hole with a depth tolerance of –7 ft to +10 ft from the 8990 ft TVD target.

Following drilling of the horizontal section Sohio ran and cemented a 7–in. liner across a gas cap. This placed the 7 in. liner shoe at 10,085 ft MD, 8970 ft TVD, and 80°. A 4¼–in. slotted liner was run to TD and hung inside the 7–in. liner, giving 1615 ft of productive zone open. The well was completed with a 4½–in. tailpipe tied into the top of the slotted liner using a seal bore receptacle and 5½–in. tubing to surface. The rig was released November 21, 1985, and production began early in December.

Prudhole Bay horizontal well JX–2 utilized sized-salt fluid technology called *Clean Bridge*. This is a patented blend of polymers and specially treated sized NaCl salt which is added to a 10.0–lb/gal NaCl brine (Stagg et al, 1986).

Lubricity studies indicate that the coefficient of friction of sized-salt fluid is similar to that of oil muds. The following observations and recommendations quoted from Stagg et al (1986) were presented following drilling of JX–2. They are useful guidelines to be followed when drilling a horizontal well with a sized-salt system:

1. **Hole Cleaning.** Drilling fluid rheology should be run at bottomhole circulating temperature to ensure the yield point is high enough for optimum hole cleaning. Reciprocating and rotating the drillpipe while circulating bottoms up is necessary for enhanced drill cuttings removal. High viscosity sweeps are suggested prior to running logs and liners. Logs and liners may encounter difficulty in being run to depth unless all drill cuttings have been removed from the low side of the hole.

2. **Solids Limit.** For Prudhoe Bay wells, the percentage of drilled solids should be limited to 40% of the total undissolved solids as indicated by the sized-salt field analysis. If the total drilled solids reach 50%, the fluid should be conditioned before drilling any producing formation to minimize formation damage caused by drilled solids. Other reservoirs should have permeability damage studies run to determine maximum drilled solids limits.

3. **Solids Control.** A mud cleaner is a prerequisite for minimizing solids buildup without stripping excess salt from the system. Screen sizes of 200 mesh provide the best solids removal in most situations.

4. **Pipe Slugging.** Sized salt should be used for all pipe slugs. Sacked salt settles to the low side of the hole, which can prevent successful running logs or liners. Consideration should be given to pulling wet pipe through the high angle before the pipe slug is pumped.

5. **Cement Drillouts.** The polymers in the sized-salt system are sensitive to cement contamination. By carefully controlling the Ph with an organic acid buffer, limited intervals of cement can be drilled without damaging the sized-salt polymers.

6. **Aeration.** Adequate surface volume should be maintained to prevent air from becoming entrained in the sized-salt system. Aerated mud can cause any MWD (measured while drilling) transmitted signal to become indecipherable. Defoamers may be required to control aerated mud.

7. **Lost Circulation.** Use various grades of sized salt for lost circulation pills. Spot and squeeze lost circulation material pills with up to 300 psi. Only in extreme situations should conventional lost circulation material be used because of potential damage to the producing formation.

Well JX–2 produced at a stable rate of about 12,400 b/d of oil for about 60 days. This compares with an average 3000 b/d for a conventional vertical well in the same part of the reservoir. Sohio indicated that a horizontal well would be able to drain twice as much reservoir area as a vertical well.

Medium Curvature Method

This technology has been developed by ARCO. The principle is to try to combine the advantages of the low and high curvature methods while avoiding some of their limitations (Dech et al, 1986). The tools and technology have now been tested successfully in many areas including the naturally fractured Austin Chalk (Texas).

Naturally fractured reservoirs possess two main characteristics considered in the development of horizontal drilling technology:

1. Many of these reservoirs are thick. Based on a survey carried out by ARCO, 65% of known fractured reservoirs in the U.S.A., have a thickness of 400 ft or more.

2. Many of the low permeability, naturally fractured reservoirs may be produced from open-hole completions. A 300 ft radius was selected by ARCO as the design curvature

of the angle-build portion to keep the entire wellbore within the producing formation. Other goals were to drill 1000 ft of horizontal hole and to use equipment not too different from conventional oil field equipment.

Drillstring Assembly. The most unconventional part of medium curvature drilling is that drillpipe is run beneath the collars in compression. The following are some advantages of this configuration:

1. The collars remain in the vertical hole in such a way that their full weight is applied axially to the drillstring rather than having some component of the weight lost to the hole wall. With this configuration, fewer collars are required for a given weight on bit, and less torque is lost from side wall contact.
2. Since the drillpipe in the curved section of the hole is in compression, the pipe is forced against the outside rather than the inside of the hole, and keyseat problems usually associated with high curvatures are eliminated. Figure 2–28 and Table 2–7 describe the generalized drillstring and drilling assemblies used during testing.

Compressive Service Drillpipe. Large compressive loads and high curvature put the following main conflicting requirements on the design of the drill pipe:

Drillstring configuration*

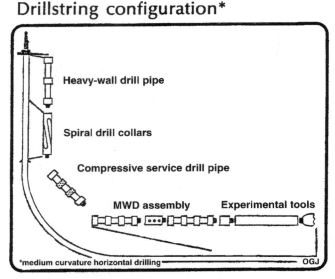

FIGURE 2–28 Medium curvature horizontal drilling drillstring configuration (after Dech et al, 1986)

TABLE 2–7 Drilling assembly, medium curvature horizontal drilling (after Dech et al, 1986)

Component	Comments
3½–in. heavy-wall drill pipe	Bring pipe to surface
4¾–in. drill collars	Provide weight; usually ran 12–20 joints
3½–in. heavy-wall drill pipe	1 joint as a transition between stiff collars and compressive-service drill pipe
3½–in. or 2⅞–in. compressive service drill pipe	Through curve and horizontal wellbore
MWD assembly	2 special collars, pulser sub, and restrictor sub
Experimental tool	Motor or rotary tool
Bit	Diamond, PDC, or mill tooth

1. The pipe must be sufficiently limber to handle the curvature without producing excessive bending movements and stresses. When the pipe is rotated in the curve, the bending stresses will be cyclic; these stresses must be smaller than those that would cause the pipe to fail from fatigue.
2. The pipe must be sufficiently stiff to both resist helical buckling and transmit torque without excessive windup. The diameter and cross-sectional area must be sufficient to carry axial and torsional loads at acceptable stress levels.
3. The pipe should be as light as possible to minimize the loss of torque from the pipe lying in the horizontal hole.

The drilling operation resulted in this design criteria for the pipe: a 6 to 6½-in. diameter wellbore; 30,000 lb compressive force; 3000 ft-lb torque; and 20°/100 ft curvature.

It was necessary to develop compressive service (CS) drillpipe since conventional drillpipe could not satisfy the structural requirements of medium curvature drilling. The cyclic bending stress limits the pipe diameter that can be run through the curvature to 3½-in. or less.

The 30-ft spacing, between tool joints of conventional drillpipe this size and smaller, was unacceptable due to the following reasons:

1. The span of pipe between the tool joints would bow out to the outside wall of the curve under the drilling forces anticipated.
2. The curvature would not be evenly distributed over the pipe, thereby concentrating the bending stresses.
3. The pipe body would contact the hole wall, leading to surface flaws in the area of the highest alternating stress.

Contact pads with the same diameter as the tool joints were introduced along the pipe body to provide lateral support. This permitted distribution of loads evenly and prevented the pipe body from contacting the wellbore.

Helical buckling of drillpipe under a given compression is a function of stiffness and radial clearance. Also associated with this buckling configuration are a bending stress and a wall contact force.

Stiffness of the pipe is limited by the curve of the wellbore. Consequently, the buckling must be controlled by placing a limit on the radial clearance. This was done by increasing the diameter of the tool joints and contact pads.

Two designs of compressive service (CS) drillpipe were tested. Each had 5-in. diameter tool joints and contact pads. The 3½-in., 13.30 lb/ft drillpipe had two contact pads at 10-ft intervals between tool joints. The 2⅞-in., 10.40 lb/ft drillpipe had three contact pads at 7.5-ft intervals between tool joints.

Of the two sizes, the 3½-in. drillpipe is stronger and stiffer. Due to the higher stiffness, there is a smaller contact force due to buckling in the vertical hole. Because of its smaller diameter, the 2⅞-in., drillpipe has lower cyclic bending stresses in the curve. Also, it has lower wall contact force in the horizontal hole due to its lighter weight.

Bottom Hole Assemblies. Ten rotary and motor bottom hole assemblies were evaluated for the angle-build portion and the horizontal portion of the hole. Motor assemblies have good steering capabilities, but are very expensive to operate. On the other hand, rotary assemblies are less expensive, but have no steering capabilities.

The idea was to design some combination of rotary and downhole–motor–drilling assemblies. The tools were designed to be used in or below 7-in. casing and would drill a nominal 6–in. hole.

Various rotary assemblies were designed for angle–build, angle–hold, and angle–drop operations. The hold and drop tools were designed mainly for use in horizontal or near–horizontal wellbores but could also be utilized in any high angle wellbore. Motor assemblies were designed for steerable angle–build, and steerable horizontal–hold/correct applications.

Drilling Fluid. The mud was a fresh-water, gel mud with extreme-pressure lubricant added to improve lubricity. No mud related problems occurred during tripping or connections, or while running and pulling the casing. However, when the rate of penetration decreased while not rotating, (perhaps due to balling), several 5–10 bbl, high viscosity gel sweeps were pumped to help clean the hole.

Accomplishments. According to Dech et al (1986) the project was successful since the original goals of developing an improved technique for horizontal drilling were accomplished and surpassed. Accomplishments include:

1. Developing new drilling tools and supporting technology to deviate controllably a 6–in. wellbore from vertical to horizontal with a constant 20°/100 ft built rate.
2. Drill more than 1000 ft horizontally, and
3. Run casing.

The project also applied an MWD surveying system using flexible nonmagnetic collars, and it kept drilling operations and tools nearly conventional.

Directional Surveys. A measurement while drilling (MWD) directional survey tool was run during the testing operations for real time directional data. The slim hole MWD tool, designed for use in 5–in. OD nonmagnetic drill collars, was run between the two joints of nonmagnetic, 3½–in. compressive service drillpipe. The plan view and the vertical section of the successful hole are shown in Figure 2–29.

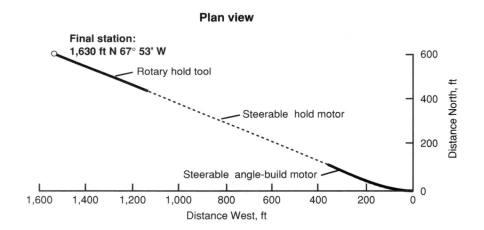

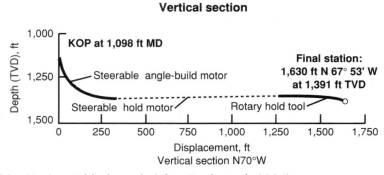

FIGURE 2–29 Horizontal hole path (after Dech et al, 1986)

Height Curvature Method

Gorody (1984) has reported on this drilling system developed by Texas Eastern Drilling Systems Inc. (TEDSI) which allows to drill a horizontal hole from a vertical well through a curve of short radius. Some of the advantages include:

1. Orientation can be set, maintained, or adjusted during drainhole drilling.
2. Several drainholes at one or more horizons can be drilled from a single wellbore.

Typical horizontal drainhole lengths have been reported to vary between 200 and 400 ft, although a 700 ft reach has been attained in a few instances. Different kinds of lithologies have been successfully drilled including sandstones and siltstones, limestones, chalks, dolomites, calcareous shales, siliceous shales, cherts, and evaporites (Gorody, 1984).

Drilling Assembly. This system uses a flexible string assembly available in two diameters, 3¾ (B tool) and 4½-in. (A tool). The assembly permits entry into any borehole with a diameter of at least 4¾-in. Each flexible string can be equipped with two separate bottom hole assemblies: one for drilling the curved portion of the hole, the other for completing the horizontal portion.

Bottom hole assemblies are mounted with the bit most compatible with the type of formation being drilled. Bit rotation is powered conventionally from the surface and any conventional drilling fluid, including air, can be circulated. Special additives or extra fluids are not required.

Internal flow passage diameters range from 1 to 1½-in. The A and B tools are capable of hauling normal drilling fluid volumes at pressure drops across the bottom hole assembly up to 1000 psi. Normal drilling parameters are 8,000–10,000 lb weight on bit at 60 rpm.

The smaller diameter curve drilling bottom hole assembly drills a 20 ft radius of curvature; the large diameter assembly, a 35 ft radius of curvature as shown on Figure 2–30. Both tools will drill a straight hole that maintains the attitude reached at the end of the curve. This attitude can be preset at any angle from vertical to up to 20° above the horizontal plane and to within an accuracy of 5°. The azimuth, or direction, of the lateral borehole can be controlled to within 20° of target as illustrated on Figure 2–31.

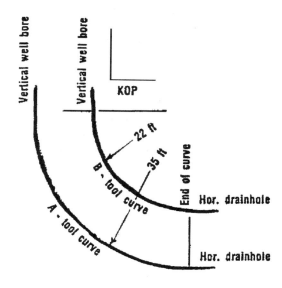

Radius of curvature
Tedsi's A and B tools

FIGURE 2–30 Horizontal drainholes drilled with tools A and B (after Gorodi, 1984)

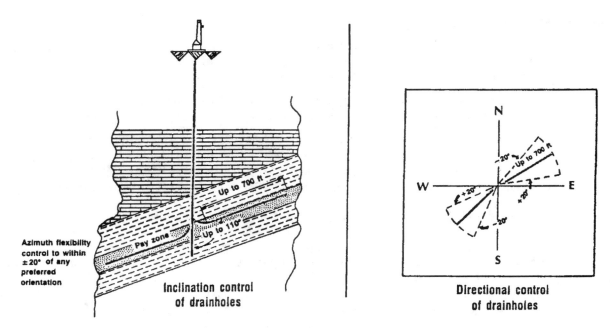

FIGURE 2–31 Drainhole control parameters (after Gorodi, 1984)

In favorable reservoir conditions, it has been possible to drill horizontally to 700 ft from the vertical wellbore. According to Gorody (1984), greater lengths can be attained if they are required.

When this method is used in the most commonly encountered situation, casing extends to a point directly above the zone of interest. The target zone or zones lie within the open hole interval below casing. The open hole is logged and kickoff point(s) selected. If orientation is required, a retrievable orienting device, is lowered, usually to the deepest target first. Conventional instrumentation is used to orient the device, which is then locked into place. After two-thirds of the curve is drilled with the *curve-cutting bottom hole assembly* (first bottom hole assembly), the hole is surveyed with a multishot camera lowered on a drillstring to confirm orientation. Azimuthal adjustment can be made, if necessary, while the final third of the curve is drilled. A second similar survey may be made after the curve is completed, but this is normally not required.

The second bottom hole assembly is then lowered through the completed curve, and the horizontal portion of the drainhole completion is drilled until the target is reached or the bit wears out. The hole can be surveyed again before re-entering with a new bit, or it can be surveyed at the end of the completed lateral.

Applications. Horizontal drainholes have been drilled in various area including the Triassic Moenkopi reservoir in Grassy Trails Field, Emery Co., Utah. Multiple zone completions in low permeability fractured reservoirs were carried out in two extremely tight (0.01 md and 4–5% porosity) siltstone beds, each varying in thickness from 10 to 60 ft.

Total length of 68 drainholes drilled from 18 wellbores was approximately 17,000 ft, for an average length of about 250 ft per drainhole. Both size tools were used to drill horizontal holes that were, generally, at 85° or 90° to the vertical wellbore.

Four of eleven exploration wells were producers. Of the seven dry exploration wells, three were distant stepouts, one was downdip, and three were updip of a "sweet" spot that was discovered and subsequently developed.

Six of seven subsequent development wells were significant producers, yet nearly all of them were dry before drilling multiple laterals.

One of the better wells in Grassy Trails field was well 11–33X. It was drilled with air to a vertical depth of 3906 ft, using a 4¾–in. bit. The well was dry and without any show of oil. The first drainhole, drilled from 3443 ft, was 220 ft long, and intersected major multiple fractures spaced approximately every 70 ft. As each fracture was approached, torque became erratic and the drilling rate slowed dramatically. Normal drilling conditions resumed as each fracture was completely crossed, and a surge of oil would fill the tanks at the end of the blooie line.

As more fractures were intersected, net oil production increased while drilling, and the background gas readings recorded on the mud logger's chromatography rose and stabilized at new levels.

Initial production was 220 bo/d and at the end of 323 days, it had decreased to 70 bo/d with cumulative production totalling 39,000 bbl.

A that time, the well was reentered and a second horizontal drainhole drilled from 3451 ft at a 55° angle to the first drainhole. This new drainhole intersected major fractures spaced approximately 35 ft apart along a lateral distance of 476 ft from the vertical wellbore. Initial production was reestablished at a rate of 280 bo/d, and the well produced an additional 35,000 bbl in a 153 day period following the second horizontal hole.

COMPLETION METHODS

Two main categories of well completions are open hole and conventional perforated completions. When a well is drilled, a decision must be made regarding completion. This decision is usually made after close analysis of shows, cores, drillstem tests, and well logs. Consideration can be given to open hole completions only in competent formations that do not present possibilities of sloughing or caving. As this is the case with most naturally fractured reservoirs, many wells in fractured media have been completed open hole. However, if it is possible to accurately detect fracture intervals, it is probably better to resort to perforated completions in naturally fractured reservoirs of Type A. In my experience, open hole completions provide better results in naturally fractured reservoirs of Type C. Open and perforated completions have been used successfully in naturally fractured reservoirs of Type B. For a discussion on characteristics of naturally fractured reservoirs of Type A, B, and C please refer to Chapter 1.

Open Hole Completions

In this type of completion, the oil string is set on top of the reservoir (Figure 2–32 [A]). Some wells of the Spraberry trend (Texas), the Altamont trend (Utah), fractured shales of Eastern United States, Palm Valley gas field (Australia), and many other naturally fractured reservoirs have been completed open hole.

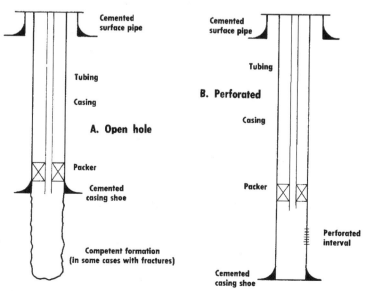

FIGURE 2–32   Schematic of A) open hole, B) perforated completions

This type of completion is used in fractured reservoirs when it is difficult to ascertain where the hydrocarbons are coming from. I especially recommend it for naturally fractured reservoirs of Type C. The basic requirement of having nonsloughing formations for open hole completions is evidently provided by highly competent, naturally fractured reservoirs.

An initial economic advantage on this completion is provided by savings in casing and perforating costs. A better possibility of production can probably be obtained in thin laminated beds and reservoirs with very poor or negligible vertical permeability (Gatlin, 1960). A key issue, invasion of the fractures by cement can be avoided. This is critical because productivity of the well can be reduced permanently once the fracture is cemented, even if the well is stimulated.

Disadvantages of open hole completions are dictated by difficulties in selective stimulation and control of gas or water entrance.

Perforated Completions

Figure 2–32(B) shows a schematic of a conventional perforated completion. The casing is set through the reservoir of interest, cemented, and perforated at desired levels. This completion for fractured reservoirs gained popularity in the Spraberry trend of Texas as early as 1952 because of sloughing problems in previous open hole completions.

At the time, the advantages and disadvantages of both types of completions were not known exactly, but satisfactory perforated completions had been already made in various Spraberry wells. With further experiences it became evident that, in some naturally fractured reservoirs, especially those of Type A, perforated completions had more advantages than disadvantages over open hole completions.

Probably the most important point to emphasize is the availability of an accurate means of detecting fracture intervals. If it seems possible to detect accurately the fracture intervals, then a perforated completion might be the best choice in reservoirs of Type A. If, on the contrary, there is no way to know where the production is coming from, it may be better to resort to open hole completions.

An important factor in perforated completions is the perforating process. The main techniques available to the industry are jets, bullets, hydraulic jetting, mechanical cutters, and permeators.

Bullet perforating has proven effective in soft formations. As naturally fractured reservoirs are made up of hard rock, the use of bullet perforating is not recommended for fracture media.

There are some special purpose bullets which allow fracturing of the formation and cement, establishing circulation in stuck tubing or drill pipe, removing casing and perforating the inner of two strings without damaging the outer string.

Table 2–8 compares bullet and jet penetration, based on studies by Lebus (1957). For soft formations, the penetration is the same using bullet and jet perforating. However, for very hard formations (naturally fractured reservoirs) the depth of penetration is larger when jet perforating is used.

Penetration of jet guns is also a function of the weight of the explosive. The larger the explosive weight, the greater the shaped charge which might create a larger hole or penetrate further.

Hydraulic jetting or hydraulic perforating is performed by forcing a high pressure stream of sand-laden fluid through a jet aimed at the casing/formation target (Pittman et al, 1961). The concentration of sand in the jet fluid ranges usually from $\frac{1}{2}$ to 1 lb per gallon of water. The main use of hydraulic jetting is in preparing a formation for fracturing. This approach does not cause casing or cement damage.

Mechanical cutters or knives are used to open windows, slots or holes between the formation and the wellbore.

Permeators are welded to windows cut in the casing before it is run into the hole (Smith, 1976). Sleeves on the permeator are extended with pump pressure to the wall of the formation of interest after the cement is in place but before it sets. Acid is used to dissolve plugs in the permeator allowing communication between the wellbore and the formation. The

TABLE 2–8 Comparison of bullet and jet penetration (after Lebus, 1957)

Cement target age, hr	Comparative formation type	Depth of penetration, in.	
		Bullet	**Jet**
24	very soft	15[a]	15[a]
48	soft	15[a]	15[a]
72	medium soft to medium	15[a]	11⅞
5 days	medium hard	12⅞	12²⁄₁₆
7 days	medium hard to hard	10½	11³⁄₁₆
10 days	hard	9⅜	10½
Other targets			
Berea sand core	very hard	4½	6¾
Steel plate	?	2	3½

[a]Shot completely through 15–in. target.

main advantage of the approach is that it does not cause damage. Disadvantages are more rig time, more expensive than jets or bullets, and lack of penetration to the formation.

Comparison of Open Hole and Perforated Completions

Two contrasting cases, one where perforated completions are preferred, and one where open hole completions provide better results are presented below.

Altamont Trend, Utah. An example of intricate completion problems is found in the fractured Altamont trend of Utah. Figure 1–38 shows a north-south stratigraphic cross section of the reservoir (see also Figures 5–1 and 5–2). A great variation in the thicknesses and stratigraphic position is evident. This, coupled with a wide variation of initial reservoir pressures including gradients of up to 0.9 psi/ft, and the presence of natural fractures, generated many problems during drilling and completion operations. Geologic aspects of the Altamont trend are found in Chapter 1. The history of the field is found in Chapter 5.

Baker and Lucas (1972) reported the history of well Shell Bleazard 1 18B4, that was drilled to 13,398 ft, completed with cemented liner to 12,223 ft, and with slotted liner to TD. Initially, the well produced 650 bo/d through the slotted liner and declined to 160 bo/d plus 115 bw/d and 75 Mscfd on 17/64–in. choke within a month. A check valve was installed above the slotted liner, and 23 zones that had shows were perforated with 1 shot/ft between 11,806 and 12,128 ft for a negligible increase in production. A reperforating job was carried out between 11,806 and 12,132 ft with two shots/ft, again for a negligible increase in production.

The perforations were stimulated with 20,000 gal of acid using wax beads as a diverting agent, and the well came with 1500 bo/d and 2.3 MMscfd on 20/64–in. choke. Production logging indicated that only a few of the net 135-ft perforated intervals were actually producing, although well log analysis indicated additional potential pay.

Production had declined to 700 bo/d after four months and cumulative production had reached over 140,000 STBO. This led Shell to the conclusion that the risk of cement entry into the fractures, while maintaining the option of stimulating the reservoir, was preferable to using uncemented slotted liner completion without the option of selective stimulation.

Palm Valley Field, Australia. Geologic aspects of this field are presented in Chapter 1. The history of Palm Valley gas field is presented in Chapter 5. A dramatic example of natural fracturing is provided by well PV2, which was drilled with air and tested over 70 MMscf/d after penetrating 1 or up most 2 ft of the Pacoota naturally fractured reservoir. This production is clearly associated with fractures as the matrix permeability from core plugs always yielded values smaller than 0.1 md (Au et al, 1991).

The well was completed open-hole and had accumulated 29.4×10^9 scf to the end of July, 1994, which represents 43% of the cumulative from the total field. The performance of this well is much better than the performance of the other wells completed with cemented casing and perforations.

Cementing Problems

If the decision is made to have a perforated completion, it is of paramount importance to pay attention to the cementation process. Natural fractures might play the role of "thief zones" which induce partial or complete circulation loss during cementation of casing. This has many detrimental effects including:

1. Plugging of natural fractures with cement which might move through the fractures far away into the formation. This can induce strong near-wellbore and deep-seated damage which in some cases might be irreparable. Appropriate care must be taken especially in reservoirs of type B and C (for naturally fractured reservoirs classification please see Chapter 1). Damage in reservoirs of type A might be due to invasion of the fractures by cement filtrates. Damage can also occur in "weak" zones which might be ruptured by the increase in hydrostatic pressure resulting from the higher density of the cement slurry.
2. Communication between intervals which are not properly sealed off with cement.
3. Probability of corrosion of exposed pipe.
4. Insufficient fill-up.

Loss circulation can be decreased or controlled by reducing the density of the slurry with light weight additives in such a way that rupture does not occur due to a large hydrostatic head. When carrying out a primary cement job in a lost circulation zone, it is important to reduce pumping pressures as low as possible.

These low pumping pressures require that turbulent flow be avoided in favor of plug flow preceded by a viscous preflush. By doing this an effective removal of the drilling mud can be achieved with a minimum of contamination, while maintaining at the same time a lower hydrostatic head than is possible with turbulent flow. Other methods of combating lost circulation include multi-stage cementing, liner cementing, and injection of nitrogen into the mud system to decrease the hydrostatic head of the fluid column. In the case of natural fractures bridging materials can be incorporated in the slurry.

Light Weight Cementing.   Various additives can be used to reduce the weight of the slurry (Dumbarld et al, 1956; Byron Jackson Inc., 1969). These include bentonite, gilsonite, diatomaceous earth, perlite, pozzolans, clay bubbles, and hydrocarbons.

Bentonite is a mixture of at least two aluminum and magnesium silicate minerals which is often added to reduce slurry weight. The concentration of bentonite (gel), however, is inversely proportional to the compressive strength and the thickening time of the set product.

Bentonite is used in (1) blended gel cements, (2) premixed or prehydrated bentonite where bentonite is added to water, (3) modified cement, a patent of Humble Oil, where 8% to 25% of bentonite and a dispersant-calcium lignosulfonate are added to regular portland cement, and (4) high-gel salt cement, a patent of Gulf Oil, where 12% to 16% bentonite, 3% to 7% inorganic salt (sodium chloride, preferably), and 0.1% to 1.5% of a dispersant-calcium lignosulfonate are added to regular portland cement (Smith, 1976).

In general, the usual amount of bentonite added is 2% to 16%. It can be used with all API classes of cement.

Diatomaceous Earth when specially graded with a high percentage of water can be used for reducing the slurry density. It provides approximately the same properties as bentonite. It presents the advantage that, when used in high percentages, it does not increase the viscosity of the slurry as does bentonite. The disadvantage of diatomaceous earth is that it is much more expensive than bentonite (Smith, 1976).

Gilsonite can be utilized as a slurry additive to decrease density mainly because its own low density (1.07) is only slightly heavier than water. Gilsonite does not absorb water under pressure. It has a higher strength at any age than other cements of the same slurry weight containing other available lightweight lost circulation additives.

Pozzolans can be used where a moderate reduction in slurry weight is adequate. Pozzolans react with the hydrated lime from the cement to yield cementation compounds. This produces a medium density slurry where bridging materials can be incorporated (Byron Jackson Inc.). Pozzolan uses up the calcium hydroxide produced during the cement-water reaction, making the set product more resistant to leaching.

Expanded Perlite particles contain open and closed pores. The open pores fill with water, some close pores crush and fill with water. The density reduction depends on the number of pores that remain closed and the amount of water that is immobilized in the open pores. Cement slurries containing perlite are mixed with what seems to be excessive water to allow the cement slurry to be pumpable at down-hole conditions.

Clay Bubbles or small, hollow clay spheres have been proposed by Coffer et al (1954) to reduce slurry density. Based on laboratory work, it is possible to obtain slurry densities as low as 10 lb/gal while maintaining satisfactory strength.

Hydrocarbons such as diesel oil and kerosene when mixed with portland cement, and a chemical dispersant have been found useful for sealing-off water-bearing zones. This material has unlimited pumping time since it does not set until brought in contact with water. It has also proved useful to prevent lost circulation.

Multi-Stage Cementing. While cementing a long casing string, the formation might not be able to withstand the hydrostatic head of drilling mud plus the required fill of cement. In these instances it might be advantageous to place cement around the casing shoe and to tie-in the string with another string of casing up the hole or to cement back to the surface.

In this method a stage collar is placed at a predetermined point in the casing string. This point occurs where the summation of hydrostatic heads of two slurry placements separated by drilling mud are not great enough to break down the formation and cause it to lose circulation.

Figure 2–33A shows a schematic of multi-stage cementing technique for the case of complete lost circulation in intermediate or production string. Plugs are not shown in the diagram. The lost circulation section might correspond to a zone of natural fractures primarily of Type B and C.

Figure 2–33B shows a diagram of multi-stage cementing technique for the case of partial lost circulation. In this case the formation might withstand a predetermined hydrostatic head. The section of partial returns might correspond to a zone of natural fractures primarily of Type A.

Nitrogen Cementing. This method was jointly developed by Nitrogen Oil Well Service Co (NOWSCO) of Houston and Livingston Oil Company of Tulsa. The method consists of (1) injecting nitrogen into the drilling mud string ahead of the cement slurry, or (2) having mud circulation established, stop the circulation and introduce a slug of fluffy, nitrified mud before cementing. The idea is to have a combined weight of the cement column, nitrified, and nitrified mud equal to, or smaller than, the hydrostatic pressure of the drilling mud used during drilling operations. This control reduces the possibility of lost circulation in "weak" formations and naturally fractured reservoirs. Another advantage of this method is that it permits a wider choice in cement types. Since the method is not dependent on low-density additives, it is possible to select a higher strength and often less expensive cement system. The nitrogen technique can also be used to advantage in cementing liners.

NOWSCO and Livingston Oil Co. have developed procedures for calculating the mud volume to be nitrified and the ratio of nitrogen to mud for obtaining the desired hydrostatic pressure. These methods have been published by Byron Jackson.

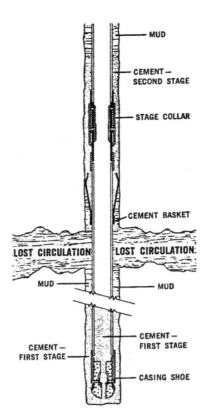

FIGURE 2–33 (A) Multi-stage cementing technique for lost circulation, with complete loss of circulation in intermediate or production string. Plugs are not shown (after Byron Jackson, Inc., 1969)

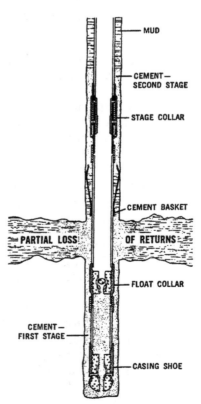

FIGURE 2–33 (B) Multi-stage cementing technique for partial lost circulation-formation will withstand predetermined hydrostatic head. Plugs are not shown (after Byron Jackson, Inc., 1969)

Figure 2–34 shows a schematic of the cemented well including nomenclature used in the equations that follow. Based on knowledge of total depth of the well in feet (L_T), height of the cement slug in feet (L_C), height of the mud capping slug in feet (L_M), annular capacity from caliper survey of circulation volume test in bbl/ft (C_A), cement column gradient in psi/ft (G_C), mud column gradient (G_M) in psi/ft, and temperature gradient in °F/100 ft (X), it is possible to run the following calculations:

1. Calculate desired bottom hole pressure (BHP) in psia from the equation:

$$BHP = G_M L_T - 200 \qquad (2\text{–}9)$$

2. Determine the average pressure (P_A) of the nitrified slug in psi with the use of:

$$P_A = (BHP - G_C L_C + G_M L_M)\,/2 \qquad (2\text{–}10)$$

3. Figure the depth from the surface to the center of the nitrified mud zone (D_A) in feet from:

$$D_A = L_M + (L_T - L_C - L_M)\,/2 \qquad (2\text{–}11)$$

4. Compute the average temperature of the nitrified slug in °F with the use of the equation:

$$T_A = 74 + X\,(D_A/100) \qquad (2\text{–}12)$$

5. Find the nitrogen volume factor (B) in scf of N_2/bbl of space at P_A and D_A, from Figure 2–35.

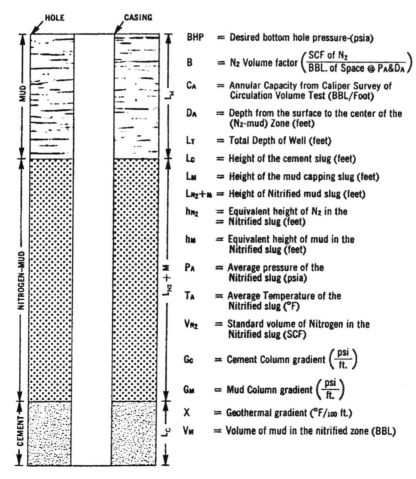

FIGURE 2–34 Nomenclature Used in Nitrogen Cementing (after Byron Jackson, Inc., 1976)

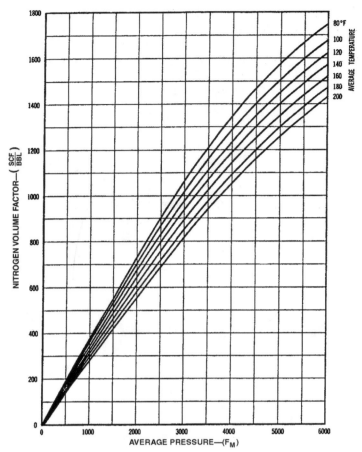

FIGURE 2–35 Nitrogen Volume Factor Chart (after Byron Jackson, Inc., 1976)

6. Calculate the equivalent height of nitrogen (h_{N2}) in the nitrified slug from:

$$h_{N2} = \frac{BHP - G_c L_c - G_M(L_T - L_c)}{0.0000908 \, B - G_M} \quad (2\text{–}13)$$

7. Determine the standard volume (scf) of nitrogen in the nitrified slug (V_{N2}) with the use of:

$$V_{N2} = h_{N2} \, C_A \, B \quad (2\text{–}14)$$

8. Figure the scf of nitrogen per bbl of mud in the nitrified zone from the equation:

$$\text{Ratio} = V_{N2} / (h_M \, C_A) \quad (2\text{–}15)$$

9. Compute the mud volume in the nitrified zone from:

$$V_M = h_M \, C_A \quad (2\text{–}16)$$

10. Calculate the nitrogen pumping rate (NPR) in scf/min from:

$$\text{NPR} = \text{Ratio X MPR} \quad (2\text{–}17)$$

where MPR is the mud pumping rate.

Example 2–3. This example is extracted from Byron Jackson Ltd.'s "Applied Engineered Cementing." Given L_T = 6665 ft, L_C = 1000 ft, L_M = 2000 ft, C_A = 0.1224 bbl/ft, G_C = 0.737 psi/ft, G_M 0.464 psi/ft and X = 0.9°F /100 ft, design a nitrogen cementing job.

Solution. The steps presented above are evaluated as follows:

1. Bottom hole pressure is calculated to be 2900 psi from equation 2–9.
2. Average pressure is found to be 1545 psia from equation 2–10.
3. Distance D_A is figured to be 3832 ft from equation 2–11.
4. Average temperature (T_A) is 108°F from equation 2–12.
5. Nitrogen volume factor (B) is 525 scf/bbl from Figure 2–35.
6. Distance h_{N2} is 1110 ft from equation 2–13.
7. Volume V_{N2} of nitrogen in the nitrified slug is 71,500 scf from equation 2–14.
8. The ratio of nitrogen volume per mud volume in the nitrified zone is 230 scf/bbl from equation 2–15.
9. The mud volume in the nitrified zone V_M is 313 bbl from equation 2–16.

Bridging Materials.   In fractured zones which require a bridging agent to prevent fluid loss during cementing operations a solid granular material such as gilsonite can provide excellent results. Gilsonite is graded in particle size from fine to ¼ in. Since only small amounts of water are required with this material, high strengths are obtained for low density slurries. Table 2–9 shows amounts of gilsonite required for any desired slurry density with Class A or H cement. Also shown are compressive strengths.

Table 2–10 shows some other materials commonly added to cement slurries to control lost circulation. Included are the nature of the various particles, amounts used, and water required.

Cementing Directional Wells.   There is the tendency for the loads acting on the string to force it against the wall of the hole in many places. These loads also create high drag, torque, and bending stress that in many cases limit pipe movement. It has been found that even with an optimum centralizer program, there are probabilities that in many parts of the hole the casing might be off center creating strong lateral forces in the eccentric annulus (Smith, 1976). Distribution of these lateral forces in directional wells is presented on Figure 2–36. In the high side of the wellbore the resistance is small and consequently mud displacement is efficient. Laboratory tests, however, indicate that excess water in the cement slurry can result in a *water channel* along the high side of the cemented annulus.

In the low side of the hole the resistance is larger, and there is solid setting from the drilling fluid which might cause a continuous *mud channel*. This is illustrated on Figure 2–37 which shows segments from a mud displacement test conducted in the laboratory at 85° deviation angle. Notice the arrows pointing to mud channels on the low side of the annulus. Water channels are observed on the high side of the annulus and also on the high side of the inner casing. This laboratory work is corroborated by field experiences which have shown that in directional wells, solids from the drilling fluid settle in the low side of the wellbore.

In a deviated hole these channels will lead to unwanted communication behind pipe through perforations which might be separated by hundreds of feet.

TABLE 2–9   Slurry weight and strength with gilsonite (D—7) (after Byron Jackson, Inc., 1976)

lbs./gal.	lbs./cu.ft.	D-7 lbs. per sk. cement	Compressive Strength, 1 day, 100°F, psi
15.7	117.4	0	2563
15.0	112.2	6.0	2085
14.0	104.7	17.0	1542
13.0	97.2	33.2	898
12.0	89.8	59.6	623
11.0	82.4	109.2	313

TABLE 2–10 Materials commonly added to cement slurries to control lost circulation (after Smith, 1976)

Type	Material	Nature of Particles	Amount Used	Water Required
Additives for Controlling Lost Circulation				
Granular	Gilsonite	Graded	5 to 50 lb/sk	2 gal/50 lb
	Perlite	Expanded	½ to 1 cu ft/sk	4 gal/cu ft
	Walnut shells	Graded	1 to 5 lb/sk	0.85 gal/50 lb
	Coal	Graded	1 to 10 lb/sk	2 gal/50 lb
Lamellated	Cellophane	Flaked	⅛ to 2 lb/sk	None
Fibrous	Nylon	Short-fibred	⅛ to ¼ lb/sk	None
Formulations of Materials for Controlling Lost Circulation				
Semisolid or flash setting	Gypsum cement	—	—	4.8 gal/100 lb
	Gypsum-portland cement	—	10 to 20% gypsum	5.0 gal/100 lb
	Bentonite cement	—	10 to 25% gel	12 to 16 gal/sk
	Cement + sodium silicate	—	—	(the silicate is mixed with water before adding cement)
Quick gelling	Bentonite-diesel oil	—	—	—

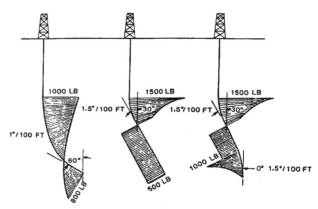

FIGURE 2–36 Distribution of lateral forces in directionally drilled holes (after Weatherford Oil Tool Co., 1974)

Laboratory research conducted by Keller et al (1983) addressing specifically the problem of cementing directional holes led the following conclusions:

1. Coherent mud channels formed by solids settling can remain at the bottom of the annulus of a deviated wellbore following mud displacement by cement.
2. Under similar conditions the size of the mud channels are significantly less at a wellbore deviation angle of 60° than at 85°.
3. Under certain conditions, barite from the mud settles to the low side of a deviated wellbore and causes the mud on the low side to be very difficult to displace.
4. The best high-angle displacement was obtained by raising the mud yield point and gel strength sufficiently high to prevent solids settling.
5. Excess water in the cement can result in a water channel at the top of the cemented annulus in a highly-deviated well.

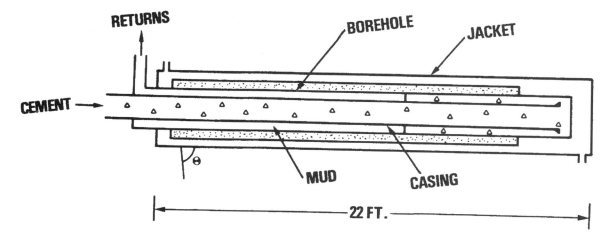

Schematic diagram of apparatus.

FIGURE 2–37 Segments from a mud displacement test conducted at 85° deviation angle. Arrows point to mud channel on low side of annulus (after Keller et al, 1983)

A summary of displacement test results from Keller et al's (1983) research is presented on Table 2–11. Included are the angle, type of formation (permeable or impermeable), density, percent solids, plastic viscosity (PV), yield point (YP), gel strength, and fluid loss for the drilling fluids. Also shown are density, PV/YP, volume, and rate for the cement slurry. The last column shows a displacement efficiency which is defined as the percent of annular volume not containing mud. These were calculated separately for the upper half and lower half of the annulus.

Smith (1976) has indicated that improved results could be obtained by using a heavier slurry in conjunction with a much lighter displacing fluid so that the casing string can float in the cemented part of the hole and be centralized off the upper side. As an example he indicates that a 9⅝-in., 40 lb/ft casing would require at least a 16.6 lb gal cement slurry in the annulus using diesel to pump the top plug into place. Mixing and pumping rates must be as high as possible in such a way that the mud in the annulus can continue to circulate without gelling.

In the case of naturally fractured reservoirs, especially those of Type B and C, the above procedure must be used carefully. In fact, the heavier slurry can lead to flow of cement into the fractures which might produce irreparable damage. Furthermore, if the natural fractures become a source of circulation loss it might be preferable to use a lighter mud and to reduce pumping pressures as low as possible. These low pumping pressures require that turbulent flow be avoided in favor of a plug flow preceded by a viscous preflush.

HYDRAULIC FRACTURING

Efforts to hydraulically fracture a naturally fractured reservoir have been carried out in many instances with varied degree of success. The idea has been that by inducing a large artificial fracture, one might be able to establish communication with the network of natural fractures increasing consequently the hydrocarbon production. In my opinion success or failure depends on (1) the ingenuity of the professionals designing and performing the job, and (2) the type of naturally fractured reservoir developed by mother nature, something that is beyond the engineer's control.

Presently, there are two main schools of thought regarding hydraulic fracturing. The first school of thought (conventional) indicates that failure by rupture occurs along a plane which is perpendicular to the least compressive stress (Howard and Fast, 1970). Based on this theory it is possible to design a hydraulic fracturing job in such a way as to obtain a certain fracture length. Well test analysis carried out following the fracturing job may tell us if the outcome of the job was as intended by the design.

A second school of thought indicates that in certain naturally fractured reservoirs there is going to be shear slippage which will result in local redistribution of stresses. This redistribution leads to branching or dendritic fractures which will improve drainage into the wellbore significantly (Murphy and Fehler, 1986; Conway, 1986). According to this theory dendritic fractures expand elliptically with the major axis of the ellipse being perpendicular to the least compressive stress. Kiel (1977) also advocated stimulation by dendritic fracturing. However, his reasoning with respect to the mechanism of fracturing was different.

Conventional Fracturing

A successful hydraulic fracturing job depends on many factors including well position with respect to the natural fractures, earth stresses, fracture height, pump rates, types of fracture fluids, leak-off controls, and proppant concentrations.

Tectonic Natural Fractures. These natural fractures are related to a tectonic event such as folding or faulting (see Chapter 1).

Naturally fractured reservoirs, especially those of the B-II and C types can be severely damaged during drilling and cementing operations. Whenever possible, these reservoirs should be drilled with air or as a second option oil-base muds. In some cases, acid jobs might prove valuable in removing the damage.

TABLE 2–11 Summary of displacement test results (after Keller et al 1983)

| Test | Angle | Drilling Fluid | | | | | | Cement | | | | Disp. Eff. top/bot % |
		Forma-tion	Den-sity (ppg)	Solids %	PV/YP @72°F cp/lb$_f$/100ft²	Gels[1] 10s/10m lb$_f$/100 ft²	Fl. Loss[2] LT/HT cm³/30 min	Den-sity (ppg)	PV/YP @180°F cp/lb$_f$/100 ft²	Vol. (bbls.)	Rate (b/min)	
1	0	imper.	12.2	20	34/9	2/2	13/34	16.8	16/32	10	1	98/98
2	0	imper.	12.1	18	31/7	2/2	14/28	16.8	18/30	10	7	98/98
3	60	imper.	16.0	30	36/4	3/5	10/13	16.8	23/29	20	4	99/97
4	60	imper.	15.9	29	91/15	3/5	8/11	16.8	23/29	10	7	88/80
5	60	imper.	16.2	30	83/24	6/10	7/20	16.8	21/33	20	7	93/87
6	60	imper.	15.9	26	61/11	3/13	11/11	16.8	20/30	10	4	99/92
7	85	imper.	15.8	22	29/2	2/2	12/25	16.8	16/30	10	1	96/31
8	85	perm.	15.7	29	53/6	2/3	11/13	16.8	24/29	10	7	95/45
9	85	perm.	15.8	27	55/8	2/4	9/15	16.8	25/30	10	4	74/48
10	85	imper.	15.7	28	56/14	4/5	8/14	16.8	15/30	10	4	99/51
11	85	imper.	15.8	28	55/11	4/4	9/26	16.8	17/29	10	4	99/55
12	85	imper.	15.9	31	72/19	5/7	7/13	16.8	20/30	10	1	97/68
13	85	imper.	15.8	26	56/19	5/9	6/13	16.8	18/29	10	7	98/80
14[3]	85	imper.	11.8	12	31/6	2/3	9/34	16.8	20/30	10	4	92/14
15	85	imper.	12.3	13	34/4	2/3	9/12	16.8	21/29	10	4	99/23
16	85	imper.	11.9	14	44/9	2/3	8/12	16.8	15/29	10	4	93/50
17	85	imper.	11.5	20	65/23	9/10	6/11	16.8	19/31	10	4	99/59
18	85	imper.	11.6	12	72/25	5/5	8/10	16.8	14/29	10	4	99/67
19	85	imper.	12.0	16	78/47	19/26	5/10	16.8	20/28	10	4	99/99
20[4]	85	imper.	12.2	20	36/12	2/4	14/29	16.8	15/30	10	4	91/37

1 At 72°F
2 LT at 72°F and 100 psi, HT at 180°F and 500 psi
3 Mud circulated at 1 bpm
4 No 24-hr. static period

A popular belief is that when natural fractures are not intercepted by the wellbore, massive hydraulic fracturing jobs are valuable because they help to establish communication between the wellbore and the natural fractures. Experience indicates, however, that in many cases when the well is poor from the beginning, it will continue being poor, in spite of massive hydraulic fracturing treatments, although these are exceptions. These poor results will happen primarily when the following two conditions are met: (1) The natural fractures have a preferential orientation. This occurs, for example, in the case of swarms of "type 2" shear and extension fractures which are fold related (see Chapter 1). (2) The contemporary least horizontal compressive stress is perpendicular to the natural fractures. These two conditions are met in the schematic of Figure 2–38A, which shows a top view

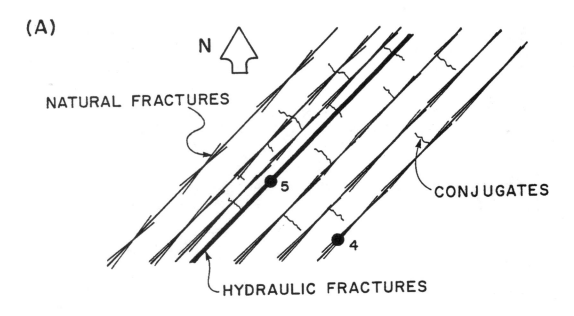

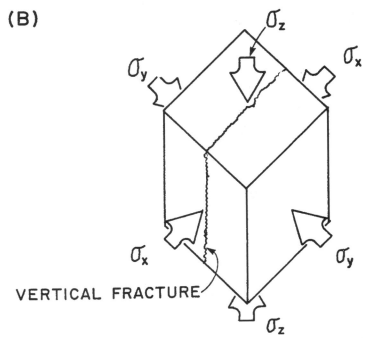

FIGURE 2–38   Schematic showing (A) a hydraulic fracture parallel to natural fractures, and (B) probable stress element and preferred plane of fracture in the Austin Chalk ($\sigma_z > \sigma_x > \sigma_y$) (after Aguilera, 1983)

of wells 4 and 5. Well 4 intercepted natural fractures and was a discovery well. The logs of well 5 were similar to those of well 4. However, when well 5 was tested it did not flow. It was decided to hydraulically fracture the well. Figure 2–38A shows that the resulting hydraulic fracture was parallel to the principal horizontal stress σ_x shown in Figure 2–38B. This has been observed some times, for instance, in the Austin Chalk. The hydraulic fracture failed then to establish communication with the main trend of natural fractures.

Of course, it can be argued that the hydraulic fracture intercepts the conjugates of the system. The problem is that the conjugates have a transmissibility much lower than the main trend of fractures. Furthermore, because the conjugates are under compressive stresses, they will probably tend to close as the pressure within the fracture is reduced and drastic decreases in production might arise.

An additional stimulation problem arises from the usual large thickness of naturally fractured reservoirs without plastic layers in between. According to Dech et al (1986) 65% of the naturally fractured reservoirs in the U.S. have a thickness of 400 ft or more. Even if only a short interval is perforated and stimulated, the artificial fracture will extend up and down until it finds a plastic barrier. This, of course, will limit the extent of the fracture away from the wellbore. Consequently, the number of conjugates that are intercepted might also be reduced.

Figure 2–39 illustrates the effect of fracture height on fracture penetration for a typical treatment in the Austin Chalk. The injection rate is 50 bbl/min, the treatment volume is 200,000 gal and all other parameters are maintained constant (Parker et al, 1982).

Assume that only the lowest portion of this chalk well showing a resistivity greater than 15 ohm-m has been perforated. An anticipated fracture height of 200 ft indicates a propped fracture length of about 975 ft (Figure 2–39). If the fracture height actually reached 250 ft (an increase of 50 ft), the fracture penetration is reduced by about 160 ft.

The above comments apply mainly to Type B-II and C reservoirs (see Chapter 1). An example of hydraulic fracturing fiascos in naturally fractured reservoirs is provided by the Austin Chalk of Texas. According to Holditch and Lancaster (1982) "even though technology has advanced tremendously over the last three to four decades, the average production from an Austin Chalk well has not really changed." This has occurred in spite of several different types of fracture treatments. In addition to the hydraulic fractures going approximately parallel to the natural fractures in the Austin Chalk, decrease in fracture penetration due to a fracture height larger than anticipated might have had something to do with this lack of success in increasing the average production from Austin Chalk wells.

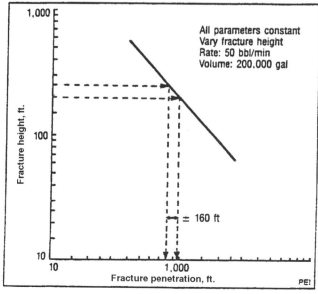

FIGURE 2–39   Effect of fracture height on fracture penetration (after Parker et al, 1982)

Some techniques have been developed by Dowell to try to control the height problem discussed above. These services are being offered under the trademarks Divertafrac and Invertafrac.

Figure 2–40 illustrates the radial configuration of typical vertical fracture growth. In stage A the fracture grows in a vertical radial pattern. In stage B the supposedly barrier rock ruptures because it is too weak. The fracture continues growing in stage C. Finally, in stage D the fracture is stopped by an upstructure barrier rock, but it continues growing downward.

In the divertafrac service an artificial lower barrier is created using a special slurry of solid diverting material which is pumped under controlled conditions to make sure that

A. *Initial fracture growth is in vertical radial pattern.*

B. *Rock strata below productive interval ruptures because it is too weak to act as barrier.*

C. *Unrestrained downward growth allows fracture to grow "out of zone."*

D. *Upward growth is restrained by barrier rock but downward growth continues.*

FIGURE 2–40 Typical fracture growth without DIVERTAFRAC Service (Courtesy of Dowell)

the diverter is deposited in the bottom of the fracture. As the diverting material reaches the bottom of the fracture, a pressure differential is created across the diverter pack with a lower pressure at point P_2 as shown on Figure 2–41. Since P_2 is smaller than P_1, the fracture is diverted upward and outward.

A sequence of events illustrating how the downward fracture growth can be restrained is presented in Figure 2–42. The initial fracture growth in a vertical radial pattern is shown on stage A. Stage B shows how an artificial barrier is deposited at the bottom of the fracture. In stage C the growth of the vertical fracture has been restrained by an artificial lower barrier of diverting materials and a natural upper barrier. Stage D shows how the vertical fracture is growing outward becoming an efficient and effective fracture.

Invertafrac operates in a similar fashion but creating an artificial upper barrier by using a special buoyant solid diverting material.

The importance of increased fracture penetration by containment of the fracture in the productive interval is illustrated with the use of Figure 2–43 published originally by McGuire and Sikora (1960). The curves represent fracture penetration (L) as a function of the drainage radius (r_e). If a good relative conductivity can be achieved, then a fracture penetration of 100% might provide as much as a 13–fold increase in production.

Nolte and Smith (1981) have published a method which allows us to identify (1) the phase during which the desired confined-height maximum extension rate can occur (Mode I), and (2) a critical pressure (Mode II) that subsequently will lead to either an undesirable pressure increase due to restricted penetration (Mode III) or an accelerated height growth (Mode IV).

The method consists of preparing a log-log plot of fracture treating pressure, above closure stress, vs. treating time (or equivalently treating cumulative fluid volume). For Mode I a straight line with a slope equal to 0.13–0.25 is obtained. The slope for Mode II is zero. A slope of 1.0 is obtained for Mode III. Finally, a negative slope would be indicative of Mode IV. Examples of this log-log plot with the different characteristic slopes are presented on Figure 2–44.

Regional Natural Fractures.   The reason for generation of these fractures is not known for sure (see Chapter 1). However, it is known that these fractures are developed over large areas with very little change in orientation. They are orthogonal and perpendicular to major bedding surfaces.

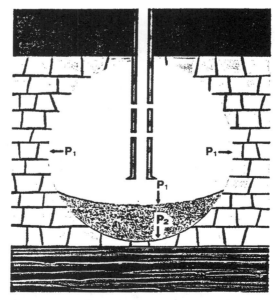

FIGURE 2–41   Pressure drop across diverter material pack results in lower pressure at point P_2. Downward fracture growth is restrained due to reduced pressure. (Courtesy of Dowell)

A. *Initial fracture growth is in vertical radial pattern.*

B. *Artificial barrier of diverting material is deposited in bottom of fracture under controlled pumping conditions.*

C. *Vertical fracture growth is restrained by artifical lower barrier and natural upper barrier.*

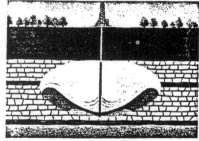

D. *Fracture is efficient and effective due to containment in productive interval.*

FIGURE 2–42 Artificial barrier created by DIVERTAFRAC Service restrains downward fracture growth (Courtesy of Dowell)

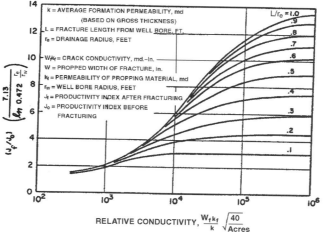

FIGURE 2–43 Increase in productivity from vertically oriented fractures (after McGuire and Sikora, 1960)

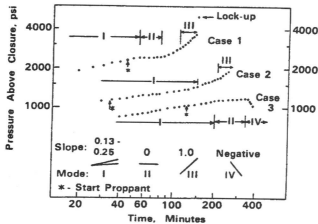

FIGURE 2–44 Examples with the different characteristic slopes (after Nolte and Smith, 1981)

Since these fractures are in many instances perpendicular to each other the induced hydraulic fracture will intercept the natural fractures no matter what is the current earth's stress state. This is a big advantage over the case of tectonic natural fractures discussed previously. However, it is likely that the hydraulic fracture will intersect the regional natural fracture with the lowest permeability. This occurs because the hydraulic fracture goes parallel to principal horizontal compressive stress. This stress in turn tends to close the regional natural fracture that goes perpendicular to it. Problems related to fracture height subsist in both tectonic and regional fractures. The idea of dendritic fracturing to be discussed later in this chapter would appear to have application in the case of regional fractures.

Contractional Natural Fractures. These fractures are associated, in general, with a reduction in volume (see Chapter 1). They are characterized by polygonal shapes. Consequently, a hydraulic fracture would stand good probabilities of intercepting contractional fractures, no matter what is the current stress state of the reservoir. A potential problem as discussed in the previous case is that the contractional fractures with the lowest permeabilities are the ones that might be intersected by the hydraulic fracture.

Dendritic Fracturing

The idea of dendritic fracturing was first advocated by Kiel (1977). In his process, a well is (1) repetitively fractured with a fluid bearing a proppant agent, (2) shut-in, and (3) vented.

According to his reasoning, the first cycle of pressurization results in spalling and self-propping of the main fracture. In subsequent cycles the proppant-bearing fluid bridges the spall-proppants producing a rise in pressure which leads to the propagation of lateral fractures perpendicular to the first one. Various case histories in oil and gas wells suggest productivity increases when this fracturing method has been used.

Another way to look at dendritic fracturing has been proposed by Murphy and Fehler (1986). Their reasoning is that when fluids are injected into natural fractures, several types of deformations can take place.

In the initial stages the pressure increase in the natural fractures is small, and there is no separation between the walls of the fracture. However, the difference between the total earth stress acting perpendicular to the fracture and the fluid pressure inside the fracture, i.e., the "effective closure stress," is reduced. As injection continues, the effective closure stress continues decreasing, and a stage might be reached in which there is not enough friction to resist shearing stresses acting parallel to the natural fracture. Slippage of the natural fracture in a shear mode will ensue, and one rough surface asperity will ride atop another producing an increase in fracture opening and permeability, which according to Murphy and Fehler (1986) is irreversible even if the fluid pressure is suddenly decreased. This permeability improvement has been termed *shear stimulation.*

If the void space increase generated by shear stimulation is not sufficient to accommodate the fluids injected, the pressure will continue to increase and might eventually reach the same level as the earth's stresses acting normal to the natural fracture. At this point the opposite walls of rock blocks, which have maintained solid to solid contact, will separate. In principle, a proppant agent trapped inside the natural fracture once the fluid pressure is released would maintain the fracture open. In this case it could be considered that the natural fracture has been effectively "stimulated."

A coupled rock mechanics/fluid flow model has been used by Murphy and Fehler (1986) to simulate the stimulation of a naturally fractured rock.

When natural fractures are aligned parallel to the principal earth stresses, a process equivalent to the conventional hydraulic fracturing is obtained.

When the natural fractures and the principal stress directions form an angle of 30° and the hydraulic fracture is formed with a fluid of low viscosity such as water, two types of stimulation patterns might occur.

The first pattern is shown on Figure 2–45. This case would occur when the void space increase or dilatancy due to shear slippage is large or when the frictional resistance to shear slippage is low. The result is that a single natural fracture is stimulated. The resolved stresses on Figure 2–45 are the result of a principal earth stress equal to 2σ applied at an angle of 30° with respect to the orientation of the natural fracture. Symbol σ represents the minimum stress acting normal to the maximum stress (2σ).

The second pattern is shown on Figure 2–46. This case would occur when dilatancy is small or when the resistance to shear slippage is high. In this situation shear slippage is accompanied by shear-stress drops. The interaction of these drops with the acting earth stresses produces opening of natural fractures which go perpendicular to the maximum stress. This leads to a dendritic pattern which tends to expand in an elliptical fashion.

Figure 2–47 allows a more detailed explanation of dendritic fracturing. The original normal stress is 1.75 σ. The main stimulated joint or natural fracture has slipped in shear and the fracture walls have separated. Since there is no solid to solid contact, there is no friction to support the initial shear stress. Consequently a stress drop occurs and the Y-direction normal compressive stress is affected as shown at the top and bottom of Figure 2–47. The original stress is reduced to as low as 1.25 σ in the lower left and upper right quadrants. This stress is now low enough to allow lateral stimulation of the natural joints as illustrated in the same quadrants. Notice that these lateral fractures are approximately parallel to each other. As injection continues, the cycle is repeated and the final product is a dendritic elliptical network as shown on Figure 2–46.

Figure 2–48 presents a comparison of conventional and dendritic fracturing. Conventional fracturing results in areal drainage which is not as effective as the volume drainage produced by dendritic fractures. These dendritic fractures have been obtained at Fenton Hill, New Mexico, by injecting a low viscosity fluid (water) at low rates into a system of

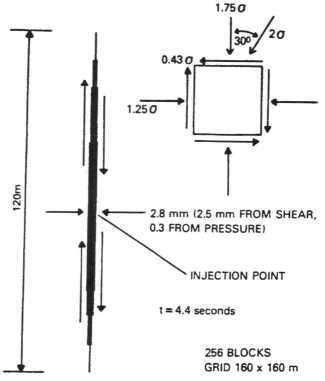

FIGURE 2–45 Single-joint stimulation induced by shear slippage when frictional resistance to shear slippage is low or the ability to open the joint in shear is high (after Murphy and Fehler, 1986)

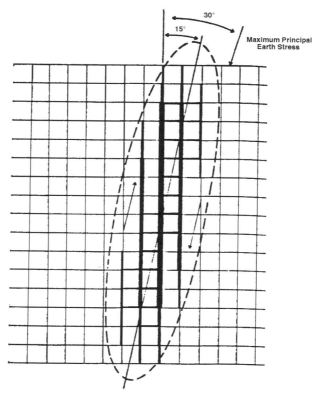

FIGURE 2–46 Multiple-joint shear stimulation that occurs when shear resistance is high or shear dilatancy is low (after Murphy and Fehler, 1986)

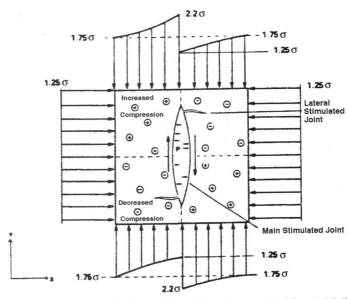

FIGURE 2–47 Stimulation of lateral joints (after Murphy and Fehler, 1986)

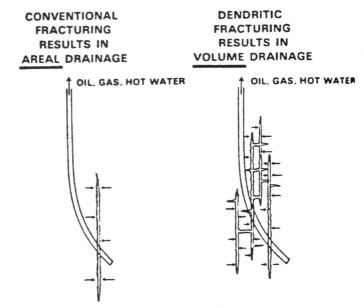

FIGURE 2–48 Volume drainage of fluids is more effective than areal drainage (after Murphy and Fehler, 1986)

natural fractures. Note that both wells in Figure 2–48 have been directionally drilled. In principle, however, dendritic fractures could be generated also in vertical and horizontal wells.

Some other examples of hydraulic fractures expanding in an elliptical fashion have been reported by Mahoney et al (1981) and Pine and Batchelor (1984).

Dynamic Gas Pulse Fracturing

This method has been marketed by Servo-Dynamics, Inc. under the name STRESSFRAC. The process is based on propellant-actuated, high pressure gas generation, which permits initiation of multiple fractures radiating in all directions from the wellbore in a pattern influenced by the casing perforation geometry.

One important advantage of the method is that the created fractures do not take the path of least resistance, as occurs in conventional fracturing, since the in-situ stress level of the formation is exceeded by a significant amount. High gas velocity, combined with fracture propagation of more than 300 ft per second erodes the fracture preventing the fracture from closing. Ideally, the tool functions when hydrostatic heads are between 250 and 6000 psi.

In a typical application the cylindrical tool is lowered to the desired zone on standard wireline under a substantial head. On ignition, the by-product of combustion of the propellant is a large quantity of CO_2 gas. Pressure is increased and within milliseconds fractures are initiated in all directions.

Horizontal Wells Fracturing

Chapter 1 indicated that the probabilities of intercepting natural fractures increase if we drill directional or horizontal wells which go perpendicular to the orientation of the fractures. Simulation work carried out by Giger et al (1983) corroborates that a horizontal well intercepting a vertical fracture will have a better productivity index than a vertical well intercepting a vertical fracture. The same principle could be anticipated to hold true for both natural and induced vertical fractures.

Figure 2–49 shows a comparison of a fractured horizontal well with a vertical well. Figure 2–49a shows the effect of the distance between vertical fractures in a crossplot of the ratio of productivity indexes between horizontal and vertical wells (J_{hf}/J_{vf}) vs. the ratio X_f/L where X_f is the half fracture length and L is distance between the well and the lateral limit of the reservoir. Figure 2–49b illustrates the effect of anisotropy and reservoir thickness in a crossplot of J_{hf}/J_{vf} vs. number of fractures. The comparison has been carried out for a hypothetical squared reservoir with dimensions $400 \times 400 \times 170$ m ($1312 \times 1312 \times 557.6$ ft) and the same horizontal and vertical permeability. The parameter in the Figure is the number of equidistant vertical fractures intercepting the horizontal well.

Figure 2–49a shows that when the number of vertical fractures is four or less than four, the productivity gain (J_{hf}/J_{vf}) is always greater than the number of vertical fractures. If the number of vertical fractures is increased, a saturation effect seems to appear and the productivity gain is not always greater than the number of vertical fractures. This is an important factor to keep in mind when dealing with economic considerations.

Figure 2–49b shows the effect of anisotropy and reservoir thickness. Comparison of the various cases (1,2, and 3) with the base case ($k_x = k_y = k$) in which permeability is isotropic leads to the following conclusions (Giger et al, 1983):

1. Deterioration of vertical permeability produces a negative effect on the productivity of the vertically fractured horizontal well.
2. Increase in vertical permeability, which might be due to vertical natural fractures, and a decrease in the height of the bed being drained results in productivity gain in the vertically fractured horizontal well.

Key Equations in Fracture Evaluation

There is a natural resistance to wellbore fracturing produced by underground stresses. Vertical stress (σ_z) at any point is given by the approximate equation:

$$\sigma_z = 1.0 \text{ D psi} \tag{2–18}$$

where 1.0 is the gross overburden gradient in psi/ft and D is depth in feet.

Horizontal strain according to Hooke's law is expressed as:

$$\varepsilon_x = \frac{\sigma_x}{E} - \frac{\mu\sigma_y}{E} - \frac{\mu\sigma_z}{E} \tag{2–19}$$

where E is Young's modulus, also called modulus of elasticity, σ_x and σ_y are horizontal stresses, and μ is Poisson's ratio. So strain is deformation caused by stress. Strain can be distortion which is a change in form, or dilation which is a change in volume, or both. The average value of E for many rocks is the order of 1.45×10^7 psi.

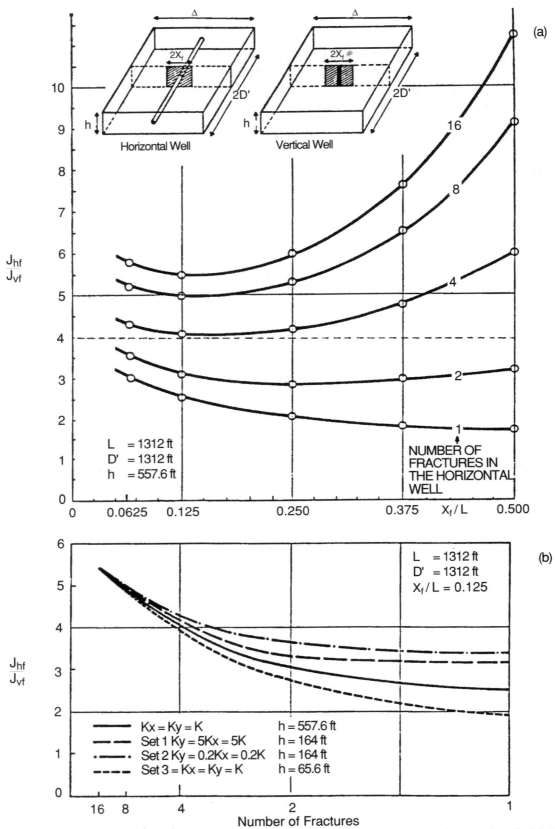

FIGURE 2–49 Advantage of a fractured horizontal well compared to a fractured vertical well. (a) Effect of the distance between fractures. (b) Effect of the anisotropy and the reservoir thickness. (after Giger et al, 1983)

Strain is also given by the equation:

$$\varepsilon = \frac{\Delta l}{l_o} \qquad (2\text{--}20)$$

where l_o is original length and Δl is change in length.

Poisson's ratio is the ratio of transverse strain to axial strain. It can be calculated with the use of the equation:

$$\mu = \frac{\Delta d/d_o}{\Delta l/l_o} \qquad (2\text{--}21)$$

where d_o is original diameter and Δd is change in diameter under tension or compression.

Rigidity modulus (G) gives the resistance to change in shape. It is expressed as:

$$G = \tau/\gamma = \frac{E}{2(1 + \mu)} \qquad (2\text{--}22)$$

where τ = shear stress and γ is shear strain. For a conventional stress-strain crossplot and discussion see Chapter 1 and Figure 1–3.

The bulk modulus (K) or inverse of compressibility is given by:

$$K = \frac{\Delta p}{\Delta V/V_o} \qquad (2\text{--}23)$$

where Δp is the change in hydrostatic pressure, V_o is original volume, and ΔV is change in volume due to a change in hydrostatic pressure.

For rocks in compression, ε_x approaches zero and Equation 2–19 becomes:

$$\sigma_x = \sigma_y = \sigma_h = \frac{\mu}{1 - \mu} \sigma_z \qquad (2\text{--}24)$$

where σ_h is horizontal stress in general. Poisson's ratio (μ) for consolidated rocks ranges from 0.18 to 0.27.

The pressure (P_f) required to fracture the borehole vertically is given by the summation of the pressure required to reduce the compressive stresses on the wall of the hole to zero plus the tensile strength of the rock (Craft et al, 1962) or:

$$P_f = 2\sigma_h + S_t = \frac{2\mu}{1 - \mu} \sigma_z + S_t \qquad (2\text{--}25)$$

where S_t is the tensile strength of the rock. Using typical tensile strengths between 0 and 500 psi and Poisson's ratio between 0.18 and 0.27 leads to the following ranges of internal pressures required to generate a vertical fracture:

$$P_f = \frac{2 \times 0.18 \times 1.0\, D}{1 - 0.18} + 0 = 0.44\, D \; psi \qquad (2\text{--}26)$$

or

$$P_f = \frac{2 \times 0.27 \times 1.0 D}{1 - 0.27} + 500 = 0.74\, D + 500 \; psi \qquad (2\text{--}27)$$

The lower limiting case where horizontal fracturing can occur will take place when

$$P_f = \sigma_z \qquad (2\text{--}28)$$

Combination of Equation 2–25 and 2–28 allows determination of the maximum depth at which horizontal fracturing can occur by assuming:

$$\frac{2\mu}{1 - \mu} \sigma_z + S_t > \sigma_z \qquad (2\text{--}29)$$

Using a vertical gross overburden stress gradient of 1.0 psi/ft, a Poisson's ratio of 0.25 and a tensile strength of 1000 psi leads to a maximum depth of 3000 ft at which horizontal fractures can be formed. This is obtained by rearranging Equation 2–29 as follows:

$$Max\ depth \cong \sigma_x \cong \frac{S_t(1-\mu)}{1-3\mu} \tag{2-30}$$

In general, it has been observed that vertical fractures are formed when the fracture gradient is 0.7 psi/ft or less. Horizontal fractures are formed when the fracture gradient is 1.0 psi/ft or larger.

Mohr diagrams (Figure 2–50) are very useful for analyzing confining forces within the elastic zone of the earth (Hubbert and Willis, 1957). For a given angle ϕ from the horizontal, it is possible to develop the following expressions for calculating normal stress σ and shearing stress τ:

$$\sigma = \frac{\sigma_z + \sigma_h}{2} + \frac{\sigma_z - \sigma_h}{2}\cos(2\phi) \tag{2-31}$$

$$\tau = \frac{\sigma_z - \sigma_h}{2}\sin(2\phi) \tag{2-32}$$

Based on the above analysis Crittendon (1959) developed the following equations for calculating fracture treating pressure (P_t):

$$p_t = \frac{P_{ob} + \dfrac{2\mu}{1-\mu}P_{ob}}{2} + \frac{P_{ob} - \dfrac{2\mu}{1-\mu}P_{ob}}{2}\cos2\phi$$

or

$$\tag{2-33}$$

$$P_t = \frac{P_{ob}}{2}\left[(1 + \frac{2\mu}{1-\mu}) + (1 - \frac{2\mu}{1-\mu})\cos2\phi\right]$$

where P_{ob} is the overburden pressure.

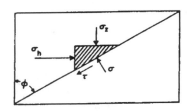

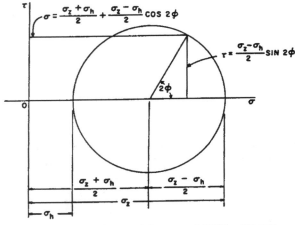

FIGURE 2–50 Mohr circle analysis (after Hubbert and Willis, 1957)

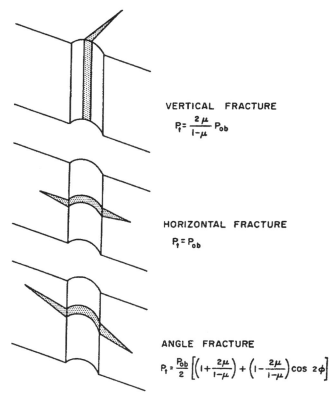

FIGURE 2–51 Fracture orientation

Various types of fracture orientation are illustrated on Figure 2–51.

Hydraulic Fracture Orientation

Knowing the orientation of hydraulic fractures is very important. This will tell us, for example, if the hydraulic fractures stand any chance of intercepting natural fractures, and this will give us information to make secondary recovery projects or infill drilling more efficient. Various methods have been developed to determine the orientation of hydraulic fractures including (1) mineback operations, (2) triaxial dynamic accelerometers, (3) passive seismic, (4) surface electrical potential, (5) active seismic methods, (6) superconducting magnetic systems, (7) solid earth tides, (8) radar imaging, (9) surface tiltmeter, (10) various logging devices such as gamma rays before and after injection of radioactive tracers, temperature logs and borehole televiewers all of which are discussed in Chapter 3, and (11) inflatable impression packers discussed in Chapter 1.

Mineback Operations. In this method, fractures formed with tagged or colored fluids are excavated to expose the decorated surfaces along which fluid penetration has occurred. The elliptical shape of hydraulic fractures presented by Mahoney et al (1981) (see dendritic fractures), was obtained by observing the colored fluorescent pigments included in the fracturing fluid. Warpinski and Teufel (1987) have also conducted mineback experiments to study the influence of geologic discontinuities on hydraulic fracture propagation. From their work they have concluded that natural fractures, faults, and bedding planes, in conjunction with the in-situ stress state, material properties, and permeability can significantly effect the overall geometry of hydraulic fractures. This can occur by (a) creating multiple secondary fractures rather than a single planar hydraulic fracture, (b) reducing fracture length due to fluid leakoff, (c) arresting lateral or vertical propagation of a fracture, and (d) limiting proppant transport and placement.

Triaxial Dynamic Accelerometers. Wittrisch and Sarda (1983) have reported on a method registered under the trade mark Simfrac. In an open-hole section, a limited hydraulic fracture is created with a given volume of fluid not containing any propping agent. The injectivity test involves various pumping and halt cycles. The breaking of the rock and the propagation of the fracture are accompanied by an emission of noise in an audible frequency range. The following data are recorded and processed to determine the orientation, in-situ stresses, and penetration of the hydraulic fracture:

1. Noise in three dimensions
2. Downhole pressure
3. Orientation of the tool to the magnetic north

The tool has been described by Wittrisch and Sarda (1983) as follows: It contains three triaxial dynamic accelerometers (Figure 2–52), a system for measuring its orientation in relation to magnetic north by a compass, a downhole pressure sensor and a hydraulic anchoring system with sidewall pads to produce proper mechanical coupling between the tool and the geological formation. This tool is connected via a 7–conductor cable to the tool support situated in the vicinity of the packer. The tool/tool-support/packer assembly is run into the well on the end of the drill string or tubing to the depth of the zone to be fractured. A standard 7–conductor electric logging cable, on the end of which is a remotely pluggable downhole connector, provides the electric and mechanical link between the tool support and the surface. This cable has the following functions:

1. Positioning the tool in the open-hole section by moving it from the surface,
2. Remote controlling the anchoring of the tool,
3. Transmitting the electric signals emitted by the accelerometers during hydraulic fracturing so that the signals can be recorded on the surface. At the wellhead, a lateral-entrance injection sub and standard sealing equipment (BOP/stuffing–box type) create a seal around the cable.

Some of the advantages of the method include that it is not limited by the depth of the well, and that it can be used offshore, a strong limitation of surface methods.

Passive Seismic.   This method allows mapping the orientation of induced hydraulic fractures by using a triaxial seismometer clamped in the treatment well to map acoustic emissions from the fractures (Dobecki, 1983, and Dobecki and Romig, 1983). The polarization of the compressional wave phase will effectively parallel the orientation of the fracture. Many of the brittle fracture signals are attributed to shear slippage (see dendritic fracturing) along planes of weakness intercepted by the fracture. It has been found that at least one fractured well has generated acoustic emissions long after it was fractured. This implies that the method could be expanded to include fracture diagnostics on old wells. The method has worked properly up to 10,900 ft in past applications. A detailed example of applications has been presented by Hart et al (1987).

Surface Electrical Potential.   This method attempts to detect the change in surface electrical potential that accompanies the generation of a hydraulic fracture. Based on the potential gradients attempts are made to map and characterize the reservoir (Bartel et al, 1976).

Active Seismic.   In these techniques an artificial seismic source is utilized to irradiate the hydraulic fracture (Stewart et al, 1981). This is opposed to the passive seismic method where the path of fluid penetration into the reservoir from the hypocentral distribution of microseismic emissions is determined as a result of increased pressure.

Superconducting Magnetic Systems.   In this approach the detection of a thin sheet magnetic anomaly allows the possibility of determining the azimuth of a hydraulic fracture (Overton, 1980). The distribution of magnetic tracer fluid injected into the fracture is determined based on the potential of oriented downhole superconducting magnetic systems (SQUIDS).

SIMFRAC PRINCIPLE

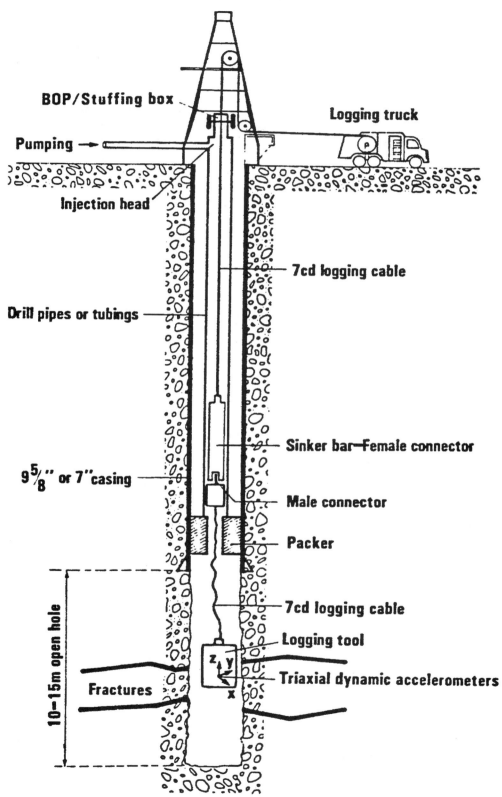

FIGURE 2–52 Simfrac principle (after Wittrisch and Sarda, 1983)

Solid Earth Tides. In this method, Bower (1982) developed a theoretical approach for calculating fracture parameters from observations of changes in borehole pressures or water head induced by the solid earth tides.

Radar Imaging. Unterberger (1980) has investigated the possibility of using radar imaging in reservoir engineering and modelling. This has involved the application of ground-penetrating radar systems as an aid to mapping natural and artificially induced fractures remote from the wellbore.

Surface Tiltmeter. This is a useful method for determining orientation of hydraulic fractures in shallow reservoirs (less than 1500 ft). The technique is also sensitive to fracture width, and consequently, the closure behavior of the fracture following shutin can be inferred. The method is based on observation of the elastic deformation field to which the fracture wall displacements give rise (Evans, 1983). The technique employs an array of continuously recording shallow borehole tiltmeters. Interpretation of the data is based on conventional strain seismology methods which allow calculating fracture parameters of buried slippage on active geological faults from observations of surface deformation (Evans et al, 1981).

The following is a typical analysis procedure of the surface tiltmeter technique as presented by Evans:

In interpreting a set of raw tilt records, the first step is to obtain a suite of fracture-related tilt waveforms by removing, as much as possible, the noise component from the records. Generally, predictive filters are used. The principal component of noise usually is the result of thermoelastic ground tilting, and an estimate of its contribution to tilting during the time window of fracturing can be derived using a least-square linear filtering technique applied to the on-site temperature record. When necessary, the earth-tide component can be estimated by least-square fitting of the principal tidal harmonics to the data. The resulting suite of fracture-related tilt waveforms then are scrutinized, and periods during which the tilt rate remains constant on the majority of records are identified. As constant tilt rate implies an unchanging pattern of surface deformation development, in-plane fracture extension is assumed to be taking place during these periods. Sudden changes in tilt rate are taken to indicate departures from this behavior. The tilt vector fields developed during each of these periods then are estimated and modeled to yield constraints on the geometry of the fractures developed during the successive time intervals.

Positive and Negative Effects of Natural Fractures on Hydraulic Fracturing

Some of the positive effects of natural fractures on well stimulation include (MacLellan, 1993):

1. Improved inflow characteristics due to more connection of the reservoir to the wellbore.
2. Improved inflow characteristics if the natural fractures themselves are properly stimulated.
3. Natural fractures may provide connection to the reservoir units.
4. Natural fractures may need minimal stimulation if not damaged during drilling operations.

Negative effects of natural fractures on well stimulation include (McLellan, 1993):

1. In some cases it is not possible to stimulate (acid or propped) the naturally fractured reservoir. This results in higher expense and lower productivity.
2. If the permeability is low this will reduce the fracture length resulting in lower productivity.
3. Increased near-wellbore fracture width leading to higher expenses.
4. Higher bottom-hole propagation pressures leading to higher expenses.
5. Increased risk of tip screenout in propped fractures leading to higher expenses and lower productivity.

6. Greater risk of severe formation damage during drilling operations leading to higher expenses and lower productivity.
7. Greater fracturing fluid leakoff leading to higher expenses and lower productivity.
8. Rapid acid leakoff in calcite-filled natural fractures leading to higher expenses and lower productivity.
9. Few intersections of natural fractures if the stresses are unfavorably oriented leading to lower productivity.
10. Possible poor packer seating in open–hole completions leading to higher expenses and lower productivity.

Re-Opening of Natural Fractures

Attempts to stimulate a depleted naturally fractured reservoir might not create a new hydraulic fracture but rather might lead to the re-opening of natural fractures. This might be recognized by the following indicators (McLelland, 1993):

1. No fracture breakdown pressure
2. Low surface and bottomhole treatment pressures
3. High overall leakoff coefficients
4. Erratic bottomhole propagation pressures
5. Nolte plot showing Mode II diagnostic slope.

Damaging Natural Fractures

Obviously we do not want this to happen. However, it is a fact that some naturally fractured reservoirs are damaged during stimulation operations. As an example, production from the low-permeability sandstones in the Mesaverde formation at the U.S. DOE Multi-well experiment (MWX) is dominated by natural fractures. Sattler et al (1991) have shown that damage to the narrow natural fractures significantly affected post-stimulation production. This was the result of the degradation of stimulation fluids. Formation of emulsions can also lead to the same result. Gall et al (1988) have also discussed permeability damage to natural fractures by fracturing fluid polymers. Sattler et al (1991) developed a controlled breaker system for use with temperature–stable biopolymer foam. This led to increased gas production and less damage in a subsequent stimulation.

Leakoff Into Natural Fractures

This phenomena has been studied by Warpinski (1988, 1991) who has indicated that leakoff into natural fractures can be either constant, stress sensitive, or accelerating. Leakoff is more severe when the natural fractures dilate and accept large volumes of treatment fluids, which might rapidly dehydrate a sand-laden slurry. Pressure decline analysis following a minifrac and evaluation of the injection pressure are required in order to make estimates of pressure-sensitive and accelerated leakoff coefficients.

Warpinski (1991) combined Darcy's law, conservation of mass, and Walsh (1981) model for stress sensitive fractures to calculate leakoff velocity. Walsh (1981) model is given by the equation:

$$k = k_o \left\{ C \ln\left(\frac{\sigma^*}{\sigma - P}\right) \right\}^3 \qquad (2\text{--}34)$$

where
k = permeability of the fracture at any imposed stress state
k_o = permeability at in-situ conditions
σ = normal stress on the fracture
σ^* = reference stress state
P = pore pressure

During hydraulic fracturing of the natural fracture, the pore pressure within the natural fracture increases leading to a reduction of the net stress. This in turn results in an increase in permeability of the natural fracture and consequently in an increase in the leakoff coefficient.

Fine-mesh sand can help control leakoff, but scheduling of fine-mesh sand is not easy. A key element is to get the fine-mesh sand to the fissures as the fissures begin to open. Some success along these lines has been obtained with 100-mesh sand.

Figure 2–53 shows the permeability of stress sensitive natural fractures calculated with the use of Walsh (1981) model. At an initial reservoir pressure of 4400 psi, $k = k_o$. Assuming that at severe drawdown conditions the pressure is reduced to 800 psi and the permeability is reduced two orders of magnitude so that $k = 0.01 k_o$, allows to solve for constants C and σ^*. Two cases are considered on Figure 2–53:

1. The natural fractures are normal to the maximum horizontal stress so that $\sigma = \sigma_{Hmax}$
2. The natural fractures are sub-parallel to the maximum horizontal stress so that $\sigma = \sigma_{Hmin}$

Notice that when nitrogen is injected at pressures below the fracture gradient but above the initial reservoir pressure, the permeability increases by one to two orders of magnitude due to the net normal stress reduction on the fractures.

Figure 2–54 shows the estimated leakoff coefficients [C(p)] calculated by Warpinski (1991) as a function of net treatment pressures for an orthogonal set of natural fractures. The leakoff values are normalized by the filtrate viscosity controlled liquid coefficient at the net closure stress value [$\sqrt{\Delta p} = \sqrt{\sigma_{Hmin} - P_\infty}$], where P_∞ is the initial reservoir pressure. In the example of Figure 2–54, there is a 1.5 to 2.8 times increase in the stress-dependent natural fracture (fissure) leakoff from the base leakoff coefficient C_o. The filtrate viscosity controlled leakoff varies from 1.0 to 1.4 C_o for the same range of net pressure change.

To estimate the type of leakoff it is necessary to carry out a minifrac and to conduct a detailed study of the injection pressure and the pressure-decline behavior. Plots of the

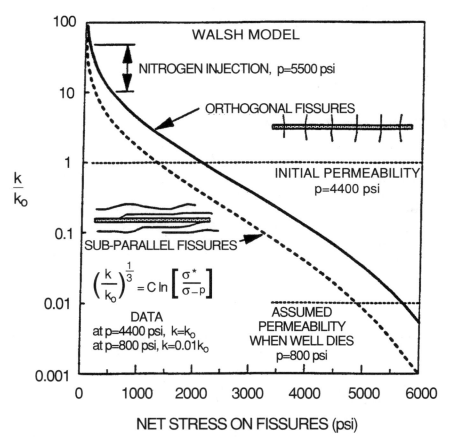

FIGURE 2–53 Permeability of stress-sensitive fissures

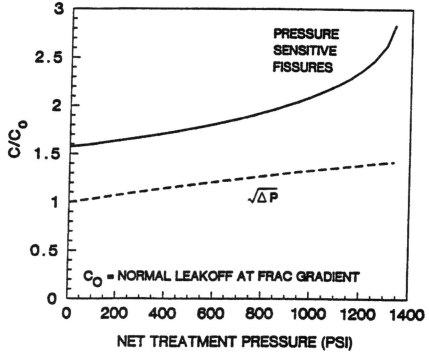

FIGURE 2–54 Leakoff into stress-sensitive fissures

derivative of the G–function, as described by Castillo (1987) can be used to assess the possibility of stress–sensitive leakoff.

Warpinski's (1991) work demonstrated the importance of evaluating the stress–sensitive leakoff coefficient in the design of induced fractures. Nolte and Smith (1981) method (Figure 2–44) allows analysis for a constant leakoff case. However, when this technique is used in the pressure–dependent leakoff situation, it can lead to inadequate designs of pad volumes, proppant dehydration and premature screenout. To alleviate this problem, Mukherjee et al (1991) have presented a method that helps in the diagnosis and evaluation of pressure–dependent leakoff. Use of this method leads to an empirical correlation between the net pressure and fluid leakoff coefficient. The reader is referred to Mukherjee et al's paper for details on this technique.

Considerations Previous to a Stimulation

The following considerations previous to stimulating a naturally fractured reservoir might save money and headaches (McLelland, 1993):

1. Determine the interaction between natural fractures and in–situ stresses.
2. Evaluate cores and well logs to assess likelihood of intersecting natural fractures.
3. Determine stress-corrected and bulk rock mass elastic properties.
4. Estimate the propagation pressures that will re-open natural fractures. Stay below these pressures to achieve long fracture lengths.
5. Examine mini-frac records to assess natural fractures effects.
6. Numerically model the treatment to help understand and predict consequences of stimulations in naturally fractured reservoirs.

Case Histories

This section reviews some cases where horizontal, vertical, inclined, and dendritic hydraulic fractures have been generated in naturally fractured reservoirs.

Austin Chalk (Vertical Fractures). Vertical hydraulic fractures, in many cases parallel to the orientation of natural fractures, have been generated repeatedly in the Austin Chalk.

A generalized fracture treatment chart (Figure 2–55) for the Austin Chalk has been presented by Parker et al (1982). As shown in Figure 2–55 a linear relationship exists between fracture height and treatment volume at a constant penetration. To determine the treatment volume in gallons enter the y–axis at the anticipated fracture height, read across to the "sand laden fluid volume" line and down to the X–axis. Based on the lb/ft² of fracture area desired, determine the total pounds of sand needed, using the same approach mentioned previously. Note the following constants: propped fracture length of 750 ft, 55 bbl/min pump rate, 0.001 CC ft √min (combined fluid loss coefficient).

Additional charts can be constructed using different fluid and formation parameters. Pad volumes are not included. Usually they vary from 30% to 40% by total volume depending on the area requirements to help avoid screenout.

As an illustration, with a propped fracture height of 200 ft, a length of 750 ft and 1.5 lb/ft², the requirements would be:

Pad volume (33%), gal = 55,000
Sand laden volume, gal = 110,000
Pounds of sand = 450,000

These parameters will vary according to area and field experience. Since they are critical, they need to be carefully addressed and defined in detail.

Another important consideration is leak-off which can be controlled using 100-mesh sand or an oil-soluble resin.

Devonian Shales (Subvertical and Horizontal Fractures). Keith et al (1982) have presented an example of fracture mapping with tiltmeters to decipher the geometry of a large scale nitrogen gas hydraulic fracture in Devonian shales. The analysis allowed to determine that in well Black No. 1, Knox County, Ohio, at about 1100 ft, a subvertical fracture of strike N62°E and dip 87° to the northwest grew bilaterally from the wellbore during the first 16 minutes of nitrogen injection. Fracture propagation toward the southwest was in the order of 650 ft, whereas that to the northwest was more than 980 ft. After 16 minutes the fracture broke out into a horizontal plane. By this time the initial vertical fracture had grown upward more than 330 ft to a point within the Bedford shale. The geometric description from the tiltmeter in this case is excellent. For a variety of stimulation methods and results see Table 4–1.

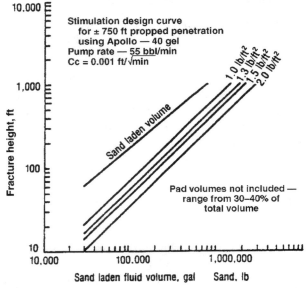

FIGURE 2–55 Generalized stimulation design graph (after Parker et al, 1982)

Monterey Shale (Vertical Fractures). A method which utilizes fluid entry surveys (Al-Khatib et al, 1984) was developed to improve fracture treatment designs at the Lost Hills Oil Field, 45 miles northwest of Bakersfield, California. The reservoir produces 25–38° API oil and large volumes of solution gas from naturally fractured silicious shales of the Upper Miocene Monterey Formation known locally as the "Cahn Zone."

A typical fracturing fluid system consisted of a 2% KCl base fluid cross-linked with a 40 PPM Guar Derivative polymer with 0.1% of clay–stabilizing polymer and water-wetting surfactant.

Many wells were fractured with this type of frac-fluid, with +/– 200 gals/ft of Sand Laden Cross–linked gel containing 6–8 lbs/gal of 20/40 sand. The sand was mixed at low concentration until the fracture was initiated and then sand concentration was increased gradually as the fracture propagated. The limited entry technique was utilized to obtain an even fluid distribution during the frac treatment. Most of the wells fractured with this design produced initially 200–500 b/d and 15–70% water cuts.

The fractures were vertical and had approximate lengths of 200 ft. Pressure buildup data indicated the existence of linear flow at early times.

Revisions to treatment procedure as presented by Al-Khatib et al (1984) included:

1. Decrease the net completion interval to 100 ft. per stage. This would necessitate two or more fracturing stages in most wells due to thick pay zones.
2. Reduce the perforation diameter from ½" to ⅜" (Figure 2–56).
3. Use a 200 psi perforation frictional pressure drop as a design criteria.
4. Use the maximum possible injection rate. This can be achieved by fracturing through casing.

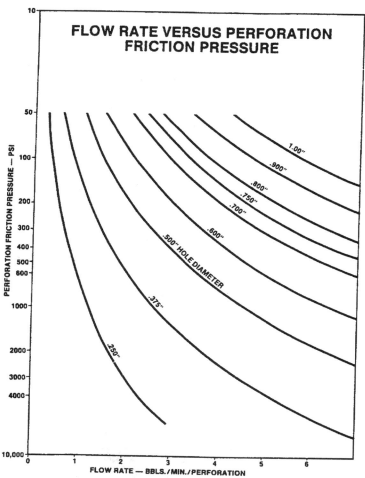

FIGURE 2–56 Flow rate vs. perforation friction pressure (after Al-Khatib et al, 1984)

5. Determine the number of holes to be perforated based on the information presented on Figure 2–56 and steps 3 and 4 above.

6. Break down all perforations before fracture treatment with a Cup Washer apparatus, instead of using ball sealers during fracturing.

These revisions to the fracture treatment design met the design requirements of the Limited Entry technique. To determine the effects of the new revisions Lost Hills One well #178 was selected for application. The well had been fractured initially using the unmodified design and was producing poorly. It flowed for only six months and then required pumping. Figure 2–57 presents the fluid entry survey for Lost Hills One #178. It shows that only 30% of the net completion interval was actually contributing to flow.

Lost Hills One #178 was refractured using the revised design criteria (high injection rate, perforations broken down with cup washers). Production was significantly improved and the well was still flowing one year after being refractured.

Lost Hills One #179 was fractured in two stages using the revised design. Initial production for this well was 350 BPD, twice the typical initial production of wells fractured with the unmodified design.

Annona Chalk (Horizontal Fractures).   Howard and Fast (1970, p.51) have reported on an experiment carried out by Pan American Petroleum Corporation in 1954 in the Pine Island Field near Shreveport, Louisiana.

Well 163 was drilled in the center of a plot of approximately 10 acres, containing four corner wells. Figure 2–58 presents the well pattern. Well 163 was an open-hole completion in the Annona Chalk. Casing was set at 1606 ft. Total depth of 1636 ft.

The fracturing treatment was cautiously controlled. It included a low-rate formation breakdown with crude oil containing a radioactive tracer followed by residual fuel oil fracturing fluid injected at a low rate. This was followed by 10,000 gal of viscous fuel oil containing an average of 1 lb of 20/40 sand per gal of oil at an average injection rate of 56 bbl/min.

WELL # 178 — FRACTURED WITH ORIGINAL TREATMENT DESIGN IN ONE STAGE

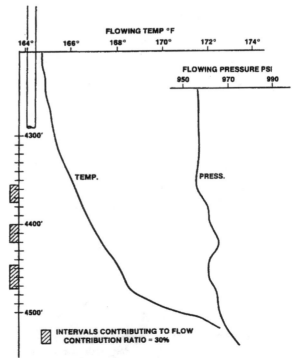

FIGURE 2–57   Well 178-fluid flow entry survey (after Al-Khatib, 1984)

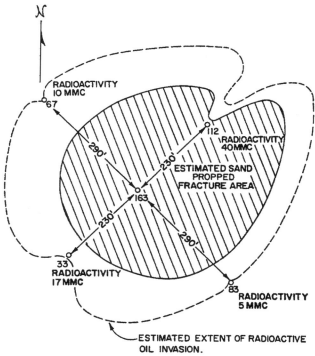

FIGURE 2–58 Areal extent of fracture system, special tests, Pine Island field, La. (after Howard and Fast, 1970)

Surface and bottom-hole temperatures and pressures were measured in the fractured well, No. 163, during the treatment. Maximum bottom-hole treating pressure gradient was 0.67 psi/ft. Average gradient was 0.61 psi/ft. Samples were taken from offset Wells 33 and 112, which were 230 ft away, and Wells 67 and 83, which were 290 ft away during and after treatment. These samples were tested for viscosity and radioactivity.

Fracturing oil and sand broke through in Well 112 during treatment. Recovery of radioactive oil from these wells indicated the proximity of Wells 33, 67, and 83 to the apparently horizontal fracture that was induced during this fracturing operation. The location of the wells, the estimated areal extent of the sand-propped fracture, and the extent of the radioactive oil invasion are shown on Figure 2–58.

ACIDIZING

Matrix and fracture acidizing have been used extensively in naturally fractured reservoirs in efforts to establish communication with the network of natural fractures.

In *matrix acidizing* a certain volume of acid is injected into the formation at a pressure below the pressure at which an artificial fracture can be generated or a natural fracture can be opened (this assumes that the natural fracture is closed). In matrix acidizing, the acid will flow preferentially into those regions with the highest permeabilities such as natural fractures, vugs, or simply the largest pores (Williams et al, 1979).

In *acid fracturing* the acid is injected into the formation at a pressure high enough that will allow generation of an artificial fracture or the opening of a natural fracture.

Matrix Acidizing

The goal of this treatment is to achieve a more or less radial penetration of the acid (usually hydrochloric acid) to improve the formation permeability around the wellbore. In some instances this treatment might establish communication with natural fractures which have not been intercepted by the wellbore.

The reaction of acid in a carbonate formation produces the formation of *wormholes* (Schechter and Gidley, 1979). These are large and highly conductive flow channels. The generation of wormholes depends on the rate of chemical reaction of the acid with the rock. Wormhole creation is favored by the high reaction rates observed between carbonates and all concentrations of hydrochloric acid. Nierode and Williams (1971) have shown that the maximum wormhole length of 10 to 100 ft. would be possible by drastically reducing the rate of fluid loss from the wormhole to the formation.

Williams et al (1979) recommended to inject from 50 to 200 gals of either 15% or 28% HCl per foot of interval perforated. Exact acid volumes and strength requirements cannot be predicted because of uncertainties in near wellbore conditions. In naturally fractured reservoirs or high permeability formations, an emulsified acid often gives better results. In wells with deep damage or with high reservoir temperatures, large volumes of acid should be used. A 28% hydrochloric acid with an effective acid fluid-loss additive should be used whenever possible in zones where acid can be injected at least at 0.25 to 0.5 bbl/min.

Example 2–4. Design a matrix acid job for a well with the following characteristics (Williams et al, 1979):

kh = 200 md ft
Perforated interval = 20 ft
Formation depth, D = 7500 ft
Fracture gradient, g_f = 0.7 psi/ft at the initial pressure of 3075 psi
Gross overburden gradient (GOG) = 1 psi/ft
Reservoir pressure, p = 2000 psi
Acid viscosity at reservoir temperature, μ = 0.4 cp
Drainage radius, r_e = 660 ft
Wellbore radius, r_w = 0.25 ft

Solution:

1. Use the fracture gradient g_f equation to calculate α:

$$g_f = \alpha + (GOG - \alpha) \times p/D \qquad (2\text{–}35)$$

α is a constant usually between 0.33 and 0.5 and other nomenclature is defined with the problem data. From Equation 2–35, α is calculated to be 0.49 at the initial pressure of 3075 psi, fracture gradient of 0.7 psi/ft and 7500 ft depth. The approximate fracture gradient at the current 2000 psi reservoir pressure is calculated from Equation 2–35 as follows:

$$g_f = 0.49 + (1 - 0.49) \times \frac{2000}{7500} = 0.63 \; psi/ft$$

2. The maximum injection rate, i_{max}, is calculated to be 0.85 bbl/min from:

$$i_{max} = \frac{4.917 \times 10^{-6} kh \; (g_f D - p)}{\mu \; ln \; (r_e/r_w)} \qquad (2\text{–}36)$$

In practice, a rate 10% lower is suggested, i.e., a maximum rate of 0.85 × 0.90 = 0.77 bbl/min would be specified.

3. Maximum surface pressure (p_{max}) without fracturing the formation would be 1500 psi from the equation:

$$p_{max} = (g_f - g_n)D \qquad (2\text{–}37)$$

where g_n is a normal gradient (0.43 psi/ft)

4. Determine the type of acid required and the volume. Williams et al (1979) recommend to inject from 50 to 200 gals of 15% or 28% HCl per foot of interval perforated. If we assume an acid volume of 100 gal/ft the matrix acidizing program could be as follows:

(i) Inject 100 gal/ft × 20 ft = 2000 gals of 28% Hcl. Do not exceed a surface pressure of 1500 psi. Reduce the rate if a surface pressure of 1500 psi is approached. Select corrosion inhibitors and other additives according to properties of the formation being acidized.

(ii) Flush the treatment with more than one tubing volume of water. Put well on production as soon as possible.

Acid Fracturing

The goal of this treatment is to inject acid into the formation at a pressure high enough to open existing natural fractures or to fracture the formation. The method is widely used in limestones and dolomites.

Initially, a fluid pad is injected at a rate higher than the matrix will accept in such a way as to create a wellbore pressure increase which becomes large enough to overcome compressive earth's stresses and rock's tensile strength. At this point the formation fails by rupture and a fracture is formed. As the injection continues, the fracture length and width increase. Acid is then injected into the fracture to react with the formation and create a flow channel that hopefully will remain open when the well is put back in production.

Figure 2–43 published by McGuire and Sikora is generally used for predicting the stimulation ratio for a fractured well after flow stabilization is achieved. Previous to stabilization, the observed stimulation ratio will exceed the stabilized ratio predicted by Figure 2–43A.

Care must be exercised when one tries to achieve deep penetration by increasing substantially the volumes of acid. Large volumes of acid will tend to remove large volumes of rock around the wellbore, and a point might be reached where the area of the rock supporting the fracture will no longer resist the closure pressures. At this moment the fracture might collapse.

Although predicting the optimum volume of acid is nearly impossible, some generalizations can be made (Molon and Fox, 1983). Since the rock volume removed from around the wellbore depends on the acid reaction rate and the contact time, it is possible to extend the contact time required to exceed the maximum conductivity by retarding the reaction rate. This will allow injecting larger volumes of acid before endangering the integrity of the fracture.

Molon and Fox (1983) have reported on the use of xantham polymer formulation as a thickening agent for acid. By increasing the viscosity of the acid, fluid efficiency is improved and the reaction rate is reduced.

Acid thickened with the xanthan polymer formulation, at a concentration of 60 lb/1000 gal of 15%HCl, has a viscosity of 41.8 cp at 170 sec^{-1} and 50°C. In contrast, ungelled 15% HCl has a viscosity of 1.5 cp (0.0015 Pa sec) at the same temperature (Novotny, 1976). The viscosity of the xanthan gelled acid is compared to ungelled acids at various temperatures in Figure 2–59.

To obtain the maximum benefit from a gelled acid, the viscosity must be stable under down-hole conditions for the duration of the treatment. The xanthan gelled acid is relatively stable at temperatures up to 105°C. At temperatures below 40°C, there is virtually no degradation of viscosity with time (Figure 2–60). At a treating temperature of 70°C, Figure 2–60 shows that there is some reduction in viscosity over a one-hour period. However, the viscosity stability is also affected by the degree of acid spending. Under acid fracturing conditions, the time that the polymer will be required to survive degradation in 15% HCl is limited. As the acid spends in the fracture or in the matrix rock, the viscosity is preserved.

Molon and Fox (1983) have published the case history of a well in Upton County, Texas. Production is from the Fusselman Formation, which is a naturally fractured limestone. The well was originally treated with 7500 gal of 15% HCl. After treatment, the well produced 42 b/d of oil and 350 Mcfd of gas through an 8.33-mm choke, with a flowing pressure of 200 psi. The well was retreated using xanthan gelled acid.

The following is a description of the well:

Depth	3467 to 3486 m (11,375 to 11,438 ft)
Natural fracture permeability	2.0 to 10.0 md
Matrix permeability	0.1 to 0.5 md
Porosity	4.5%
BHT	82°

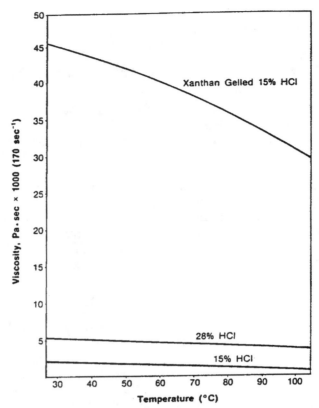

FIGURE 2–59 Viscosity profiles for xanthan gelled 15% HCl (@ 60lb/1000 gal) and ungelled acids (after Molon and Fox, 1983)

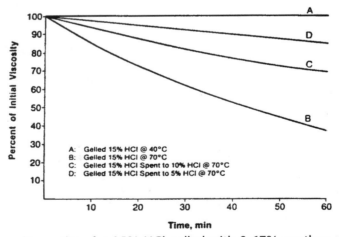

FIGURE 2–60 Viscosity vs. time for 15% HCl gelled with 0.67% xanthan polymer formulation (after Molon and Fox, 1983)

Treatment Design was as follows:

4000 gal Pad
6000 gal Gelled 15% HCl
4000 gal Pad
6000 gal Gelled 15% HCl
4000 gal Pad
8000 gal Gelled 15% HCl
Flush

After treatment with the xanthan gelled acid, production was increased to 145 b/d of oil and 1.4 MMcfd of gas, with a flowing pressure of 675 psi through an 8.33–mm choke. This represents a 3.5–fold increase in oil production and a 4.0–fold increase in gas production. Open–flow calculations indicate a potential rate of 200 b/d of oil and 4.6 MMcfd of gas. This is now one of the better wells in the field.

Favorable results with gelled acids have also been reported in some naturally fractured Austin Chalk wells.

A detailed step by step acid fracturing treatment design has been published in Chapter 7 of SPE Monograph Volume 6 by Williams, Gidley, and Schecter (1979). Input data required for the analysis include net and gross formation thickness, permeability, porosity, depth, fracturing gradient, Poisson's ratio, sonic travel time, formation and fluid injection temperatures, reservoir pressure, and viscosity, compressibility, and density of reservoir fluids.

EXPLOSIVE FRACTURING

A chemical explosive can be initiated by a number of ways including impact, heat, friction, or shock (Hill and Daggett, 1976). Following initiation, the explosive undergoes a quick, self-propagating decomposition yielding gases which exert huge pressures as they expand at high temperatures. This rapid release of energy generates a shock wave composed of compressional and shear waves. When the energy of any of these components exceeds the rock strength under dynamic loading, the rock will fail by rupture, generating a network of fractures.

Hydraulic fracturing and acidizing have displaced oil and gas shooting down to a minimum. One big problem is safety. In the past, several companies have experienced predetonations.

REFERENCES

Aguilera, R., J.S. Artindale, G.M. Cordell, M.C. Ng, G.W. Nicholl, G.A. Runions. *Horizontal Wells, Formation Evaluation, Drilling and Production, Including Heavy Oil Recovery*, Gulf Publishing Company, Houston, Texas (1991), 401 p.

Aguilera, R., and H.K. van Poollen "Naturally Fractured Reservoirs–Choice of Completion Method Affects Productivity." *Oil and Gas Journal* (February 19, 1979) 106–108.

Al-Khatib, A.M., A.R. King, and M.S. Wilson. "Hydraulic Fracturing Design and Evaluation: A Case History." SPE paper 12482 presented at the California Regional Meeting held in Long Beach, Ca (April 11–13, 1984).

Angel, R.R. "Volume Requirements for Air or Gas Drilling." *Transactions.* AIME 210, (1957), 325.

Au, A.D., L.N. Franks and R. Aguilera. "Simulation of a Gas Naturally Fractured Reservoir." SPE paper 22920 presented at the 66th Annual Technical Conference and Exhibition held in Dallas, Texas (October 6–9, 1991).

Baker, D.A., and P.T. Lucas. "Strat Trap Production May Cover 280+ Square Miles." *World Oil* (April 1972), 65–68.

Baron, G., and C. Wittrisch. "Une Nouvelle Method pour la Realisation des Diagraphies dans les Puits Horizontaux: SIMPHOR." *Petrole et Techniques* (December 1982) No.294, 42–44.

Bartel, L.C. et al. "Use of Potential Gradients in Massive Hydraulic Fracture Mapping and Characterization." SPE paper 6090 presented at Annual Technical Conference and Exhibition, New Orleans (1976).

Beck, R.W., W.F. Nuss, and T.G. Dunn. "The Flow Properties of Drilling Muds." *API drilling and Production Practices* (1947), 9.

Bentsen, N.W. and J.N. Veny. "Preformed Stable Foam Performance in Drilling and Evaluating Shallow Gas Wells in Alberta." *Journal of Petroleum Technology* (October 1976), 1237–1240.

Bower, D.R. "Fracture Parameters from the Interpretation of Well Tides." Submitted to *J. Geophys. Res.* (1982).

Butler, R.M. *Horizontal Wells for the Recovery of Oil, Gas and Bitumen.* Petroleum Society of CIM Monograph No. 2, Calgary, Alberta (1994), 227 p.

Byron Jackson, Inc. "Applied Engineering Cementing." A Byron Jackson, Inc. Publication (1969).

Castillo, J.L. "Modified Fracture Pressure Decline Analysis Including Pressure-Dependent Leakoff." SPE 16417, proceedings, SPE/DOE Low Permeability Reservoirs Symp., Denver, Colorado, (May 18–19, 1987), 273–281.

Coberly, C.J. "Selection of Screen Openings for Unconsolidated Sands." *API Drilling and Production Practices* (1937), 189.

Coffer, H.F., J.J. Reynolds, and R.C. Clark, Jr. "A Ten-Pound Cement Slurry for Oil Wells." *Transactions* AIME (1954) v. 201, 146.

Conway, M.W., Personnal Communication (October, 1986).

Cook, W.H. "Offshore Directional Drilling Practices Today and Tomorrow." SPE paper 783-G presented at New Orleans (February 1957).

Craft, B.C., W.R. Holden, E.D. Graves, Jr. *Well Design, Drilling and Production.* Prentice Hall, Inc. Englewood Cliffs, New Jersey (1962).

Cranfield, John. "French Refine Horizontal Drilling Techniques." *Oil and Gas Digest* (October 1984), vol.6, no.17.

Critendon, D.C. "The Mechanics of Design and Interpretation of Hydraulic Fractures, In–Situ Observations." Presented at 36th Annual Meeting of the Midwest Society of Exploration Geophysicists, Denver, Colorado (March 8, 1983).

Daniel, W.L. and W.H. Fertl. "Well Logging Operations in High Angle, Long Reach Boreholes—The Slant Hole Express." A Dresser Atlas Publication (1984), 241–248.

Davies, A.P. and O.M. Knight. "Tools Run on Drill Pipe Reduce Logging Problems in High Angle Wellbores." *World Oil* (April 1983)

Dawson, R., and P.R. Pasley. "Drillpipe Buckling in Inclined Holes." *Journal of Petroleum Technology* (October 1984), 1734–1738.

Dech, A.J., D.D. Hear, F.J. Schuh, and B. Lenhart. "New Tools Allow Medium–Radius Horizontal Drilling." *Oil and Gas Journal* (July 14, 1986) 95–99.

Dobecki, T.L. "Hydraulic Fracture Orientation Using Passive Borehole Seismics." SPE paper 12110 presented at the 58th Annual Technical Conference and Exhibition held in San Francisco, California (October 5–8, 1983).

Dobecki, T.L. and P.R. Romig. "Acoustic Signals Generated by Hydraulic Fractures, In–Situ Observations." Presented at 36th Annual Meeting of the Midwest Society of Exploration Geophysicists, Denver Colorado (March 8, 1983).

Dorel, Marc. "Horizontal Drilling Methods Proven in Three Test Wells." *World Oil* (May 1983) 127–135.

Dowell. "Divertafrac and Invertafrac, Services to Control Vertical Fracture Growth." Dowell Publication TSL-2578, Houston, Texas 77042 (circa 1985).

Dumbarld, G.K. et al. "A Lightweight, Low Water-Loss, Oil-Emulsion Cement for Use in Oil Wells." *Journal of Petroleum Technology* (May 1956) 99–104.

Dyna-Drill, 1976–1977 Catalog, P.O. Box 327, Long Beach, California.

Eastman. *Introduction to Directional Drilling*, an Eastman Whipstock Publication, P.O. Box 14609, Houston, Texas 77021.

Eckel, J.R., and W.J. Bielstein. "Nozzle Design and its Effect on Drilling Rate and Pump Operation." *API Drilling and Production Practices* (1951), 28.

Escaron, P.C. "A Technique to Evaluate Deviated Wells with Standard Logging Tools." SPE paper 12180 presented at the 58th Annual Technical Conference and Exhibition Held in San Francisco, Ca. (October 5–8, 1983).

"Esso Drills World's Longest Horizontal Well." *Oil Week*, Drilling Report (December 3, 1984), 27–28.

Evans, K. "On the Development of Shallow Hydraulic Fractures as Viewed Through the Surface Deformation Field: Part 1—Principles." *Journal of Petroleum Technology* (February 1983), 406–410.

Evans, K. "On the Development of Shallow Hydraulic Fractures as Viewed Through the Surface Deformation Field: Part 2—Case Histories." *Journal of Petroleum Technology* (February 1983), 411–420.

Evans, K. et al. "Propagating Episodic Creep and Aseismic Slip Behavior of the Calaveras Fault North of Hollister, California." *Journal of Geophysical Research* (May 1981), V.86, 3721–3735.

Evans, K. et al. "The Geometry of Large-Scale Nitrogen Gas Hydraulic Fracture Formed in Devonian Shale. An Example of Fracture Mapping with Tiltmeters." *Society of Petroleum Engineers Journal* (October 1982) 755–763.

Findley, L.D. "Why Uinta Basin Drilling is Costly and Difficult." *World Oil* (April 1972), 77–91.

Forsyth, V.L. "A Review of Gun-Perforating Methods and Equipment." *API Drilling and Production Practices* (1950).

Franklin, L.O. "Drilling and Completion Practices, Spraberry Trend." *API Drilling and Production Practices* (1952), 128–137.

Gall, B.L. et al. "Permeability Damage to Natural Fractures Caused by Fracturing Fluid Polymers." SPE 17542 presented at the SPE Rocky Mountain Regional Meeting held in Casper, Wyoming (May 11–13, 1988).

Gates, R.L. and G. Schwab. "Specialized Drilling Systems Set New World Records in High-Angle Holes." *Journal of Petroleum Technology* (February 1984), 241–248.

Gatlin, Carl. *Petroleum Engineering—Drilling and Well Completions*. Prentice–Hall, Inc. (1960).

Gearhart Industries. "Toolpusher, High Angle Logging System." A Gearhart Publication, P.O. Box 1258, Fort Worth, Texas 76101 (1983).

Giannesini, J.F. "Horizontal Wells. A Today's Technology." *PetroAsia* (January–February, 1984), 50–56.

Giger, F., J. Combe and L.H. Reiss. "L' Intéret du Forage Horizontal Pour l'exploitation de Gisements D'Hydrocarbures." *Revue de l'Institut Francais du Petrole* (May–June 1983) v.38, 329–350.

Gorodi, A.W. "TEDSI Develops Horizontal Drilling Technology." *Oil and Gas Journal* (October, 1984) 118–126.

Hart, C.M. et al. "Fracture Diagnostics Results for the First Multiwell Experiments Paludal Zone Stimulation." *SPE Formation Evaluation* (September 1987), 320–326.

Hill, W.L. and W.K. Daggett. "A New Look at Chemical Explosive Fracturing." *Petroleum Engineer* (Nov. 1976).

Holditch, S.A., and D.E. Lancaster. "Economics of Austin Chalk Production." *Oil and Gas Journal* (August 9, 1982) 183–189.

Horwell. *Horizontal Well Technology*, a Horwell Publication, 177, Avenue Napoleon Bonaparte, 92500 Rueil—Malmaison, France (1984).

Howard, G.C., and P.P. Scott. "An Analysis and the Control of Lost Circulation." *Transactions*. AIME, (1951), 192, 171.

Howard, G.C., and C.R. Fast. *"Hydraulic Fracturing."* SPE Monograph Volume 2, Henry L. Dohery Series (1970).

Hubbert, M.K. and D.G. Willis. "Mechanics of Hydraulic Fracturing." *Transactions*. AIME (1957) v. 210, 154.

Hutchison, S.O. and G.W. Anderson. "What to Consider When Selecting Drilling Fluids." *World Oil* (October 1974).

Johnston, David. "Discussion of Drilling and Completion Practices, Spraberry Trend." API (1952) 134–135.

Jourdan, P. and Guy Baron. "Horizontal Well Proves Productivity Advantages." *Petroleum Engineer International* (October 1984), 23–25.

Keller, S.R., R.J. Crook, R.C. Haut, and D.S. Kulakofsky. "Problems Associated with Deviated—Wellbore Cementing." SPE paper 11979 presented at 58th Annual Technical Conference and Exhibition held in San Francisco, Ca (October 5–8, 1983).

Kiel, O.M. "The Kiel Process—Reservoir Stimulation by Dendritic Fracturing." Paper SPE 6984 Presented at the 52nd Annual Conference held in Denver, Colorado (1977).

Knight, O.M. "Toolpusher Logging—an Aid in Advanced Directional Drilling Programs." First Northern European Drilling Conference (October 24, 1983).

Lebus, J. "Should We Use Bullets or Jets?" *World Oil* (March 1957).

Lincicome, J.D. "Using Foam to Drill Deep Underpressured Zones." *World Oil* (February 1984).

Littleton, Jeff. "Sohio Studies Extended-Reach Drilling to Prudhoe Bay." *Petroleum Engineer International* (October 1985) 28–34.

Littleton, J.H. "Standard Oil Applies Extended-Reach Drilling to Prudhoe Bay." *Petroleum Engineer International* (April 1986).

Loxam, D.C. "Texaco Canada Complete Unique Horizontal Drilling Program." *Petroleum Engineer International* (September 1982) 40–52.

Lubinski, A., W.S. Althouse, and J.L. Logan. "Helical Buckling of Tubing Sealed in Packers." *Journal of Petroleum Technology* (May, 1956) 99–104.

Lubinski, Arthur. "Maximum Permissible Dog–Legs in Rotary Boreholes." *Journal of Petroleum Technology* (February 1961), 175–188.

Mahoney, J.V. et al. "Effects of a No-Proppant Foam Stimulation Treatment on a Coal-Seam Degasification Borehole." *Journal of Petroleum Technology* (November 1981) 2227–2235.

Martin, P., and E.B. Nuckols. "Geology and Oil and Gas Occurrence in the Devonian Shales, Northern and West Virginia," Proceedings of the Seventh Appalachian Petroleum Geology Symposium, Morgantown, West Virginia (March 1–4, 1976), 20.

McLellan, P.J. "Designing Stimulation Treatments to Account for Natural Fractures." Presented at the Petroleum Society of CIM Calgary Section Monthly Technical Meeting, Calgary, Alberta (March 10, 1993).

McGuire, W.J., and V.J. Sikora. "The Effect of Vertical Fractures on Well Productivity." *Transactions*. AIME (1960) v. 219, 401–403.

Melrose, J.C., et al. "A Practical Utilization of the Theory of Bingham Plastic Flow in Stationary Pipes and Annuli." *SPE Petroleum Transactions Reprint Series No 6, Drilling*. 204 (Circa 1969).

Molon, J.P. and K.B. Fox. "Increased Fracture Penetration and Productivity Using Xanthan Gelled Acid in Massive Carbonate Formations." SPE paper 11502 presented at the Middle East Oil Technical Conference held in Manama, Bahrain (March 14–17, 1983).

Moore, P.L. *Drilling Practices Manual*. Petroleum Publishing Company (1974).

Mud Engineering. Dresser Magcobar (1968).

Mukherjee, H., S. Larkin, and W. Kordziel. "Extension of Fracture Pressure Decline Curve Analysis to Fissured Formations." SPE 21872 presented at the Rocky Mountain Regional Meeting held in Denver, Colorado (April 15–17, 1991).

Murphy, H.D. and M.C. Fehler. "Hydraulic Fracturing of Jointed Formations." SPE paper 14088 presented at the International Meeting on Petroleum Engineering held in Beijing, China (March 17–20, 1986).

New Hughes Simplified Hydraulics. Hughes Tool Company (1968).

Nicholson, K. "Air and Gas Drilling, Fundamentals of Rotary Drilling." *The Petroleum Engineer* (1954), 86.

Nierode, D.E. and B.B. Williams. "Characteristics of Acid Reaction in Limestone Formations." *Society of Petroleum Engineers Journal* (December 1971) 406–418.

Nolte, K.G. "Discussion of Influence of Geologic Discontinuities on Hydraulic Fracture Propagation." *Journal of Petroleum Technology* (August 1987), 998.

Nolte, K.G. and M.B. Smith. "Interpretation of Fracturing Pressures." *Journal of Petroleum Technology* (September 1981) 1767–1775.

Novotny, E.J. "Prediction of Stimulation from Acid Fracturing Treatments Using Finite Fracture Conductivity." SPE paper 6123 presented at the Annual Technical Conference and Exhibition, New Orleans (Oct. 3–6, 1976).

Ormsby, G.S. "Calculation and Control of Mud Pressures in Drilling and Completion Operations." *API Drilling and Production Practices* (1954), 14.

Overton, W. "Detection of a Thin Sheet Magnetic Anomaly by SQUID Gradiometer Systems: Possibility of Hydrofracture Azimuth Determination." *Proc., Soc. Explor. Geophys. Res.*, Tulsa (1980).

Parker, C.D., D. Weber, D. Garza, and S. Swaner. "Austin Chalk Stimulation—Techniques and Design." Austin Chalk Oil Recovery Conference, Texas A&M University, college Station (April 28, 1982).

Paslay, P.R., and D.B. Bogy. "The Stability of Circular Rod Laterally Constrained to be in Contact with an Inclined Circular Cylinder." *Journal of Applied Mechanics* (1964) v. 31, 605–610.

Petzet, G.A. "Prudhoe Bay Horizontal Well Yields Hefty Flow." *Oil and Gas Journal*. (October, 1984) 42–43.

Pine, R.J. and A.S. Batchelor. "Downward Growth of Hydraulic Stimulation by Shearing in Jointed Rock." *Int. J. Rock Mech. Min. Sci. and Geomech. Abstr.*, (1984).

Pittman, F.C. et al. "Investigation of Abrasive Laden Fluid Method for Perforation and Fracture Initiation." *Journal of Petroleum Technology* (May 1961) 489–495.

Reed, R.M. "Air Drilling with Foam Combats Water Influx." *The Petroleum Engineer* (May 1958), B–57.

Sattler, A.R. et al. "Stimulation Fluid Systems for Naturally Fractured Tight Gas Sandstones; A General Case Study." *SPE Production Engineering* (August 1991), 313–322.

Schecter, R.S. and J.L. Gidley. "The Change in Pore Size Distribution from Surface Reactions in Porous Media." *AICHE Journal* (May 1969) vol. 15, 339–350.

Directional Drilling. School of Drilling Technology. Wilson Downhole Service. Sperry Sun Catalog, 767.

Smith, D.K. *Cementing.* SPE Monograph volume 4, Henry L. Doherty Series (1976).

Stagg, T.O., J.W. Powell, R.L. Dewess, and M.P. Stephens. "Horizontal Drill-In Project Uses Sized-Salt Fluid." *Petroleum Engineer International* (July, 1986) 31–36.

"Standard Procedure for Evaluation of Well Perforators."API RP 43 (October 1962).

Stearns, G. "Engineering Fundamentals in Modern Drilling." *Oil and Gas Journal.* (1953), 69–81.

Stewart, et al. "Study of a Subsurface Fracture Zone by Vertical Seismic Profiling." *Geophys. Res. Lett.* (November 1981) v.8, 1132–1135.

"Stressfrac—For Enhanced Well Stimulation With Multiple Dynamic Fractures." A Servo–Dynamics, Inc. Publication (Circa 1985).

Turbodrill Handbook. Eastman Oil Well Survey Company (May 19, 1969).

Unterberger, R. "Reservoir Engineering and Modeling: Radar Imaging." Report to Los Almos National Laboratory, in Hot Dry Rock Geothermal Energy Development Program Annual Report, LA–8280–HDR (August 1980).

Walsh, J.B. "Effect of Pore Pressure and Confining Pressure on Fracture Permeability." *International Journal of Rock Mechanics, Minerals, Science & Geomechanics Abstract*, vol. 18 (1981) 429–475.

Walstrom, J.E., A.A. Brow, and R.P. Harvey. "An Analysis of Uncertainty in Directional Surveying." *Transactions.* AIME, (1969), 246, 515–523.

Warpinski, N.R. and L.W. Teufel. "Influence of Geologic Discontinuities on Hydraulic Fracture Propagation." *Journal of Petroleum Technology* (February 1987), 209–220.

Warpinski, N.R. "Dual Leakoff Behavior in Hydraulic Fracturing of Tight, Lenticular Gas Sands." SPE paper 18259 presented at the 63rd Annual Technical Conference and Exhibition held in Houston, Texas (October 2–5, 1988).

Warpinski, N.R. "Hydraulic Fracturing in Tight, Fissured Media." *Journal of Petroleum Technology* (February 1991), 146–152.

Weatherford Oil Tool Co. "Stresses on a Centralizer." Houston (1974).

Williams, B.B., J.L. Gidley, and R.S. Schechter. *"Acidizing Fundamentals."* SPE Monograph Volume 6, Henry L. Doherty Series (1979).

Witthrisch, C. "Horizontal Well Logging by SIMPHOR." Paper 5, SPWLA Symposium, London (March 1982).

Witthrisch, C. and G. Baron. "SIMPHOR The First Horizontal Well Logging System." in *Horizontal Well Technology*, a Horwell Publication, 177, Avenue Napoleon Bonaparte, 92500 Rueil—Malmaison, France (1984).

Witthrisch C. and J.P. Sarda. "New Method for Determining the Direction of a Hydraulic Fracture–Simfrac." *Oil and Enterprise* (October 1983) 56–59.

Witthrisch C. and J.P. Sarda. "Simfrac-Systeme d'Instrumentation et de Mesure des fractures." Ninth International Formation Evaluation Transaction, Paris (October 24–26, 1984).

Woods, H.B., and Arthur Lubinski. "Charts Solve Hole-Deviation Problems and Help Determine Most Economical Hole size." *Oil and Gas Journal.* (May 31, 1954), 34–36.

Woods, H.B., and Arthur Lubinski. "Practical Charts for Solving Problems in Hole Deviation." *API Drilling and Production Practices* (1954), 56.

Woods, H.B., and Arthur Lubinski. "Use of Stabilizers in Drill Collar String to Spud Drilling in Crooked-Hole Country," *Oil and Gas Journal* (April 4, 1955), 215–232.

CHAPTER 3

Formation Evaluation by Well Log Analysis

Logs have been an integral part of the formation evaluation effort of fractured media since 1951 when Mardock and Myers, and Lyttle and Ricke published techniques for evaluating the fractured Spraberry trend of Texas using radioactive and induction logs. These pioneering efforts provided reasonable results for distinguishing lithologies. However, quantitative analysis was not possible. Since then, many interpretation techniques have been developed. For completeness, this chapter discusses modern tools and interpretation techniques, but also old logs that are the only ones available in wells drilled more than 15 years ago.

SONIC AMPLITUDE LOGS

These logs are frequently used in attempts to detect fractures. Pickett indicated in 1963, and it is still recognized today, that the recordings of acoustic velocity generated by a logging tool identify four wave types:

The first type is a compressional wave, which travels from transmitter to formation as a fluid pressure wave, is refracted at the well bore, travels through at the compressional wave velocity of the formation, and arrives at the receiver as a fluid pressure wave.

The second type, a shear wave, travels from transmitter to formation as a fluid pressure wave, goes through the formation at the shear wave velocity of the formation, and comes back to the receiver as a fluid pressure wave. Particle motion of a shear wave is perpendicular to the borehole axis.

A fluid or water wave, the third type, travels from transmitter to receiver at the compressional-wave velocity in the borehole fluid.

The final type is a low velocity wave. It travels from transmitter to receiver at less than the compressional-wave velocity in the borehole fluid. Stonely waves usually occur here.

Morris et al (1964) have shown that the compressional wave amplitude is generally more attenuated by vertical and high angle fractures, while the shear wave amplitude seems to be more attenuated by horizontal and low angle fractures.

The attenuation due to fractures appear to be mostly the result of drastic changes in permeability rather than porosity. Theoretically, this can be shown with the use of Biot's (1956) equation for the shear wave.

$$a_s = 2\pi^2 \left(\frac{M_2}{M_2 + M_1} \right) \frac{f\rho_f k}{\phi\mu_f} \cong \frac{2\pi^2_f k}{\rho_b \mu_f} \qquad (3\text{--}1)$$

where
$\quad a_s$ = shear wave attenuation
$\quad M_1$ = mass of dry rock
$\quad M_2$ = mass of fluid
$\quad f$ = frequency, cycles/second
$\quad \rho_f$ = fluid density, gr/cc
$\quad k$ = permeability
$\quad \phi$ = porosity, fraction
$\quad \mu_f$ = fluid viscosity
$\quad \rho_b$ = bulk density, gr/cc

181

The above equation shows that shear wave attenuation is directly proportional to permeability and inversely proportional to bulk density. Since a small open crack can have Darcys of permeability even if its porosity is very small it can be concluded that permeability is the controlling factor in shear wave attenuation. A similar but more complex type of equation can be used for the compressional wave attenuation. In general, the reduction in amplitude of the compressional wave is not as large as the attenuation of the shear wave.

Figure 3–1 illustrates the reduction caused by fractures on the amplitude of all wave trains. Core analysis at 11,970 ft indicated the presence of fractures filled with pyrobitumen. The rest of the section was indicated to be unfractured. Notice in Figure 3–1 that the section at 11,970 ft shows that all wave types are reduced, especially the shear wave. The unfractured section does not show any amplitude reduction.

This example illustrates that in practice it might be possible to detect fractures by making a log of the amplitude of the shear wave. A decrease in shear amplitude may indicate the possible presence of fractures.

Figure 3–2 shows the idealized model of a horizontal fracture presented by Pickett and Reynolds (1969). When the fracture is located between transmitter (*Tx*) and receiver (*R*), the refracted compressional wave travelling along the borehole (A_o) is reflected in a great amount (A_r) at the fractured interface. This results in an amplitude drop of the compression wave. The effects on the shear wave are even stronger, and the shear wave amplitude is almost completely vanished due to lack of acoustic coupling between the solid and the fluid.

The phenomenon of cycle skipping in sonic logs sometimes develops in the presence of fractures. In general, the compressional wave corresponds to the first arrival in the receiver. When fractures are present, the amplitude of the compressional wave may be reduced due to reflection in the fracture. Consequently, the first arrival of the compressional wave may not be detected but rather some later arrival. If travel times are bigger than those normally recorded the cycle skipping of sonic log develops. Figure 3–3 shows an example of cycle skipping in a fractured calcareous shale of the lowlands of Quebec.

In many instances a good correlation exists between shear amplitude reduction and fractures as determined from other sources. However, Pickett (1969) does not recommend using this log alone, rather as a part of the formation evaluation package for several reasons.

First, there might be solid-to-solid contact in the fracture face which reduces the degree of acoustic discontinuity. Consequently, the amplitude of the shear wave may be only slightly reduced. Second, most fractures at depths below 3000 ft are vertical or of high angle inclination. In some cases, this might reduce the effectiveness of the log in detecting fractures. Finally, amplitude reductions might be generated in the presence of lithology variations, acoustic wave attenuation in the formation, uncentralized tool, rough borehole, and/or porosity variation.

Four wave types

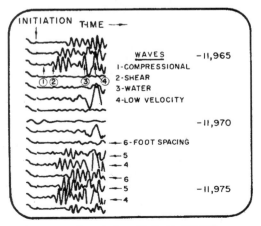

FIGURE 3–1   Multispacing character log (after Pickett, 1963)

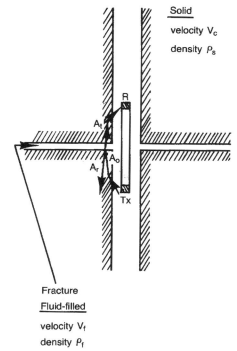

FIGURE 3–2 Idealized model, horizontal fracture (after Pickett & Reynolds, 1969)

Figure 3–4 shows a cross plot of reciprocal compressional velocity ($1/V_c = \Delta t_c$) vs. reciprocal shear velocity ($1/V_s = \Delta t_s$) using, as a base laboratory measurements on unfractured limestones, dolomites, and sands. From this cross plot, it appears that variations in compressional and shear wave velocities might be used to distinguish lithologies rather than fractures.

Acoustic array sonic waveforms can be recorded from which Stoneley (fluid) arrivals can be easily identified to evaluate reflection coefficients which might indicate the presence of natural fractures. Schlumberger conducts this type of analysis using a program called STE-FRAC. Another program, slow time coherence (STC) permits to find the slowness of all coherent non-dispersing propagating waves exited in the borehole as they pass sonic receivers.

VARIABLE INTENSITY LOGS

These logs have gained popularity for detecting fractures. They are presented commercially as a recording of depth vs. the time after the initiation of an acoustic pulse at the transmitter (t). Amplitude changes are indicated by a succession of varying shades of gray across the film track. The darkest areas correspond to the largest positive amplitudes, while the lightest areas correspond to the largest negative amplitudes.

Figure 3–5 shows an example of a variable intensity log (or variable density log or microseismogram) in a limestone cored well. A 1–ft vertical fracture is present in the core at depth indicated by an arrow. This corresponds to drastic band breaks in the log. When no fractures are present, the log gives the impression of banding.

Since the reflected/refracted waves arrive later than the compressional wave, they might create interference patterns that are usually referred to as Chevron patterns.

The Stoneley (Tube) wave has a low frequency (≈ 2 Khz), high amplitude which varies with hole size, and a velocity slower than the fluid. These properties are related to the Shear Modulus of the formation. Generally, Stonely waves occur after the fluid wave arrival. In small boreholes and fast formations, however, they can arrive together with the fluid wave.

Cycle skipping

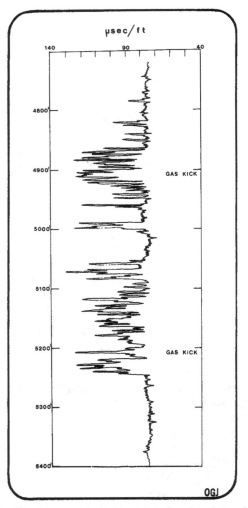

FIGURE 3-3   Cycle skipping in fractured shale of Lowlands of Quebec, Canada

While moving along the borehole the Stoneley wave is able to exchange energy with the formation through acoustic flow. This represents relative mobility of flows. Consequently, the Stoneley wave attenuation is correlatable with formation permeability, and can provide a qualitative scale of pseudo-permeability where low attenuation corresponds to low permeability and high attenuation corresponds to high permeability. Integration of these pseudo-permeabilities with production logs can prove very powerful in naturally fractured reservoirs (Dennis and Standon, 1988).

The variable intensity log may provide qualitative information regarding the presence of fractures. However, Beck et al (1977) reported that the resolution of the log is also affected by other factors such as tool centralization, receiver/transmitter spacing, hole size, lithology changes, rough boreholes, fracture and bedding plane orientation, and presence of vuggy or crystalline porosity.

LONG SPACING SONIC

This tool has improved the accuracy of transit time measurements in formations which have been affected by the drilling process. The longer the spacing between transmitter and receiver, the greater the separation of the various waves from each other. Furthermore, the

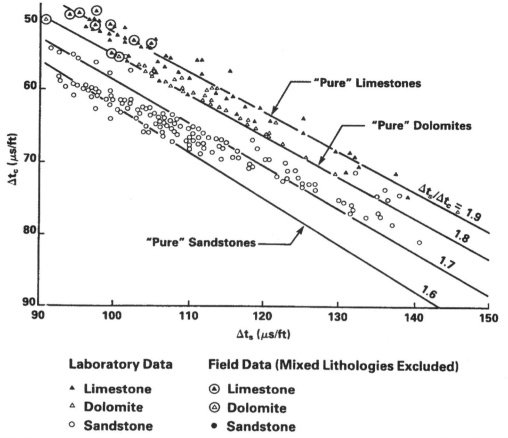

FIGURE 3–4 Reciprocal compressional velocity (Δt_c) vs. reciprocal shear velocity (Δt_s) including laboratory and field data (after Pickett, 1963)

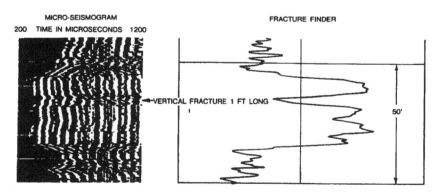

FIGURE 3–5 Micro-seismogram/Fracture Finder (Courtesy of Welex)

greater the spacing the larger the attenuation of the wave train. Spacings for this tool are for example 8, 10, or 12 ft.

Present acquisition and processing technology (Schlumberger) allows computation of velocities and energies of the shear and compressional waves. It has been found that the ratio between shear and compressional wave energies provides a useful fracture indicator. Figure 3–6 shows an example of Delta–T shear processing. The first track presents amplitude and the energy ratio (RSCE). The second track shows compressional and shear wave energies (CEWl and SEWl). The last track shows the computed Δt compressional and Δt shear (DTC and DTSM), and the long spacing sonic Δt (DTC).

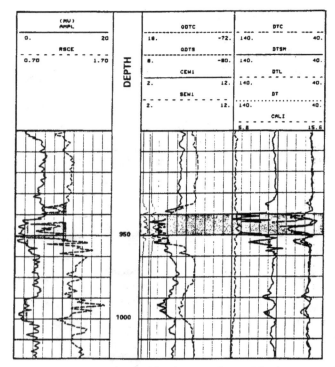

FIGURE 3–6 Long Spacing Sonic Delta T (DTL), together with the computed Delta T Compressional (DTC) and Delta T Shear (DTSM). Compressional and Shear Wave Energies (CEW1 and SEW1), and the ratio of these energies (RSCE) are also displayed (after Schlumberger)

Note that when both waves are strongly attenuated, Δt_c and Δt_s are not reliable as cycle skipping occurs. The ratio of energies and the other curves suggest the presence of fractures in the shaded areas. It must be stressed, however, that other characteristics such as lithology changes, caving, borehole rugosity and bed boundaries produce similar energy reductions even when no fractures are present. Consequently, these possibilities must be ruled out before any conclusions with respect to the presence of fractures are drawn.

SHEAR-WAVE SPLITTING

A substantial body of information from several major oil companies provides evidence that shear-wave splitting is produced by propagation through stress-aligned, fluid-filled inclusions, fractures, microfractures and preferentially oriented pore space that exist in most rocks in the crust (Crampin et al, 1989). It is now recognized that seismic shear waves contain much more information than compressional (P) waves about the internal structure of a rock along the ray path. This can be interpreted in terms of fractures and stress geometries within the rock. The approach uses shear-wave vertical seismic profiles (VSP's), where a surface source is recorded on downhole geophones. From the data, fracture orientation and reservoir description can be obtained.

Figure 3–7 is a schematic of shear-wave splitting in the crust. The greek letter σ represents stress. The fractures are aligned perpendicular to the minimum compressional stress (σ_{min}). The rock contains vertical fractures aligned east-west. Shear wave propagation is presented for three initial polarizations: (a) the initial shear wave is polarized parallel to the strike of the fractures; (b) the initial shear wave is polarized perpendicular to the strike of the fracture; and (c) the initial shear wave has intermediate polarization.

In cases (a) and (b) there is no splitting and no change in the polarization of the shear wave, although the waves in case (a) travel faster and are less attenuated than the waves in case (b). The shear wave splits in case (c) when it has an intermediate polarization. The leading faster-travelling split shear wave is polarized parallel to the strike of the fractures,

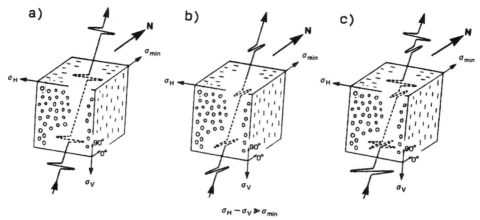

FIGURE 3–7 Shear-wave propagation through a typical rock with vertical cracks aligned east/west for three initial polarizations: (a) initial shear wave polarized parallel to the strike of the cracks; (b) initial shear wave polarized perpendicular to the strike; and (c) initial shear wave with intermediate polarization (after Crampin et al, 1989)

and the slower-travelling shear wave is polarized perpendicular to the fractures. The polarization of the leading shear wave is parallel to the horizontal direction of maximum compressional stress. The time delay between the split shear waves is proportional to the fracture density for each ray path through the fractured rocks. (Crampin et al, 1989; Yale and Sprunt, 1989).

Figure 3–8 shows comparisons of synthetic (boxed) and observed (open) horizontal polarization diagrams (PD) for six multioffset VSP's in the Paris basin. The synthetic PD's were drawn with a computer program developed by Taylor (1987). Figure 3–8 (a) shows

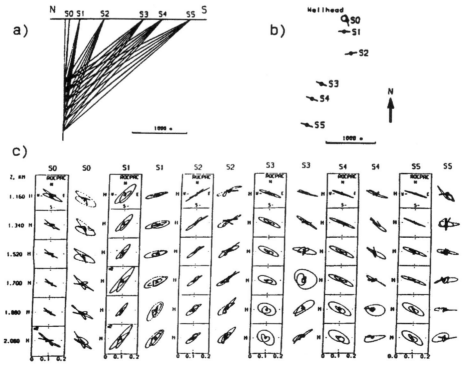

FIGURE 3–8 Multioffset VSP's in the Paris basin: (a) straight-line ray paths from surface source to borehole receivers; (b) plan of wellhead (open circle), source of offsets (solid circles), and orientation of the shear-wave vibrators (bars); and (c) comparison of synthetic (boxed) and observed (open) patterns of PD's for the six offsets (after Crampin et al, 1989)

straight-line ray paths from surface source to borehole receivers. Figure 3–8 (b) shows the plan of wellhead (open circle), source offsets (solid circles), and orientation of the shear wave vibrators (bars). Figure 3–8 (c) shows the comparison of synthetic and observed PD's. The modelling provides strong evidence that the sedimentary sequences above the geophones have an internal structure of vertical parallel EDA (extensive-dilatancy anisotropy) fractures striking N30°W, parallel to the measured horizontal compressional stress, with a fracture density of 0.03, equivalent to a fracture diameter of approximately 0.6 in. each unit cube.

Shear-wave splitting has been observed in many areas including the Austin chalk in Texas, Pennsylvania, California and the Niobrara chalk in Wyoming.

INDUCTION LOGS

In some instances, induction logs indicate the presence of fractures. This is possible when there are resistivity anomalies in an interval and sufficient resistivity contrast with adjacent beds (Timko, 1966).

Figure 3–9 presents gamma ray and induction logs for a Gulf Coast well run at three different times. Figure 3–9A shows the gamma ray and induction logs run between 15,850 and 16,020 ft following drilling with 17.1 lb/gal inverted oil emulsion mud. When the well was deepened to 16,294 ft, a pressure kick occurred. As a result the mud density was increased to 17.5 lb/gal. This new density allowed the mud to flow in the formation, losing 275 b of mud. Figure 3–9B shows the logs obtained between 15,850 and 16,020 ft following the mud density transition from 17.1 to 17.5 lb/gal.

Then the mud density decreased to 17.3 lb/gal and the mud flowed back from the formation. Figure 3–9C shows the logs obtained following the reduction in mud weight.

Various interesting features are to be noted from this example. First, there is a general increase in resistivity ($R_{\parallel}$) in Figure 3–9B compared with Figure 3–9A especially between 15,910 and 15,925 ft. This can be interpreted by saying that the formation was fractured (or maybe an existing hairline fracture was opened) as a result of increasing the mud density from 17.1 to 17.5 lb/gal. The 275 b of nonconductive inverted oil emulsion mud lost to the formation were placed in the fractures, thus reducing the induction log current paths, and consequently increasing the induction log apparent resistivity.

Second, the reduction of mud weight from 17.5 to 17.3 lb/gal allowed the fracture to close and the mud to flow back from the fracture into the well bore. Consequently, the induction logs's apparent resistivities should go back to lower values, which in fact occurred (Figure 3–9C).

This is an excellent example to illustrate that fractures do indeed open and close at depth. Consequently any forecasting of reservoir performance should be conducted by taking into account possible fracture closure, and hence fracture permeability and fracture porosity reductions, as the reservoir is being depleted. Not taking this into account might lead to forecast very optimistic recoveries.

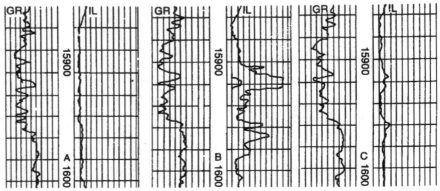

FIGURE 3–9 Resistivity logs, Gulf Coast Well (after Timko, 1966)

DUAL INDUCTION LATEROLOG 8

These logs indicate the presence of fractures if the laterolog 8 reads less than the induction curve (Schlumberger). The laterolog 8 is a vertically focused, short-spacing resistivity tool that may respond to thin bedded formations and/or to vertical fractures when they are filled with filtrate of resistivity lower than that of the formation.

The induction log reads horizontal conductivity, which is presented in the induction log as the inverse, i.e., as resistivity (R_{ll}). Consequently the presence of fractures in the wall of a borehole might be detected if the laterolog 8 indicates less resistivity than the induction log.

Figure 3–10 shows an example of a fractured zone located with dual induction laterolog 8. Since the induction log depends on induced currents which, in general flow in horizontal loops, it can be concluded that the induction-log curve is affected negligibly by the conductive fluid-filled vertical fracture. This method must be used with care because R_{mf}/R_w ratio; resistivity of fluid in the fractures; fracture width, length, configuration, and lateral extent; lithology and porosity; shaliness; and borehole size affect the resolution of the method (Beck et al, 1977).

SONIC, NEUTRON AND DENSITY LOGS

The combination of sonic, neutron and density logs might indicate the presence of fractures (Schlumberger). In this method, it is assumed that the sonic log provides matrix porosity. The difference between neutron (or density or the combination neutron-density) and sonic porosity is interpreted as fracture porosity, if it is known that there is no other kind of secondary porosity.

Figure 3–11 shows sonic, neutron, and density logs of a well in the Auquilco formation, Neuquen basin of Argentina. The section contains anhydride with fractured porosity. Notice that the sonic transit time (Δt) remains approximately constant over the entire section, while the bulk density (ρ_B) decreases from 2.97 to 2.83 g/cc and the neutron porosity (ϕ_N) increases from 0% to 4%.

According to Beck et al (1977), there are four major problems in using the sonic-neutron and/or density combination. First, this combination provides values of total secondary porosity. Fractured porosity can be estimated only if it is known that there are no other types of secondary porosity. Second, total porosity can be underestimated, as it is usually derived from a measurement on only one side of the hole. Consequently, if the fracture happens to be in the opposite side of the borehole, its porosity will not be accounted for. Third, this method may indicate fracture porosity which does not really exist due to variations in shaliness. Fourth, total porosity can be overestimated because of borehole irregularities.

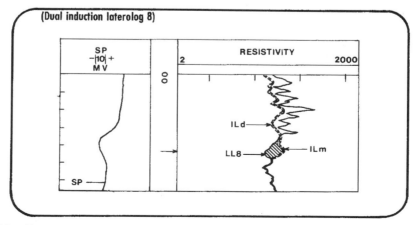

FIGURE 3–10 Fracture response (at level of arrow) on Dual Induction Laterolog 8 log (after Beck et al, 1977)

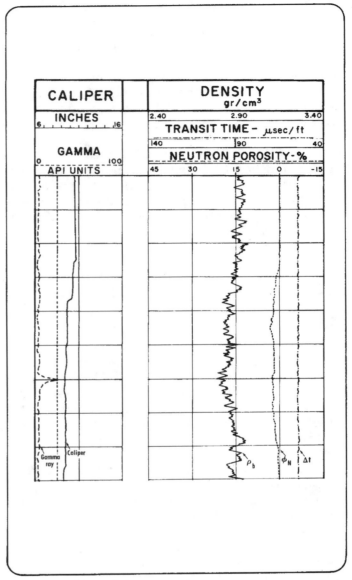

FIGURE 3–11 Logs of Auquilco Formation, Neuquen Basin (Argentina) (after Schlumberger)

Difficulties 3 and 4 might sometimes be accounted for using computer programs. Usual input data for these programs include the proper resistivity tool for calculating R_t; neutron, density and sonic logs for calculating total, matrix and secondary porosity and for accounting for variations in lithology; proximity log or microlaterolog for determination of R_{xo}; gamma ray and SP curves to aid in calculations of clay content; and the caliper log to indicate enlarged or rugose hole.

COMPARISON OF POROSITY ESTIMATES FROM DIFFERENT SOURCES

Pickett and Reynolds (1969) introduced a statistical method for evaluating fractured reservoirs. Initially, it is assumed that the neutron log (or density or the combination neutron-density) provides total porosity and that core porosity provides matrix porosity.

Figure 3–12 (A) shows an idealized schematic of neutron response vs. core porosity, with no uncertainty in measurements, i.e., both log and core measurements have perfect precision. Data points representing unfractured zones would be located in segment AA', where total porosity equals the matrix porosity. Points D and F would represent fractured zones,

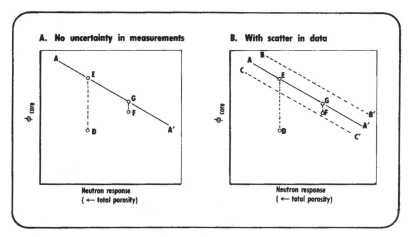

FIGURE 3–12   Neutron response vs. core porosity, idealized case (after Pickett & Reynolds, 1969)

i.e., zones where total porosity is greater than matrix porosity. Measures of fracture porosity would be provided by the distances DE and FG.

Figure 3–12 (B) shows a diagram of neutron response vs. core porosity for a more realistic case where there is scatter in data. Here, the data points representing the unfractured zones are contained between BB' and CC'. Point D represents a fractured zone, and the distance DE is a measure of fracture porosity.

Some fracture zones could fall in the area of scatter for unfractured zones, as in point F. The evaluation from there depends on the presence of a normal distribution about the average correlation of the response when no fractures are present. Experience indicates that unfractured porosity usually has a normal distribution.

BOREHOLE TELEVIEWER

The borehole televiewer (BHTV) produces an acoustic picture with a rotating ultrasonic transducer or scanner (Zemanek et al, 1969). The transducers can be focused or unfocused. Typically they rotate 3 to 16 times per second. The transducers frequencies range between 200 KHz and 1 MHz. The log inspects the borehole and evaluates the formation as it reveals induced and natural fractures, vugs, distribution of perforations in casing, and casing failures. Table 3–1 shows a summary of characteristics of borehole imaging devices including the borehole televiewer and the formation micro-scanner discussed later.

Figure 3–13 presents a BHTV of a Spraberry (Texas) well. One readily concludes that the well has a high angle fracture which intercepts the borehole from 7,040 to 7,072 ft. This fracture is only 5 in. tilted over the 32–ft interval, i.e., the fracture is nearly vertical.

Some of the requirements for obtaining a good borehole picture are: near-perfect centralization, low solids content in the borehole fluid, and a slow, constant logging speed. In some cases, damage may make the fracture appear wider than it actually is.

A borehole televiewer tool that has been reported to provide good results in horizontal wells drilled in the Austin chalk is the Western Atlas circumferential borehole imaging log (CBIL) (Western Atlas, 1990). The tool is pipe-conveyed and a combination of centralizers and knuckle joints prevent eccentric positioning of the CBIL instrument.

Figure 3–14 shows a combination of CBIL, gamma ray, and spectralog run in a horizontal hole. The spectralog was recorded while going into the hole. The CBIL and gamma ray measurements were made coming out of the hole through the entire 1500 ft of horizontal section. Logging operations were carried out at a standard speed of 600 ft/hr.

The left side of Figure 3–14 shows a natural fracture intercepted by the horizontal wellbore. This is confirmed by the spectralog which displays an increase in uranium counts.

The right side of Figure 3–14 shows vee pattern in the CBIL image indicating that the wellbore exited the chalk and drifted into a marl. The spectralog verifies once again the presence of the chalk-mark interface. Halliburton's CAST is also a type of borehole televiewer.

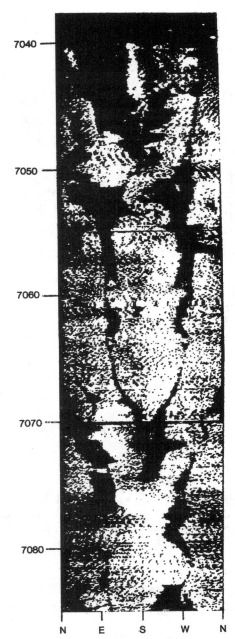

FIGURE 3–13 BHTV log of hydraulically induced fracture in Spraberry formation of West Texas (after Zemanek, 1969)

DIPMETER AND FRACTURE IDENTIFICATION LOG (FIL)

The FIL is a high resolution dipmeter. The dipmeter has been used extensively to locate fractures since the early 1960s when it was noted that the three-arm, continuous dipmeter log correlated with vertical fractures. Further progress has lead to today's high-resolution tools.

Figure 3–15 (A) shows how the 4-curve presentation of a high-resolution dipmeter (FIL) can help obtain the direction of a vertical fracture. The azimuth always gives the bearing of curve 1. The other curves are numbered sequentially and clockwise.

For example, if pad 1 is recording on the north side of the hole, then curve 2 will be recorded on the east, curve 3 will be recorded on the south, and curve 4 will be recorded on the west side of the hole.

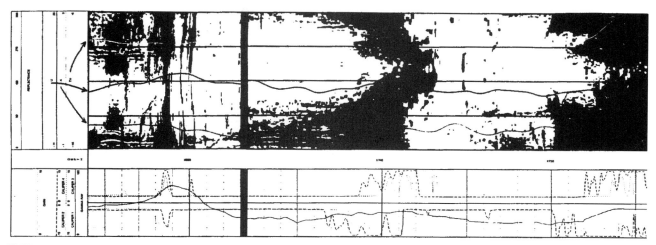

FIGURE 3–14 Circumferential borehole imaging log (CBIL) of horizontal well in Austin chalk (Courtesy of Western Atlas, 1990)

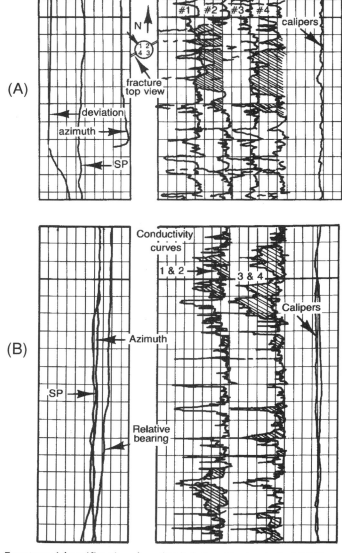

FIGURE 3–15 Fracture identification log (FIL) (after Beck et al, 1977)

Sometimes, vertical fractures can be detected in opposite curves, i.e., in curves 1 and 3 or curves 2 and 4. Notice in Figure 3–15 (A) that the fractures are indicated in curves 2 and 4. This type of response depends on invasion of the fracture by a conductive fluid.

Figure 3–15 (A) also shows a schematic top view of the borehole indicating that the fracture is oriented in a NE/SW direction, the usual orientation in Gulf Coast fractured carbonates.

The previous statement is based on the assumption that the fracture orientation is given by the larger of the caliper two diameters. There are cases, however, in which the smaller diameter might indicate the fracture orientation as shown by Cox (1983). In fact, he states that "the correlation of natural fractures with borehole elongation is not strong except in the Cotton Valley and Austin Chalk" and he recommends to search for a relationship first, based on additional information, rather than assuming that natural fractures orientation is given by the larger diameter of the caliper.

Figure 3–15 (B) shows a dual two-curve overlay which is useful for indicating zones of vertical fractures. The curve separation gives a qualitative indication of fracture quality. This is very important because reasonable estimates of fracture pay can be made by adding the intervals which show curve separation.

Care must be exercised when interpreting the high resolution dipmeter curves because the presentation is sometimes scrambled by sedimentary features. Furthermore, in some cases, shaded areas such as those shown on Figure 3–15 (A) might represent layering and apparent dips rather than fractures.

Another device, Schlumberger's SHDT Dual Dipmeter tool, has four dual electrodes which register eight microconductivity curves, permitting a large density of dip results to be computed for interpretation. A mechanical devise allows good centralization in deviated wells. For good correlation, an SP tool or a gamma ray curve can be run in combination with the dual Dipmeter tool.

Under some conditions the conventional dipmeter may detect fractures associated with faults. Figure 3–16 shows a dipmeter log run in the St. Lawrence lowlands of Quebec. The dipmeter clearly shows the zones of thrusting. In the St. Lawrence lowlands, gas shows are present where thrusting occurs.

FORMATION MICROSCANNER AND FULLBORE FORMATION MICROIMAGER

Schlumberger's Formation MicroScanner (FMS) is similar to the SHDT Dual Dipmeter tool in all respects except that in one of the configurations it has two arrays of electrodes on pads three and four (Standen, 1986). Currently on the market is another version of the tool with four imaging pads perpendicular to each other.

In one configuration the arrays consist of three rows of seven buttons and one row of six. Each of the buttons is 2/10 of an in. in diameter with the entire array 2.7 in. wide by 1.4 in. deep. The configuration allows for a 60% overlap of the electrical signals. With arrays on two pads, a 20% coverage of the well bore is obtained in an 8 1/2 in. diameter hole. With the four pad configuration, a 45% circumferential coverage is obtained in an 8 1/2 in. diameter hole.

Increases in borehole coverage have been advancing at a rapid pace. Table 3–1 shows a summary of characteristics of the FMS tool and the BHTV discussed in a previous section.

In addition to the array electrodes, the tool also has 10 standard SHTD electrodes (8 measure electrodes plus 2 speed electrodes) as well as a directional cartridge containing accelerometers and magnetometers (GPIT). Thus the total number of electrodes in this set-up is 27 + 27 + 10 = 64 electrodes. Acquisition of the data is by two modes; either as an SHDT without image data or in imaging mode with all 64 electrodes being recorded.

The image processing uses all electrodes from the arrays and goes through programs for depth and speed correction, repair of missing electrodes, equalization of electrode responses and normalization of the image over a sliding window. These programs allow for the enhancement of fine stratigraphic features.

The plotting program allows the presentation of the images as either grey-scale images from white (high resistivity) to black (low resistivity), wiggle-traces of the electrodes or

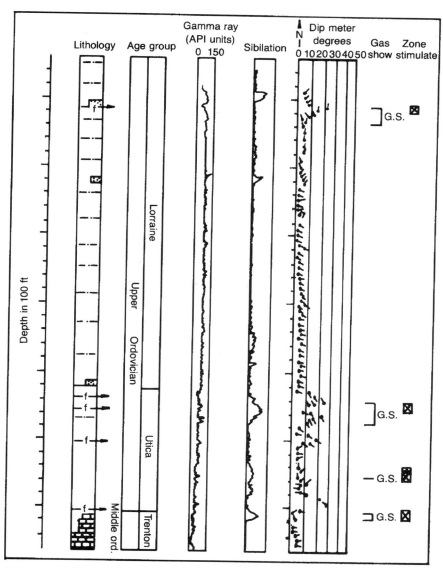

FIGURE 3–16 Dipmeter shows zones of fracturing corroborated by sibilation log and gas shows (after Beiers, 1976)

color images (similar to grey-scale). Another program can also display multiple passes of the images in an azimuthal presentation. This gives the images their proper orientation as a wrap-around picture of the wellbore from 0° to 360°.

Figure 3–17 shows an example of a fractured sandstone reservoir with recementation. Note that an excellent repeat of the picture was obtained. Good results with the FMS for locating natural fractures have also been reported in horizontal wells drilled in the Austin chalk, Texas (Fett and Henderson, 1990).

The Fullbore Formation MicroImager (FMI) is Schlumberger's latest generation electrical imaging device. The FMI generates electrical images which are reported to be almost insensitive to borehole conditions. It is particulary useful for analysis of fractures and can provide quantitative information such as fracture width and fracture porosity. A computer program known as FracView is used for the analysis. In my experience, however, the user should assign quantitative information from the FMS and the FMI a relative rather than an absolute value.

For example, an interval with a larger fracture aperture could be anticipated to have better productivity than an interval with a smaller aperture only if the calculations are carried out in the same well and the same formation. However, I have found that a calculated

TABLE 3–1   Summary of characteristics of borehole imaging devices (after Luthi, 1992)

Characteristic	Borehole Televiewer	Formation MicroScanner
Typical logging speed	800 ft/hr (245 m/hr)	1600 ft/hr (490 m/hr)
Borehole wall coverage	Full (360°)	Partial (depending on number of pads and runs)
Borehole size	6–12 in. (15–30 cm)	6–18 in. (15–46 cm)
Borehole muds	Best results in light muds	Conductive muds only

Typical Sampling Rates:
    Vertical
    Horizontal
	0.2–0.5 in. (5–12.5 mm)
	0.1 in. (2.5 mm)
		0.1 in. (2.5 mm)
		0.1–0.2 in. (2.5–5 mm)

Properties influencing measurement	Acoustic impedance contrasts; borehole geometry	Rock microresistivity
Shortcomings	Sensitive to borehole rugosity and drillmarks; sensitive to tool eccentering and borehole ellipticity	Resolution decreased by pad stand-off; irregular tool speed causes layer misalignment
Main applications	Fracture detection (orientation); borehole geometry; breakout detection (orientation)	Bedding types; fracture detection (orientation and width); fault detection (orientation)

larger fracture aperture in a well does not necessarily provide better productivity than a smaller fracture aperture in an offset well. Thus care must be exercised with the quantitative information from these tools. A potential reason for this problem is that the ratio of mud to flushed zone resistivity (R_m / R_{xo}) plays an important role in computing the fracture opening, and that the model assumes that the fractures are planar with parallel walls of infinite extent. Another point to keep in mind is that a large number of case histories have shown that a few, and in some cases only one macro-fracture, can lead to bigger flow than a large number of smaller fractures.

The apparent area of the fractures from the FracView computer program is the sum of the fracture apertures over an area of the borehole. The apparent fracture porosity is estimated as the ratio of the area of fractures to the total borehole wall area.

The borehole coverage of the fullbore FMI is quite significant. The coverage in an 8½ in. borehole is close to 80%. This compares with a 45% coverage by the FMS. The following table summarizes coverage as a function of hole diameter for the FMI tool:

Hole Diameter in.	Coverage %	Hole Diameter in.	Coverage %
6¼	93	14	47
8	80	16	41
10	63	18	37
12	53	20	33

The number of sensors used, the coverage in a 8 1/2 in. hole and the maximum logging speed of each FMI operating mode is summarized in the following table:

Operating Mode	Number of Sensors	Coverage in 8½-in. borehole	Maximum Logging Speed
Fullbore	192	80%	1800 ft/hr
Four-Pad	96	40%	3600 ft/hr
Dipmeter	8	—	5400 ft/hr

FRACTURED SANDSTONE RESERVOIR WITH RECEMENTATION

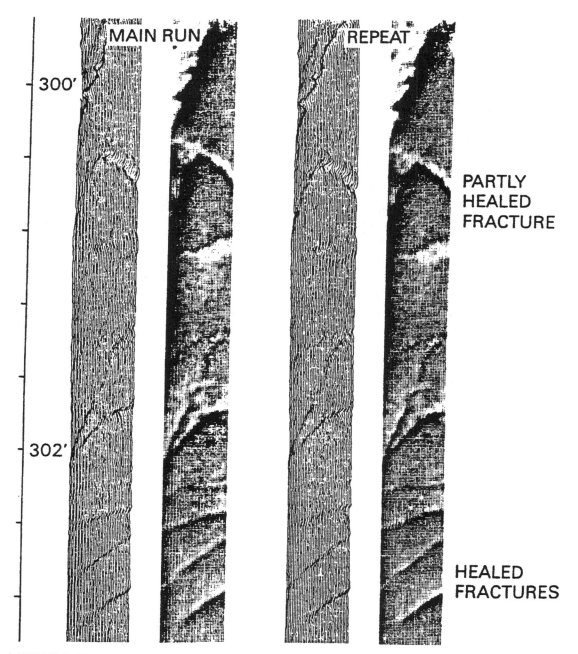

VERTICAL AND HORIZONTAL SCALE REDUCTION OF 1/5

FIGURE 3–17 Electric picture of fractured sandstone reservoir using the Formation Microscanner (after Standen, 1986)

The FMS/FMI allow two main types of fractured interpretations, (i) fracture characterization, and (ii) fracture analysis (Paauwe, 1994). The characterization includes the identification of the fracture type, fracture morphology and orientation. These are derived from a workstation by visual inspection of the images. The fracture analysis is conducted with computer programs (one of them is called FracView) that calculate parameters such as fracture aperture, fracture intensity and fracture porosity.

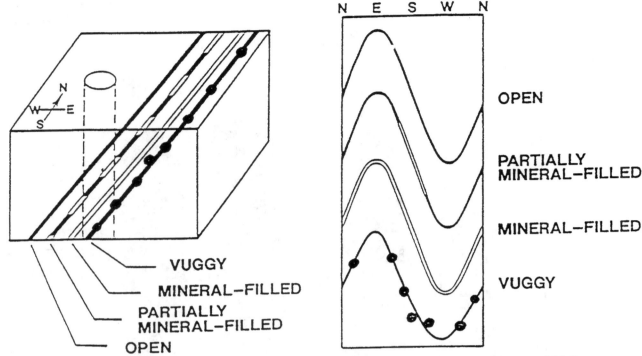

FIGURE 3–18 (A)   Fracture morphology (open, mineral-filled and vuggy) (after Paauwe, 1994)

I must emphasize that the image is simply the result of an interpretation and should never be accepted at face value. Some of the factors that will affect the quality of the image include:

- Resistivity of the mud (R_m) at formation temperature
- Resistivity of the flushed zone (R_{xo}) at formation temperature
- Fracture aperture (width)
- Experience of the evaluator. Geological input is critical for a proper interpretation of the image.

The quality of all the above factors cannot be overemphasized. Figure 3–18A shows a schematic of how the FMS/FMI can identify fracture morphology. For a discussion of fracture morphology (vuggy, mineral-filled, partially mineral-filled, and open fractures) please refer to Chapter 1. Figure 3–18B presents schematics of a viewing perspective of vertical, 45° deviated, and horizontal wellbores that penetrate high inclination natural fractures and horizontal bedding planes. In the case of the vertical well, average depth of fracture (ADF), fracture dip angle (α), and fracture strike are calculated as follows:

$$ADF = (A + B)/2$$

$$\alpha = \tan^{-1}\left[\frac{|B - A|}{Borehole\ Diameter}\right]$$

Strike = Dip azimuth at x $\pm\ 90°$

As indicated in Chapter 1, fluid flow through a fracture (Figure 3–19a) can be calculated from the equation:

$$q = \frac{w^2 A(P_1 - P_2)}{12\,\mu L}$$

Where: q = flowrate, cm³/sec
             w = fracture width or aperture, cm
             Y = fracture trace length, cm
             A = cross-sectional area of fracture, cm²

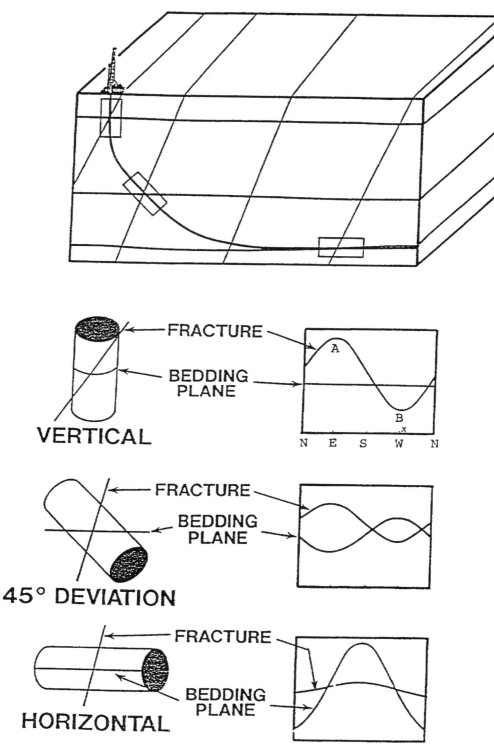

FIGURE 3–18 (B)   Viewing perspective (after Paauwe, 1994)

$(P_1 - P_2)$ = pressure differential between end points of the fracture length, dynes/cm²

L = fracture length, cm

μ = fluid viscosity, poises

In order to look at the effects of fracture size and shape, various terms in the above equation are lumped together as a constant (Paauwe, 1994):

$$C = \frac{(P_1 - P_2)}{12\,\mu L}$$

and,

$$q = Cw^2\,A = Cw^3\,Y$$

Figures 3–19 b and c present two fractures of different shape. The area of both fractures is equal to 22. The flow rates through fractures #1 and #2 can be calculated as follows:

$$q_1 = C\sum_{y=1}^{10} w^3\,Y = C[3(1.0^3 \times 1) + 3(5.0^3 \times 1) + 4(1.0^3 \times 1)]$$

$$= C\,[3 + 375 + 4]$$

$$= 382\,C$$

$$q_2 = C\sum_{y=1}^{10} w^3\,Y = C[10(2.2^3 \times 1)]$$

$$= C\,[10(10.648)]$$

$$= 106.48\,C$$

The ratio of flow rate in fracture #1 to that in fracture #2 is:

$$q_1/q_2 = \frac{382\,C}{106.48\,C} = 3.588$$

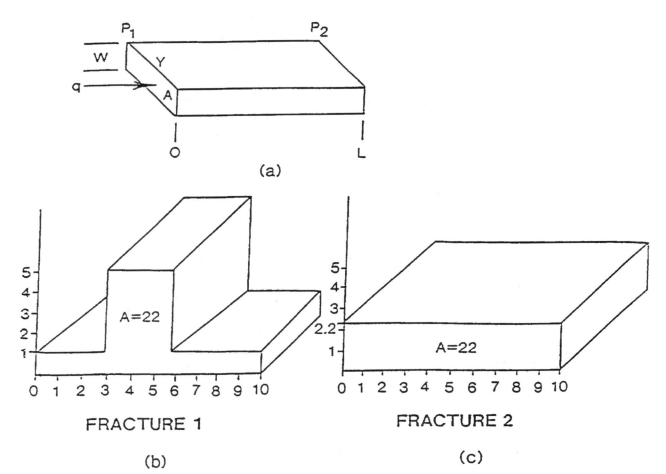

FIGURE 3–19 Fracture schematics (after Paauwe, 1994)

Consequently, although the areas of the two fracture faces are the same, fracture #1 will have a flow rate more than three times that of fracture #2. This illustrates the fact that the wider fractures will be the main conduit for fluid flow in the formations.

Two types of fracture aperture calculations are performed in Schlumberger's program FracView: 1) main fracture aperture, F_A and 2) hydraulic fracture aperture, F_{AH}. The mean aperture is a simple arithmetic average of the fracture width whereas the hydraulic aperture is a cubic mean of the fracture width, i.e., cube root of the sum of the cubes of each width element.

Mean fracture apertures for fractures #1 & #2 are:

$$F_{A1} = \frac{\Sigma w_1 i}{n} = \frac{3(1.0) + 3(5.0) + 4(1.0)}{10} = 2.2$$

$$F_{A2} = \frac{\Sigma w_2 i}{n} = \frac{10(2.2)}{10} = 2.2$$

and

$$\frac{F_{A1}}{F_{A2}} = \frac{2.2}{2.2} = 1.0$$

Consequently fracture #1 and fracture #2 have the same mean apertures. Hydraulic fracture aperture calculations are carried out as follows:

$$F_{AH1} = \sqrt[3]{\frac{\Sigma w^3_1 i}{n}} = \sqrt[3]{\frac{3(1.0)^3 + 3(5.0)^3 + 4(1.0)^3}{10}} = \sqrt[3]{\frac{382}{10}} = 3.368$$

$$F_{AH2} = \sqrt[3]{\frac{\Sigma w^3_2 i}{n}} = \sqrt[3]{\frac{10(2.2)^3}{10}} = \sqrt[3]{\frac{106.48}{10}} = 2.2$$

$$\left(\frac{F_{AH1}}{F_{AH2}}\right)^3 = \left(\frac{3.368}{2.2}\right)^3 = (1.531)^3 = 3.588$$

Therefore the cube of the ratio of the hydraulic fracture apertures produces the same value as the ratio of flow rates through the two fractures.

This indicates that fracture aperture gives a measure of the physical size of the fracture opening, whereas hydraulic aperture gives a measure of the relative flow capacity of the fracture. This is the reason Schlumberger's FMS and FMI normally use the hydraulic aperture values when comparing different fractures.

In addition to evaluating fracture intervals the FMI has other applications that include evaluation of other types of secondary porosity, determination of net/gross ratio in sand/shale sequences, depth matching with core data, structural and textural analysis, and characterization of sedimentary bodies.

Halliburton has recently introduced an Electric Micro Imaging (EMI) tool that is also reported to provide fracture identification and orientation. The EMI images are obtained from a 6–arm (rather than 4–arm) high resolution dipmeter (Eubanks, 1994).

SPONTANEOUS POTENTIAL

The SP curve develops in some naturally fractured reservoirs and can be used qualitatively to determine the presence of fractures. In some cases the SP development is smooth and hardly noticeable with respect to a base line. In other cases the SP curve tends to be hachured (Figure 3–20) where each pick most likely corresponds to a streaming potential effect, resulting from mud filtrate into the fractures (Pirson, 1967).

CORRECTION CURVE ON THE COMPENSATED DENSITY LOG

Fracture indications may be obtained from the correction curve ($\Delta\rho$) in the compensated density log. Since the $\Delta\rho$ curve corrects the density logs for effects of rough borehole and

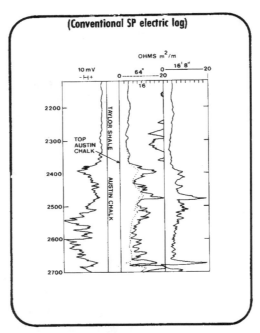

FIGURE 3–20 Conventional SP-Electric log in the fractured Austin chalk, Salt Flat-Tenney Creek Field, Caldwell County, Texas (after Pirson, 1970)

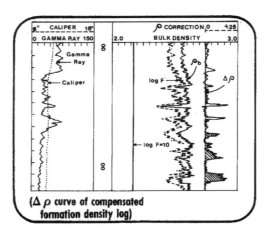

FIGURE 3–21 Fracture indications (cross-hatched excursions) on $\Delta\rho$ curve of compensated formation density log (after Beck et al, 1977)

mud cake, the $\Delta\rho$ curve may be affected by the mud in the fracture and indicate a correction even if the hole is in gauge.

A disadvantage of this method is that it might indicate the presence of fractures in only one side of the borehole and miss the possible fractures on the other side of the borehole. Figure 3–21 is an example of how the correction curve ($\Delta\rho$) on the compensated density log helps detect fractures.

COMPARISON OF SHALE VOLUME TO URANIUM INDEX

Fractures may be indicated by a comparison between the values of uranium index, as determined from the spectralog (Dresser Atlas), and the volume of shale in a zone (Heflin et al, 1976). Since uranium is very soluble in water it is commonly contained in ground waters. A volume of shale is calculated that is independent of the natural formation radioactivity.

For normal depositional environments the uranium index and the shale volume have the same value. When a fracture exists the uranium index may be larger than the shale volume. The main limitation of this technique is that it cannot indicate if the fracture is open or healed.

LITHOPOROSITY CROSSPLOT

This technique was introduced to help interpret formations with complex lithologies (Burke et al, 1969). The method handles the data from neutron, density and sonic logs simultaneously. From the readings of these logs, two porosity-independent parameters, M and N, can be determined as follows:

$$M = \frac{\Delta t_f - \Delta t}{\rho_B - \rho_f} \times 0.1 \qquad (3-2A)$$

$$N = \frac{(\phi_N)_f - \phi_N}{\rho_B - \rho_f} \qquad (3-2B)$$

Where Δt is sonic transit time (μsec/ft), ρ is density (gr/cc), ϕ_N is neutron porosity and the subscripts B and f stand for bulk and fluid, respectively.

In the crossplot of M vs. N each pure mineral is represented by a unique point regardless of porosity. For complex lithologies the position of the data points on the M–N plot helps identify the various minerals in the formation and the approximate percentage of each one. In addition, the lithoporosity crossplot may help detect secondary porosity without differentiating between vugs and fractures.

By assuming that the sonic log responds to only matrix porosity, and the neutron and density logs respond to total porosity notice in Equation 3–2B that secondary porosity does not affect the value of N. Analysis of Equation 3–2A however reveals that the value of M increases as the secondary porosity increases. Consequently the crossplot of M vs. N allows the detection of vuggy and/or fractured zones.

Figure 3–22 shows a generalized lithoporosity crossplot for fresh mud. The secondary porosity areas are located above the dolomite-limestone-silica line. They are indicated as areas B, C, D and E.

Table 3–2 shows matrix coefficients and M and N values for some common minerals. The following nomenclature is used in the table: Δt = sonic transit time in μsec/ft, ρ = density, and the subscripts ma and f stand for matrix and fluid, respectively. These M and N values have been used in the generalized lithoporosity crossplot presented in Figure 3–22.

Example 3–1. Figure 3–23 shows a lithoporosity crossplot which exhibits an extensive secondary porosity development, indicated by the large amount of data points above the dolomite-$CaCO_3$ line.

The selected lithology triangles are dolomite—$CaCO_3$—secondary porosity and dolomite—$CaCO_3$—silica. The points to the left of the dolomite-secondary porosity line are due to the mud cake effect on the SNP log. When no corrections are made for mud cake thickness before generating the M–N crossplot, each ¼ inch of mud cake reduces the N value by about 0.01.

PRODUCTION INDEX LOG

Experience in Devonian shales indicates that some anomalies occur in gamma ray, density, and induction logs (Myung, 1976). Usually in fractured zones, the gamma ray and resistivity increase while density decreases. The increase in the gamma ray intensity is attributed to zones rich in organic matter. The increase in resistivity is attributed to gas-filled fractures and/or increased kerogen content. The decrease in bulk density is attributed to increase in porosity due to the presence of fractures and/or to lower density of kerogen.

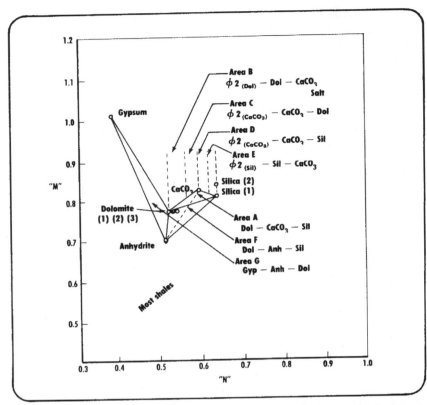

FIGURE 3–22 M–N Generalized lithoporosity crossplot for fresh mud (after Burke et al, 1969)

TABLE 3–2 Matrix coefficients and M & N values for some common minerals

Mineral	Matrix Coefficients			Salt Mud $\Delta t_f = 185.$ $\rho_f = 1.10$		Fresh Mud $\Delta t_f = 189.$ $\rho_f = 1.00$	
	Δt_{ma}	ρ_{ma}	$(\phi\,SNP)_{ma}$	M	N	M	N
Silica (1) (V_{ma} = 18,000)	55.5	2.65	−.035	.835	.669	.810	.628
Silica (2) (V_{ma} = 19,500)	51.2	2.65	−.035	.862	.669	.835	.628
CaCO$_3$	47.6	2.71	0.00	.854	.621	.827	.585
Dolomite (1) ϕ = 5.5% to 30%	43.5	2.87	.035	.800	.544	.778	.513
Dolomite (2) ϕ = 1.5% to 5.5% & > 30%	43.5	2.87	.02	.800	.554	.778	.524
Dolomite (3) ϕ = 0.0% to 1.5%	43.5	2.87	.005	.800	.561	.778	.532
Anhydride	50.0	2.98	0.00	.718	.532	.702	.505
Gypsum	52.0	2.35	0.49	1.060	.408	1.015	.378
Salt	67.0	2.05	0.04	1.240	1.010	1.16	.914

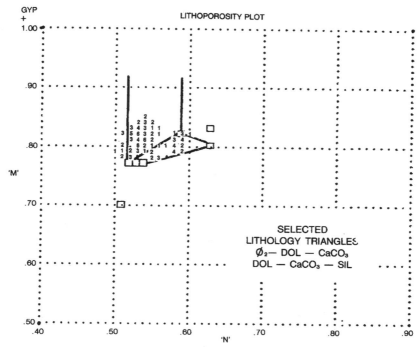

FIGURE 3–23 Lithoporosity plot (after Burke et al, 1969)

Myung (1976) proposed an equation that combined the gamma radiation, bulk density, and resistivity as follows:

$$Production\ index = \frac{G + R}{\rho_B}$$ (3–3)

where: $G = \dfrac{\text{API units from gamma ray log}}{\text{average API units of shale}}$

$R = \dfrac{\text{Resistivity from induction log}}{\text{average resistivity of shale}}$

ρ_B = Bulk density from a density log

The average resistivity and API units of shale must be established based on local conditions. As an example, values of 20 ohm–m and 200 API units have been used in some cases for Devonian shales.

Figure 3–24 is a plot of the production index log from a West Virginia fractured well. Interval 3,406–3,656 ft which shows a high production index, proved to be highly fractured by other means. Various cases like the one presented in Figure 3–24 suggest that the production index is closely related to the degree of fracturing of the formation. Figure 3–24 also shows the kerogen content of the rock in gallons/ton. This is important because, although kerogen is not considered of commercial value, it is believed to be in the vicinity of gas source beds. An estimate of kerogen content or yield can be obtained from:

$$Yield = 496.325(\rho_B)^{-0.6} - 285.176$$ (3–4)

where the kerogen yield is in gal/ton.

TEMPERATURE LOG

This a valuable tool for determining gas entry into the borehole. When gas enters the well bore, there is a drastic deflection of the curve toward lower values of temperature. Figure 3–25 shows an example of a temperature log run in the lowlands of Quebec. The zone of low temperature tested at initial rates as high as 3 MMscfd.

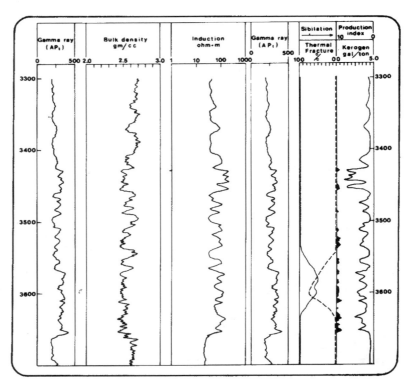

FIGURE 3-24 Production index log in a fractured West Virginia well (after Myung, 1976)

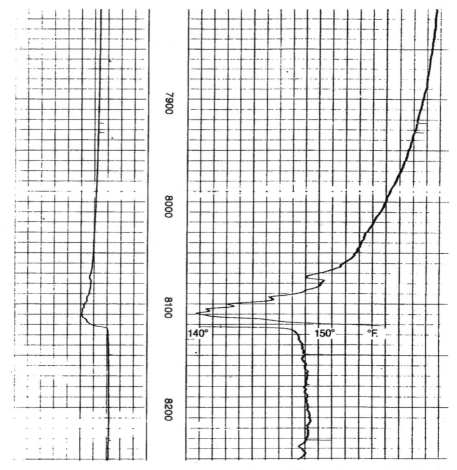

FIGURE 3-25 Temperature log showing fractured interval in St. Lawrence lowlands of Quebec.

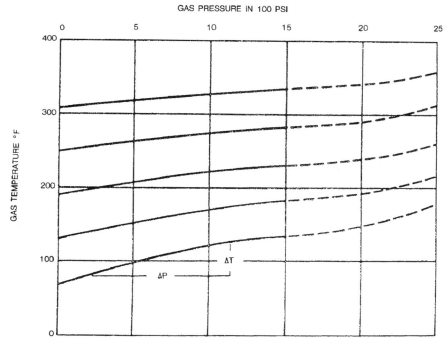

FIGURE 3–26 Joule-Thompson effect: gas temperature change due to expansion (after Myung, 1976)

Myung (1976) has used the concept of Joule-Thompson expansion temperature to determine the thermal fracture index of a formation from a temperature log. Figure 3–26 shows the theoretical Joule-Thompson effect, i.e., the gas temperature change due to expansion. The zero in the thermal fracture index scale is determined by the gas temperature due to expansion. The 100% in the thermal fracture index scale is the gradient temperature. The higher the thermal fracture index, the higher the degree of fracturing of the formation.

Figure 3–27 shows two temperature logs run in the same hole before and after hydraulic fracturing of the formation. The temperature in the zone of interest is higher following stimulation due to heating of expanding gas in the increased fracture system. A zero thermal fracture index is assigned to the theoretical Joule-Thompson temperature (50°F.) determined from Figure 3–26. A 100% thermal fracture index is assigned to the normal geothermal gradient at the depth of interest. This indicates that the thermal fracture index before fracing was about 30% and after fracing, 80%.

In the case of oil reservoirs, if open fractures are present and mud losses have occurred, the temperature log will show a cooling effect due to mud invasion in the fracture interval. The log is recorded going down soon after the last calculation so that the temperature profile is not disturbed. If only small anomalies are present, it is worthwhile to run a log later on that hopefully will be closer to thermal equilibrium. Comparison of the two temperature logs will indicate where the fractures are located.

SIBILATION LOG

This log, also known as noise log, has been extended from the location of collar leaks, channelling, and other types of leaks in gas storage wells to the location of small, potential gas-producing zones in fractured systems (Myung, 1976). A sibilation log consists of two piezoelectric transducers, one of high sensitivity and one of low sensitivity. Their frequency response is on the order of 40Khz. This allows the instrument to detect gas movements at the borehole wall, while ignoring most noises caused by the instrument and gas movement in the borehole past the instrument.

Figure 3–28 indicates a potential gas-bearing zone at 3278 ft. Notice that the temperature log does not show any anomaly at this depth. This might be an indication that gas is

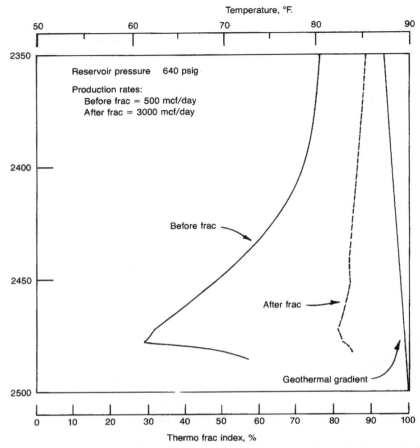

FIGURE 3–27 Temperature logs, before and after fracturing (after Myung, 1976)

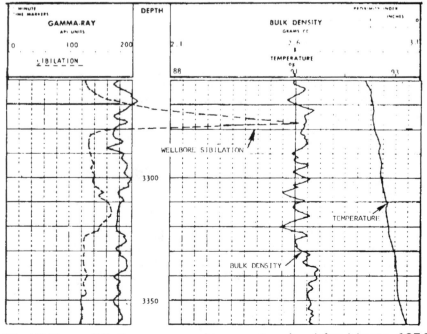

FIGURE 3–28 Wellbore sibilation survey and temperature log (after Myung, 1976)

not actually entering the well bore, but rather that gas is moving in the formation beyond the borehole.

KEROGEN ANALYSIS

This analysis can be used for the evaluation of Devonian shales. The idea is to account for the volume of kerogen present in the formation and to come out with smaller, but more realistic, porosities. The logging program leading to kerogen analysis consists of gamma ray, FDC density, CNL (on SNP) neutron, BHC sonic, and dual laterolog or dual-induction laterolog (Hilton, 1977).

For evaluation, matrix characteristics of kerogen, silt, and clay must be known or assumed based on experience in the area. Based on Hilton's research, some typical values used in kerogen analysis are:

Device	Silt	Clay	Kerogen
Sonic (μsec/ft)	55.50	85.00	174
Density (gm/cc)	2.68	2.82	1.10
SNP ϕ (Lm)	–0.02	0.27	0.67
CNL ϕ (Lm)	0.14	0.315	0.67

With the values of the previous table, it is possible to generate various crossplots as in Figure 3–29. Volumes of kerogen are determined from these crossplots by dividing the distance between the data point (x) and the silt-clay line by the total distance from the kerogen point to the silt-clay line, or

$$\text{kerogen value} = \frac{\text{distance from data point (x) to silt–clay line}}{\text{distance from kerogen point to silt–clay line}}$$

As different kerogen values may be obtained from crossplots similar to the ones presented in Figure 3–29, select the lower of the two kerogen values. The data point can then be shifted to the silt-clay line along the line connecting the kerogen point and the silt-clay line to obtain kerogen-free data that can then be analyzed by more conventional computer programs.

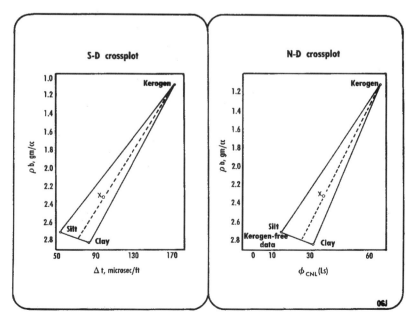

FIGURE 3–29 Sonic-density and neutron-density crossplots in Kerogen-bearing shale (after Hilton, 1977)

GAMMA RAY

In some cases radio isotopes might help to locate fractured intervals, including the number of fractures intercepted, fracture spacing and the vertical extent of a vertical fracture. In this procedure a base gamma ray of the formation of interest is run having in the hole the fluid used to drill the well. Then a radio-active tracer is introduced in the mud and this is followed by circulation. The idea is that the fractures will be invaded by radioactive mud. This is followed once again by circulation with a conventional mud and by a new gamma ray log. The logging is carried out with the tool coming up the hole.

An increase in gamma ray activity above the base log is interpreted as a fracture invaded by the radioactive mud. The same approach has been used in the case of hydraulic fractures. Based on field and model studies, the following criteria has been given by Mihram (1964) to distinguish vertical from horizontal fractures in the case of hydraulic fracturing jobs:

1. A horizontal fracture exists when a is less than two ft, and b and c are each less than three ft in Figure 3–30.
2. A vertical fracture exists when a is larger than two ft. Figure 3–31 shows an example of a base gamma ray log and a new gamma ray log run after a hydraulic fracturing job where radioactive beads were used as tracers. The shaded area shows additional radiation due to the beads injected with the fracturing materials.

Another technique along the same lines presented above using spectral gamma log analysis for determination of fracture height has been presented by Anderson et al (1986).

CIRCUMFERENTIAL ACOUSTICAL LOG

Acoustical devices which use sound waves which propagate circumferentially in a horizontal plan include the circumferential acoustic device (Vogel and Herolz, 1981), and circumferential microsonic tool (Koerperich, 1975; Setser, 1981).

In the case of a microsonic tool, transducers mounted in pads on four arms of a sonde are applied to the wall of the borehole. Two of the pads, 180° apart, house the transmitters. The other two pads house the receivers. Measurements include an amplitude for the received signal in each quadrant, transit times, tool orientation and borehole diameters along the two perpendicular directions. The circumferential acoustical device uses transmitters and receivers which are cylinders several wavelengths long.

Wave forms are recorded on tape for example every three in. This can be presented individually or side-by-side in variable density log or waveform format. The log appears to be valuable for detection of vertical or high angle fractures. However, care has to be exercised because it is difficult to say what are the amplitude alterations which are due to

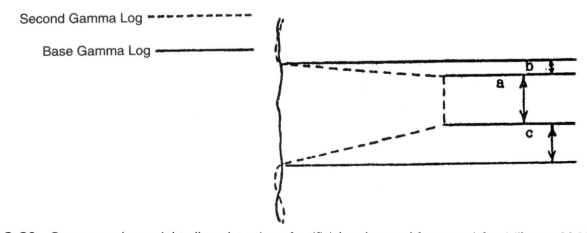

FIGURE 3–30 Gamma ray logs might allow detection of artificial and natural fractures (after Mihram, 1964)

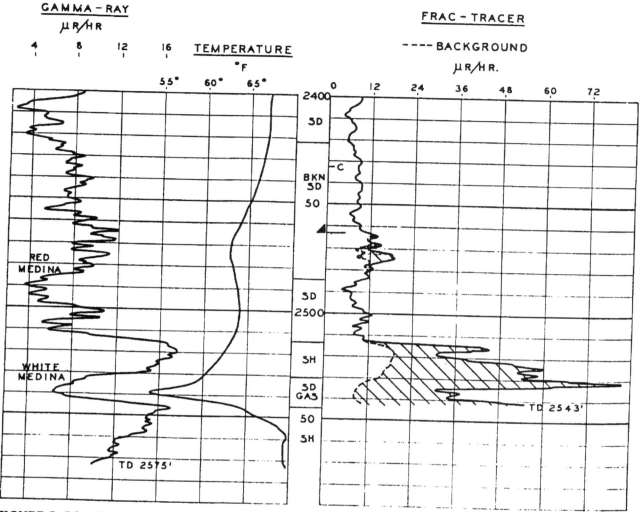

FIGURE 3–31 Comparison of gamma ray logs following a hydraulic fracturing job where radioactive beads were used as tracers (after Demprey and Hickey, 1958)

non-fracture effects. Delhomme et al (1982) have indicated, that in the average, the circumferential acoustic tool seems to yield more fracture indications than the high resolution dipmeter. Figure 3–32 shows an example of circumferential acoustical logs (CAD) obtained in a West Texas well containing vertical saw cuts which are shown by the reductions in amplitude of the CAD guided fluid wave and the CAD shear wave. Q1, Q2, Q3, and Q4 in the log heading represents the quadrant number. A borehole televiewer on the left hand side of the figure shows the saw cuts. The numbers inside circles allow correlation of the logs.

DUAL LATEROLOG-MICROSPHERICALLY FOCUSED LOG

This combination has proven very useful in various naturally fractured reservoirs around the world, as in many instances it is possible to see large cross-overs and separations between the two laterologs in front of the fractured intervals detected by low resistivity readings in the microspherically focused log.

Qualitative Estimates of Fracture Intensity

Results presented by Rasmus (1983) indicate that a crossplot of laterolog deep resistivity (R_{LLd}) divided by laterolog shallow resistivity (R_{LLs}) versus (R_{LLd}) might give qualitative

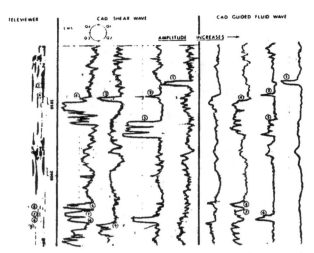

FIGURE 3–32 CAD logs from a well with vertical saw cuts (after Vogel and Herolz, 1981)

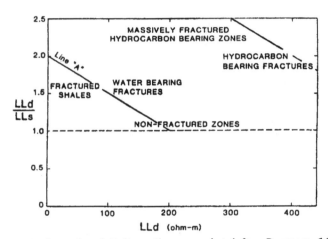

FIGURE 3–33 Interpretation of resistivity-ratio crossplot (after Rasmus, 1982)

information with respect to water-bearing and hydrocarbon-bearing intervals and degree of fracturing. The method is limited to reservoirs with very low primary porosities and permeabilities and to reservoirs where the matrix water saturation is close to 100%. This would be the case of reservoirs of type C or B–II as discussed in Chapter 1.

The principle behind the method is illustrated with the use of Figure 3–33. The water-bearing fractured intervals would fall along line "A". Non-fractured zones would be characterized by equal deep and shallow laterolog resistivities, i.e., $R_{LLd}/R_{LLs} = 1$, as shown by the dashed line. Massively fractured hydrocarbon-bearing zones would plot to the northeast of the water-bearing line, and hydrocarbon-bearing fractures with a lesser degree of fracturing would plot to the east of the water-bearing trend. Fracture shales plot to the left side of line "A". A master plot would have to be generated empirically for each reservoir. The values of R_{LLd} shown on the abscissa are presented for illustration purposes only, and the position of the water-bearing line will change from reservoir to reservoir.

Figure 3–34 shows a dual laterolog for the Twin Creek limestone, overthrust belt, Wyoming. The laterolog shallow reads a smaller resistivity in the intervals thought to be fractured. This requires to have a mud of relatively low resistivity in the wellbore which hopefully has invaded the fracture. Also shown is a gamma ray in a scale from 0 to 160 API units, a caliper (scale six to 16 in.), a ratio or R_{LLd}/R_{lls} curve (scale minus 1.5 to 3.5), and a weight curve (scale 10 to zero).

The weight curve is generated empirically based on the plot presented on Figure 3–35. A weight of zero is assigned to points falling on line "A" and a weight of 10 is assigned to

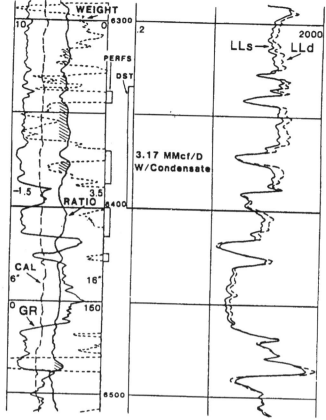

FIGURE 3–34 DLL with crossplot weight, and ratio (after Rasmus, 1982)

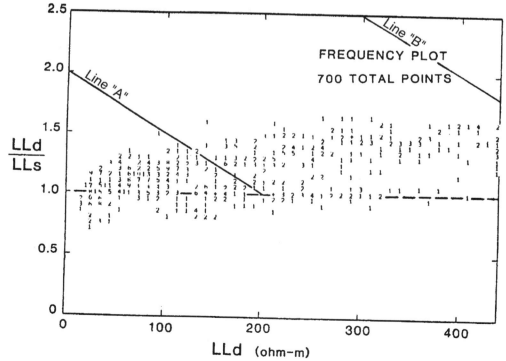

FIGURE 3–35 Fractured, clean, 0% to 3% porosity. Initial production of 2.2 MMcf/d, 1410 bopd (after Rasmus, 1982)

points falling on line "B". The weight of intermediate points is prorated and the result is presented as the weight curve on Figure 3–34. Fracture hydrocarbon-bearing intervals have a weight greater than zero and a resistivity ratio greater than one. Intervals where the weight curve crosses over the ratio curve are potentially highly permeable hydrocarbon-bearing fractured zones (One example is interval 6375–80 ft in Figure 3–34).

During a DST this well produced 3.17 MMscf/d of gas with condensate and mud. Following completion and perforation the well produced 2.20 MMscf/d with 1416 barrels of condensate per day. This production decrease might be the result of damage induced during the completion process (for example cement going into the fractures) or a perforating job which did not contact properly all the fractured intervals. This last explanation might be reasonable as interval 6344–6365 which is fractured according to the dual laterolog was not perforated.

Quantitative Estimates of Fracture Intensity

Under some conditions it is possible to extend the qualitative method presented above to the calculation of fracture porosity. This can be done by assuming that the two laterologs respond to matrix and fracture systems as if they were connected in parallel. If furthermore, it is assumed that only the fractures (not the matrix) are subjected to mud invasion, the following equations can be written (Boyeldieu and Winchester, 1982):

Deep Laterolog:

$$\frac{1}{R_{LLd}} \geq \frac{\phi_b^{m_b} S_{wf}^{n_f}}{R_w} + \frac{\phi_2^{m_f} S_{wf}^{n_f}}{R_w} \qquad (3–5)$$

Shallow Laterolog:

$$\frac{1}{R_{LLs}} \leq \frac{\phi_b^{m_b} S_{wb}^{n_b}}{R_w} + \frac{\phi_2^{m_b} S_{xof}^{n_f}}{R_{mf}} \qquad (3–6)$$

where $\quad \phi \quad = \quad$ porosity, fraction
$\qquad S \quad = \quad$ saturation, fraction
$\qquad R \quad = \quad$ resistivity, ohm-m
$\qquad m \quad = \quad$ porosity exponent
$\qquad n \quad = \quad$ water saturation exponent

and the subscripts are LLd = laterolog deep, LLs = laterolog shallow, b = matrix block, 2 and f = fracture, w = water and mf = mud filtrate.

Subtracting Equation 3–5 from Equation 3–6 we obtain:

$$\frac{1}{R_{LLs}} - \frac{1}{R_{LLd}} \leq \phi_2^{m_f} \left[\frac{S_{xof}^n}{R_{mf}} - \frac{S_{wf}}{R_w} \right] \qquad (3–7)$$

If the assumption is made that $S_{wf} = 0$ and $S_{xof} = 1$, Equation 3–7 can be solved to calculate fracture porosity as follows:

$$\phi_2 = \left[R_{mf} \left[\frac{1}{R_{LLs}} - \frac{1}{R_{LLd}} \right] \right]^{1/m_f} \qquad (3–8)$$

If the assumption is made that $S_{wf} = 1$ and $S_{xof} = 1$, Equation 3–7 can be solved to calculate fracture porosity as follows:

$$\phi_2 = \left[\frac{\left(\frac{1}{R_{LLs}} - \frac{1}{R_{LLd}} \right)}{\left(\frac{1}{R_{mf}} - \frac{1}{R_w} \right)} \right]^{1/m_f} \qquad (3–9)$$

Fracture porosity from Equations 3–8 and 9 can be considered as a lower limit. The porosity exponent of the fracture, m_f, is usually close to 1.0. In some instances, however, it might be slightly larger. As an example, Aguilera (1974) and Towle (1962) have reported values of m_f ranging between 1.1 and 1.2 for a theoretical model made out of cubes with spaces in between, where the spaces represent the fractures and the cubes represent the matrix. Draxler and Edwards (1984) have determined in the laboratory m_f values of 1.0 at room conditions and 1.25 at a net overburden of 8000 psi for carboniferous fractured reservoirs of northern Europe. Boyeldieu and Winchester (1982) have cited statistical evaluations which suggest m_f values in the order of 1.3 and going all the way up to 1.5.

Figure 3–36 shows gamma ray, microspherically focused, and dual laterolog of a tight carbonate reservoir. The log comes from a hard limestone containing gas. The mud filtrate resistivity was equal to 0.18 ohm-m at reservoir temperature. Fracture porosity of zone A was calculated to be ≥1.27% from Equation 3–8 using R_{LLd} = 115 ohm–m, R_{LLs} = 60 ohm–m, and a fracture porosity exponent, m_f, of 1.5. This larger value of m_f comes from an observation of larger tortuosity than usual due to recrystallized elements within the fractures. Care must be excersized with respect to the selected value of the fracture porosity exponent as small changes in m_f lead to drastic changes in fracture porosity. This is illustrated in the following table with the use of the same data discussed above.

m_f	$\phi_2(\%)\geq$
1.0	0.14
1.1	0.26
1.2	0.43
1.3	0.65
1.4	0.93
1.5	1.27

Unless there is some additional supporting evidence I recommend to use an m_f of 1.0 to avoid ending up with very high and unrealistic values of fracture porosity. Shankland and Waff (1974) have indicated on the basis of a single cubic array of 2340 resistors that a value m equal to one can be associated with open cracks. Note in Figure 3–36 that the fracture intervals are clearly indicated by the low resistivity readings in the microspherical focused log and the separation between the two laterologs. The above laterolog methods are valid only in the case of vertical or high inclination fractures. In horizontal or low angle fractures the separation might be reversed with the shallow giving higher resistivities than the deep laterolog.

P_e CURVE ON THE LITHO-DENSITY LOG

The P_e curve is an index of the effective photo-electric absorption cross section of the formation. The P_e of barite is 267 as opposed to low values of P_e for quartz (1.81), calcite (5.08), dolomite (3.14), water (0.734 for 200,000 ppm N_aCL water), oil (0.119) and gas (0.095). This makes the P_e curve on the litho-density log an excellent fracture indicator in wells drilled with mud containing barite. Figure 3–37 is an example of a litho-density log including a caliper, a compensated density log with $\Delta\rho$ correction and a P_e curve. A fracture interval is clearly delineated by the increase in P_e reading. Note that the borehole is in reasonably good shape. However, there is a deflection of the $\Delta\rho$ curve which provides support to the interpretation regarding the fracture.

SONIC POROSITY GREATER THAN NEUTRON-DENSITY POROSITY

The usual wisdom and some of the methods discussed in this chapter assume that neutron and density logs read total porosity, while sonic logs give only matrix porosity. There are some instances, however, in which the sonic log might give higher porosities than the neutron and density. This can happen, for example, in the heavily fractured Monterey Shale of California. A plausible explanation for the larger transit times is the presence of low angle or horizontal fractures, with multiple orientations, uncemented or uncompacted sands and clayey silts.

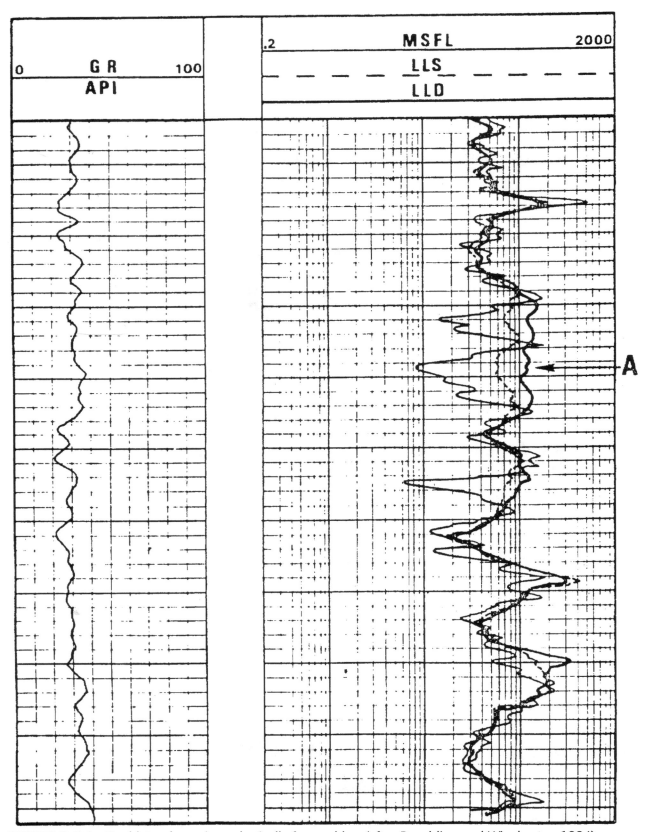

FIGURE 3–36 Dual laterolog-microspherically focused log (after Boyeldieu and Winchester, 1984)

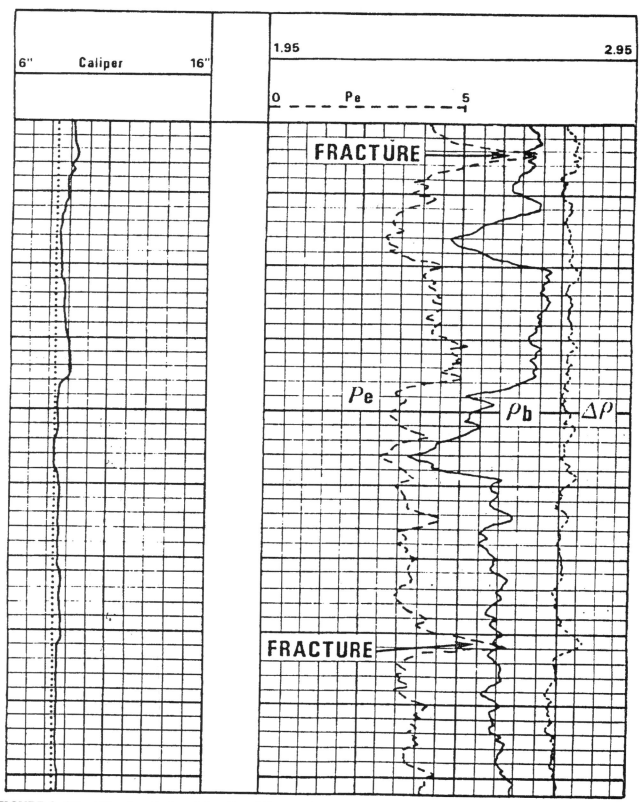

FIGURE 3–37 Lithodensity log (after Boyeldieu and Winchester, 1984)

In these cases it is possible to identify the fractured intervals by using the following procedure (Cannon, 1979):

1) Calculate a total porosity from neutron and density logs. This can be done from crossplots presented on Figure 3–66 or it can be approximated from the equation:

$$\phi = \left(\frac{\phi_N + \phi_D}{2} \right) \tag{3–10}$$

for the case of oil, and

$$\phi = \sqrt{\frac{\phi_N^2 + \phi_D^2}{2}} \tag{3–11}$$

for the case of gas. ϕ_N and ϕ_D in Equations 3–10 and 3–11 stand for neutron and density porosity, respectively.

2) Calculate the sonic transit time equivalent (Δt_{me}) to the above total porosity from the equation:

$$\Delta t_{me} = \frac{\Delta t - \phi \Delta t_f}{1 - \phi} \tag{3–12}$$

where Δt is sonic transit time and the subscript f stands for fluid. If the Δt_{me} value calculated from Equation 3–12 is larger than the matrix sonic time for the lithology under consideration probably there is a fracture at the depth under investigation. Another way to see the resolution of the problem is by saying that the sonic porosity calculated with the use of a conventional Δt_m is larger than the total porosity from neutron and density logs.

MUD LOG

A good record of hydrocarbon evidence in mud and cuttings and a log of drilling rate can provide valuable information for locating natural fractures. During a drilling operation, a small or large degree of flushing can occur depending on various properties including rock porosity and permeability, depth, hole size, mud density, reservoir fluid pressures and volume of mud being circulated. In many cases, the cuttings will retain some of the interstitial fluids. However, during the trip from the bottom of the hole to the surface some of the fluids originally within the cuttings will flow into the mud stream due to the mud pressure being reduced from hydrostatic to atmospheric. In cases where permeability is high most of these fluids will flow into the mud within the last 600 ft. There are consequently two sources that can provide information with respect to the presence of hydrocarbons: the cuttings and the mud.

The cuttings when analyzed will release fluids contained within the pore space by the chopping and beating action of the blender. At this stage probably only the heavier liquid hydrocarbons will remain in the cuttings. A "cuttings gas detector" will usually be efficient to detect hydrocarbons up to about C_{18}, giving thus information with respect to oil content. Analyses under ultra-violet light to detect fluorescence, leaching, color and staining of the cutting are put together on the mud log to give qualitative estimates of oil content.

The "hot wire log" of gas hydrocarbons and the "gas percent log" obtained conventionally from the aeration gas trap and the gas chromatograph are useful for indicating potential hydrocarbon reservoirs. In general, a large amount of methane will be a positive qualitative indicator. In some instances, however, abnormally high values of C_1 might be a result of salt water-bearing formations.

Results presented by Ferrie et al (1981) indicate that it is the composition rather than the magnitude of the gas show the key parameter in terms of hydrocarbon productivity. An empirical useful approach for determining the potential productivity involves the calculation of the ratios C_1/C_2, C_1/C_3, C_1/C_4, and C_1/C_5 (net excluding background gas).

If the ratios are smaller than two, the formation can be anticipated to be non-productive. If the C_1/C_2 ratio is between about two and 15 and there is a continuous increase in

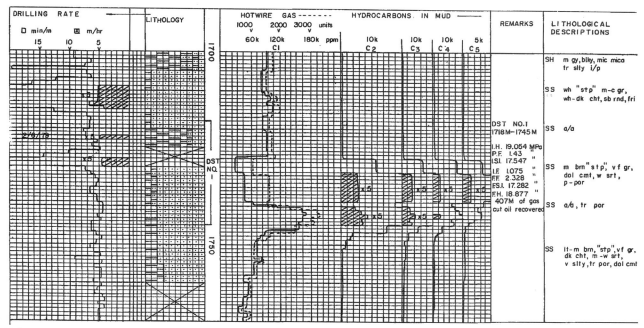

FIGURE 3–38 Section of PPM log used in conjunction with Figure 3–39. The ppm log is obtained through a quantitative analysis of the hydrocarbons in the drilling fluids, specifically parts per million of the lighter hydrocarbons in the fluids. The total ppm log is the summation of the following factors: a drilling rate column which plots penetration rate inversely to the gas, on a linear scale, a visual porosity column, a lithology column which shows either an interpretive or percentage lithology; five columns showing C_1 through C_5 in ppm; a remarks column; and finally a column detailing lithological descriptions (after Ferrie et al, 1981)

the successive ratios, the formation is probably oil-bearing. If the ratio is lower than the preceding ratio, for example, if C_1/C_3 is lower than C_1/C_2 the formation is probably water-bearing (an exception occurs with C_1/C_5 when oil is used in the mud). A C_1/C_2 ratio between 15 and 60 increasing continuously for successive ratios might be indicative of a gas zone. Finally, abnormally high values of C_1/C_2 might be indicative of a non-productive zone.

Figure 3–38 shows a mud log including drilling rate in meters/hour, lithology, hot wire gas, hydrocarbons in mud in parts per million of the lighter hydrocarbons, remarks with respect to a DST which tested an oil-bearing zone and some lithological descriptions. The evaluation of this mud log was carried out by Ferrie et al (1981) as shown on Figure 3–39.

The dashed lines on Figure 3–39 show the empirical ratio limits for non-productive, gas and oil zones. For this example, plot #1 was generated using data from the mud log. The results indicate probable oil production which was corroborated during the DST as 407 meters of gas-cut oil were recovered. Plot #2 on Figure 3–39 is based on experiments carried out on some of the gas obtained during the DST. The comparison with Plot #1 is good.

In some naturally fractured reservoirs a good interval might be detected by the combination of an increase in gas shows and an increase in drilling rates. Mud losses also occur frequently in fractured intervals.

VELOCITY RATIO PLOT

This method allows in some instances the location of natural fractures in formations which have solid to solid contact in the fracture face. The velocity ratio (VR) is calculated with the use of the equation:

$$VR = \frac{\Delta t_s}{\Delta t_c} \tag{3–13}$$

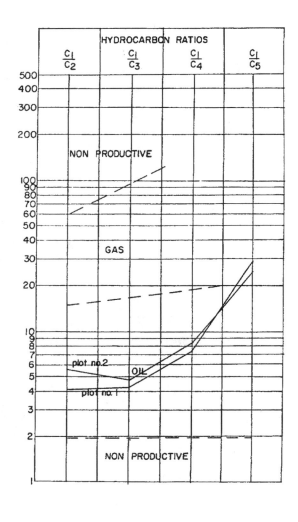

MUD GAS LOG
SHOW EVALUATION REPORT

COMPANY _____

WELL ___EXAMPLE NO. I_____

LOCATION ___72-7 W6M_____

FORMATION ___BASAL GETHING_____

DEPTH ___plot no.I (1743m)___ plot no.2 (1718m - 1745m)___

PROBABLE PRODUCTION ___OIL_____

MUD ANALYSIS

	BACKGROUND	NET INCREASE	
		PLOT I	PLOT 2
		DEPTH_____	DEPTH_____
METHANE – kC$_1$		170k	63.7 %
ETHANE – kC$_2$		42	11.4
PROPANE – kC$_3$		40	13.3
BUTANE – kC$_4$		23	7.6
PENTANE + – kC$_5$+		6k	2.5 %
METHANE % _ UNITS _			
TOTAL % _ UNITS _			
CHLORIDES			

PLOT I

C_1/C_2 = _4.1_ C_1/C_3 = _4.3_ C_1/C_4 = _7.4_ C_1/C_5 = _28.3_

PLOT 2

C_1/C_2 = _5.6_ C_1/C_3 = _4.8_ C_1/C_4 = _8.4_ C_1/C_5 = _25.5_

This plot illustrates the relationship between a chromatographic analysis taken from two different sources. Plot no. I is the ratio taken from the mud stream, on wellsite, while drilling is in operation. Plot no. 2 is the ratio, taken from the formation, of the gas sampled from the Drill Stem Test. The two plots are taken from the lower section of the Basal-Gething. The interesting point to note, is that plot no I was obtained while drilling was in operation. Within reason, the quality of gas, oil, salt water, etc. can be determined before an expensive testing operation is run. Lithological data is determined, at the same time, by the wellsite geologist; therefore, if money or time is a factor, in well completion, this method might be a viable alternative to expensive testing or logging runs, while drilling

FIGURE 3–39 Mud gas log, show evaluation report (after Ferrie et al, 1981)

where Δt is transit time in μsec/ft and the subscripts s and c stand for shear and compressive, respectively. Pickett (1963) has shown that VR is constant for a given lithology independent of porosity. This ratio is 1.6–1.7 for sandstones, 1.8 for dolomite and 1.9 for limestone.

Compressive Transit Time

Compressive transit time is determined preferentially from a sonic log. When this log is not available Δt_c can be estimated from a microseismogram or VDL with the use of the equation:

$$\Delta t_c = \frac{t_c - \frac{\Delta t_f(d - d_t)}{12} - TD}{TS}$$

(3–14)

where t_c = time to compressive wave, μ sec
d = borehole diameter, inches
d_t = tool diameter, inches
TD = tool delay, μsec
TS = tool spacing, feet
Δt_f = fluid transit time, μsec/ft

Shear Transit Time

Shear transit time is given by:

$$\Delta t_s = \frac{t_s - t_c}{TS} + \Delta t_c$$

(3–15)

where t_s is time to shear wave in μsec and other nomenclature has been defined previously.

Figure 3–40 shows a typical four ft spacing response of a microseismogram in the Mancos B formation (Leeth and Holmes, 1978). Compressional (Δt_c) transit time was determined to be 75 μsec/ft at 3200 ft from a sonic log. Shear transit time is calculated to be 125 μsec/ft from Equation 3–15 by using TS = 4 ft, and by reading from the microseismogram $t_c = 450$ μsec and $t_s = 650$ μsec. Finally, the velocity ratio is calculated from Equation 3–13 to be VR = 125/75 = 1.67. The cutoff for productivity has to be determined empirically for each area. The smaller the value of VR, the larger the probabilities of success. It must be stressed that there are cases in which natural fractures without acoustic continuity are recognized because there are drastic breaks in the bandings presented by the dark (positive half cycles) and light (negative half cycles) shades in a microseismogram. In cases where solid to solid contact in the fracture face is suspected, however, the present method is worthy of testing.

Shear or Rigidity Modulus

Another approach that can be introduced consists of calculating shear or rigidity modulus, i.e., the resistance to change in shape (G), and the bulk compressibility (C_b) from the equations (Tixier et al, 1973):

$$G = 1.34 \times 10^{10} \frac{\rho_b}{\Delta t_s^2}$$

(3–16)

and

$$C_b = \frac{1}{1.34 \times 10^{10} \rho_b \left[\frac{1}{\Delta t_c^2} - \frac{4}{3 \Delta t_s^2} \right]}$$

(3–17)

where ρ_b is bulk density in gr/cc and other nomenclature has been defined previously. For the same example described above G is calculated to be 2.27×10^6 psi based on $\rho_b = 2.65$ gr/cc and $\Delta t_s = 125$ μsec/ft. Bulk compressibility is determined to be 0.3×10^{-6} psi^{-1}.

Weak formations might show a rigidity modulus (G) as low as 0.4×10^6 psi, whereas, well compacted formation may have a value in excess of 1.6×10^6 psi. Bulk compressibility (C_b) may range between 1.3×10^{-6} psi for weak formations and 0.25×10^{-6} for very strong ones. Equations 3–16 and 3–17 have been used mainly in reservoirs with poor consolidation to try to predict whether or not a well will have sanding problems. In some reservoirs it might be possible to detect natural fractures based on variation in G and C_b along the borehole.

PRODUCTION LOGS

These tools are designed for measuring the performance of producing and injecting wells. In very tight reservoirs with natural fractures this log might be instrumental in determining where the hydrocarbons are coming from, the distance between fractures and the fraction of the total production contributed by each fracture. Some of the instruments usually

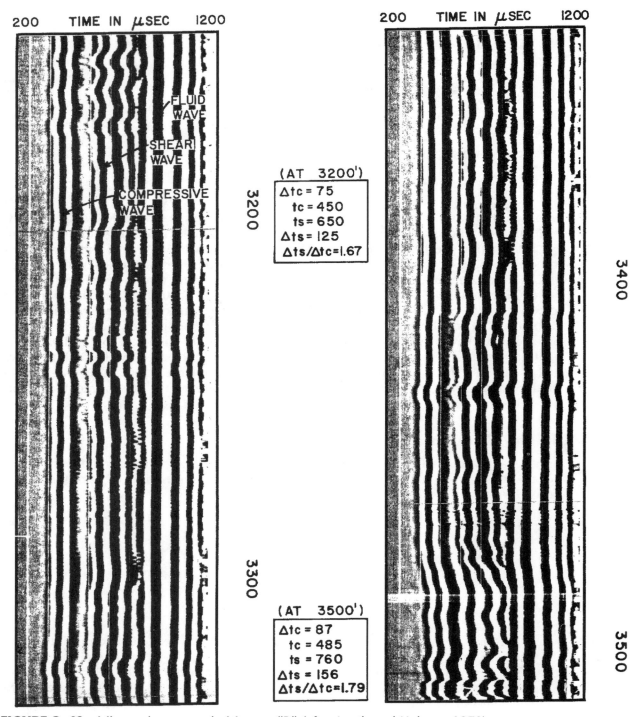

FIGURE 3–40 Micro-seismogram in Mancos "B" (after Leeth and Holmes, 1978)

considered as production logging tools include a thermometer, gradiomanometer, full-bore-spinner flowmeter, continuous flowmeter, packer flowmeter, manometer, caliper, water-holdup meter and radioactive-tracer survey (Schlumberger, Production Log Interpretation, 1973). Figure 3–41 shows an example of a production log in a naturally fractured reservoir. Table 5–7 shows results of flowmeter surveys in the Agha Jari field in Iran. Results indicate that oil enters through many fractured zones in this limestone reservoir. Results presented by Baker an Lukas (1972) (See Chapter 2, Comparison of Open Hole and Perforated Completions) show how production logging indicated that only a few of the

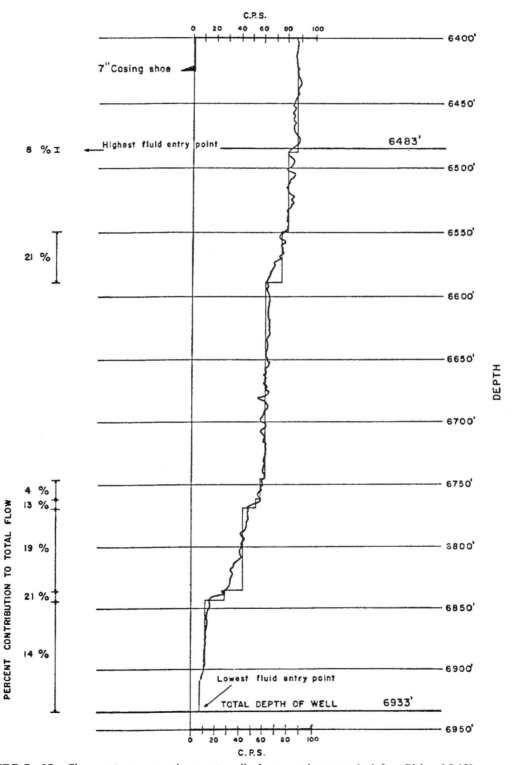

FIGURE 3–41 Flowmeter survey in a naturally fractured reservoir (after Birks, 1963)

net 135 ft perforated intervals in well Bleazard 1–18 B4, Altamont trend, Wasatch Formation, were actually producing although conventional well log analysis indicated additional potential pay.

Thermometer. It measures temperatures of the fluids in the borehole under either dynamic (flowing) or static (shutin) conditions. Temperature logs are powerful for identifying fracture intervals when run in empty hole. Some examples are presented in Figures 3–25 and 3–27.

Gradiomanometer. It measures a pressure differential over two ft of borehole and relates it to the mean density of the fluids in the borehole (ρ_f), a kinetic component (k), and a friction component (F), through the use of the equation:

$$\rho_{gr} = \rho_f (1 + k + F) \qquad (3\text{--}18)$$

where ρ_{gr} is the gradiomanometer reading in gr/cc. In addition to the ρ_{gr} curve the gradiomanometer log includes an unscaled amplified measurement with a 0.2 gr/cc sensitivity per log track, i.e., five times the sensitivity of the ρ_{gr} curve. In the case of deviated holes, an additional correction must be introduced as follows:

$$\rho_{corr} = \rho_{gr}/\cos\theta \qquad (3\text{--}19)$$

where ρ_{corr} is the corrected density in gr/cc and θ is the hole deviation of the logged section. This tool is very useful for computing water holdup.

Fullbore-Spinner Flowmeter. It gives readings of fluid velocities in revolutions per second (rps) which can be translated into volumetric flow rates (q) in barrels per day with the use of the equation:

$$q = 1.40\, vd^2 \times (VCF) \qquad (3\text{--}20)$$

where v is fluid velocity in ft/sec, d is diameter in inches read from a caliper log and VCF is a velocity correction factor usually in the order of 0.83. The velocity calibration for water based on laboratory experimentation is 4.70 rps per 100 ft/min.

This tool is useful down to low rates when the flow is monophasic. For polyphasic flow, larger rates are required. Some of the lower limits given by Schlumberger are:

Casing Size (in)	b/d		
	W	W/O/G	W/G
5½	20	200	300
7	30	400	600
9⅝	60	800	1200

W, O and G in the above table stand for water, oil and gas, respectively. Equation 3–20 is not valid where the total flow rate is below the limit values indicated in the above table.

Continuous Flowmeter. It gives fluid velocities which can be translated into volumetric rates with the use of Equation 3–20. It can be run in smaller casing diameters than the spinner flowmeter. However, higher rates are required for sound evaluation as shown in the following table:

Casing Size (in)	b/d		
	W	W/O/G	W/G
3½	40	1,500	3,000
5	100	2,500	5,000
6⅝	150	5,000	10,000

The tool has found its main applications in monophasic flow regimes, including waterfloods, high flow-rate gas wells, and high flow-rate oil wells.

Packer Flowmeter. It has found application for rates between 10 and 1900 b/d. In this approach all fluids are forced through the spinner assembly following inflation of a packer to fill the annulus between the tool and casing. The fluid velocity measured by the spinner is translated into a volumetric rate with the use of Equation 3–20.

Monometer. Usually it is available for ranges between 0 and 5000 psi or 0 and 10,000 psi. The manometer measures the pressure of the fluid in the borehole. It is valuable for open flow potential in gas wells and productivity index test in oil wells.

Caliper. It measures the diameter of the hole or the casing. The accuracy of the tool is ± 0.2 in. Hole enlargements in some cases are indicative of fractures. If the hole is in gauge and there are drastic corrections in the $\Delta\rho$ curve in a compensated density log, there might be some natural fractures in that horizon.

Radioactive-Tracer Survey. It provides good information up to 6000 psi and 300°F. This survey allows determination of fluid velocity in flow regimes where there is only one phase flowing. In some cases it permits detection of fluid movement outside casing or tubing.

FRACTURE PLAUSIBILITY

In this method, tool responses to fractures are divided into five broad categories (Boyeldieu and Martin, 1984):

1. Electrical category which includes dual laterolog, induction, microlaterolog, invasion diameter, and microspherical log.
2. Multi-pad category which includes high resolution dipmeter (HDT) and stratigraphic high resolution dipmeter (SHRD).
3. Radioactivity category which includes gamma ray and natural gamma ray spectroscopy.
4. Rugosity category including bulk density, $\Delta\rho$ correction, P_e curve, calipers and the electromagnetic propagation tool (EPT).
5. Acoustic category which includes secondary porosity index, compressional and shear waves and amplitude analysis.

A fracture "criterion" is selected for each one of the fracture indicators. This "criterion" is defined by an expression, a threshold, a median and a maximum plausibility. As an example, a dual laterolog might pinpoint a vertical or high angle fracture which has been invaded by a mud of relatively low resistivity. The larger the separation between the deep and shallow laterologs, the wider is going to be the opening of the fracture. Consequently, the dual laterolog "criterion" might be represented by the expression:

$$\log_{10} (R_{LLd}/R_{LLs})$$

When deep and shallow resistivities are equal, the assumption is made that no fractures are present and the above expression becomes equal to zero.

A threshold value for the above expression might be 0.05, indicating that this is the minimum value at which the expression is indicating a fracture plausibility. This corresponds to a resistivity ration R_{LLd}/R_{LLs} equal to 1.12. A median plausibility might be 0.10 indicating a resistivity ratio of 1.26. This median of 0.10 corresponds to a fracture plausibility which is equal to one half of the maximum plausibility ($P_{max}/2$) as shown on Figure 3–42.

A maximum plausibility might be 0.4 corresponding to a resistivity ratio R_{LLd}/R_{LLs} equal to 2.51. This maximum plausibility is reached when the difference between the expression and the threshold is infinite.

Some plausibility correction coefficients might be introduced in the analysis. As an example, for a larger value of porosity the fracture plausibility might become smaller. So a correction might be introduced to reduce P_{max} when porosity is larger than approximately 10%.

Another correction coefficient might be introduced in the case of resistivities. When R_t is small, the separation between R_{LLd} and R_{LLs} is not very significant. So for R_t values smaller than approximately 20 ohm-m, the maximum plausibility is zero. When R_t is larger than 100 ohm-m more conventional plausibilities, such as 0.05, 0.1 and 0.4 for threshold, media and maximum, respectively, might be introduced.

A correction coefficient might be used also in the case of fracture porosity as determined with the use of Equation 3–8. As an example, if fracture porosity is less than 0.1 percent, the maximum plausibility is reduced.

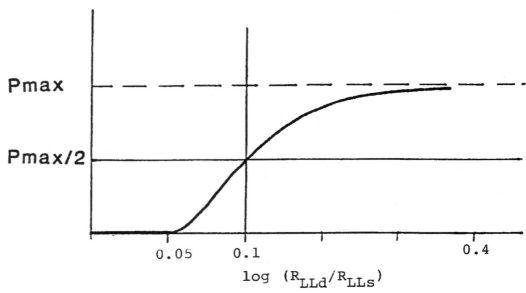

FIGURE 3–42 Fracture plausibility based on dual laterolog (after Boyeldieu and Martin, 1984)

Bayesian logic is used to combine the various plausibilities. For the case of two criterions having individual plausibilities P_1 and P_2, the combined plausibility will be:

$$P = 1 - (1 - P_1)(1 - P_2) \qquad (3\text{--}21)$$

Equation 3–21 is associative and can be used to combine any number of plausibilities. Boyeldieu and Martin (1984) have used it to obtain "category plausibilities" based on "criterion plausibilities" and to obtain a "final plausibility" based on "category plausibilities". Results are presented in a DETFRA (detection of fractures) computed log as shown on Figure 3–43. The first track shows fracture porosity as calculated from Equation 3–8 and total porosity. Track 2 shows plausibilities for the five different categories. Final fracture plausibility is shown on track 3 and formation analysis by volume is shown on track 4. Notice that there is good correlation between fractures as indicated in tracks 1 and 3.

FRACTURE EVALUATION IN CASED HOLES

In a few instances, acoustical methods such as the variable density log or microseismogram can be used for detection of fractures in cased holes. The method, however, is difficult to use with a good level of certainty due to its dependency on excellent cement bonding. In some areas, it has been possible to use spectral gamma ray logging devices such as the spectrolog with reasonably good success. However, because of its dependency upon uranium deposits in the fractures the method cannot be used in all areas.

A method that has provided reasonable results in Texas (Austin Chalk) and Michigan (Trenton Black River) is the combination of pulsed neutron (TDT) and gamma spectometry (GST) logs. The principle behind this method (Butsch, 1982; Sutton, 1984) indicates that the gamma ray will increase with shale content or radioactive precipitates in the fractures. As the shale content increases the total porosity will also increase. On the other hand, the value of sigma is corrected for background radiation and consequently it is not affected by any radioactive precipitates. By setting sigma equal to gamma ray in an unfractured interval it is possible to carry out the following qualitative analysis:

(1) In a shale both gamma ray and sigma values will increase.
(2) In clean intervals with low porosity both the gamma ray and sigma readings will decrease.

D E T F R A*

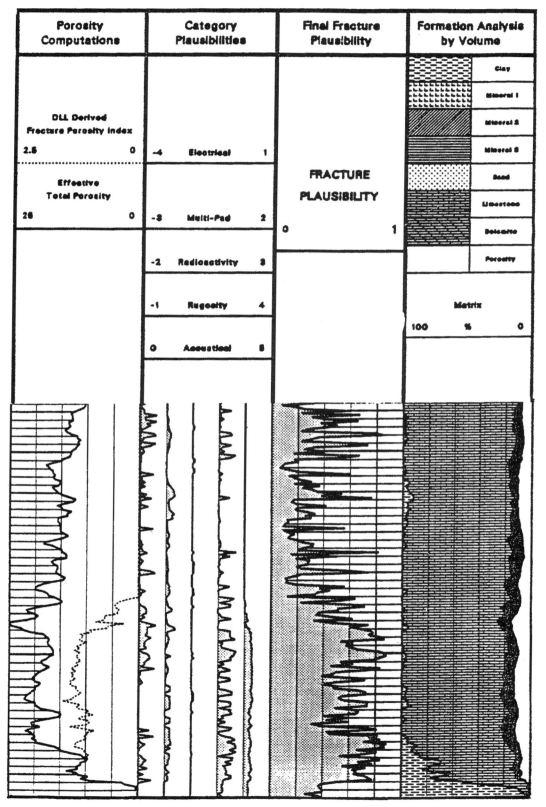

FIGURE 3–43 Computed plausibility log (after Boyeldieu and Martin, 1984)

(3) In a naturally fractured reservoir with radioactive precipitates the gamma ray would increase but sigma would be equal to its value in an unfractured interval.

(4) In a naturally fractured reservoir with hydrocarbons, sigma would decrease due to low Σ values of hydrocarbons and the gamma ray reading would increase.

A procedure to carry out this analysis would be to plot Σ vs. the response of the gamma spectometry tool on cartesian paper as shown on Figure 3–44. A straight line is drawn through the data points. From this plot the following key parameters are obtained:

$$
\begin{array}{ccc}
\text{Gamma ray} = 12 & \text{at} & \text{Sigma} = 10 \\
\text{Gamma ray} = 62 & \text{at} & \text{Sigma} = 30
\end{array}
$$

These end points are used to construct the scales presented on Figure 3–45. The overlay indicates potentially fractured intervals when the gamma ray (dashed line in second track) shifts to the left of the sigma curve. The comparison with perforated intervals results is good. Other curves in Figure 3–45 are the shale volume and gamma ray in track 1, and water saturation, water volume and total effective porosity in track 3. Eighty barrels of fresh water were flowed into the well while logging.

The method is valuable for intervals with reasonably constant lithology. When lithology varies it is important to go through a detailed zonation process.

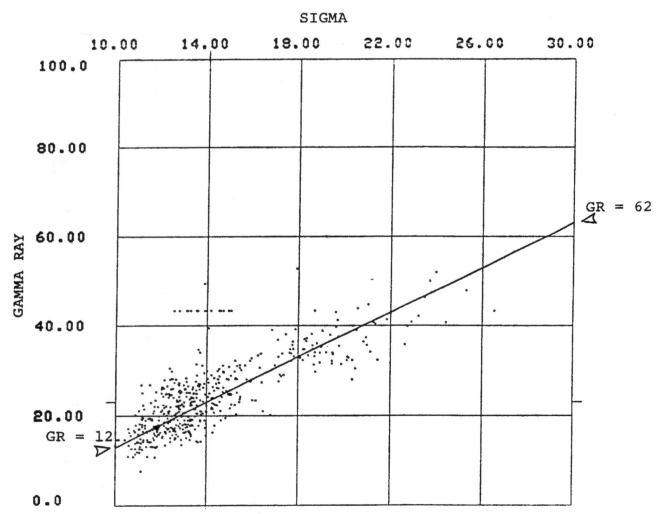

FIGURE 3–44 Gamma Ray vs. Sigma. Cased well log interpretation. The calibration, GR = 12 at sigma = 10 and GR = 62 at sigma = 30, is used as scale in the second track of Figure 3–45 to determine the intervals that are potentially naturally fractured.

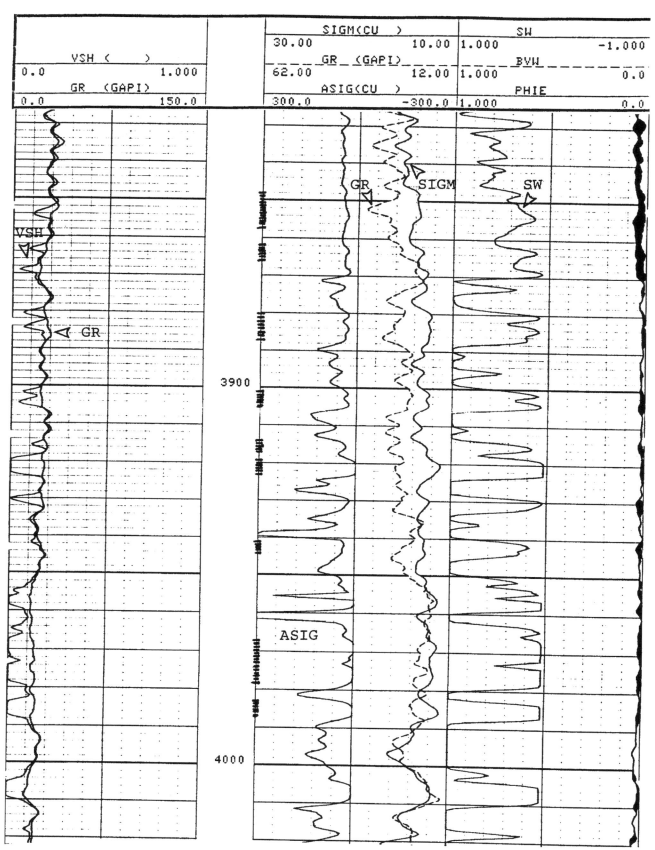

FIGURE 3–45 Cased borehole fracture identification. Increases in GR (dashed line in the second track) and decreases in sigma pinpoint intervals that are potentially naturally fractured.

Another approach to fracture detection in cased holes uses a pulsed neutron and a porosity log (Engelke and Hilchie, 1971) In this method the mud must be saltier than the formation water. This leads to a sigma mud much larger than sigma water. The mud might be normal salt mud or a fresh mud with boron added. This increases the capture cross-section to a salt mud level. If invasion is deep (at least two or three ft) the "porosity" calculated from the pulsed neutron log, using Equation 3–53 and assuming S_w = 100% would be larger than the porosity determined from neutron, sonic and density logs.

An example illustrating this method in the Ellenburger Formation, Pecos County, Texas, is presented in Figure 3–46. This well was drilled using fresh mud to about 22,700 ft. From there on borax was added as drilling continued. The logs presented in Figure 3–46 suggest possible fractures below 22,800 ft. This could not be corroborated from production test.

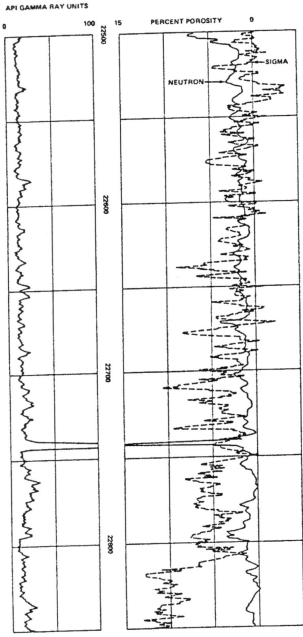

FIGURE 3–46 Apparent Porosities from sigma and Neutron-Ellenburger Formation, Pecos County, Texas (after Engelke and Hilchie, 1971)

Notice that in this case the porosity control is given by a gamma ray-neutron log run through casing.

QUANTITATIVE ANALYSIS

This section describes techniques for quantitative analysis of naturally fractured reservoirs from conventional well logs. The methods are not claimed to be perfect. However, they have provided reasonable results in evaluating orders of magnitude of fracture, total, and matrix porosities; and water saturations in matrix, fractures, and the composite matrix-fractures system.

Porosity Exponent, m

The porosity exponent (m) of a matrix-fracture system is smaller than the porosity exponent (m_b) of the matrix if the fracture is not healed. The value of m for a fracture can be shown theoretically to be equal to 1.0. Figure 3–47A shows the schematic of a rock of length L, with a flow path length equal to L_A. If the water saturation is 100%, the resistance (R) of the rock can be calculated as follows:

$$R = \frac{R_w L_a}{\phi A} \tag{3–22}$$

where: R_w = water resistivity
ϕ = porosity, fraction
A = section area
L = length of rock
L_a = length of flow path

By definition, the resistance of a system is given by:

$$R = R_o \frac{L}{A} \tag{3–23}$$

where: R_o = Rock resistivity when it is 100% saturated with water.

Combination of Equation 3–22 and 3–23 leads to:

$$R_o = R_w \frac{L_a}{L\phi} \tag{3–24}$$

Rock tortuosity, τ

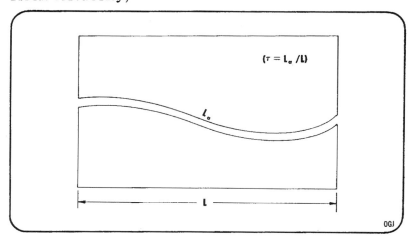

FIGURE 3–47 (A)   Schematic of rock tortuosity, $\tau = L_a/L$

The tortuosity of the system (τ) is defined as $\tau = L_a/L$. Consequently, Equation 3–24 can be written as:

$$R_o = \frac{R_w \tau}{\phi} \tag{3–25}$$

and,

$$F = \frac{R_o}{R_w} = \frac{\tau}{\phi} \tag{3–26}$$

where F is the formation factor. For an open fracture, the tortuosity is equal to 1. Consequently, $F = 1/\phi$. This is significant because for the theoretical case of a planar fracture, the value of m in Archie's equation, $F = \phi^{-m}$, would equal 1. This of course, is quite different from the values of m normally used in formation evaluation by well log analysis. Based on my experience, naturally fractured reservoirs are characterized by values of *m* which are smaller than usual. The larger the degree of natural fracturing the smaller the value of *m* of the composite system. Boyeldieu and Winchester (1982) have cited statistical evaluations which suggest m values for the fractures (m_f) in the order of 1.3 and going all the way up to 1.5.

With the premise that the value of m for a planar open fracture is smaller than usual (for example 1.0), an equation was searched which could handle fracture porosity alone, matrix porosity alone, and the combination of both matrix and fractures. This equation was found by considering a double-porosity model connected in parallel. Pirson (1975) presented the following relationship for a case in which matrix and fractures are 100% saturated with water:

$$\frac{1}{R_{fo}} = \frac{v\phi}{R_w} + \frac{1-v}{R_o} \tag{3–27}$$

Where R_{fo} is resistivity of the composite system and R_o is resistivity of the matrix system.

The partitioning coefficient represents the fraction of total pore volume made up of fractures and is defined by:

$$v = \frac{\phi - \phi_b}{\phi\,(1.0 - \phi_b)} = \frac{\phi - \phi_m}{\phi} \tag{3–28}$$

where ϕ is total porosity, ϕ_b is matrix porosity attached to matrix bulk volume and ϕ_m is matrix porosity attached to bulk volume of the composite system (matrix bulk volume + fracture bulk volume). Fracture porosity (ϕ_2) attached to bulk volume of the composite system is given by:

$$\phi_2 = v\,\phi \tag{3–29}$$

This fracture porosity is small, in many instances smaller than 1% and is strongly scale-dependent.

Figure 3–47B illustrates the scale dependency of the various porosities mentioned above ϕ_m, ϕ_b and ϕ_2. It also introduces a fracture porosity, ϕ_f, that in the case of open uncemented fractures without any secondary mineralization would be equal to 100%. If there is some secondary mineralization the fracture void space, $(VS)_f$, decreases and ϕ_f also decreases.

From a practical point of view we are always interested in ϕ_2 rather than ϕ_f. As indicated previously ϕ_2 is in many instances smaller than 1% although there are exceptions. If for example, we have a 1–ft drilling brake, the value of ϕ_2 within that particular foot at that specific location is 100%. Other exceptions are provided in some cases by vuggy fractures and karst topography.

Rearranging Equation 3–27 results in:

$$R_{fo} = \frac{R_w R_o}{v\phi R_o + (1-v)R_w} \tag{3–30}$$

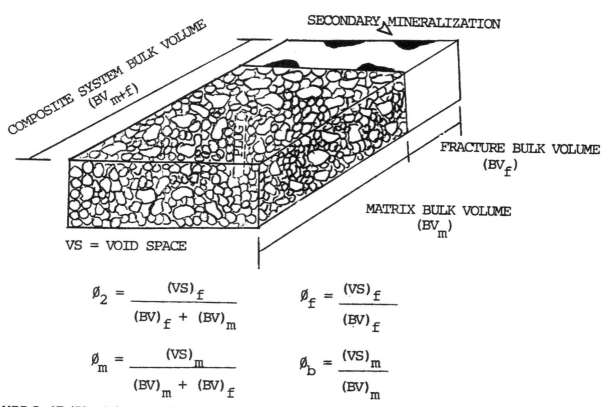

$$\phi_2 = \frac{(VS)_f}{(BV)_f + (BV)_m} \qquad \phi_f = \frac{(VS)_f}{(BV)_f}$$

$$\phi_m = \frac{(VS)_m}{(BV)_m + (BV)_f} \qquad \phi_b = \frac{(VS)_m}{(BV)_m}$$

FIGURE 3–47 (B) Schematic illustrating strong scale dependency of different types of porosity in naturally fractured reservoirs. Notice that the matrix porosity of an unfractured plug is equivalent to ϕ_b, not ϕ_m. ϕ_2 is fracture porosity attached to bulk properties of the composite system. ϕ_f is attached to bulk volume of the fracture system.

The total system (matrix plus fractures) formation factor (F_t) is defined as the relation between the resistivity of the system 100% saturated with water (R_{fo}) and the resistivity of the connate water (R_w). Substituting Equation 3–30 into this definition leads to

$$F_t = \left[\frac{R_w R_o}{v\phi R_o + (1-v)R_w} \right] / R_w \tag{3–31}$$

Equation 3–31 is valid for three different cases: only matrix porosity is present in the system; only fracture porosity is present in the system; and both matrix and fracture porosities are present in the system.

Case 1. If only matrix porosity is present in the system, the partitioning coefficient (v) is equal to zero and $F_t = F$. Therefore, Equation 3–31 can be rewritten as:

$$F = \frac{R_w R_o}{R_w} / R_w = \frac{R_o}{R_w} \tag{3–32}$$

Equation 3–32 represents the relationship defining the formation factor (F) in intergranular media. Consequently, it can be concluded that Equation 3–31 holds for this particular case.

Case 2. If only fracture porosity is present in the system, the partitioning coefficient (v) is equal to 1.0. Equation 3–31 can be rewritten as:

$$F_t = \frac{R_w R_o}{\phi R_o} / R_w = \frac{1}{\phi} = \frac{1}{\phi^m} \tag{3–33}$$

The inferred m in Equation 3–33 is equal to 1, indicating that for a fracture the porosity exponent should be 1 as discussed previously. Consequently, Equation 3–31 holds for this case.

Case 3. This case, which considers that both matrix and fracture porosities are present in the system, is represented by Equation 3–31. Simplification of this equation leads to:

$$F_t = \frac{R_o}{v\phi R_o + (1-v)R_w}$$

Taking R_o as a common factor in the denominator results in:

$$F_t = \frac{R_o}{R_o[v\phi + (1-v)R_w/R_o]} \tag{3–34}$$

Since $F = R_o/R_w$, Equation 3–34 can be written as:

$$F_t = \frac{1}{v\phi + (1-v)/F} \tag{3–35}$$

Finally, Archie's equation can be inserted in Equation 3–35 to give

$$\phi^{-m} = \frac{1}{v\phi + (1-v)/\phi_b^{-mb}} \tag{3–36}$$

Remember that ϕ is total porosity, ϕ_b is matrix block porosity, v is the partitioning coefficient, m is the porosity exponent of the composite system, and m_b is the porosity exponent of the matrix blocks.

Equations 3–28 and 36 represent the relationships that can be used to study all possible combinations of two-porosity systems and their influence on the double-porosity exponent, *m*. Computer runs were made to study the following ranges.

m_b : between 1.6 and 2.4

ϕ_b : between 0.02 and 0.30

v : between 0.05 and 0.90

Figures 3–48 through 3–52, respectively, present the results for the cases in which the matrix-porosity exponent (m_b) is equal to 1.6, 1.8, 2.0, 2.2 and 2.4. These figures were generated as indicated in the flow diagram of Figure 3–53.

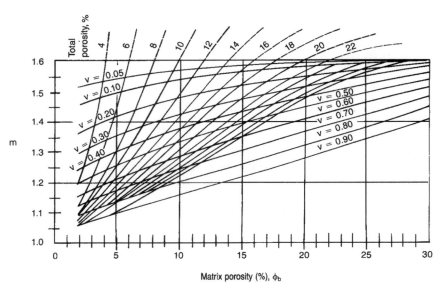

FIGURE 3–48 Chart for evaluating naturally fractured reservoirs (m_b = 1.6) (after Aguilera, 1976)

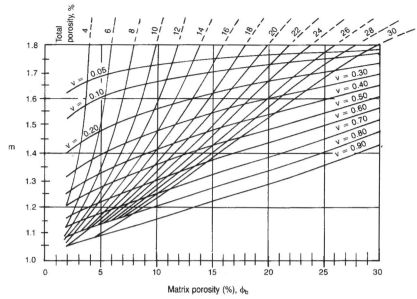

FIGURE 3–49 Chart for evaluating naturally fractured reservoirs ($m_b = 1.8$) (after Aguilera, 1976)

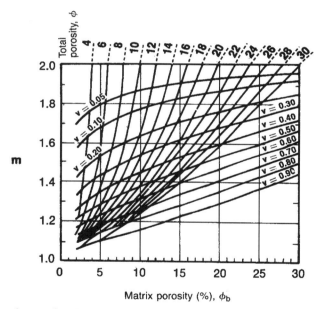

FIGURE 3–50 Chart for evaluating naturally fractured reservoirs ($m_b = 2.0$) (after Aguilera, 1976)

Based on the previous information, Equation 3–31 is considered valid for Case 3 (both matrix and fracture porosities are present in the system) for the following reasons.

1. The equation holds for Cases 1 and 2 as indicated above, that is, it holds for matrix and fractured systems when considered independently.
2. If $m = m_b$, then $\phi = \phi_b$. In other words, the model provides consistent information.

In a strict sense, Figures 3–48 through 3–52 are only valid for the theoretical model that assumes fractures and matrix are connected in parallel (Equation 3–27). However, cautious use of the charts provides reasonable answers. Results from these charts have been corroborated with other models developed by Rasmus (1983), Draxler and Edwards (1984), Crary et al (1987), Edmunson (1988), and Schlumberger (1991).

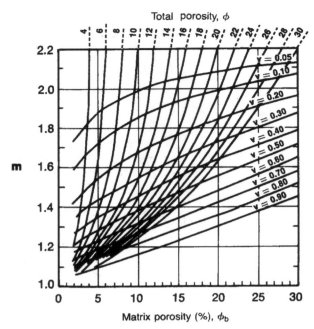

FIGURE 3–51 Chart for evaluating naturally fractured reservoirs (m$_b$ = 2.2) (after Aguilera, 1976)

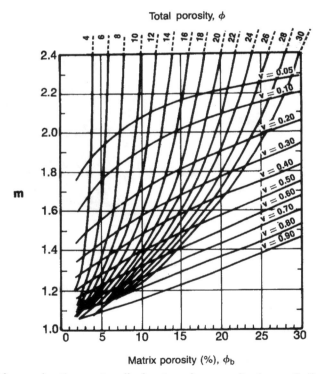

FIGURE 3–52 Chart for evaluating naturally fractured reservoirs (m$_b$ = 2.4) (after Aguilera, 1976)

Water Saturation Exponent, *n*

In log formation evaluation, *n* is assumed many times to be approximately equal to 2.0 in water-wet systems. In general, this assumption yields reasonable results. It appears, however, that this assumption might lead to pessimistic values of total water saturation in naturally fractured reservoirs. My experience is that more reasonable values of S$_w$ may be obtained by assuming that *m* equals *n*. This implies that the values of n should be smaller

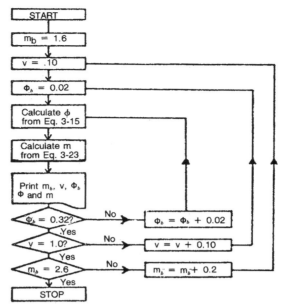

FIGURE 3–53 Flow chart for generating Figures 3–48 to 3–52 (after Aguilera, 1976)

than usual for naturally fractured reservoirs. This implication is reasonable by assuming that a fractured system can be assumed to be equivalent to a bundle of tubes, as suggested by Hilchie and Pirson (1961). In fact, Fatt (1956) showed that for a simple bundle of tubes, a log-log plot of resistivity index (I) vs. water saturation (S_w) resulted in a straight line with a slope (n) equal to –1 (Figure 3–54). Consequently, for a fracture the equality m = n = 1.0 seems to hold, and it is reasonable to speculate that, as in the case of the porosity exponent, the water saturation exponent may vary between 1 and the n value of the matrix depending on the degree of fracturing. This reinforces the assumption that m equals n, especially when laboratory analysis indicates that for unfractured plugs the matrix porosity exponent (m_b) is approximately equal to the matrix water saturation exponent (n_b).

Uses of Interpretation Charts

Figures 3–48 through 3–52 can estimate the following:

1. Total (ϕ) and fracture (ϕ_2) porosities and the partitioning coefficient (v) as a function of ϕ_b, m, and m_b.
2. Matrix (ϕ_b) and fracture (ϕ_2) porosities and the partitioning coefficient (v) as a function of ϕ, m, and m_b.
3. The double-porosity exponent (m) as a function of ϕ, ϕ_b, and m_b. This a very important application for the evaluation of fractured reservoirs where ϕ and ϕ_b can be evaluated independently. The values of m_b and ϕ_b can be obtained in the laboratory from the analysis of intergranular unfractured cores. Total porosities can be determined from neutron and density logs. Sonic logs can provide matrix and in some cases total porosity.

Mathematically, the value of m can be determined by re-writing Equation 3–36 as follows:

$$m = \log[v\phi + (1 - v)/\phi_b^{-mb}]/\log \phi \qquad (3\text{–}37)$$

Some other equations have been proposed for calculation of the double-porosity exponent, m. One of them was developed by Rasmus (1983) and includes potential changes in fracture tortuosities (τ_f) as follows:

$$m = \frac{\log[(1 - (\phi - \phi_m)) \, \phi_m^{mb} + (\phi - \phi_m)/\tau_f^2]}{\log\phi} \qquad (3\text{–}38)$$

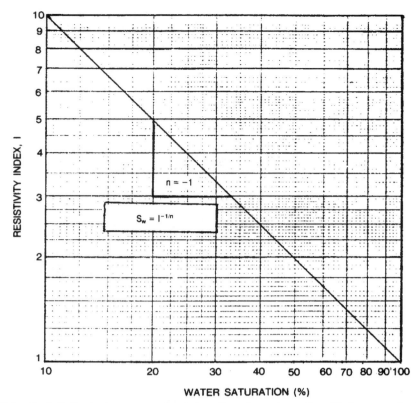

FIGURE 3–54 Resistivity index vs. water saturation for a bundle of tubes results in a water saturation exponent, n, equal to 1.0 (after Fatt, 1956)

Where ϕ_m is matrix porosity attached to bulk volume, of the composite system and ϕ is total porosity. For fractured carbonates with fenestral fracture/vug porosity, Rasmus (1983) recommends to calculate total porosity from a neutron-density crossplot and matrix porosity from a sonic log.

Another equation for calculating m of the composite system includes a fracture cementation exponent, m_f, and can be written as follows (Draxler and Edwards, 1984):

$$m = \frac{\log (\phi_m^{mb} + \phi_2^{mf})}{\log \phi} \tag{3–39}$$

When $m_f = \tau_f = 1$ the values of m calculated from Equations 3–37, 3–38 and 3–39 are approximately equal as shown on Table 3–3 and Figure 3–55. Consequently the formation factors of the composite system are also approximately the same. Values of m_f and τ_f equal to 1.0 could be anticipated to be valid for many reservoirs with open unhealed fractures.

When m_f is different from 1.0, Equation 3–39 might be valuable but must be used carefully. For example if $m_b = 2$, $m_f = 1.5$, $\phi = 0.10$, $\phi_2 = 0.01$, and $\phi_m = 0.09$, Equation 3–39 yields an m of 2.041 which is larger than the m_b of the matrix. This not reasonable as m for a fractured system should be smaller than the m_b of the matrix. However, values of m larger than m_b do occur in dual-porosity reservoirs made out of non-connected vugs and matrix as shown originally by Towle (1962). This is discussed in more detail in the next section.

Vuggy Reservoirs

In the case of a vuggy reservoir the value of m of the composite system is larger than the porosity exponent of the matrix system (m_b). This finding was published originally by Towle (1962), who indicated that as the vuginess increases so does the value of m. Figure 3–56 shows a crossplot of vuginess vs m for tube-vug and plane-vug systems.

TABLE 3-3 Comparison of cementation exponents for naturally fractured reservoirs using different models ($m_b = 2.0$)

φ (total)	$φ_m$	$φ_b$	$φ_2$	F_R	F_D	F_A	m_R	m_D	m_A
			Partitioning Coefficient, v = 10%						
.02	.018	.0180361	.002	430.31	430.29	436.15	1.550	1.550	1.554
.05	.045	.0452261	.005	142.55	142.35	146.18	1.656	1.655	1.664
.10	.090	.0909091	.010	55.50	55.25	57.35	1.744	1.742	1.759
.15	.135	.137056	.015	30.35	30.10	31.34	1.799	1.795	1.816
.20	.180	.183573	.020	19.32	19.08	19.86	1.840	1.832	1.857
			Partitioning Coefficient, v = 30%						
.02	.014	.0140845	.006	161.42	161.39	162.90	1.300	1.300	1.302
.05	.035	.035533	.015	61.70	61.63	62.96	1.376	1.376	1.383
.10	.070	.0721649	.030	28.77	28.65	29.72	1.459	1.457	1.473
.15	.105	.109948	.045	18.01	17.85	18.70	1.542	1.519	1.544
.20	.140	.148936	.060	12.75	12.56	13.24	1.582	1.572	1.606
			Partitioning Coefficient, v = 50%						
.02	.010	.010101	.010	99.02	99.01	99.49	1.175	1.175	1.176
.05	.025	.025641	.025	39.05	39.02	39.48	1.223	1.223	1.227
.10	.050	.0526316	.050	19.09	19.05	19.46	1.281	1.280	1.289
.15	.075	.081081	.075	12.47	12.40	12.77	1.330	1.327	1.342
.20	.100	.111111	.100	9.17	9.09	9.42	1.377	1.371	1.393

m_R = dual-porosity exponent from Eq. 3-38
m_D = dual-porosity exponent from Eq. 3-39
m_A = dual-porosity exponent from Eq. 3.37

F_R = formation factor from Rasmus (1983) model
F_D = formation factor from Draxler and Edwards (1984) model
F_A = formation factor from Aguilera (1976) model

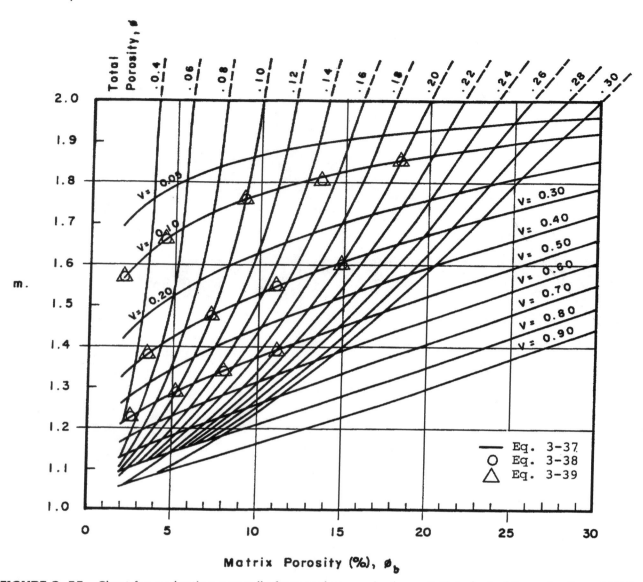

FIGURE 3–55 Chart for evaluating naturally fractured reservoirs (m$_b$ = 2.0) (adapted from Aguilera, 1976)

The values of m for a vuggy reservoir can be calculated based on the equations (Schlumberger, Middle East Well Evaluation Review, January 1991):

$$R_t = R_w/\phi^m \qquad (3\text{–}40)$$

and,

$$R_t = \frac{R_w}{(\phi - \phi_{nc})^{m_b}} \qquad (3\text{–}41)$$

where ϕ =   total porosity, fraction
ϕ_{nc} =   porosity of non-connecting vugs, fraction
m_b =   porosity exponent of the matrix blocks
m =   porosity exponent of the composite system

and other nomenclature has been defined previously.

Equations 3–40 and 3–41 can be combined to yield the value of m for the composite system of matrix porosity and non-connecting vugs:

$$m = m_b \log (\phi - \phi_{nc})/\log \phi \qquad (3\text{–}42)$$

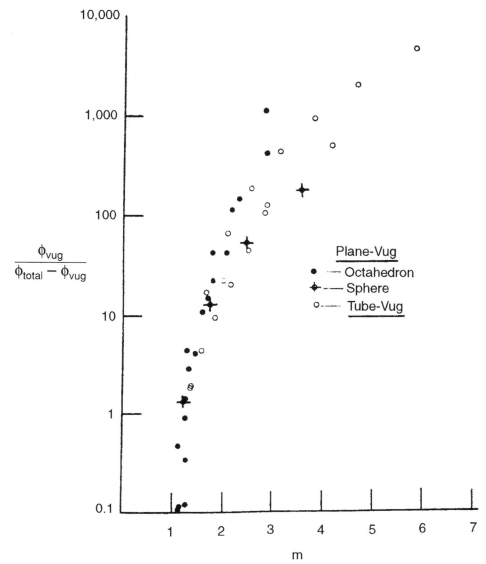

FIGURE 3–56 Cementation factor "m" vs. "vugginess" for tube-vug and plane-vug systems (after Towle, 1962)

The same Schlumberger publication uses Equation 3–39 for calculating m of a naturally fractured reservoir. As discussed previously, Equation 3–39 might give erroneous values of m when the cementation exponent of the fracture, m_f, is greater than 1.0.

The results from these equations are presented in Figure 3–57, which was developed for the case in which the porosity exponent of the matrix (m_b) is equal to 2.0. Notice that in the case of the naturally fractured reservoir the porosity exponent of the composite system (m) is smaller than m_b. On the other hand in the case of the reservoir with non-connecting vugs the value of m of the composite system is larger than m_b.

The solutions presented in Figures 3–50 and 3–57 provide the same results. For example if total porosity (ϕ) is 5% and fracture porosity (ϕ_2 using this book's nomenclature or ϕ_f using Schlumberger's nomenclature) is 1% the value of m for the composite system from Figure 3–57 is equal to 1.5. To use Figure 3–50 calculate first the partitioning coefficient ($v = \phi_2/\phi = 1/5 = 0.20$). Enter total porosity (5%) at the top of Figure 3–50, go down parallel between the lines for 4 and 6% until reaching v = 0.20. At that point read the value of m = 1.5 from the y axis. The answer is the same one obtained from Figure 3–57.

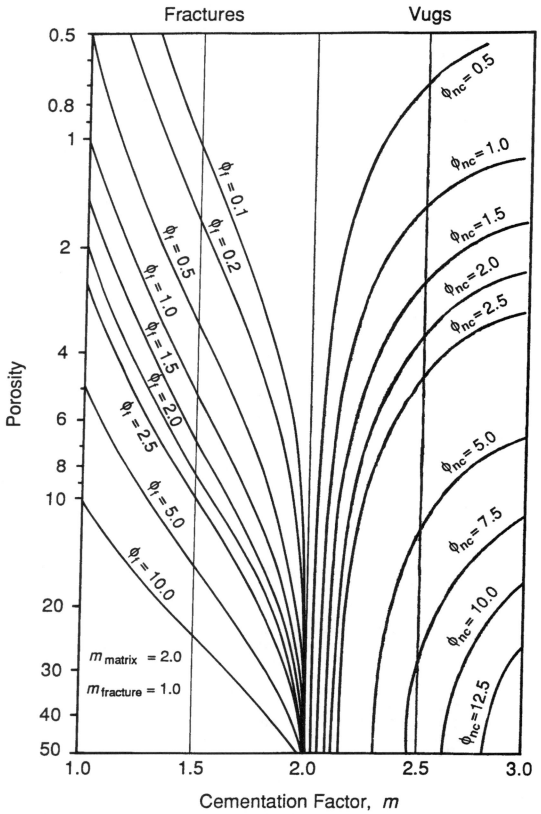

FIGURE 3–57 Chart for evaluating dual-porosity systems from well logs (ϕ_f = fracture porosity. ϕ_{nc} = porosity of non-connecting vugs) (after Schlumberger, 1987)

The fact that various dual-porosity models developed independently by Aguilera (1976), Rasmus (1983), Draxler and Edwards (1984), and Schlumberger (1991) provide approximately the same results give confidence with respect to the solutions provided by the careful use of these equations.

Determination of m from Pickett Plots

According to Archie (1942), the basic relationships in formation evaluation from logs are:

$$S_w = I^{-1/n} \tag{3-43}$$

$$I = R_t/(FR_w) = R_t/R_o \tag{3-44}$$

$$F = a\phi^{-m} = R_o/R_w \tag{3-45}$$

where: S_w = water saturation, fraction
  I = resistivity index
  R_t = true formation resistivity
  a = constant usually equal to 1

Other parameters are as defined previously.

Manipulating Equations 3–44 and 3–45 leads to the mathematical relationship representing Pickett's crossplot:

$$\log R_t = -m \log \phi + \log (aR_w) + \log I \tag{3-46}$$

or,

$$\log \phi = -\frac{1}{m} \log R_t + \frac{1}{m} \log (aR_w) + \frac{1}{m} \log I \tag{3-47}$$

Equation 3–47 fits better the standard notation for slope (y/x) than Equation 3–46. Equation 3–47 indicates that a log-log crossplot of ϕ vs. R_t should result in a straight line with a slope equal to $-1/m$ and intercept $(1/m) \log(aR_w)$ at a resistivity of 1.0 when the water saturation is 100% ($I = 1$). In this case the slope is the inverse of the cementation exponent (m).

Equation 3–46 is more widely used in the oil industry and thus it has been chosen as standard throughout this chapter.

Equation 3–46 tells that a plot of log R_t vs. log ϕ should result in a straight line with a slope of $-m$ for zones with constant I and aR_w. For fractured reservoirs, the resulting slope (porosity exponent m) should be smaller than the matrix porosity exponent (m_b) determined in the laboratory. The application of this method implies that Archie's equations also apply to naturally fractured reservoirs.

Figure 3–58 shows schematic diagrams of this crossplot for homogeneous and naturally fractured reservoirs. Notice that the water zones, i.e., zones with I equal to 1, form linear trends with slopes equal to 2.0 and 1.3, respectively.

When I = 1 and ϕ = 1, Equation 3–46 reduces to log R_t = log (aR_w). Consequently, the value of aR_w can be determined directly from this crossplot by extrapolating the 100% water-bearing trend to a porosity of 100% as indicated by the dashed line of Figure 3–58(a). If the value of constant "a" is 1.0, the R_w can be determined from this kind of crossplot alone. This is valuable because it permits sound comparisons of R_w values determined from various sources.

If the value of "a" is different from 1, extrapolation of the 100% water-bearing trend to a porosity of 100% would indicate the value of (aR_w) as shown by the dashed line of Figure 3–58(a). Knowing one of these two parameters would permit immediate calculation of the other one.

When hydrocarbon zones are present, they can be recognized using this kind of plot as shown in the schematic of Figure 3–58. Further, the value of the resistivity index (I) for the hydrocarbon zone can be determined by dividing its resistivity (R_t) by the resistivity of the water-bearing trend (R_o) at the same level of porosity.

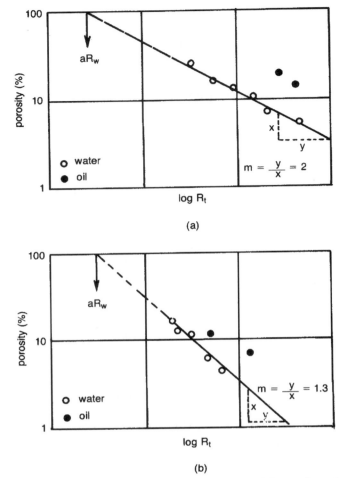

FIGURE 3-58 Crossplots for a) homogeneous and b) fractured formations (after Aguilera, 1974)

Finally, the value of S_w can be calculated from the relationship $S_w = I^{-1/n}$ (Equation 3–43) by assuming that m equals n. The advantages of this type of analysis up to this point are self-evident and include:

1. The value of water resistivity (R_w) and the constant (a) do not have to be known previously to end up with a value of water saturation.
2. The value of the porosity exponent (m) does not have to be known in advance.
3. Values of the above unknown parameters can be determined from this kind of crossplot.

The main limitation of the method is that a statistically significant number of zones of constant a, R_w, and m must be available. Since m is taken as a constant it is implicitly assumed that the interval being analyzed has a relatively uniform fracture distribution.

The value of water saturation can be also determined from the equation:

$$S_w = \left[\frac{\phi_w}{\phi_{hy}} \right]^{m/n} \tag{3–48}$$

where ϕ_{hy} is porosity of the hydrocarbon interval for which S_w is going to be calculated and ϕ_w is porosity in the 100% water saturation line at the same resistivity of ϕ_{hy}. Since the assumption that m equals n seems to be valid in many cases in naturally fractured reservoirs, water saturation can be calculated directly from $S_w = \phi_w/\phi_{hy}$.

Example 3–2. This example is designed to provide a basic working knowledge of this method. Only four zones are considered in this example for ease and simplicity. The following information is available.

Zone	R_t	ϕ (Total)
1	55	0.015
2	25	0.026
3	10	0.047
4	56	0.030

Find:

1. The porosity exponent, m
2. The water resistivity (R_w) if a = 1
3. The resistivity index (I) of the hydrocarbon zone
4. Is this reservoir likely to be fractured? The value of the matrix porosity exponent (m_b) is 2.0.
5. The water saturation (S_w) of the hydrocarbon interval (zone 4) assuming that m equals n
6. If the reservoir is fractured, estimate matrix porosity (ϕ_b), the fraction of total porosity made up of fractures or partitioning coefficient (v) and the fracture porosity (ϕ_2).
7. Draw lines of S_w equal to 25, 50 and 75%.

Solution.

1. A crossplot of log R_t vs. log ϕ is shown in Figure 3–59. Notice that zones 1, 2, and 3 form a straight line with a slope of –1.5. Consequently, the porosity exponent (m) is equal to 1.5.
2. Extrapolation of the 100% water-bearing trend to a porosity of 100% yields an aR_w value of 0.1. Since a = 1, the water resistivity is 0.1 Ω-m.
3. The resistivity index (I) of zone 4 is 2.87 by dividing the R_t value of zone 4 by R_t of the 100% water-bearing trend (or R_o) at the same level of porosity, i.e., 56/19.5 = 2.87.
4. Comparison of the porosity exponent of the matrix, m_b = 2.0, with the porosity exponent from the Pickett plot, m = 1.5, suggests that the reservoir may be naturally fractured.
5. Water saturation of the composite system made out of matrix and fractures (zone 4) is calculated to be 49.5% from Equation 3–43 by assuming that m = n = 1.5. The same value of water saturation is calculated from Equation 3–48 by using S_w = 0.01485/0.030 = 0.495. It is emphasized that because the reservoir is naturally fractured, S_w from Equation 3–43 corresponds to water saturation of the composite system.
6. An estimate of matrix porosity (ϕ_b) and the fraction of total porosity made up of fractures can be made using Figure 3–50. For example, for zone 3, total porosity ϕ = 0.047 and m = 1.5. Consequently, from Figure 3–50, v = 0.19 and matrix porosity (ϕ_b) = 0.038. Fracture porosity in this case is ϕ_2 = vϕ = 0.19 × 0.047 = 0.0089 = 0.89%.

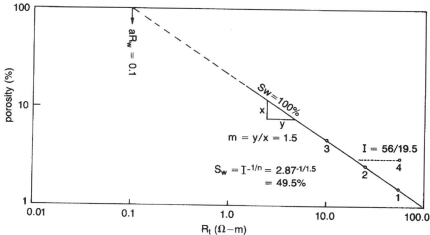

FIGURE 3–59 R_t vs. porosity for hypothetical example showing calculation of m and S_w

7. Lines of water saturation for the composite system equal to 25%, 50%, and 75% can be drawn by calculating the resistivity index (I) from Equation 3–44 by assuming m = n = 1.5. The following values of I are calculated:

S_w	I
0.25	8.00
0.50	2.83
0.75	1.54

Parallel lines to the water-bearing trend (S_w = 100%) are drawn for each "I" of interest by considering the relationship $R_t = IR_o$. For example, the line of water saturation equal to 25% is drawn by considering any R_o determined from the water bearing trend and multiplying this R_o by 8. The R_t, so determined, is plotted at the same level of porosity from which the R_o was selected and a line is drawn parallel to the 100% water saturation line (Figure 3–60).

Thus it is possible to determine values of water saturation without previous knowledge of the double-porosity exponent (m) and the water resistivity (R_w). On the contrary, under favorable conditions it is possible to determine m and R_w by the crossplotting technique discussed in this example.

Porosity Response of Conventional Well Logs

Sometimes the value of ϕ is not available for preparing the log-log crossplot of R_t vs. ϕ. In these cases, it is still possible to make the same type of analysis by considering the porosity response equations. For example, for the sonic log the porosity response equation can be written as:

$$\Delta t = A + B \phi \tag{3–49}$$

Wyllie et al (1956) indicated in their original study that the sonic log sees intergranular or matrix porosity but ignores secondary porosity. Their equation for matrix porosity can be written as:

$$\Delta t = \Delta t_m + (\Delta t_f - \Delta t_m)\phi_b \tag{3–50}$$

where:
Δt = sonic log response, μsec/ft
Δt_m = sonic log response at zero porosity, μsec/ft
Δt_f = fluid transit time, usually taken as 189 μsec/ft

Experience with double-porosity systems, however, indicates that in some cases the sonic log might provide reasonable estimates of total porosity. In these cases, parameters

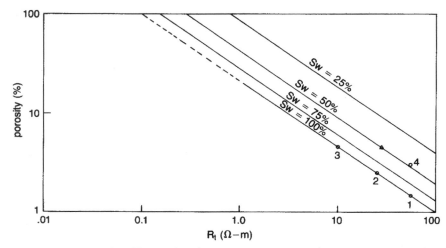

FIGURE 3–60 R_t vs. porosity, illustrating how to generate various water saturation lines

A and B may have little physical significance but provide a means of calculating total porosity. Notice that for intergranular reservoirs $A = \Delta t_m$ and $B = \Delta t_f - \Delta t_m$.

Equation 3–49 can be rewritten as:

$$\phi = \frac{\Delta t - A}{B} \tag{3–51}$$

Inserting Equation 3–51 in Equation 3–46 results in:

$$\log R_t = -m \log (\Delta t - A) + m \log B + \log (aR_w) + \log I \tag{3–52}$$

Equation 3–52 indicates that a log-log crossplot of R_t vs. $(\Delta t - A)$ must result in a straight line of slope $-m$ for zones with constant A, B, a, R_w, and I. This crossplot is powerful, as it permits estimates of water saturation values with no previous knowledge of A, B, a, R_w, m, and porosity. On the contrary, under favorable conditions, it is possible to estimate values for these unknowns from this type of crossplot.

The same type of analysis can be carried out by crossplotting:

Density: $\log (\rho_s - \rho_B)$ vs. $\log R_t$, where ρ_s is grain density and ρ_B is bulk density
Epithermal Neutron: $\log \phi_N$ vs. $\log R_t$, where ϕ_N is neutron porosity
Thermal Neutron: Neutron deflection (API or cps) in linear scale rather than logarithmic vs. $\log R_t$

Older wells with very limited log suites are prime candidates for evaluations using the above types of crossplots. Most recent wells with modern logs can also be evaluated using the same approach. Probably the biggest advantage of the Pickett crossplot is that is encompasses most tools and interpretation techniques available as of this writing.

Pulsed neutron logs can be evaluated by crossplotting $\log (\Sigma - \Sigma_m)$ vs. $\log R_t$ for zones with $S_w = 100\%$. In this case Σ is the neutron capture cross section in capture units (c.u.) from the well log and Σ_m is the matrix neutron capture cross section. However, for zones with $S_w < 100\%$ the analysis changes, as discussed by Aguilera (1979). An example of this type of analysis is presented in Figure 3–61. In the case of pulsed neutron logs porosity (ϕ) is calculated from the equation:

$$\phi = \frac{\Sigma - \Sigma_m}{(1 - S_w) \Sigma_{hy} + S_w \Sigma_w - \Sigma_m} \tag{3–53}$$

where:

$$
\begin{aligned}
\Sigma &= \text{neutron capture cross section from log, c.u. (capture units)} \\
\Sigma_m &= \text{matrix neutron capture cross section, c.u.} \\
\Sigma_{hy} &= \text{hydrocarbon neutron capture cross section, c.u.} \\
\Sigma_w &= \text{water neutron capture cross section, c.u.}
\end{aligned}
$$

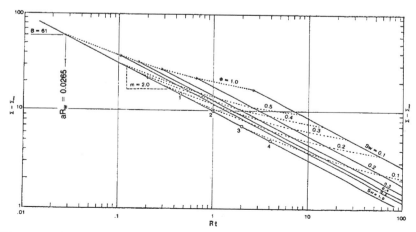

FIGURE 3–61 R_t vs. $(\Sigma - \Sigma_m)$ for various water saturations and porosities (after Aguilera, 1979)

Example 3–3. This example is designed to provide a working knowledge of the previous approach. As in the previous example, only four zones are considered.

The following information is available:

Zone	R_t	Δt
1	55	58.0
2	25	60.0
3	10	64.2
4	56	61.0

It is desired to determine or discuss:

1. The porosity exponent (m) and the Constant A
2. The resistivity index (I) of the hydrocarbon zone
3. The water saturation (S_w) of the hydrocarbon zone assuming that m equals n
4. The value of the matrix porosity exponent (m_b) is 2.0. Is this reservoir likely to be fractured?
5. Draw lines for S_w equal to 25%, 50%, and 75%

Solution.

1. The porosity exponent (m) can be calculated by trial and error by assuming values of A until a straight line is obtained (Figure 3–62). Notice that only when A equals 55 is a straight line obtained. If, for example, A is equal to 53 or 57, curves are obtained. With a straight line, it is possible to determine the value of m by measuring the slope of the straight line. For this case m = 1.5 (Figure 3–63).
2. The resistivity index (I) of zone 4 is calculated by taking the ratio R_t/R_o at the same porosity level. For this case $I = R_t/R_o = 56/19.5 = 2.87$.
3. Water saturation is calculated to be 49.5% from Equation 3–43 by assuming that m = n = 1.5. The same value of water saturation can be calculated with the use of the equation:

$$S_w = \left[\frac{(\Delta t - A)_w}{(\Delta t - A)_{hy}}\right]^{m/n}$$

Where $(\Delta t - A)_{hy}$ refers to the interval for which S_w is going to be calculated and $(\Delta t - A)_w$ is read in the water-bearing line at the same value of R_t. For zone 4, $(\Delta t - A)_{hy}$ equals 6 and $(\Delta t - A)_w = 2.97$ μsec/ft. Consequently $S_w = 2.97/6 = 0.495$.
4. Comparison of the matrix porosity exponent, $m_b = 2.0$, with the porosity exponent from Figure 3–63, m = 1.5, suggests that the reservoir is naturally fractured. Thus S_w calculated in step 3 corresponds to water saturation of the composite system.

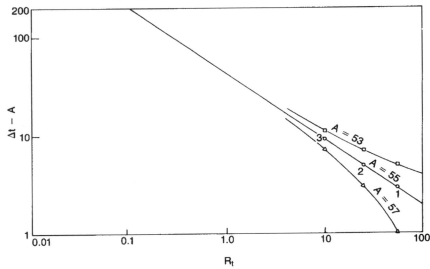

FIGURE 3–62 R_t vs ($\Delta t - A$) for hypothetical example. Calculation is done by trial and error, A = 55

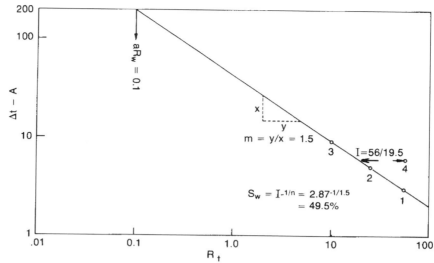

FIGURE 3–63 Hypothetical example showing calculation of m, R_w and S_w

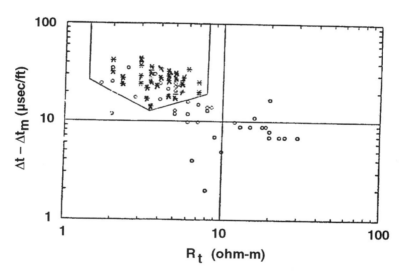

FIGURE 3–64 Pickett plot for Council Grove Formation, Panoma Field, Kansas. Asterisks represent naturally fractured intervals from core data (after Etnyre, 1981)

5. Water saturation lines equal to 25%, 50%, and 75% are drawn exactly as in the previous example.

Etnyre (1981) has shown that even a modest program including sonic and resistivity logs might give reliable fracture determinations. He has indicated that in the Permian Council Grove Formation of the Panoma Field (Kansas), chicken wire fractures fall in a distinctive area in a Pickett plot as shown on Figure 3–64, where the asterisks represent fracture intervals from core data. A matrix travel time of 48 μsec/ft was used in this plot. One hundred percent of the fractured zones fall within the closed area and the percentage of zones without fractures in the area is equal to 28%. Fracture intervals in this Formation are characterized by slow sonic travel time and lower resistivities. The formation is composed of thin beds of heterogeneous mixtures and laminations of siltstone, limestone, dolomite, anhydrite and significant clay volumes. The evaluation problem is compounded by the absence of hydrocarbon "shows" during drilling. No hydrocarbons are produced until after perforations and a significant fracture treatment.

Calibration of Logs with Core Porosities

Logs can be calibrated by comparing their responses with core porosities from whole core analysis. For example Equation 3–49 indicates that a crossplot of Δt vs. porosity on cartesian coordinates must result in a straight line with a slope equal to B and interception at zero porosity equal to A. This type of crossplot permits a double check with the value of A calculated by trial and error in the previous example. Knowing B enables determining the value of aR_w in the R_t vs. (Δt – A) log–log crossplot by extrapolating the 100% water saturation line to Δt – A = B, i.e., to the value of (Δt – A) when φ = 100%. Finally, the calibration allows the estimation of porosities from the log reading.

Example 3–4. Assume that whole core analysis is available in the previous Example 3–3 as follows:

Δt(μsec/ft)	Porosity from whole core analysis (%)
58.5	2.0
59.0	1.9
61.0	2.8
63.0	4.0
65.0	5.0
67.0	6.0

Complete the evaluation of the previous problem (Example 3–3) by determining:

1. The value of the constants A and B
2. The value of water resistivity, assuming that "a" equals 1.0
3. The total porosity (φ) of each zone
4. The matrix porosity ($φ_b$), partitioning coefficient (v), and fracture porosity ($φ_2$) of each zone.

Solution.

1. A crossplot of core porosity vs. sonic transit time is prepared on cartesian coordinates as shown in Figure 3–65. The data points form an approximate straight line with a

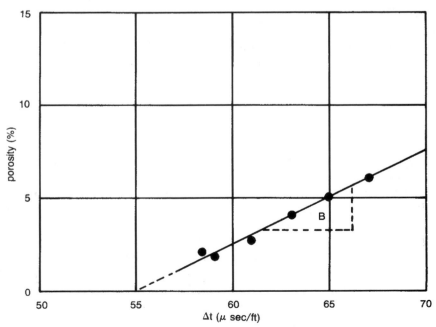

FIGURE 3–65 Calibration of sonic log with core porosity

slope (B) equal to 200 and interception at zero porosity (A) equal to 55. The calibrated equation is thus $\Delta t = 55 + 200\ \phi$. I emphasize that the response of the sonic log in this naturally fractured reservoir is non-conventional. In fact, in the paragraph above Equation 3–51 I indicated that $B = \Delta t_f - \Delta t_m$. If this was a conventional unfractured reservoir it could be anticipated that $B = 189 - 55 = 134\ \mu sec/ft$. However, as indicated previously this is a naturally fractured reservoir and B is 200 $\mu sec/ft$. Values of B vary considerably depending on the type of fracture reservoir and the fracture intensity.

2. The water resistivity (R_w) is calculated by extrapolating the 100% water saturation line to B = 200 as shown in Figure 3–63. From this analysis it is found that R_w equals 0.1.

3. Total porosity of each zone is readily calculated from the calibration equation obtained from Figure 3–65 i.e., $\phi = (\Delta t - 55)/200$.

Zone	Calculated ϕ
1	0.015
2	0.025
3	0.047
4	0.030

4. Matrix porosity (ϕ_b), the partitioning coefficient (v) and fracture porosity (ϕ_2) are determined from Figure 3–50 as a function of ϕ, m, and m_b as discussed in Example 3–2.

LITHOLOGY DETERMINATION

Crossplots of porosity responses from logs are used for determination of porosity and for investigating lithologies. If the formation contains only two minerals, the crossplot of neutron response versus density or sonic will provide information with respect to porosity and volume of each lithology. If the lithology is more complex it is necessary to use the combination neutron-density-sonic and in some instances the gamma ray and some other logs.

Use of Two Porosity Logs

Figure 3–66 shows a plot that allows porosity and lithology determination from density and neutron logs in water-filled holes. It must be remembered that neutron and density logs tend to give total porosities in naturally fractured reservoirs. The chart is entered by assuming that the matrix has the same properties as water-saturated limestone. As an example, a zone P has a sidewall neutron porosity (limestone) of 19% and a density porosity (limestone) of 15%. Assuming a mixed lithology of limestone and dolomite, the percentage of each lithology is estimated based on the distance between the two curves. For this case there is about 60% of limestone and 40% of dolomite. Porosity is approximately 18%.

It is interesting to see that an error in choosing the pair of lithologies will not result in large porosity errors provided that the choice is limited to limestone, dolomite, anhydrite (excluding gypsum) and sandstones or cherts. From the point of view of porosity estimates this is fortunate as most naturally fractured reservoir occur in dolomites, limestones, cherts and sandstones.

As an example, if the lithologies of zone P mentioned above were known to be quartz and dolomite instead of limestone and dolomite, a porosity of 18.3% rather than 18% would be calculated. The mineral proportions would be 55% of dolomite and 45% or quartz.

A similar approach is used for determination of lithology from sonic-density (Figure 3–67) and sonic-neutron crossplots (Figure 3–68 and 3–69). The resolution with the sonic-density combination is not very good for the group quartz-limestone-dolomite, and an error in choosing the pair of lithologies might result in a large error in the calculated porosity. This combination is very useful for distinguishing salt, gypsum and anhydrite.

Sonic-neutron crossplots have a good resolution for the quartz-limestone-dolomite group, and an error in choosing the pair of lithologies will not drastically affect the porosity results.

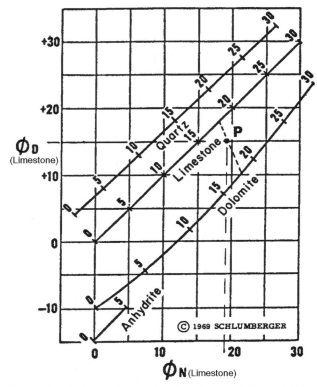

FIGURE 3–66 Porosity and lithology determination from FDC and SNP Logs in water-filled holes (after Schlumberger, 1969)

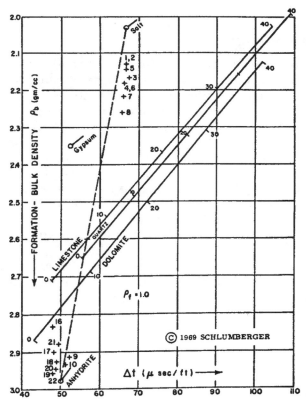

FIGURE 3–67 Porosity and lithology determination from FDC and Sonic Logs (after Schlumberger, 1969)

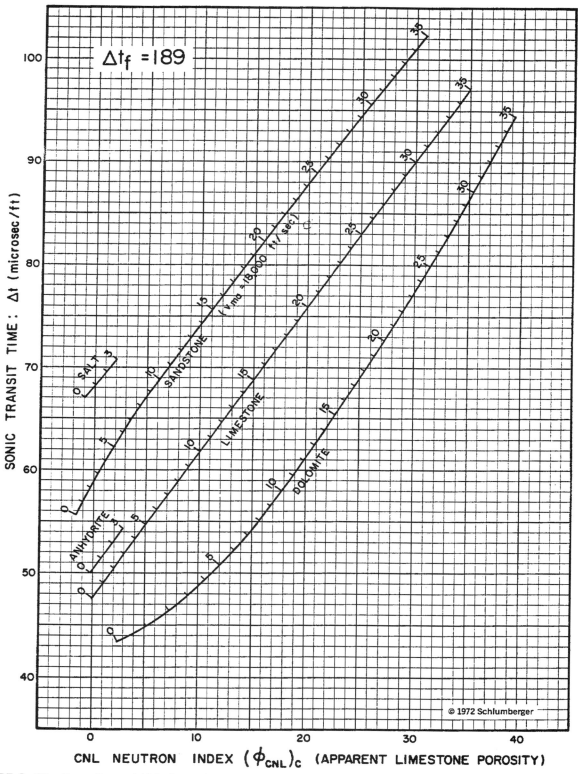

FIGURE 3–68 Porosity and Lithology determination from sonic and compensated neutron log CNL (after Schlumberger, 1972)

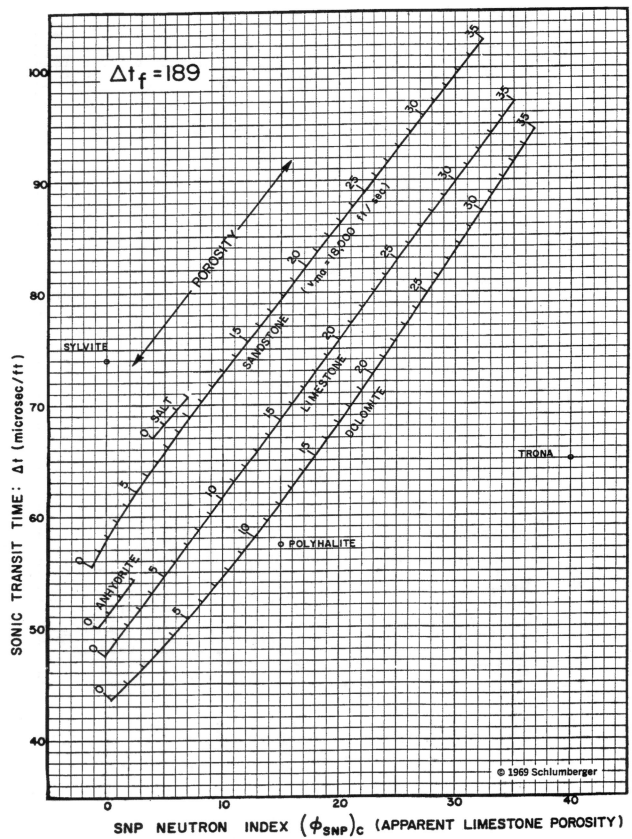

FIGURE 3–69 Porosity and lithology determination from sonic and sidewall neutron porosity log SNP (after Schlumberger, 1969)

In some reservoirs with secondary porosity it is possible to calculate a "secondary porosity index" (SPI) from the equation:

$$SPI = \frac{\phi_{xp} - \phi_s}{\phi_{xp}} \tag{3-54}$$

Where
ϕ_{xp} = porosity from the neutron-density crossplot, fraction
ϕ_s = sonic porosity, fraction

If the sonic log is reading only matrix porosity, SPI is the same as the partitioning coefficient v, discussed previously.

The effect of gas must be taken into account as it will produce a decrease in the readings of the density log and an increase in the neutron log. The average porosity determined from the crossplot will not be very far from the correct one but the lithology volumes can be completely wrong. In some cases gas might produce cycle skipping in the sonic log.

Use of Three Porosity Logs

There are some instances in which the neutron and density logs give total porosity and the sonic log provides only matrix porosity. For these cases it is possible to estimate the approximate volume of two minerals (for example limestone and dolomite), and matrix and secondary porosity by solving simultaneously the following equations (Burke et al, 1967):

$$\rho_b = \rho_{fm} \phi_m + \rho_{f2} \phi_2 + \rho_{dol} V_{dol} + \rho_{lm} V_{lm} \tag{3-55}$$

$$\phi_N = \phi_{Nfm} \phi_m + \phi_{Nfm} \phi_2 + \phi_{Ndol} V_{dol} + \phi_{Nlm} V_{lm} \tag{3-56}$$

$$\Delta t = \Delta t_f \phi_m + \Delta t_{\phi2} \phi_2 + \Delta t_{dol} V_{dol} + \Delta t_{lm} V_{lm} \tag{3-57}$$

$$1 = \phi_m + \phi_2 + V_{dol} + V_{lm} \tag{3-58}$$

Where
ρ = density, gr/cc
ϕ = porosity, fraction
V = volume, fraction
Δt = sonic transit time, μsec/ft

and the subscripts b = bulk, f = fluid, m = matrix, 2 secondary, dol = dolomite, lm = limestone and N = neutron. Matrix and fluid coefficients of various minerals and types of porosity are shown on Table 3–4. The value of $\Delta t_{\phi2}$ is calculated based on an estimate of the lithology mixture. For example, a 50 – 50% mixture of dolomite and limestone results in a $\Delta t_{\phi2}$ equal to (0.5) (43.5) + (0.5) (47.6) = 45.5 μsec/ft. This is, in general, a good average to use in the carbonates.

Example 3–5. Determine lithologies and calculate volume of each mineral, and primary and secondary porosity for an interval with the following average characteristics: $\rho_b = 2.42$ gr/cc, $\Delta t = 71$ μsec/ft, ϕ_N 0.219.

Solution. Values of M and N are calculated to be 0.83 and 0.55 with the use of Equations 3–2A and 3–2B, respectively. In addition to the average characteristics mentioned above, the following parameters were input into these equations: $(\phi_N)_f = 1$, $\rho_f = 1$ gr/cc, $\Delta t_f = 189$ μsec/ft.

These values of M and N are entered in Figure 3–22. The resulting data point plots in Area B indicating an interval which probably contains dolomite, limestone, primary and secondary porosity.

Quantification of minerals and porosities is accomplished with the use of Equation 3–55 through 3–58 as follows:

1) $2.42 = 1\phi_m + 1\phi_2 + 2.87 V_{dol} + 2.71 V_{lm}$
2) $0.219 = 1\phi_m + 1\phi_2 + 0.035 V_{dol} + 0 V_{lm}$
3) $71 = 189\phi_m + 45.5\phi_2 + 43.5 V_{dol} + 47.6 V_{lm}$
4) $1 = \phi_m + \phi_2 + V_{dol} + V_{lm}$

TABLE 3–4 Matrix and fluid coefficients of several minerals and types of porosity (after Schlumberger)

Minerals		Δt_{ma}	ρ_{ma}	$(\phi_{SNP})_{ma}$
Sandstone (1) (v_{ma} = 18,000)		55.5	2.65	–.035
Sandstone (2) (v_{ma} = 19,500)		51.2	2.65	–.035
Limestone		47.5	2.71	0.000
Dolomite (1) (ϕ = 5.5% to 30%)		43.5	2.87	.035
Dolomite (2) (ϕ = 1.5% to 5.5% & > 30%)		43.5	2.87	.020
Dolomite (3) ϕ = 0.0% to 1.5%		43.5	2.87	.005
Anhydrite		50.0	2.96-3.00	0.000
Gypsum		52.0	2.35	0.490
Salt		67.0	2.03	0.040
Siderite*		40.8	3.90	0.120
Fluids		Δt_{ma}	ρ_{ma}	$(\phi_N)_f$
Primary Porosity (Liquid-Filled):	Fresh Mud	189.0	1.00	
	Salt Mud	185.0	1.10	1.00
Secondary Porosity (In dolomite):	Fresh Mud		1.00	
	Salt Mud	43.5	1.10	1.00
(In Limestone):	Fresh Mud		1.00	
	Salt Mud	47.5	1.10	1.00
(In Sandstone):	Fresh Mud		1.00	
	Salt Mud	55.5	1.10	1.00

* After O'Brien (1983)

There are many ways of solving the above four equations with four unknowns. One possibility is:

    a. Subtract Equation 2 from Equation 1, and Equation 4 from Equation 1 to obtain:

5) $2.201 = 2.835 V_{dol} + 2.71 V_{lm}$

6) $1.42 = 1.87 V_{dol} + 1.71 V_{lm}$

    b. Multiply Equation 6 by 1.5848 and subtract Equation 6 from Equation 5 to obtain $V_{dol} = 0.3798$

    c. Insert V_{dol} into Equation 5 and calculate $V_{lm} = 0.4149$

    d. Multiply Equation 4 by 189 and subtract Equation 4 from Equation 3 to obtain ϕ_2 = 0.02834

    e. Calculate a matrix porosity, ϕ_m, of 0.17696 from Equation 4

    f. Total porosity is consequently 0.2053. This figure compares with 0.219 given by the sidewall neutron porosity log.

Use of Three Porosity Logs and the Gamma Ray

In some instances, it is possible to calculate primary and secondary porosity and the contributing fractions of 3 lithologies, including clay, with the use of density, neutron, sonic and gamma ray logs. The basic equations for the case in which we have three lithologies (for example quartz, siderite and clay) can be written as follows (O'Brien, 1983):

$$\rho_b = \phi_t \rho_f + V_q \rho_q + V_s \rho_s + V_{cl} \rho_{cl} \tag{3–59}$$

$$\phi_N = \phi_t \phi_{Nf} + V_q \phi_{Nq} + V_s \phi_{Ns} + V_{cl} \phi_{Ncl} \tag{3–60}$$

$$\Delta t = \phi_m \Delta t_f + V_q \Delta t_q + V_s \Delta t_s + V_{cl} \Delta t_{cl} \tag{3–61}$$

$$GR \; \rho_b = V_{cl} \; GR_{cl} \; \rho_{cl} + (V_q \rho_q + V_s \rho_s) \; GR_{qs} \tag{3–62}$$

$$1 = \phi_t + V_q + V_s + V_{cl} \tag{3–63}$$

$$\phi_t = \phi_2 + \phi_m \tag{3–64}$$

where GR = gamma ray reading in API units, and the subscripts t = total, q = quartz, s = siderite, cl = clay, and the rest of nomenclature is as in the previous section on the use of three porosity logs.

The above equations assume that the various clay minerals (for example, illite and glauconite) can be treated as a single lithology. Furthermore it is assumed that total porosity is given by neutron and density logs while matrix porosity is given by the sonic log.

O'Brien (1983) has reported on the successful use of the above equations in the Kurparuk reservoir, North Slope, Alaska. For proper use of this method it is necessary to define the clay parameters using core data as a calibration tool. After matching core to log porosity in the Kurparuk reservoir, the following parameters were defined:

ϕ_{Ncl} = 0.5, GR_{qs} = 10 API units

Δt_{cl} = 100 μsec/ft, GR_{cl} = 150 – 180 API units

ρ_{cl} = 2.5 gr/cc

Mineral coefficients for siderite were found to be as follows:

ρ_s = 3.9 gr/cc, ϕ_{Ns} = 0.12, Δt_s = 40.8 μsec/ft.

The subscripts cl and s stand for clay and siderite, respectively.

ESTIMATE OF WATER SATURATION EXPONENT (n) FROM LOGS

Under favorable conditions, i.e., high water salinities and porosities greater than 15%, it is possible to estimate the water saturation exponent (n) from logs alone with the use of pulsed neutron and resistivity logs.

For this purpose a crossplot of Σ vs. porosity is prepared as shown in Figure 3–70. The value of Σ_m is obtained at zero porosity, and the Σ_w and Σ_{hy} values are plotted at 100% porosity. The $\Sigma_m - \Sigma_w$ line represents a water saturation of 100%. The $\Sigma_m - \Sigma_{hy}$ line represents a water saturation of zero. A family of water saturation lines can be easily generated between the two boundary lines. From this type of plot it is possible to calculate S_w for a given zone as a function of Σ and porosity. Having the S_w value, one can calculate n using Archie's equations.

Example 3–6. Assume a zone with ϕ = 19%, Σ = 13, and R_t = 20 ohm m. Calculate the value of the water saturation exponent, n, if a = 1, R_w = 0.0265 ohm-m, m = 2, Σ_m = 9, Σ_{hy} = 21, and Σ_w = 70.

Solution. Construct the graph shown in Figure 3–70, with Σ_m = 9, Σ_{hy} = 21, and Σ_w = 70. From Figure 3–70, a zone with ϕ = 19% and Σ = 13 would have a water saturation of 18.48%. If aR_w = 0.0265 and m = 2,

R_o = $\phi^{-m} aR_w$ = $0.19^{-2} \times 0.0265$ = 0.73

I = R_t/R_o = 20/0.73 = 27.40

S_w = $I^{-1/n}$

n = $- \log I / \log S_w$

n = $- \log 27.40 / \log .1848$ = 1.96

Water saturation can be calculated rather than read from Figure 3–70 by using the following procedure:

1) Calculate B_n from:

$$B_n = (\Sigma - \Sigma_m)/\phi = (13 - 9)/0.19 = 21.053$$

2) Calculate S_w from:

$$S_w = (B_n - \Sigma_{hy} + \Sigma_m) / (\Sigma_w - \Sigma_{hy})$$
$$= (21.053 - 21 + 9) / (70 - 21) = 0.1848$$

Based on this S_w, the value of n is calculated to be 1.96. The same type of information was used for constructing the plot shown on Figure 3–61.

In some cases it is possible to calculate values of n from the crossplot of R_t vs. ϕ or the response of a porosity tool on log-log coordinates. The method that follows is strictly applicable to intervals at irreducible water saturation and was developed originally to be used in intergranular porosity. It appears, however, that since the irreducible water saturation of the fractures is close to zero, the method could be used in some instances for determining n values in naturally fractured reservoirs.

Empirically, it has been observed in some cases that for zones at irreducible water saturation:

$$\phi S_w \cong C \tag{3–65}$$

Where C is a constant. From Equations 3–44 and 3–45:

$$R_t = a\phi^{-m} R_w I \tag{3–66}$$

Since $I = S_w^{-n}$, Equation 3–66 can be written as:

$$R_t = a\phi^{-m} R_w S_w^{-n} \tag{3–67}$$

Inserting Equation 3–65 into 3–67 and taking logarithm in both sides of the equation results in:

$$\text{Log } R_t = (n - m) \log \phi + \log \left(\frac{aR_w}{C^n}\right) \tag{3–68}$$

Equation 3–68 indicates that a crossplot of R_t vs. ϕ on log-log coordinates should result in a straight line with a slope equal to n – m for intervals at irreducible water saturation as

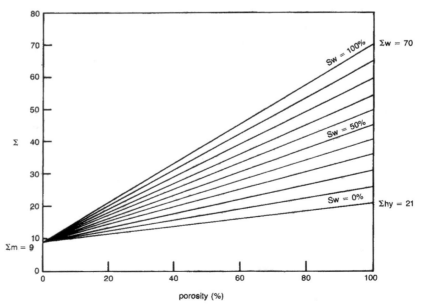

FIGURE 3–70 $\Sigma - \phi$ crossplot allows calculating water saturation S_w

shown on Figure 3–71. Since m is determined from the 100% water saturation line, the value of n is readily calculated. For improved results, Sanyal and Ellithorpe (1978) have suggested the following procedure:

1) Determine the value of n
2) Extrapolate the irreducible water saturation line to 100% porosity. Read at this point an R_t which is equal to aR_w/C^n. Calculate C.
3) Extrapolate the irreducible water saturation line until it intercepts the 100% water saturation line. If the porosity read at the intercept is equal to constant C calculated in step 2, the selected irreducible water saturation line is correct. If it is not equal, shift the irreducible water saturation line to a new position and repeat the procedure.

Daniel and Hvala (1982) have reported on the successful use of this method for analysis of naturally fractured volcanic rocks in Argentina.

FLUSHED FRACTURES

Some of the methods presented in previous sections are based on Pickett log-log crossplots of R_t vs. ϕ or the response of a porosity log. Pickett plots can also be used to analyze flushed intervals. The equation controlling the analysis can be written as follows:

$$\log R_{xo} = -m \log \phi + \log (aR_{mf}) + \log I_{xo} \qquad (3–69)$$

where I_{xo} = S_{xo}^{-n} = resistivity index of the flushed zone
        R_{xo} = resistivity of the flushed zone, ohm–m
        R_{mf} = resistivity of mud filtrate at reservoir temperature, ohm–m

and other nomenclature has been defined previously.

Equation 3–69 indicates that a crossplot of R_{xo} vs. ϕ on log-log paper should result in a straight line with a negative slope equal to m provided that a, R_{mf} and I_{xo} are constant. If m is smaller than the porosity exponent of the matrix blocks (m_b) the interval being analyzed is probably fractured. Extrapolation of the straight line representing $I_{xo} = 1$ or $S_{xo} = 100\%$ to a porosity of 100% should result in an R_{xo} equal to aR_{mf}. Since the value of R_{mf} is always known, this type of crossplot is very useful for estimating "a" values of naturally fractured reservoirs. The same type of analysis can be carried out by crossplotting responses of porosity logs such as $(\Delta t - \Delta t_m)$ or $(\rho_s - \rho_b)$ vs. R_{xo} on log-log paper. In some cases, a similar kind of evaluation can be performed as discussed under the heading "Electromagnetic Propagation (EPT)-Resistivity Combination".

In addition to the determination of m, the Pickett plot allows direct calculation of I_{xo} from $I_{xo} = (R_{xo})_t/(R_{xo})_o$, where $(R_{xo})_t$ is the resistivity of a partially flushed zone and $(R_{xo})_o$ is a resistivity read in the 100% S_{xo} line. Both of these resistivities are read at the same porosity. Finally, S_{xo} can be calculated from:

$$S_{xo} = I_{xo}^{-1/n} \qquad (3–70)$$

From here an estimate of the movable oil saturation (MOS) can be made with the use of MOS = $S_o - S_{or} = S_{xo} - S_w$, where S_w is calculated as explained in previous sections or with the use of the $P^{1/2}$ statistical analysis discussed next, S_o is oil saturation, and S_{or} is residual oil saturation.

WATER SATURATION OF THE COMPOSITE SYSTEM

A previous section showed how to calculate S_w of the composite system directly from the log-log crossplot of R_t vs. porosity response. The method involved the evaluation of I, then the calculation of S_w from Equation 3–43. Another approach involved taking the ratio between porosity at 100% water saturation and porosity of the interval being evaluated.

Sometimes, the analysis is not as straightforward due to uncertainty in establishing the position of the 100% water-bearing trend. In these cases, a statistical analysis which uses a

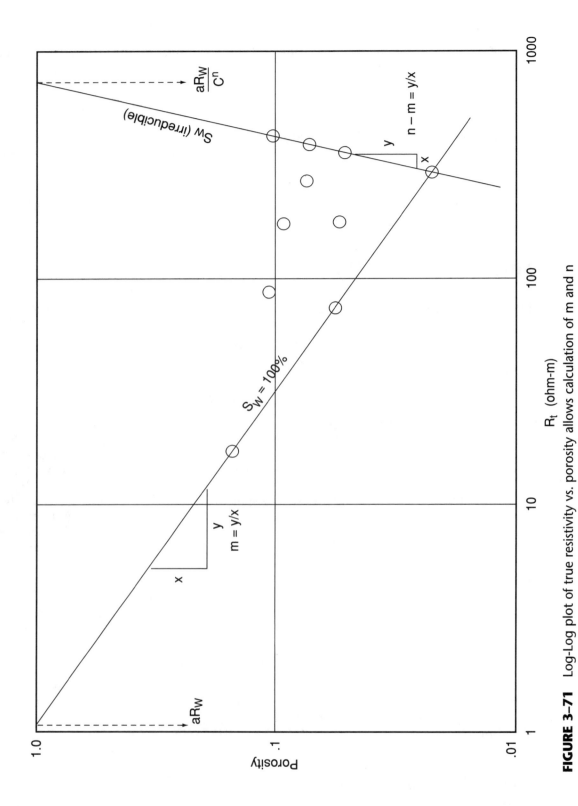

FIGURE 3–71 Log-Log plot of true resistivity vs. porosity allows calculation of m and n

parameter ($P^{1/2}$) becomes a significant evaluation tool (Porter et al, 1969). $P^{1/2}$ is a function of formation resistivity, porosity tool response, and the porosity exponent m.

The statistical behavior of $P^{1/2}$ was investigated previously by Porter et al (1969) for intergranular porous media in 13 wells, including sections from the Cretaceous sands in the Illinois Basin, Miocene sands from the U.S. Gulf Coast, Cretaceous sands from northwest Montana, Silurian carbonates in the Williston Basin, and Devonian sands from the Great Artesian Basin of Australia. Burial depths ranged from a few hundred feet to more than 12,000 ft. Hydrocarbon production ranged from zero to hundreds of barrels of oil per day. Matrix porosity ranged from a few percent to more than 30%. There were also significant ranges in water resistivity, grain density, water saturation, and porosity exponent.

Porter et al (1969) found that $P^{1/2}$ was a parameter with a normal distribution for 100% water saturation zones. When this technique was extended to naturally fractured reservoirs (Aguilera, 1974), $P^{1/2}$ also had a normal distribution for 100% water saturation zones.

The parameter P leading to the $P^{1/2}$ statistical analysis can be obtained in the case of the sonic log by combining Equations 3–44, 3–45 and 3–50:

$$P = R_t (\Delta t - \Delta t_m)^m = aR_w B^m I \qquad (3\text{–}71)$$

Notice that in Equation 3–71 the constants aR_w, B, and m, and the resistivity index, I, have been placed on one side of the equation. Consequently, P should be a constant for zones with 100% water saturation if the measurements of R_t and Δt were perfect, and if the values of aR_w, B, m, and Δt_m were perfectly constant. Since this is highly idealistic, P was investigated and found to have a square-root-normal distribution for zones with 100% water saturation. This distribution accounts for deviations of the ideal conditions indicated previously.

From the preceding analysis, it follows that:

$$P^{1/2} = [R_t(\Delta t - \Delta t_m)^m]^{1/2} \qquad (3\text{–}72)$$

As in the previous cases, a constant A rather than Δt_m can yield:

$$P^{1/2} = [R_t(\Delta t - A)^m]^{1/2} \qquad (3\text{–}73)$$

Similar treatment for a density log results in:

$$P^{1/2} = [R_t(\rho_s - \rho_B)^m]^{1/2} \qquad (3\text{–}74)$$

If porosity is known, the $P^{1/2}$ equation can be written as

$$P^{1/2} = [R_t \phi^m]^{1/2} \qquad (3\text{–}75)$$

This last equation is particularly useful in the case of mixed lithologies which are changing continuously with depth.

Since P had a square root normal distribution for zones with 100% water saturation, a plot of $P^{1/2}$ vs. cumulative frequency (which includes total number of samples for values of $P^{1/2}$ within a particular range) on probability paper should result in an approximately straight line. Hydrocarbon zones should deviate from this approximately straight line.

Figure 3–72 shows schematics associated with the $P^{1/2}$ statistical analysis. The familiar bell shape typical of intervals with 100% water saturation is presented in the upper part of the figure. Notice that hydrocarbon-bearing intervals fall to the right of the bell shape. The lower part of Figure 3–72 shows the probability crossplot where the bell-shape normal distribution becomes a straight line. Dots representing hydrocarbon-bearing intervals deviate from the straight line.

Once the hydrocarbon zones have been recognized, values of water saturation can be calculated as follows:

1. Consider the 100% water saturation zones as a single distribution. This should result in a straight line on probability paper.
2. Determine the median value of $P^{1/2}$ at a cumulative frequency of 50%. Since the distribution gives a straight line, i.e., it is symmetric or non-skewed, the median is also the mean and the mode.

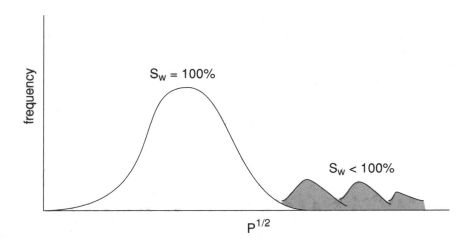

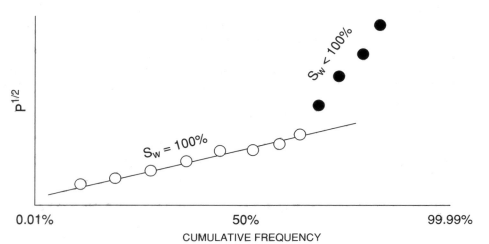

FIGURE 3–72 Schematics showing normal distribution for intervals that are 100% saturated with water. Hydrocarbon-bearing intervals deviate from the normal distribution.

3. Calculate the resistivity index I from the relationship:

$$I = \left[\frac{P_h^{1/2}}{P_{100}^{1/2}} \right]^2 \tag{3–76}$$

where $P_h^{1/2}$ is the value of $P^{1/2}$ for a hydrocarbon-bearing zone and $P_{100}^{1/2}$ is the median value of $P^{1/2}$ determined in Step 2.

4. Calculate water saturation from Equation 3–43. To accomplish this, assume that the porosity exponent (m) equals the saturation exponent (n). Based on previous experience, this assumption does not appear unreasonable.

A faster way to calculate water saturations consists of anchoring the median value of $P^{1/2}$ at a water saturation of 100% and drawing a straight line with a slope of –n/2 on log-log paper. This is valid because Equation 3–76 can be written as:

$$S_w^{-n} = \left[\frac{P_h^{1/2}}{P_{100}^{1/2}} \right]^2 \tag{3–77}$$

or,

$$S_w^{-n/2} = \frac{P_h^{1/2}}{P_{100}^{1/2}} \tag{3–78}$$

Consequently, $-n/2 \log S_w = \log P_h^{1/2} - \log P_{100}^{1/2}$. Values of water saturation for any $P^{1/2}$ can be determined from this kind of plot.

Example 3–7. This is a simplified example designed to provide a working knowledge of the $P^{1/2}$ technique. In most practical situations a statistically significant number of zones should be used for sound analysis.

Assume that the information of columns two and three is available:

(1) Zones	(2) Δt	(3) R_t	(4) $P^{1/2}$
1	58.0	55	16.91
2	60.0	27	17.37
3	64.2	11	17.52
4	61.0	29	20.64
5	59.0	49	19.80
6	60.0	35	19.78
7	61.0	23	18.39
8	62.0	19	18.76
9	63.0	15	18.42
10	63.0	100	47.57
11	62.0	115	46.15

Values of $P^{1/2}$ were calculated in column 4 using Equation 3–73 by taking A = 55 and m = 1.5. An estimate of these values was obtained by crossplotting Δt – 55 vs. R_t in log-log coordinates. Grouping the values of $P^{1/2}$ presented in column 4 resulted in the following distribution:

Range of $P^{1/2}$	No. of samples	Frequency	Cum. Frequency
16–17	1	0.091	0.091
17–18	2	0.182	0.273
18–19	3	0.273	0.546
19–20	2	0.182	0.728
20–21	1	0.091	0.819
46–47	1	0.091	0.910
47–48	1	0.091	1.001

Figure 3–73 shows a crossplot of $P^{1/2}$ vs. cumulative frequency for the ranges of $P^{1/2}$ indicated above. Notice that there is a trend for $P^{1/2}$ ranges 16 to 21 that breaks when ranges 46 to 48 are reached. This indicates that zones with $P^{1/2}$ ranges 46 to 48 are hydrocarbon-bearing.

Considering only the zones which are 100% water saturated leads to the following distribution:

Range	No. of samples	Frequency	Cum. Frequency
16–17	1	0.111	0.111
17–18	2	0.222	0.333
18–19	3	0.333	0.666
19–20	2	0.222	0.888
20–21	1	0.111	0.999
	9	0.999	

Figure 3–74 shows a crossplot of $P^{1/2}$ vs. cumulative frequency for the above distribution. A straight line is clearly delineated. The median value of $P^{1/2}$ for water-bearing zones is 18 at a cumulative frequency of 50%.

Values of resistivity index (I) for the hydrocarbon zones and water saturation (S_w) can be determined from Equation 3–76 as follows:

Zone	$P^{1/2}$	I	$S_w = I^{-1/n}$
10	47.57	6.98	0.27
11	46.15	6.57	0.29

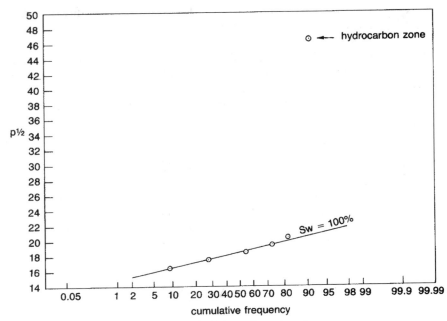

FIGURE 3–73 Cumulative frequency vs. P$^{1/2}$ for oil and water zones

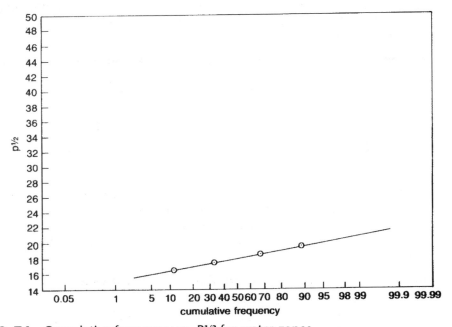

FIGURE 3–74 Cumulative frequency vs. P$^{1/2}$ for water zones

The above values of S_w were calculated assuming m = n = 1.5. Notice that we have arrived at values of S_w without having previous knowledge of aR_w, m, n, and porosity.

PRACTICAL APPLICATION

Figure 3–75 shows a log-log crossplot of Δt – A vs. R_a (LL8) for a well drilled in the naturally fractured Wasatch formation, Uinta Basin, Utah. The laterolog–8 was selected for resistivity due to the thin-bedded characteristics of the formation. The value of A was determined to be 50 by trial and error. The crossplot reveals a definite trend of points with

Log-log plot for Utah well

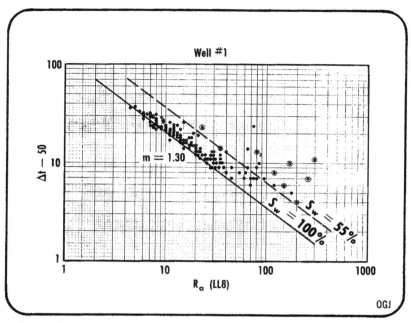

FIGURE 3–75 R_a vs. $(\Delta t - A)$, Well No. 1. R_a is an apparent resistivity from the LL8 proportional to true resistivity (after Aguilera, 1974)

a negative slope m of 1.3. This suggests the presence of fractures, since the porosity exponent of the matrix m_b as determined by a commercial laboratory was 2.0. The reservoir was known to be naturally fractured from other sources of information including cores and well testing data (Aguilera, 1973).

The evaluation of water saturation from Figure 3–75 is rather straightforward. As an example for zone 3, the resistivity index $(I = R_t/R_o)$ is given by $I = 80/18 = 4.44$; and $S_w = I^{-1/n} = 4.44^{-1/1.3} = 32\%$.

The $P^{1/2}$ statistical approach was used to corroborate S_w values due to the uncertainty of locating the 100% water-bearing line.

Figure 3–76 presents plots on probability paper of $P^{1/2}$ vs. cumulative frequency. The upper plot includes all zones analyzed. The plot shows a linear trend $(S_w = 100\%)$ until a cumulative frequency of about 83% is reached, at which time there is a drastic change in the trend indicating that the hydrocarbon zones have been reached.

The lower plot in Figure 3–76 shows a plot of $P^{1/2}$ vs. cumulative frequency on probability paper for the 100% water saturated zones. Notice that an approximate straight line is obtained. From this plot the median value of $P^{1/2}$ was determined to be 23.3.

Values of S_w as determined from Pickett's crossplot and the $P^{1/2}$ statistical analysis compare reasonably well.

As an example, for zone 3, $P_h^{1/2} = [R_t(\Delta t - A)^m]^{1/2} = [80(63 - 50)^{1.3}]^{1/2} = 47.4$. Consequently from Equation 3–76, $I = (47.4/23.3)^2 = 4.14$. Finally, S_w is found to be 34% from $S_w = I^{-1/n}$ by assuming that m equals n = 1.3. This compares with 32% from the Pickett plot.

Figure 3–77 shows a crossplot of S_w vs. $P^{1/2}$, prepared by anchoring the mean value of $P^{1/2}$ (23.3) at $S_w = 100\%$ and drawing a straight line with a slope of $-n/2$ $(-1.3/2 = -0.65)$. Water saturation for all intervals considered in this case can be read directly from this plot by knowing the correspondent $P^{1/2}$. For example for zone 3, $P_h^{1/2} = 47.4$ and from Figure 3–77 we read $S_w \simeq 34\%$.

This well was perforated at the intervals represented by the double-circle in Figure 3–75 and produced initially at oil rates exceeding 1000 b/d.

Cumulative frequency vs. P$^{1/2}$

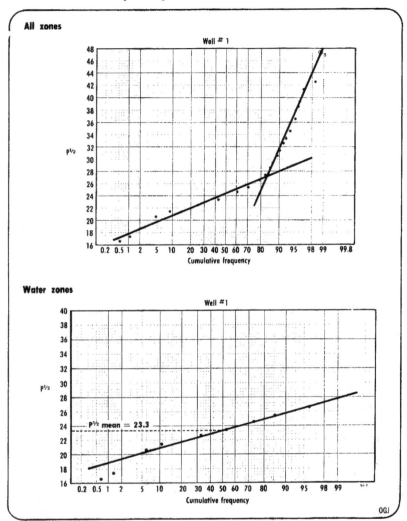

FIGURE 3–76 Cumulative frequency vs. P$^{1/2}$, Well No. 1 (after Aguilera, 1974)

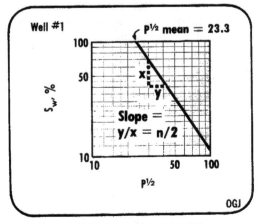

FIGURE 3–77 S$_w$ vs. P$^{1/2}$, Well No. 1 (after Aguilera, 1979)

ESTIMATES OF WATER SATURATION IN MATRIX AND FRACTURES

When a naturally fractured reservoir is discovered, there are always serious doubts as to the commerciality of the reservoir even if high initial hydrocarbon rates are obtained. The deep concern for this situation is obvious, as many fractured reservoirs produce at high initial rates that soon drop to noncommercial levels. These initially high rates come from hydrocarbons stored in the fractures. Thus, it is very important to have an estimate of hydrocarbon saturation in the fractures.

The procedure presented here for estimating water saturation is neither exact nor perfect, but in my experience provides reasonable orders of magnitude.

Fatt (1956) has shown that for a bundle of tubes, the relative permeability curves are straight lines with 45° angles (Figure 3–78). By assuming, as proposed by Hilchie and Pirson (1961), that a fracture system is approximately equivalent to a bundle of tubes, it is possible to estimate the hydrocarbon saturation in fractures and matrix as follows:

1. Measure carefully the initial water cut. Make sure you are dealing with formation water, not water lost during drilling operations.
2. Determine oil and water viscosities (μ_o and μ_w) and the oil formation factor (B_o) at reservoir conditions.
3. Calculate the fractured water saturation (S_{wf}) from the equation:

$$S_{wf} = \frac{\mu_w WOR}{B_o \mu_o + \mu_w WOR} \tag{3–79}$$

where WOR = water oil ratio

4. Calculate water saturation in the matrix (S_{wm}) from the equation:

$$S_{wm} = \frac{S_w - v S_{wf}}{(1 - v)} \tag{3–80}$$

Where S_{wm} is a matrix water saturation attached to bulk properties of the matrix-fracture composite system, S_w is the average water saturation of the composite system determined, for example from a Pickett plot or the $P^{1/2}$ statistical analysis, and v is the partitioning coefficient calculated as explained previously. Water saturation attached to matrix bulk properties (S_{wb}) is given by:

$$S_{wb} = \frac{S_w - v S_{wf}}{(1 - v)(1 - v\phi)} \tag{3–81}$$

Where ϕ is total porosity. S_{wb} should compare with water saturations from capillary pressures conducted in unfractured plugs. If the initial water cut is zero, it is not unreasonable to speculate that the water saturation in the fracture is zero. Conversely, the oil saturation is 100%. This is usually the case above the water-oil contact.

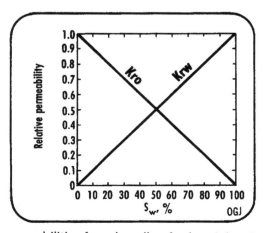

FIGURE 3–78 Relative permeabilities for a bundle of tubes (after Fatt, 1956)

Assume that the initial water cut of a well is 40% (or WOR = 0.67), the oil viscosity (μ_o) is 0.26 cp, the water viscosity (μ_w) is 0.50 cp, and the oil formation volume factor (B_o) is 1.5 bbl/STB. Using Equation 3–79, the value of water saturation in the fractures is 26.48%. Value of water saturation in the matrix (S_{wm}) can be calculated from Equation 3–80 as function of the partitioning coefficient (v), the water saturation in the fractures (S_{wf}), and the water saturation in the matrix-fracture system as found from a Pickett plot or the $P^{1/2}$ statistical analysis. If for example, v = 0.20, S_w = 0.50, and ϕ = 0.05, then S_{wm} = 0.559 and S_{wb} = 0.564. Thus in this case the difference between S_{wm} and S_{wb} is negligible.

FRACTURE COMPLETION LOG

The Pickett plot and $P^{1/2}$ statistical analysis have been combined into a computer program (Aguilera and Acevedo, 1982) called Fracture Completion Log (FCL). The program allows handling single and mixed lithologies, shaly formations, dual-water models and silty formations with chert, Opal A and Opal CT components. This process allows reasonable estimates of primary and secondary porosity, and water saturation in the primary, secondary and composite systems. The FCL utilizes conventional well logs such as resistivity, density, neutron and/or sonic. In some cases it might detect fractured systems even when the fractures are not intercepted by the wellbore, provided that the fractures are within the radius of investigation of the tools.

Figure 3–79 shows the relationship between vertical resolution and depth of investigation for some Schlumberger tools. Seismic would fall about an inch above the upper right corner, and scanning electron micrography about an inch diagonally below the origin. The borehole televiewer has a depth of investigation close to zero and vertical resolution of about 1 in. (2.5 cm) given optimal borehole conditions. The bar next to some measurements represents the range of the depth of investigation. Depth of investigation typically varies with diameter of invasion, drilling fluid resistivity, and resistivities in the invaded and virgin zones. Vertical resolution also varies with these parameters and, for some tools, with the effect of shoulder beds. In homogeneous formations, the High Resolution Dipmeter (HDT), Dual Dipmeter and Formation MicroScanner measurements can penetrate as deep as the shallow laterolog (LLS), but in the usual setting they are very shallow.

Figure 3–80 shows the fracture completion log of a well in the Williston Basin of Montana drilled through Carbonates.

The plot shows in the first track the gamma ray and the partitioning coefficient (v), i.e., the fraction of the total porosity that is made out of fractures. Next are depth and a black vertical line that defines the intervals, which based on a $P^{1/2}$ cutoff, are worthy of testing.

The second track shows matrix, fracture and total porosities. The last track shows total and matrix water saturation.

The well was perforated as indicated by the boxes on Figure 3–80, i.e., following very closely the pay flag generated by the FCL. The well came with an initial rate equivalent to more than 4000 bo/d. A water coning study using Birk's method (1970), however, suggested that the maximum efficient rate (mer) to avoid sucking water into the perforations was less than 500 bo/d.

SHALY FORMATIONS

In most situations it is not unreasonable to anticipate that the degree of fracturing decreases as shaly volumes increase. However, there are exceptions. The techniques presented in this section address this issue.

Many models have been published in the log interpretation literature to handle shaly formations. In general, they can be grouped into laminar, dispersed, and total shale models.

Laminar shales are those that exist in the form of laminae between layers of sand. This type of shales does not affect porosity or permeability of the sand layers. Structural shales—that is, those shales that might exist as grains or nodules in the formation matrix—are considered to have properties similar to those of laminar shales.

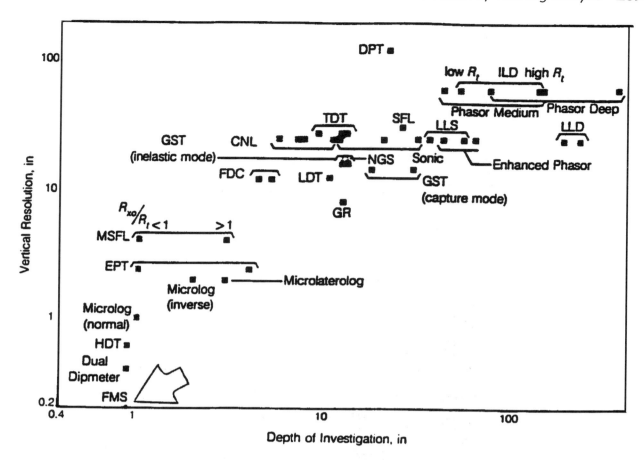

FIGURE 3–79 Vertical resolution and depth of investigation of various logging tools (after Schlumberger)

CNL*– Compensated Neutron tool
DPT*– Deep Propagation tool
Dual Dipmeter* tool– Also known as the Stratigraphic High Resolution Dipmeter [SHDT] tool.
EPT*– Electromagnetic Propagation tool
FDC*– Compensated Density tool
FMS– Formation MicroScanner* tool
GR– Gamma ray
GST*– Gamma Ray Spectrometry tool (capture and inelastic modes)
HDT*– High Resolution Dipmeter tool
ILD– Dual Induction Laterolog (DIL*) tool, deep

LDT– Litho-Density* tool
LLD– Dual Laterolog (DLL*) tool, deep
LLS– Dual Laterolog (DLL*) tool, shallow
MSFL– Microspherically focused log (MicroSFL)
NGS*– Natural Gamma Ray Spectrometry tool
Phasor*– Induction log (high vertical resolution)
SFL*– Spherically Focused Resistivity log
Sonic– Borehole compensated sonic tool
TDT*– Thermal Decay Time tool

Dispersed shales are those that might be dispersed throughout the sand, filling partially the intergranular porosity or coating the sand grains. Dispersed shales can reduce the formation permeability significantly.

Total shale models are those that apply reasonably well for many shaly formations independently of the distribution of shales.

Figure 3–81 shows the above forms of shale classified by manner of distribution in the formation. The upper part of the figure shows pictorial representations, whereas the lower part shows volumetric representations.

The log-log crossplot methods presented previously can be extended to the case of naturally fractured reservoirs in shaly formations (Aguilera, 1988, 1990). This section demonstrates that laminar, dispersed and total shale models can be analyzed using Pickett log-log

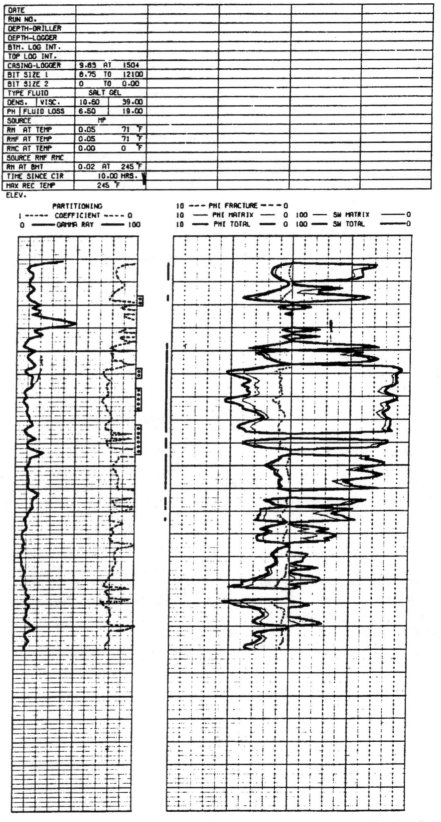

DATE					
RUN NO.					
DEPTH-DRILLER					
DEPTH-LOGGER					
BTM. LOG INT.					
TOP LOG INT.					
CASING-LOGGER	9.63 AT 1504				
BIT SIZE 1	8.75 TO 12100				
BIT SIZE 2	0 TO 0.00				
TYPE FLUID	SALT GEL				
DENS. | VISC.	10.60 39.00				
PH | FLUID LOSS	6.50 19.00				
SOURCE	MP				
RM AT TEMP	0.05 71 °F				
RMF AT TEMP	0.05 71 °F				
RMC AT TEMP	0.00 0 °F				
SOURCE RMF RMC					
RM AT BHT	0.02 AT 245 °F				
TIME SINCE CIR	10.00 HRS.				
MAX REC TEMP	245 °F				
ELEV.					

FIGURE 3–80 Fracture completion log of Williston Basin well (after Aguilera and Acevedo, 1979)

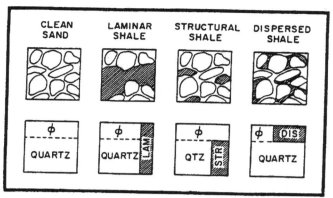

FIGURE 3–81 Types of shale classified by manner of distribution-pictorial representations above and volumetric representations below (after Schlumberger)

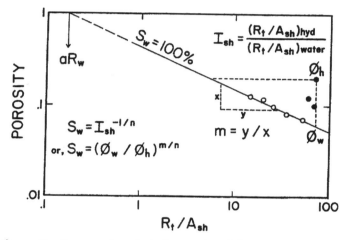

FIGURE 3–82 Schematic diagram for shaly formation (after Aguilera, 1990)

crossplots of porosity vs. true resistivity as affected by a shale group (A_{sh}). This is illustrated in the schematic diagram shown on Figure 3–82.

Laminated Shale Model

Poupon et al. (1954) have used the following parallel-conductivity model for the evaluation of laminated shaly formations:

$$S_w^n = \left[\frac{a(1 - V_{lam}) R_w}{\phi^m} \left(\frac{1}{R_t} - \frac{V_{lam}}{R_{sh}} \right) \right] \tag{3–82}$$

where V_{lam} = bulk volume of shale, distributed in laminate, each of uniform thickness, fraction

          R_{sh} = resistivity of the shale laminate, ohm-m

Other nomenclature is more conventional and has been provided previously.

Equation 3–82 can be rewritten as follows (Aguilera, 1988, 1990):

$$\frac{R_t}{A_{lam}} = aR_w \, \phi^{-m} S_w^{-n} \tag{3–83}$$

where A_{lam} is a laminar shaly group given by (Aguilera, 1988, 1990):

$$A_{lam} = \frac{(R_{sh} - R_t V_{lam}) \, (1 - V_{lam})}{R_{sh}} \tag{3–84}$$

Taking logarithm in both sides of Equation 3–84 leads to (Aguilera, 1988, 1990):

$$\log\left(\frac{R_t}{A_{lam}}\right) = -m\log\phi + \log(aR_w) + \log S_w^{-n} \tag{3–85}$$

Equation 3–85 indicates that a crossplot of (R_t/A_{lam}) vs. ϕ on log-log coordinates should result in a straight line with a slope equal to $-m$, provided that aR_w, S_w, and n are constants. If the line of 100% water saturation is extrapolated to a porosity of 100%, the value of aR_w could be read from the log-log plot on the (R_t/A_{lam}) scale.

If there are intervals at irreducible water saturation which meet the criteria $S_{wi}\phi =$ constant $= C$, the water saturation exponent, n, can be calculated by inserting $S_w = C/\phi$ into Equation 3–83, as follows:

$$\left(\frac{R_t}{A_{lam}}\right) = aR_w\,\phi^{-m}\left(\frac{C}{\phi}\right)^{-n} \tag{3–86}$$

Taking logarithm on both sides of the equation results in:

$$\log\left(\frac{R_t}{A_{lam}}\right) = (n-m)\log\phi + \log\left(\frac{aR_w}{C^n}\right) \tag{3–87}$$

Equation 3–87 indicates that a crossplot of (R_t/A_{lam}) vs. ϕ on log-log coordinates should result in a straight line with a slope equal to $(n-m)$, for intervals at irreducible water saturation. Since the value of m has been already determined as the slope of the 100% water saturation line, it is possible to readily calculate the water saturation exponent n.

Knowing m, n and aR_w, values of S_w are determined with the use of Equation 3–82 or 3–83.

Another possibility is to determine S_w directly from the Pickett plot. This is accomplished by calculating a resistivity index of the laminated shale hydrocarbon interval, I_{lam}, from the equation (Aguilera, 1988, 1990):

$$I_{lam} = \frac{(R_t/A_{lam})_{hyd}}{(R_t/A_{lam})_{water}} \tag{3–88}$$

and water saturation from

$$S_w = I_{lam}^{-1/n} \tag{3–89}$$

Example 3–8. Calculate water saturation in a formation with laminar shales using (1) Equation 3–82 and (2) a Pickett plot based on the following data: $a = 1$, $R_w = 0.3$, $\phi = 0.20$, $m = n = 2$, $R_t = 14$, $V_{lam} = 0.2$, $R_{sh} = 4$.

Solution.
1. S_w is calculated to be 35.9% with the use of Equation 3–82.
2. The shale group, A_{lam}, is found to be 0.24 with the use of Equation 3–84. Then $R_t/A_{lam} = 58.33$. A Pickett plot for these data is shown on Figure 3–83. Resistivity index, I_{lam}, is 58.33/7.5 from Figure 3–83 and S_w is 35.86% from Equation 3–89. For the case of a naturally fractured reservoir the porosity exponent m of the composite system on a Pickett plot should be smaller than m_b of the matrix blocks, and Figures 3–48 through 3–52 could be used to quantify the degree of fracturing.

Dispersed Shale Model

In some cases, shales might be dispersed throughout a sand, partially filling the smaller pores or coating the sand grains. Schlumberger have presented the following equation to calculate water saturation in the case of dispersed shale reservoirs:

$$S_w = \sqrt{\frac{aR_w}{\phi^2 R_t} + \left(\frac{V_{dis}(R_{dis}-R_w)}{2\phi R_{dis}}\right)^2} - \left[\frac{V_{dis}}{\phi}\frac{(R_{dis}+R_w)}{2R_{dis}}\right] \tag{3–90}$$

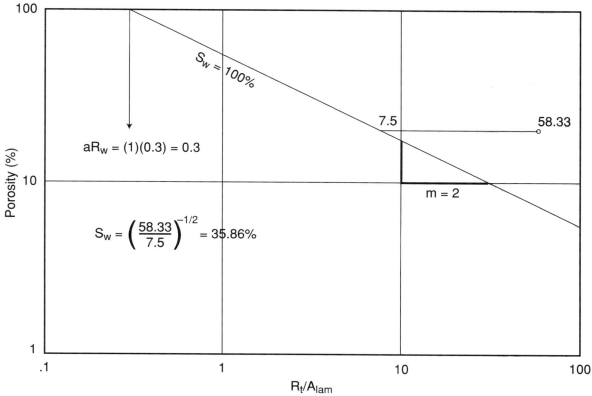

FIGURE 3–83 Pickett plot for laminated sand shales, example 3–8.

where the subscript dis stands for dispersed, V is fractional bulk volume and other nomenclature has been defined previously.

The above equation has been proved valuable in several areas where all the required input data are available. For example, it has given good results in low resistivity pay sands in the Gulf Coast. The problem is that in many cases there is a high degree of uncertainty with respect to the constant a, the water resistivity, R_w, the exponent m and the fractional bulk volume of the dispersed shale, V_{dis}. To reduce this uncertainty Equation 3–90 has been rewritten as follows (Aguilera, 1988, 1990):

$$\frac{R_t}{A_{dis}} = aR_w \, \phi^{-m} \, S_w^{-2} \tag{3–91}$$

where a cementation exponent, m, rather than 2.0 has been introduced to allow handling of dual-porosity systems, and A_{dis} is a dispersed shale group given by:

$$A_{dis} = \left[1 + \frac{\phi^m R_t}{aR_w} \left(B_{dis} - 2\sqrt{\frac{aR_w}{\phi^m R_t} + B_{dis} \, C_{dis} + C_{dis}^2} \right) \right] \tag{3–92}$$

$$B_{dis} = \left[\frac{V_{dis}(R_{dis} - R_w)}{2\phi \, R_{dis}} \right]^2 \tag{3–93}$$

and

$$C_{dis} = \frac{V_{dis}(R_{dis} + R_w)}{2\phi R_{dis}} \tag{3–94}$$

Taking logarithm in both sides of Equation 3–91 results in:

$$\log\left(\frac{R_t}{A_{dis}}\right) = -m \log \phi + \log(aR_w) + \log S_w^{-2} \tag{3–95}$$

Equation 3–95 shows that a crossplot of log (R_t/A_{dis}) vs. log ϕ should result in a straight line with a slope equal to –m for intervals with constant aR_w and constant S_w.

Analysis of a dispersed shale reservoir using a Pickett plot would involve the following steps:

1. Calculate B_{dis} and C_{dis} at each depth of interest from Equation 3–93 and 3–94, respectively.
2. Calculate A_{dis} from Equation 3–92. Note that in order to calculate A_{dis}, it is required to know aR_w and m. If these parameters are not known make an educated estimate of them. Assumed incorrect aR_w and m values would show up later in the analysis.
3. Prepare a crossplot of R_t/A_{dis} vs. ϕ on logarithmic paper. This should result in straight line for intervals of constant water saturation.
4. Extrapolate the 100% water saturation straight line to a porosity of 100%. At this point read the product aR_w in the R_t/A_{dis} logarithmic scale. Also determine the value of the slope –m. If aR_w and m are different from the values input in step 2, assume new values and repeat the Pickett plot. The assumption that m is equal to 2.0 is valid for many reservoirs with dispersed shale that do contain any natural fractures or where the degree of fracturing is very small. However, if the degree of fracturing is medium to intense m should be smaller than 2.0.
5. Calculate a resistivity index, I_{dis}, from the equation (Aguilera, 1988, 1990):

$$I_{dis} = \frac{(R_t/A_{dis})_{hyd}}{(R_t/A_{dis})_{water}} \qquad (3\text{–}96)$$

Where both numerator and denominator can be read directly from the Pickett plot.
6) Calculate water saturation of the dispersed shale hydrocarbon interval from the equation:

$$S_w = I_{dis}^{-1/2} \qquad (3\text{–}97)$$

Equation 3–97 implies the assumption that n equals 2.0 in formations with dispersed shales. In naturally fractured reservoir, however, the assumption that m equals n provides in my experience better results.

When Equation 3–90 is used in a conventional fashion, it is necessary to have a very good grasp of aR_w and m in order to come out with reasonable values of water saturation. In the present method m does not have to be known in advance. On the contrary it can be determined directly from the Pickett plot, provided that there are intervals with constant water saturation. This implies trial and error because a value of m has to be input in Equation 3–92 to prepare the Pickett crossplot.

Since the method allows to calculate the porosity exponent, m, from logs, it is possible to use the present analysis for evaluation of naturally fractured reservoirs with dispersed shales.

Waxman and Smits (1968) have presented the following equation to calculate water saturation in sands having dispersed clays:

$$S_w^{-n\star} = \frac{R_t}{\phi^{-m\star}R_w}(1 + R_wBQ_v/S_w) \qquad (3\text{–}98)$$

where n* and m* are saturation and porosity exponents of the dispersed clay, respectively, B is the equivalent conductance of clay-exchange cations in (ohm-m)$^{-1}$ (meq/ml)$^{-1}$, and Q_v is the volume concentration of clay exchange cations, in meq/ml.

Equation 3–98 has proven valuable in shaly reservoirs including the unusual situation of high water resistivities. For example, it has been used by Koerperich (1975) for evaluation of shaly reservoirs J, M, and Q in the Kern River field of California where the water resistivities at bottom hole temperatures were found to be 5.6, 6.4 and 6.5 ohm–m, respectively.

For the cases in which n* is equal to 2.0. Equation 3–98 can be written as follows (Aguilera, 1988, 1990):

$$\left(\frac{R_t}{A_{cec}}\right) = \phi^{-m\star}R_wS_w^{-2} \qquad (3\text{–}99)$$

where the group A_{cec} is given by:

$$A_{cec} = \left[1 + \frac{R_t}{2\phi^{-m^*}R_w} \left(B_{cec}^2 - B\sqrt{B^2 + \frac{4\phi^{-m^*}R_w}{R_t}} \right) \right] \tag{3–100}$$

$$B_{cec} = B\, R_w\, Q_v \tag{3–101}$$

The equivalent conductance of clay-exchange cations, B, is usually determined from correlations published by Waxman and Thomas (1974) or estimated from the equation:

$$B = 4.6 \left[1 - 0.6 \exp\left(-0.77/R_w\right) \right] \tag{3–102}$$

where R_w is a water resistivity at 25°C.

The volume concentration of clay exchange cations, Q_v, has been found to correlate with gamma ray responses in certain areas including sandstones in the Mackenzie Delta in Canada (Johnson and Linke, 1978). In other areas this correlation has not been possible. According to Koerperich (1975) direct laboratory measurements is the only reliable means of determining Q_v. The problem is that conventional cores and suitable sidewall samples are not usually available throughout the whole section in all wells. As a consequence, an approach by which the cation exchange capacity is correlated to a certain log response appears the most reasonable from a practical point of view, although recognizing that the approach is imperfect.

Taking logarithm of Equation 3–99 in both sides of the equation leads to:

$$\log\left(\frac{R_t}{A_{cec}}\right) = -m^* \log\phi + \log R_w + S_w^{-2} \tag{3–103}$$

Equation 3–103 indicates that a log-log crossplot of (R_t/A_{cec}) vs. ϕ should result in a straight line with a slope equal to $-m^*$ for intervals with constant R_w and S_w. If the water-bearing straight line is extrapolated to 100% porosity, the water resistivity, R_w, can be read in the R_t/A_{cec} scale. A resistivity index, I_{cec}, is calculated with the use of the equation (Aguilera, 1988, 1990):

$$I_{cec} = \frac{(R_t/A_{cec})_{hyd}}{(R_t/A_{cec})_{water}} \tag{3–104}$$

and water saturation is given by:

$$S_w = I_{cec}^{-1/n} \tag{3–105}$$

where the water saturation exponent is assumed to be equal to 2.0.

Example 3–9. Calculate water saturation using (1) Equation 3–90 and (2) a Pickett crossplot based on the same data presented on Example 3–8. Note that shale volume and resistivity are for a laminar model in Example 3–8 as opposed to this example where a dispersed shale model is being considered.

Solution.
1. Water saturation is calculated to be 32.8% with the use of Equation 3–90 for dispersed shales. This S_w compares with 35.9% determined by considering a laminar shale model in Example 3–8.
2. Calculation of R_t/A_{dis} involves the following steps:
   B_{dis} = 0.2139 from Equation 3–93
   C_{dis} = 0.5375 from Equation 3–94
   A_{dis} = 0.2011 from Equation 3–92
   R_t/A_{dis} = 14/0.2011 = 69.62

A Pickett plot for the previous data is presented on Figure 3–84. Resistivity index, I_{dis}, is 69.62/7.5 = 9.28 from Figure 3–84. Water saturation is calculated to be 32.8% from Equation 3–97. For a naturally fractured reservoir the slope m, should be smaller than m_b, in which case Figures 3–48 through 3–52 could be used to quantify degree of fracturing. Furthermore, it has been found empirically that the assumption m = n, tends to give reasonably good results in naturally fractured reservoirs.

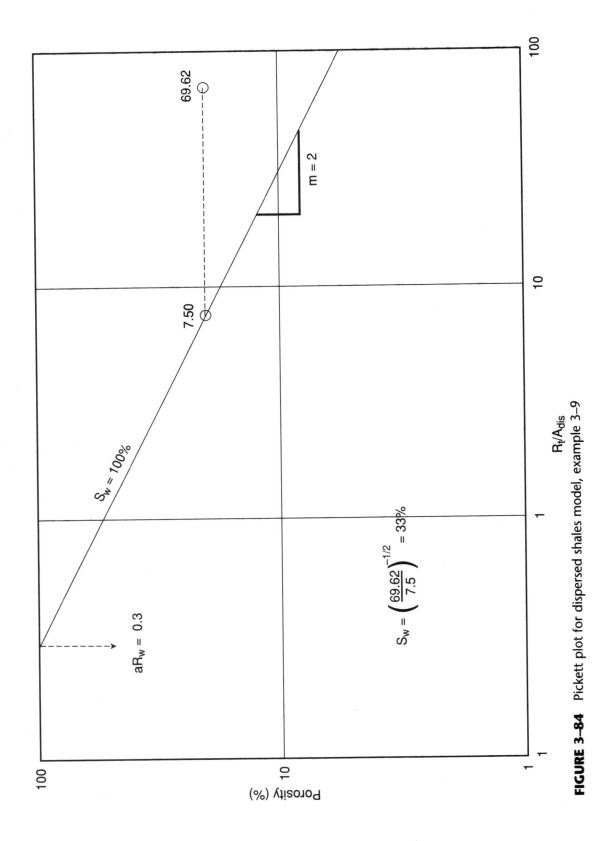

FIGURE 3–84 Pickett plot for dispersed shales model, example 3–9

Example 3–10. Calculate water saturation using (1) Waxman and Smits Equation 3–98 and (2) a Pickett plot based on the following data: $R_w = 6.4$, $Q_v = 0.36$, $R_t = 21$, $\phi = 0.29$, B = 2.15, $m^* = 1.6$, and $n^* = 2.0$.

Solution.

1. Water saturation is found to be 41.3% by using the following steps:
   Determine $B_{cec} = R_w B\, Q_v = 6.4 \times 2.15 \times 0.36 = 4.95$ (equation 3–101)
   Calculate water saturation by rearranging Equation 3–98 as follows:

$$S_w = \frac{B_{cec} + (B_{cec}^2 + (4\phi^{-m^*} R_w/R_t))^{1/2}}{2} \qquad (3\text{–}106)$$

2. Calculation of R_t/A_{cec} involves the following steps:
   B_{cec} = 4.95 from Equation 3–101
   A_{cec} = 0.0778 from Equation 3–100
   R_t/A_{cec} = 21/0.0778 = 269.85

A Pickett plot of the previous data is presented in Figure 3–85. Resistivity index, I_{cec}, is calculated to be 269.85/45.97 = 5.87 from Figure 3–85. Finally S_w is determined to be 41.3% with the use of Equation 3–105.

Total Shale Model

Simandoux (1963) and Poupon et al (1954) have shown that in some cases it is possible to use the following equation to calculate water saturation, independently of the distribution of shale:

$$S_w = \sqrt{\frac{aR_w}{\phi^2 R_t} + \left(\frac{aR_w}{2\phi^2}\frac{V_{tsh}}{R_{tsh}}\right)^2} - \left(\frac{aR_w}{2\phi^2}\frac{V_{tsh}}{R_{tsh}}\right) \qquad (3\text{–}107)$$

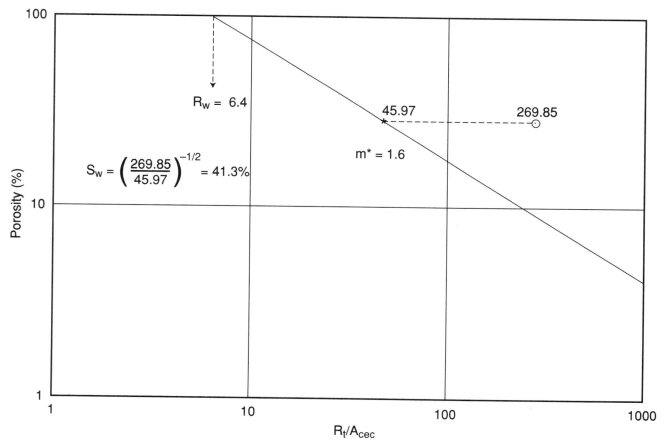

FIGURE 3–85 Pickett plot for Waxman-Smits model, example 3–10

where the subscript tsh stands for total shale and other nomenclature has been defined previously.

The above equation has been used with a good level of success in several places, including the U.S.A., Nigeria, Libya, and Argentina. Equation 3–107 can be written as follows (Aguilera, 1988, 1990):

$$\left(\frac{R_t}{A_{tsh}}\right) = aR_w\phi^{-m} s_w^{-2} \qquad (3\text{–}108)$$

where a cementation exponent m, rather than 2.0, has been introduced to allow handling a naturally fractured reservoir. The total shale group A_{tsh} is given by:

$$A_{tsh} = \left[1 + \frac{\phi^m R_t}{aR_w}\left(2B_{tsh}^2 - 2B_{tsh}\sqrt{\frac{aR_w}{\phi^m R_t} + B_{tsh}^2}\right)\right] \qquad (3\text{–}109)$$

and,

$$B_{tsh} = \frac{aR_w}{2\phi^m}\frac{V_{tsh}}{R_{tsh}} \qquad (3\text{–}110)$$

Taking logarithm in both sides of Equation 3–108 results in:

$$\log\frac{R_t}{A_{tsh}} = -m\log\phi + \log(aR_w) + \log S_w^{-2} \qquad (3\text{–}111)$$

Equation 3–111 indicates that a crossplot of R_t/A_{tsh} vs. ϕ should result in a straight line with a slope Equal to $-m$, provided that aR_w and S_w are constant.

Resistivity index of a total shale hydrocarbon zone is calculated from (Aguilera, 1988, 1990):

$$I_{tsh} = \frac{(R_t/A_{tsh})_{hyd}}{(R_t/A_{tsh})_{water}} \qquad (3\text{–}112)$$

Water saturation is determined with the use of the equation:

$$S_w = I_{tsh}^{-1/n} \qquad (3\text{–}113)$$

Where n is assumed to be equal to 2.0 for unfractured systems. Empirically, it has been found that the assumption m = n provides good results in many naturally fractured reservoirs.

Example 3–11. Calculate water saturation using (1) Equation 3–107 and (2) a Pickett plot based on the same data presented on Example 3–8. Note that shale volume and resistivity are for a laminar model in Example 3–8 as opposed to this example where a total shale model is being considered.

Solution.
1. Water saturation is calculated to be 56.81% with the use of Equation 3–107. This S_w compares with 35.9% determined by considering a laminar shale model in Example 3–8 and 32.8% determined with the use of a dispersed shale in Example 3–9.
2. Calculation of R_t/A_{tsh} involves the following steps:
   B_{tsh} = 0.1875 from Equation 3–110
   A_{tsh} = 0.6024 from Equation 3–109
   R_t/A_{tsh} = 14/0.6024 = 23.24

A Pickett crossplot based on the previous data is presented on Figure 3–86. Resistivity index, I_{tsh}, is 23.24/7.5 = 3.009 from Figure 3–86. Water saturation is calculated to be 56.81% with the use of Equation 3–113. For a naturally fractured reservoir the slope m should be smaller than m_{bi} in which case Figures 3–48 through 3–52 could be used to quantify degree of fracturing. Furthermore, it must be kept in mind that the assumption m = n usually provides reasonable results in fracture media.

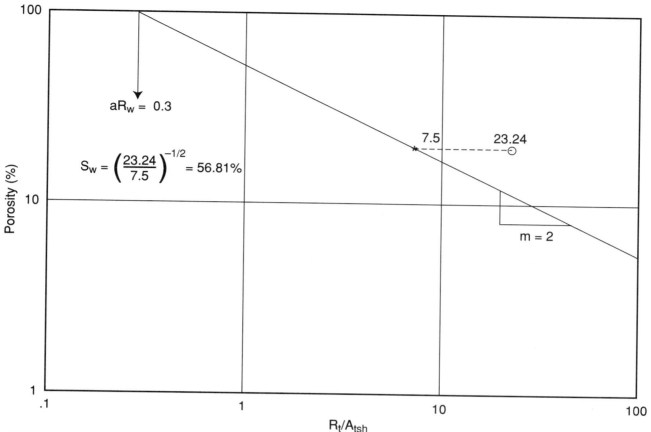

FIGURE 3–86 Pickett plot for total shale model, example 3–11

Pattchet and Rausch Shaly Model

In this approach water saturation is calculated with the use of the equations (Pattchet and Rausch, 1967):

$$S_w^n = \frac{a\phi^{-m}R_{mf}R_w}{R_t(R_{mf} - R_w)} \times A_{sh}$$

(3–114)

$$A_{sh} = 1 - 10^{-SP/(60 + 0.133T)}$$

(3–115)

where SP is corrected spontaneous potential in millivolts and T is formation temperature in degrees Fahrenheit. The above equation can be written as follows (Sanyal and Ellithorpe, 1978):

$$\log\left(\frac{R_t}{A_{sh}}\right) = -m \log \phi + \log\left(\frac{aR_wR_{mf}}{R_{mf} - R_w}\right) - \log S_w^{-n}$$

(3–116)

Equation 3–116 indicates that a crossplot of R_t/A_{sh} vs.ϕ on log-log paper should result in a straight line with a slope Equal to –m for intervals with constant S_w, a, R_w, and R_{mf}. The straight line for a water-bearing interval should give an R_t Equal to $(aR_wR_{mf}/(R_{mf}-R_w))$ at 100% porosity. From this plot it is possible to determine n using methods discussed previously. Finally water saturation is calculated from the equation:

$$S_w = [(R_t/A_{sh})_{hyd} / (R_t/A_{sh})_{water}]^{-1/n}$$

(3–117)

My experience indicates that the assumption m = n in naturally fractured reservoirs usually provides reasonable results.

Other Shaly Models

Similar procedures to those described above have been published by Aguilera (1988, 1990) for other shaly models, including the Indonesian model, Dual-Water model, Simandoux model, and Hossin model.

NUCLEAR MAGNETIC–RESISTIVITY COMBINATION

The nuclear magnetic log (NML) measures the earth's field proton free induction decay of formation fluids. Herrick et al (1979) have discussed the use of the tool for determination of irreducible water saturation, porosity in unusual lithologies, residual oil saturation and permeability. Aguilera (1990) has extended the technique to evaluations using Pickett log-log crossplots.

The NML gives free fluid porosity such as free oil when the formation is at irreducible conditions of water saturation. As such, it is a good indicator of permeability and indirectly a potential indicator of natural fractures.

Irreducible water saturation (S_{wi}) is given by the Equation:

$$S_{wi} = \frac{\phi - \phi_{fl}}{\phi} \tag{3–118}$$

where ϕ_{fl} is free fluid porosity and ϕ is effective porosity as determined, for example, from a neutron-density crossplot. The above equation can be inserted into Equation 3–67 to yield (Aguilera, 1990):

$$R_t = a\phi^{-m}R_w\left(\frac{\phi - \phi_{fl}}{\phi}\right)^{-n} \tag{3–119}$$

Assuming that m equals n, reduces Equation 3–119 to:

$$R_t = aR_w \, (\phi - \phi_{fl})^{-n} \tag{3–120}$$

Taking logarithm in both sides of the equation, leads to (Aguilera, 1990):

$$R_t = -n \log (\phi - \phi_{fl}) + \log (aR_w) \tag{3–121}$$

Equation 3–121 indicates that a crossplot of $\phi - \phi_{fl}$ vs. R_t on log-log paper should result in a straight with a slope equal to $-n$ and intercept equal to aR_w at $\phi - \phi_{fl} = 1$, provided that the zones are at irreducible water saturation. Intervals with movable water should fall below the straight line. Gas bearing intervals should fall above the straight line. Since n of the composite system is smaller than n_b of the matrix blocks in naturally fractured reservoirs, this crossplot has the potential of detecting natural fractures uniformly distributed.

Matrix Permeability

An empirical equation that sometimes gives reasonable estimates of formation matrix permeability when S_{wi} is irreducible has the form (Schlumberger):

$$k^{1/2} = c\phi^3/S_{wi} \tag{3–122}$$

where   k = formation matrix permeability, md
            c = constant whose value depends on the density of the hydrocarbon in the formation. For a medium gravity oil, $c \approx 250$; for a dry gas at shallow depth, $c \approx 79$.

Combing Equations 3–43, 3–44, 3–45 and 3–122 leads to (Aguilera, 1990):

$$\log Rt = (-3n -m) \log \phi + \log [aR_w \, (c/k^{1/2})^{-n}] \tag{3–123}$$

Equation 3–123 is telling us that a log-log crossplot of R_t vs ϕ should result in a straight line with a slope equal to $(-3n-m)$ for intervals at irreducible water saturation with constant aR_w and constant k. When m equals n, the slope would be equal to $-4m$. Extrapolation of this straight line to $\phi = 100\%$ would yield the product $[aR_w \, (c/k^{1/2})^{-n}]$.

Note that if Equation 3–122 is used, then the matrix permeabilities are valid. However, if Equation 3–123 is used, the permeabilities will be valid only if R_t is under irreducible water saturation conditions.

Because the essence of a "Pickett plot" is a crossplot of R_t vs ϕ on log-log paper, Equation 3–123 indicates that lines of constant matrix permeability can be superimposed on the Pickett plot to make it a more complete formation evaluation tool. For this technique to be valid in naturally fractured reservoirs the matrix porosity has to be approximately equal to the total (matrix + fracture) porosity, and the matrix water saturation has to be approximately equal to the water saturation of the composite system (Type A reservoirs).

Example 3–12. Evaluate the log of a high-porosity sand-shale sequence presented on Figure 3–87.

Solution. Values of porosity and resistivity are presented in columns 2 through 5 of Table 3–5. Determine if the eight zones of interest contain irreducible or movable water. Make estimates of formation permeability.

Effective porosity, ϕ, was calculated conventionally based on neutron and density crossplots. Free fluid porosity was read from the nuclear magnetic log. A crossplot of $\phi - \phi_{fl}$ vs. R_t on log-log paper is shown in Figure 3–88. Zones 1, 3, 6, 7, and 8, fall in the same linear trend indicating that they are irreducible conditions of water saturation. The slope or water saturation exponent, n, is 2.0. Extrapolation of the straight line to $\phi - \phi_{fl} = 1$ results in $aR_w = 0.036$. Water saturation can be calculated for these zones with the use of Equation 3–118. Results are shown in column 6, Table 3–5. Zones 2, 4 and 5 fall below the straight line indicating that they have some movable water. Gas zones should fall above the straight line. Calculations of S_w using conventional Archie's Equations (3–43 through 3–45), $aR_w = 0.036$, and the assumption that m equals n are shown in column 7. The comparison is good for the intervals at irreducible water saturation. When S_w is bigger than S_{wi} the interval contains movable water.

A major contribution of the plot presented in Figure 3–88 is that it allows determination of aR_w from intervals which produce clean oil with no water. This should help to alleviate the problem regarding lack of water resistivities in those reservoirs from which water samples are not available, reservoirs with poor SP development, or reservoirs for which water resistivity catalogues are not available.

An advantage of Archie's equations is that they calculate S_w for intervals which are or are not at irreducible water saturation.

An advantage of the NML log is that it calculates irreducible water saturation from Equation 3–118 without previous knowledge of R_t, m, n, a and R_w.

Figure 3–89 shows a Pickett plot using the same data in Table 3–5. On the basis of $aR_w = 0.036$ and m = 2 (m is assumed to be equal to n in the development of this technique) determined from Figure 3–88, the 100% water saturation straight line was drawn. From this plot the same values of water saturation shown in column 7 of Table 3–5, can be determined. For example, for Zone 1, moving horizontally along the dotted line, $S_w = (R_t/R_o)^{-1/n} = (5.5/0.3)^{-1/2} = 0.231$, or moving vertically along the dashed line $S_w = \phi_w/\phi = 0.081/0.35 = 0.231$, where ϕ_w is porosity read in the 100% water saturation line at the same resistivity of Zone 1. It is interesting to note that intervals 2 and 3 have approximately the same water saturation. Interval 2, however, contains movable water, while interval 3 does not.

Permeability is calculated from Equation 3–122. As an example for a medium gravity oil, c = 0.026, and using $S_{wi} = 0.231$ and $\phi = 0.35$ (Zone 1, Table 3–5), formation permeability is calculated to be 2190.9 md. Other values of permeability are presented in column 9, in Table 3–5. These permeabilities compare well with core data reported by Herrick et al (1979). Note that permeability decreases as water saturation (S_{wi}) increases. Furthermore, based on the analysis of Equation 3–123, the slope $(-3n-m)$ is equal to 8 because m = n = 2. If we want to draw a line of constant permeability equal to 1 md, we calculate R_t from:

$$R_t = a \, \phi^{-3n-m} \, R_w \, (c/k^{1/2})^{-n}$$

using $aR_w = 0.036$ and any porosity. For example, for $\phi = 0.10$ we calculate $R_t = 57.6$. Plot a control point corresponding to $\phi = 0.10$, $R_t = 57.6$ and draw a straight line

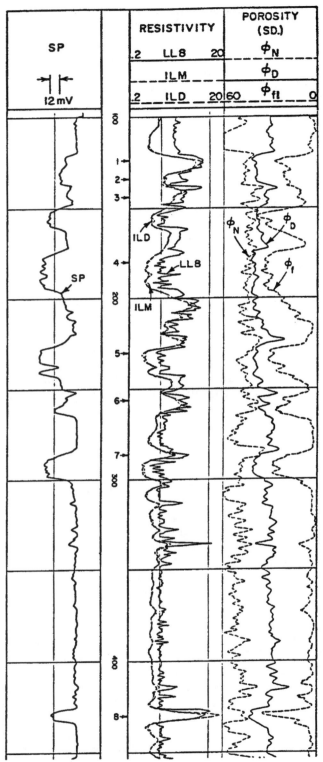

FIGURE 3–87 Logs of Well in high-porosity sand-shale sequence (after Herrick et al)

TABLE 3–5 Evaluation of log data presented in Figure 3-87

(1) Zone	(2) ϕ	(3) ϕ_{fl}	(4) $\phi - \phi_{fl}$	(5) R_t	(6) S_{wi}	(7) S_w	(8) ϕS_{wi}	(9) k (md)	(10) Water
1	0.35	0.27	0.08	5.5	.229	.231	.080	2190.9	Irreducible
2	0.30	0.19	0.11	1.3	.367	.555	.110	338.3	Movable
3	0.25	0.10	0.15	2.0	.600	.537	.150	42.4	Irreducible
4	0.32	0.26	0.06	0.5	.187	.839	.060	1919.1	Movable
5	0.30	0.23	0.07	0.4	.233	1.000	.070	839.3	Movable
6	0.25	0.14	0.11	3.0	.440	.438	.110	78.8	Irreducible
7	0.30	0.16	0.14	1.9	.467	.459	.140	208.9	Irreducible
8	0.35	0.29	0.06	10.9	.171	.171	.060	3921.1	Irreducible

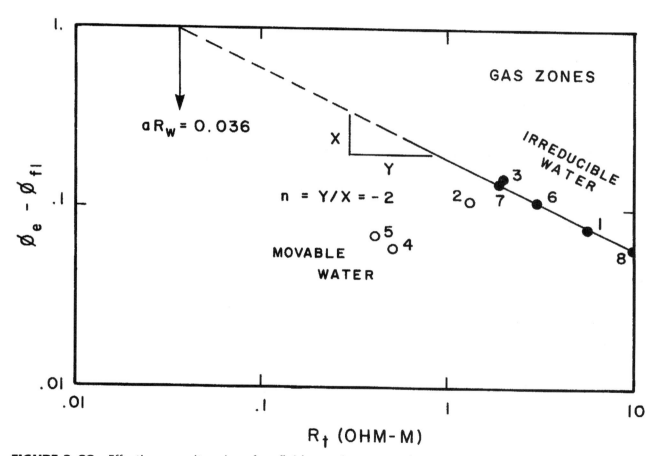

FIGURE 3–88 Effective porosity minus free fluid porosity vs. true formation resistivity. Data from well in high porosity sand-shale sequence (after Aguilera, 1990)

through this point with a negative slope equal to 8. The same procedure is followed for other permeabilities of interest. Note that the above equation is the same as Equation 3–123.

This process was utilized for generating Figure 3–90. The point corresponding to $\phi = 0.10$ and $R_t = 57.6$ discussed above, is highlighted by a triangle in Figure 3–90. It must be stressed that these estimates of formation matrix permeability are based strictly on irreducible water saturations, not saturations which include movable water.

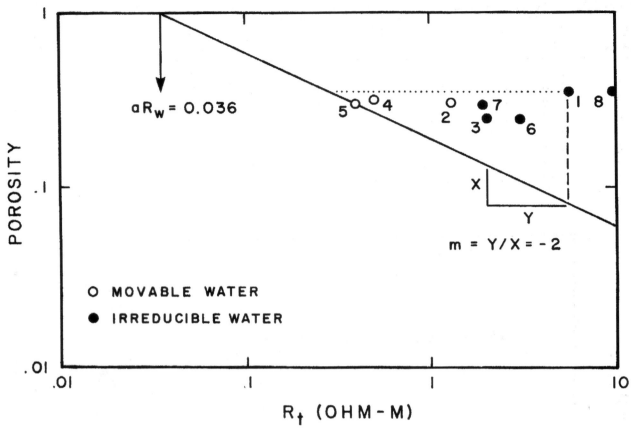

FIGURE 3–89 Pickett plot. Data from well in high porosity sand-shale sequence (after Aguilera, 1990)

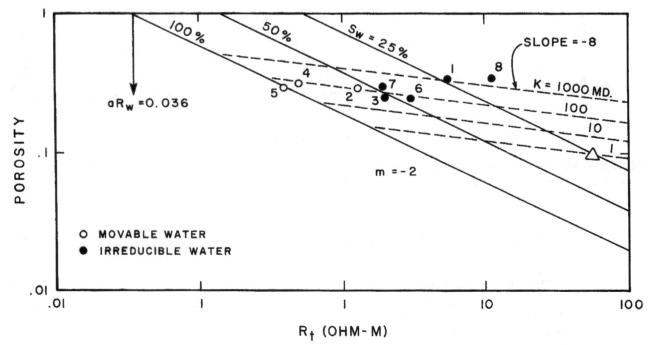

FIGURE 3–90 Pickett plot incorporating formation permeability. Data from well in high porosity sand-shale sequence (after Aguilera, 1990)

A Word of Caution

The previous example raises a word of caution with respect to irreducible water saturation. In many methods, some of which are presented in this chapter, the assumption is made that the product $\phi\, S_{wi}$ is approximately constant for intervals with the same lithology at irreducible water saturation. Example 3–12, based on irreducible water saturation determined from NML log, illustrates that the value of this product is changing as shown in column 8, in Table 3–5.

In fact, $\phi\, S_{wi}$ is equal to $\phi - \phi_{fl}$ based on Equation 3–118, and the crossplot of porosity vs. water saturation (Figure 3–91) does not show the coherent hyperbolic pattern sometimes seen in intervals at irreducible water saturation. It can be noticed that each one of the zones at irreducible water saturation has its own hyperbolic relationship ($\phi\, S_{wi} \approx$ constant) in spite of a constant lithology as shown by the dashed lines. This application suggests that, in some cases, it will be necessary to resort to additional information provided by cores in order to obtain meaningful irreducible water saturations.

LITHO DENSITY-RESISTIVITY COMBINATION

The P_e curve is a good qualitative indicator of natural fractures as discussed in the section "P_e Curve on the Litho-Density Log." The Litho-density can be combined with resistivity

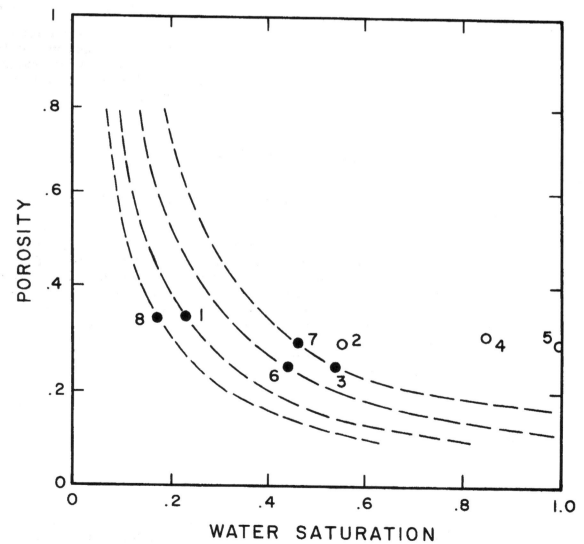

FIGURE 3–91 Porosity vs. water saturation. Data from well in high porosity sand-shale sequence (after Aguilera, 1990)

logs for quantitative analysis. In this case, porosity is given by (Gardner and Dumanoir, 1980):

$$\phi = \frac{U_m - U}{U_m - U_f} \qquad (3\text{--}124)$$

where U is a volumetric cross section, or the effective photoelectric absorption cross section index per unit volume in barns/unit volume, and the subscripts m and f stand for matrix and fluid, respectively. The volumetric cross section, U, is given by:

$$U = P_e\, \rho_e \qquad (3\text{--}125)$$

where P_e is an index of the effective photoelectric absorption cross section of the formation given by the litho-density log, in barns/electron and ρ_e is a measure of the number of electrons per unit volume. The value of ρ_e can be calculated with the use of the equation:

$$\rho_e = \frac{\rho_b + 0.1883}{1.0704} \qquad (3\text{--}126)$$

where ρ_b is bulk density in gr/cc.

Equations 3–44, 3–45, and 3–124 can be combined to yield:

$$\log R_t = -m \log (U_m - U) + m \log (U_m - U_f) + \log (aR_w) + \log I \qquad (3\text{--}127)$$

Equation 3–127 indicates that a crossplot of log R_t vs. log $(U_m - U)$ should result in a straight line with a slope equal to –m provided that U_m, U_f, aR_w and I are constant. From this kind of crossplot it is possible to estimate the value of U_m by trial and error and to calculate S_w as explained in the previous sections. A slope m smaller than m_b of the matrix blocks would suggest a possible naturally fractured reservoir.

ELECTROMAGNETIC PROPAGATION (EPT)— RESISTIVITY COMBINATION

A Pickett plot prepared with these tools responses might give indications with respect to the presence of natural fractures.

The EPT "is a downhole microwave instrumentation device which channels 1.1 GHz energy into the formation and measures the travel time and attenuation of the microwave signal as it propagates through the invaded zone" (Wharton et al, 1980). The EPT measures bulk volumes of water, including bound and movable water and mud filtrate. It was originally developed to run in fresh water environment. However, more recently it has been used as a water salinity estimator and in oil-base mud logs in the Middle East (Chardac, 1985). The EPT allows to differentiate hydrocarbon and water intervals even in formations with varying lithologies and water resistivities including fresh waters. The tool has a shallow depth of investigation (1 to 6 inches).

Water saturation (S_{xoe}) from the EPT tool is given by:

$$S_{xoe} = \frac{t_{po} - t_{pm} + \phi(t_{pm} - t_{ph})}{\phi(t_{pwo} - t_{ph})} \qquad (3\text{--}128)$$

Where
t_{po} = loss free propagation time
t_{pm} = loss free propagation time for matrix
t_{ph} = loss free propagation time for hydrocarbons
t_{pwo} = loss free propagation time for water
ϕ = total porosity, fraction

In cases of heavy crude oil where oil is not displaced by mud filtrate, S_{xoe} is equal to the original water saturation of the undisturbed reservoir. The same might happen in very tight zones such as those found in the matrix of naturally fractured reservoirs. Intervals of high permeability or with fractures will show values of S_{xoe} equivalent to the conventional

S_{xo}, or water saturation of the flushed zone. This concept allows estimates of m for the fractured intervals. In fact, Equation 3–128 can be rewritten as follows:

$$\phi = \frac{t_{po} - t_{pm}}{S_{xoe}t_{pwo} + (1 - S_{xoe})\,t_{ph} - t_{pm}} = \frac{t_{po} - t_{pm}}{B} \qquad (3\text{–}129)$$

Equation 3–129 can be inserted into Equation 3–69 for flushed zones to end up with:

$$\log R_{xo} = -m \log (t_{po} - t_{pm}) + m \log B + \log (aR_{mf}) + \log I_{xo} \qquad (3\text{–}130)$$

Thus a crossplot of R_{xo} vs. $(t_{po} - t_{pm})$ on log-log paper should result in a straight line with a slope equal to $-m$ provided that B, a, R_{mf}, and I_{xo} are constant. Intervals with $S_{xoe} = 100\%$ when extrapolated to B, should give an $R_{xo} = aR_{mf}$. Note that t_{pm} can be determined from this analysis by trial and error. Another possibility is to use the following propagation times (t_{pm}) for various minerals:

Mineral	t_{pm} (nanosec/m)	Mineral	t_{pm} (nanosec/m)
Sandstone	7.2	Halite	7.9 – 8.4
Dolomite	8.7	Gypsum	6.8
Limestone	9.1 – 10.2	Petroleum	4.7 – 5.2
Anhydrite	8.4	Shale	7.45 – 16.6
Dry colloids	8.0	Fresh water at 25°C	29.5

Propagation times for dry colloids, halite and gypsum were estimated by Wharton et al (1980) from published literature.

Equation 3–129 indicates that a crossplot of ϕ vs. t_{po} on Cartesian paper should yield a straight line with a slope equal to B from which it is possible to calculate S_{xoe} based on knowledge of t_{pwo}, t_{ph} and t_{pm}.

The EPT tool can also be used to estimate variable values of m in different rock types including fracture media. With S_{xo} from the EPT log (corrected for salinity effects), ϕ from the porosity logs, R_{xo} from a shallow resistivity tool, and R_{mf} from analysis of the mud filtrate, it is possible to calculate m from Equation 3–69. Focke and Munn (1987) have presented applications of this technique to Middle East carbonate reservoirs. Crossplots of formation factor (F) vs. porosity for fracture wells suggest values of m in the order of 1.6 for the fractured carbonate wells considered in Focke and Munn (1987) study.

BOREHOLE GRAVIMETER—RESISTIVITY COMBINATION

The borehole gravimeter (BHGM) appears to be a useful tool in highly fractured reservoirs as shown by Schmoker (1977). A radius of investigation of at least 50 ft provides excellent average bulk properties of the formation and, thus, the advantage of being highly insensitive to mud filtrate invasion and wellbore conditions. In addition it can be run through casing.

The BHGM can be calibrated when combined with a resistivity log. Under some conditions this combination allows determination of the cementation exponent (m) and water saturation.

Porosity from a BHGM derived bulk density is given by the equation:

$$\phi = \frac{\rho_s - \rho_b}{\rho_s - S_w\rho_w - \rho_h(1 - S_w)} \qquad (3\text{–}131)$$

where ρ is density in gr/cc and the subscripts s, b, w, and h stand for grain, bulk, water and hydrocarbon, respectively.

Equation 3–131 can be combined with Equations 3–44 and 3–45 to give:

$$\log R_t = -m \log (\rho_s - \rho_b) + \log (aR_w) + m \log B + \log I \qquad (3\text{–}132)$$

where,

$$B = \rho_s - S_w\rho_w - \rho_h (1 - S_w) \qquad (3\text{–}133)$$

Equation 3–132 tells us that a crossplot of R_t vs. $(\rho_s - \rho_b)$ on log-log paper should result in a straight line with a negative slope equal to m for intervals with constant B, aR_w, ρ_s and

I. An improved value of ρ_s can be determined by trial and error. In fact, different values of ρ_s can be tried in the log-log crossplot of $(\rho_s - \rho_b)$ vs. R_t until the desired straight line is obtained. Extrapolation of the 100% water-bearing line to $\rho_s - \rho_w = \rho_s - \rho_b$ permits determination of the product aR_w.

Example 3–13. The following data are available from a water-bearing formation:

Zone	ρ_b(gr/cc)	R_t (ohm)
1	2.13	1
2	2.28	2
3	2.39	4
4	2.50	12

Determine: (1) The value of ρ_s; (2) the porosity exponent, m; (3) the product of aR_w; (4) a family of straight lines for water saturations equal to 75, 50 and 25%, assuming that $\rho_h = 0.15$ and $\rho_w = 1.0$; (5) a family of lines for porosities equal to 10, 20 and 30%; (6) what are the values of porosity and water saturation for a zone with $\rho_b = 2.21$ gr/cc and $R_t = 21$; (7) if the value of m_b is 2.0, is the reservoir likely to be fractured?

Solution

1. Grain density, ρ_s, is calculated to be 2.65 gr/cc by trial and error. In fact, a crossplot of $2.65 - \rho_b$ vs. R_t on log-log paper results in a straight line as shown on Figure 3–92. Values different from 2.65 would yield curves rather than a straight line.
2. The porosity exponent, or slope of the 100% water saturation line is 2.0.
3. The product aR_w as obtained by extrapolating the water-bearing line to $\rho_s - \rho_b = \rho_s - \rho_w = 1.65$ is 0.1.

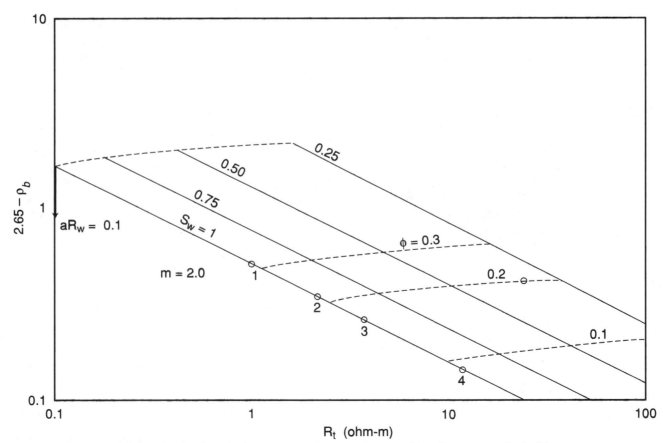

FIGURE 3–92 Pickett plot for borehole gravimeter-resistivity combination, example 3–13.

4. The family of constant water saturations is generated by previously calculating values of B from Equation 3–133, I from Equation 3–43 and the product IaR_w.

The following table shows calculation for each water saturation:

S_w	B	I	$I(aR_w)$
1.00	1.650	1.00	0.100
0.75	1.863	1.78	0.178
0.50	2.075	4.00	0.400
0.25	2.288	16.00	1.600

The $I(aR_w)$ values are plotted in the R_t axis against the values of B in the $\rho_s - \rho_b$ axis. Parallel lines to the 100% water line are drawn through these points, as shown by the black solid lines in Figure 3–92.

5. The family of porosity lines is generated with the use of Equation 3-131, rearranged in such a way as to calculate $\rho_s - \rho_b$. Results are summarized in the following table:

		$\rho_s - \rho_b = \mathbf{B} \phi$		
S_w	B	$\phi = 0.10$	$\phi = 0.20$	$\phi = 0.30$
1.00	1.650	0.165	0.330	0.495
0.75	1.863	0.186	0.373	0.559
0.50	2.075	0.208	0.415	0.623
0.25	2.288	0.229	0.446	0.686

The above numbers were used to draw the dashed lines representing 10%, 20%, and 30% porosity in Figure 3–92.

6. From Figure 3–92 a zone with $R_t = 21$ and $\rho_b = 2.21$ gr/cc ($\rho_s - \rho_b = 2.65 - 2.21 = 0.44$) would have a porosity of about 20% and a water saturation of approximately 35%.

7. For this example $m_b = m$. This method would suggest that the reservoir probably is not naturally fractured.

Equation 3–131 indicates that a crossplot of ρ_b vs. ϕ on Cartesian coordinates should result in straight lines for constant water saturations as shown on Figure 3–93. The intercept at zero porosity should be equal to ρ_s. The 100% water line intercept at 100% porosity gives the value of ρ_w. The zero water line intercept at 100% porosity gives ρ_h. Note that without the assistance of the resistivity log, there is a high uncertainty in determining S_w. For example, an interval with $\phi = 20\%$ and $\rho_b = 2.21$ was shown previously to have a water saturation of 35%. Figure 3–93 shows approximately the same value. However, there is uncertainty because the S_w lines are very close. The smaller the porosity the larger the uncertainty.

EFFECT OF VARIATIONS IN WATER RESISTIVITIES

In some cases, it is possible to calculate R_w from the SP log response. This has proven successful in some sand-shale sequences. In carbonates, quantification of R_w from the SP response is more unreliable.

Pickett (1972) has suggested the use of the following equation relating SP and R_w:

$$SP = -K \log (R_{mf}/R_w) \tag{3–134}$$

where K is a coefficient given by:

$$K = 61 + 0.133\,T \tag{3–135}$$

and,

R_{mf} = mud filtrate resistivity at reservoir temperature, ohm-m
T = reservoir temperature, °F

Equation 3–134 can be inserted into Equation 3–44 and 3–45 to yield:

$$R_t = \phi^{-m}a\left(\frac{R_{mf}}{10^{-SP/K}}\right)I \tag{3–136}$$

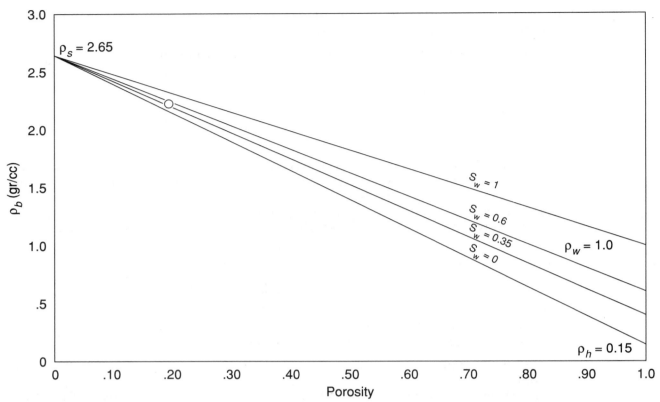

FIGURE 3–93 Borehole gravimeter density vs. porosity, example 3–13.

By making $A = 10^{SP/K}$ we can write:

$$\frac{R_t}{A} = \phi^{-m}aR_{mf}I \qquad (3\text{--}137)$$

Taking logarithm in both sides of the equation yields:

$$\log\left(\frac{R_t}{A}\right) = -m\log\phi + \log(aR_{mf}) + \log I \qquad (3\text{--}138)$$

Equation 3–138 indicates that a crossplot of R_t/A vs. ϕ on log-log paper should result in a straight line with a slope equal to –m provided that a, R_{mf} and I are constant. In a naturally fractured reservoir m should be smaller than the porosity exponent m_b of the matrix blocks. Extrapolation of the 100% water line to 100% porosity yields the product aR_{mf}. Data points representing hydrocarbon intervals should fall to the right of the water-bearing line. Water saturation is calculated from:

$$S_w = [(R_t / A)_h / (R_t/A)_w]^{-1/n} \qquad (3\text{--}139)$$

where the subscripts h and w stand for hydrocarbon and water, respectively. Experience indicates that the assumption m = n usually yields reasonable results in naturally fractured reservoirs.

Example 3–14. The following data are available from a formation which shows variations in water resistivities (K = 81 mv):

Zones	ϕ	R_t	–SP	A	R_t/A
1	0.168	0.74	62	0.1716	4.312
2	0.078	8.00	32	0.4027	19.866
3	0.120	2.60	42	0.3030	8.581
4	0.110	3.00	40	0.3208	9.352
5	0.120	2.30	95	0.0672	34.226

Determine (1) the value of m; (2) the product aR_{mf}; (3) water saturation for zone 5 assuming m = n; (4) is the reservoir likely to be fractured if m_b of the matrix = 2.0?; (5) It is possible to carry out the analysis with a plot of ϕ vs. R_t on log-log paper?

Solution

1. Values of A were calculated from $A = 10^{SP/K}$. Note the SP values in the above table are negative. Next a crossplot of ϕ vs. R_t/A was prepared on log-log paper as shown on Figure 3–94. Zones 1, and 2, 3, and 4 form a straight line with m =2.
2. The product aR_{mf} = 0.12 ohm-m at reservoir temperature is read at 100% porosity. If a equals 1, R_{mf} = 0.12. This value should compare with R_{mf} shown in the log heading once it is corrected to reservoir temperature.
3. Water saturation is calculated from Equation 3–139. The ratio $(R_t/A)_h$ = 34.226 for zone 5. A value of $(R_t/A)_w$ = 8.5 is read at the same porosity of zone 5 in the 100% water saturation line. Thus from Equation 3–139, $S_w = (34.226/8.5)^{-1/2}$ = 49.8%.
4. Probably the reservoir is not fractured as m = m_b
5. Figure 3–95 shows a conventional Pickett crossplot of log R_t vs. log ϕ. All data points tend to fall in same linear trend suggesting a constant water saturation. This is not correct as zone 5 has 50% water and the other intervals are 100% water saturated. Thus the analysis cannot be carried out with a conventional plot of ϕ vs. R_t in log-log paper.

HORIZONTAL WELLS

Logging of horizontal wells technology has advanced very rapidly (Aguilera et al, 1991). Good quality results can be anticipated in horizontal wells when running spontaneous potential, natural and spectral gamma ray, dual induction, high resolution dipmeter with

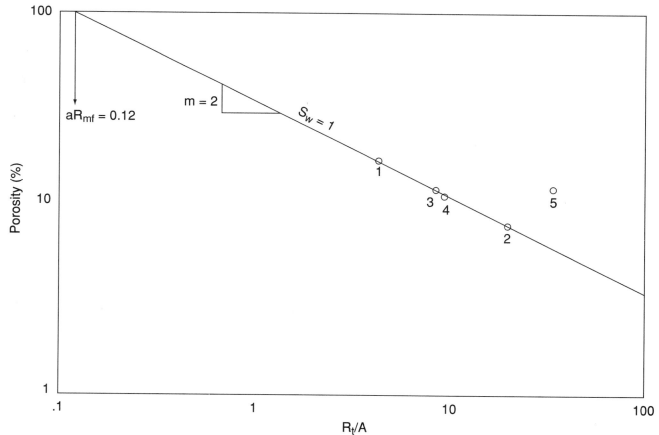

FIGURE 3–94 Pickett plot for reservoir with variations in water resistivities, example 3–14.

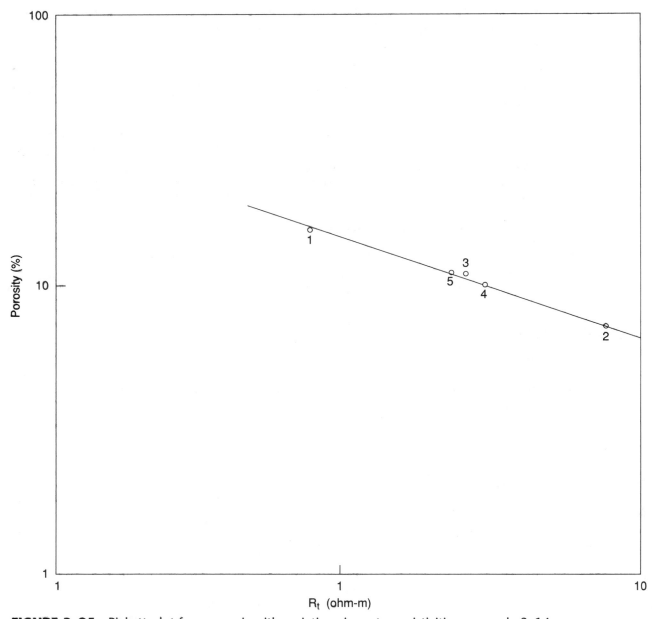

FIGURE 3–95 Pickett plot for reservoir with variations in water resistivities, example 3–14.

up to six hands, formation microscanner, formation microimager, shallow focused resistivity, dual induction, sonic (including log spacing), neutron, and density logs. Many of these tools are rated to over 400°F and more than 20,000 psi. The determination of porosity and water saturation can be conducted with conventional equations developed throughout the years. However, corrections must be introduced to the input data due to reasons outlined in the following sections. Most of the information on logging equipment is extracted from Halliburton literature (1990).

Pipe Conveyed Logging System

The tools can be transported downhole using pipe conveyed logging systems (PCLS). Once the tools are accurately positioned at the bottom of the zone to be logged the collection of logging data can be started (Halliburton, 1990). At all times the logging tools remain attached to the transport assembly allowing easy retrieval after the logging has been completed.

The pipe-conveyed logging system (PCLS) presents several advantages over the more conventional wireline-conveyed methods which utilize pumpdown techniques and gravity. One of the key advantages is that the PCLS allows circulation at almost any time during the operation (Halliburton, 1990; Western Atlas, 1990). This in turn permits hole conditioning whenever it is required and allows continuous control of the well. The probabilities of a successful latch are increased by circulating just prior to latching the wet connectors, as this cleans the connectors' mating assemblies.

A downhole navigational system provides wellbore trajectory and monitors the position of the logging tools in the borehole. Problems associated with doglegs, key seating, depth correlations, and high deviation angles are largely eliminated. Severe doglegs can be traversed by using special flex joints. Multiple logs can be made in a single run of horizontal wells. Stresses applied to the logging tools and pipe drag can be monitored through a load cell. This cell is very useful to indicate if the tools are working under extreme tension or compression. This is particularly important when the logging tools are run without protective housings.

When logging openhole intervals that are larger than the cased section of the well, a multiple latching feature becomes very useful, as the wet connector assemblies can be latched and unlatched as many times as necessary without removing the logging tools and the protective assembly from the wellbore.

Coiled or Reeled-Tubing-Conveyed Wireline Systems

In this approach standard wireline tools are attached at the surface to reeled-tubing through which a wireline cable has been run (Halliburton, 1990).

No side entry subassemblies or wet connectors are required (as in the case of the pipe-conveyed logging systems) since the electrical connection between the logging tools and the cable is made previous to going into the hole. The injector assembly of the reeled-tubing unit pushes the tubing and tool string into the wellbore and later retrieves them. Electrical power and commands are sent downhole through the wireline cable, and logging data is transmitted uphole.

This system presents several advantages including that many logging, perforating and auxiliary services can be performed, some of which are outlined in Table 3–6. Full-size standard logging tools can be employed in the operation. The special flex-joints between

TABLE 3–6 Reeled-tubing-conveyed wireline services (after Halliburton, 1990).

Logging	Perforating
Casing Collar Locator	Strip Guns
Natural Gamma Ray	Hollow Carrier Through-Tubing Guns
Spectral Natural Gamma Ray	Hollow Carrier Casing Guns
Single Detector Neutron	Tubing Punchers
Dual Detector Neutron	
Pulsed Neutron Capture	Tubing Cutters
Multiple Radioactive Tracer	Casing Cutters
Fluid Travel	Drillpipe Cutters
Temperature	Severing Tools
Pressure	Auxiliary Services
Spinner	Junk Baskets
Fluid Density	
Fluid Capacitance	Gauge Rings
Conventional Cement Bond	Bridge Plugs
Ultrasonic Cement Bond	Free Point/Pipe Recovery
Casing Inspection	
Caliper	
Depth Determination	

logging tools and the flexibility of the reeled-tubing allows the assembly to traverse small radius deviations. Furthermore, rollers in the tools facilitate transport. Because the tools are attached to reeled-tubing which is pushed and pulled by a tubing injector attached to the wellhead, no rig is required. Logging can be carried out in new or existing wells.

Centralization of the tools is possible. The tools can be oriented with respect to the low side of the wellbore, or they can be permitted to ride the low side of the borehole. This radial positioning is very important when orientation of perforating guns is desired.

Circulation can be maintained during wireline operations. This allows lower borehole temperature and cleaner fluids, which lead, in turn, to more reliable tool performance. Should the operation require it, the tool string can be quickly released from the tubing and cable. This would leave a standard fishing neck exposed at the top of the tool string.

Measurement While Drilling (MWD)

These systems provide measurements of downhole data from sensors mounted in collars directly above the drill bit. The data are transmitted to the surface using a mud pulse telemetry system. MWD data are very useful while drilling horizontal wells because they allow efficient steering of downhole motors, provide early information related to basic geological and reservoir engineering parameters, and give critical information conditions such as overpressures (Halliburton, 1990).

MWD sensors are placed in nonmagnetic collars. These include the gamma-directional and the resistivity-gamma-directional combinations. The directional sensors measure toolface orientation azimuth and inclination. Other sensors commonly available in MWD systems include conventional and focused gamma ray, resistivity, porosity, temperature, and pressure sensors.

Lateral and bit resistivity measurements can give an indication with respect to overpressuring, and when combined with porosity give estimates of water saturation from conventional equations.

An interval temperature sensor is used to correct wellbore directional measurements for temperature. An annulus sensor is used to correct resistivity measurements for temperature. Local geothermal gradients and potentially overpressured zones are indicated by the temperature sensors. Downhole pressure is measured directly by sensitive pressure transducers. This allows determination of borehole-assembly and swab-surge pressures, pressure drop across the bit, and equivalent circulating density.

Log Interpretation

Evaluation of horizontal wells is different from that of deviated or vertical wells (Struyk and Poon, 1990; Gianzero et al; 1990). The problem arises from the design of conventional logging tools which assumes symmetry in the formation around the wellbore.

This assumption is reasonable in most vertical wells. In horizontal wells, however, this is not necessarily the case because the formation above, below, and on the sides of the wellbore can change considerably. Thus, there is lack of symmetry around the wellbore and the use of logging devices that are radially averaged is cumbersome. Focused devices such as sidewall neutron, density, and dipmeter are very useful for evaluating horizontal wells if the location of the pads is known (Struyk and Poon, 1990).

Differences Between Vertical and Horizontal Wells. The discussion presented in this section follows very closely the original work of Struyk and Poon (1990) and has also been documented by Aguilera et al (1991).

Natural Fractures. The formation microscanner (FMS), formation microimager (FMI), and the borehole televiewer (CBIL) have been used successfully in some cases to locate natural fractures intersected by horizontal wells. Since most natural fractures tend to be vertical or of high inclination, they show up in the FMS and FMI as low angle or horizontal events (Bourke et al, 1989). A schematic diagram of this situation is presented in Figure 3–18B.

The response of sonic and waveform logs is different in vertical and horizontal wells. Usually, in vertical wells, the shear waves are more attenuated than the compressional waves. In horizontal wells, however, both waves are significantly attenuated, because the acoustic energy is more exposed to the vertical and high angle fractures.

If the compressional wave is sufficiently diminished, cycle skipping might occur. Keep in mind, however, that other things such as the presence of gas or road noise might also produce cycle skipping. On the other hand, natural fractures might be present even if cycle skipping does not develop. This might be the result of solid to solid contact across the natural fracture, in which case the degree of acoustic discontinuity is diminished. Furthermore, some modern software processing is designed to mask cycle skipping.

Invasion. Historically, the invasion profile of a vertical well has been modeled by assuming a cylinder that surrounds the wellbore. This does not necessarily apply to horizontal wells. In single-porosity conventional reservoirs without natural fractures, the value of horizontal permeability is generally larger than the vertical permeability. For this situation the invasion profile in the horizontal well is probably elliptical about the axis of the borehole.

In the case of bed boundaries, fractures and changes in permeability, complex invasion profiles are probably created due to the anisotropic nature of the formation around the wellbore. This is illustrated in Figure 3–96, which presents the probable invasion profile in a horizontal well at different times. The invasion is probably elliptical and increases when in contact with an aquifer.

Mud Cake. Solids settled on the floor of the wellbore, also referred to as a "cuttings bed," can easily be confused with mud cake along the floor of the hole. Under these circumstances, it is difficult to obtain good qualitative indications of permeability from microlog tools. In principle, this problem could be solved by directing the pads in such a way that they read along the sides of the borehole. This is reasonable because the sides are not as affected by rugosity or the "cuttings bed".

Borehole Rugosity. Most likely, the ceilings of horizontal wells are more rugose than the floor or the sides. Consequently, better results can be obtained if pad devices are directed to the floor and preferentially to the sides. Single pad devices tend to read along the low side of the wellbore due to their weight distribution. Thus, a single caliper registers hole diameters between the floor and ceiling. The borehole rugosity shown by this caliper consequently does not provide a valid check on the quality of the pad reading.

Anisotropy. The effects of natural fractures and anisotropy on the logging responses can be unexpected and difficult to interpret because the readings are directly influenced by the vertical components of the formation parameters.

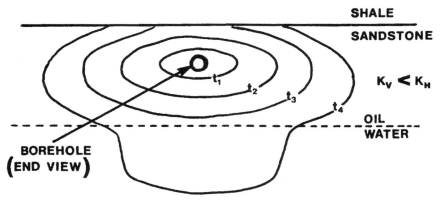

FIGURE 3–96 Perceived invasion profile in a horizontal well at time t_1, t_2 t_3 t_4 (after Struyk and Poon, 1990)

Bed Boundaries.   It is important to properly determine the location of the bed boundaries to qualify the success of the horizontal well with respect to the actual production area.

Bed Boundaries Intersecting the Borehole.   Logging measurements in a horizontal well do not adequately reveal low angle boundaries and fluid contacts. The boundary indicated in a log depends on the tool's depth on investigation, measurement bias, measurement resolution, and direction of the tool reading. Thus, logs might give the impression of being off depth with each other. For proper log correlation it is important to run a base log such as a gamma ray on each logging run.

A tool that seems to accurately locate bed boundaries in horizontal wells at this writing is the four-or six-arm dipmeter.

Bed Boundaries Away From the Wellbore.   Having good information about bed boundaries that do not intersect the wellbore is quite important in horizontal boreholes. Keep in mind that the generalized assumption that the formation tops form a straight line between two vertical wells is not necessarily correct. A Horizontal well permits to keep track of the bed boundaries, allowing determination of their position with respect to nearby fluid contacts or shale zones. This can probably be accomplished with a dipmeter tool by taking oriented resistivity measurements with various depths of investigation.

True Vertical Depth.   It is common practice to redisplay logs of deviated wells to true vertical depth to allow sound correlation with other wells and other types of geological data. This does not appear to be very meaningful in horizontal wells. What does appear worthwhile is redisplaying horizontal well logs into true horizontal well logs, because the horizontal well is in many cases sinusoidal. With this type of work, the lateral extent of the penetrated zone can be more easily determined.

Measurements Conducted by Logging Tools

Logging tools can conduct (1) directionally focused measurements and (2) radially averaged measurements. The principles behind these types of measurements for horizontal wells as presented by Struyk and Poon (1990), and documented by Aguilera et al. (1991) are discussed next.

Directionally Focused Measurements.   The density and epithermal neutron (pad-type) tools are directionally focused. The pads of these tools are heavy, and in many cases cause the tool to read along the floor of the horizontal hole. The "cuttings bed" probably does not affect the readings because the plow-shaped design of the pad pushes the cuttings and mud cake aside.

Usually, the exact direction in which the pad is reading is not known, but is assumed to be downwards. A directional device run in combination with these tools could be used to indicate the exact pad orientation. It may also prove useful to design a directionally focused gamma ray that reads in the same direction as the density or pad-type neutron tools. Under these circumstances it would be possible to correct the density or pad-type neutron more accurately for shale effects.

One type of dipmeter consists of four directionally focused resistivity readings each at ninety degrees to the adjoining one. Six-arm dipmeters are also available. These resistivity readings can be made to approximate a shallow resistivity with a similar depth of investigation to that of a spherically focused log. Under this approximation, a dipmeter could provide resistivity information along the top, bottom, and sides of the well.

This directional resistivity log could give an indication of the proximity and location of bed boundaries. The curves along the horizontal axis could be considered as the "normal" curves, while the curves from the high and low sides of the well could be compared to these normal curves. It would be important to record tool rotations with the corresponding resistivity readings. Bed boundaries intersecting the wellbore could thus be identified by the dipmeter. Other important information from the dipmeter would include a directional survey and dip information.

The dipmeter could prove to be a very valuable resistivity tool in horizontal wells. It could prove to be even more useful if the resistivity readings could have various depths of investigation (Struyk and Poon, 1990).

Radially Averaged Measurements. The radially averaged measurement takes an average reading over a plane perpendicular to and radiating out from the borehole, as illustrated in Figure 3–97. The induction log, for example, provides this type of measurement. Radial averaging provides excellent readings as long as the formation is homogeneous and isotropic. This probably happens in most vertical wells because the plane over which an averaged reading is taken is parallel to the bedding planes.

In horizontal wells, however, radially averaged measurements are in a plane perpendicular to the bedding planes. The measurements are therefore carried out in a nonhomogeneous, nonisotropic medium, and the radial averaging might provide unreasonable results.

On the other hand, a directionally focused measurement is normally recorded by a pad-type tool that takes readings in a particular side of the borehole, as illustrated in the schematic in Figure 3–98. The density tool, for example, provides directionally focused measurements readings in one side of the borehole, while the dipmeter takes four separate directionally focused measurements, each from a different side of the borehole. Because of their characteristics, directionally focused tools identify bed boundaries in horizontal wells much more distinctly than radially averaged measurements.

Induction and laterologs provide radially averaged measurements for which some resistivity modelling has been done in horizontal wells (Gianzero et al, 1990). This modelling indicates that the laterolog and induction tools can be used in some instances to identify nearby beds that are parallel to the borehole. However, only beds within one or two feet of the borehole can be identified. Furthermore, it is not possible to determine if the bed is located above or below the borehole.

The gamma ray, pulsed neutron, compensated neutron (mandril-type neutrons) and MWD tools are examples of radially averaged logging tools. Radially averaged measurements can be used to give an idea of the angle between the borehole and the intersecting bed, although a dipmeter would be the preferred tool for this purpose.

If a boundary is reasonably sharp on a radially averaged tool, one can assume the boundary intersects the borehole at approximately right angles. The more gradual the bed boundary appears on radially averaged logs in horizontal wells, the more parallel are the bed boundary and the borehole.

Some of the radially averaged tools tend to ride along the floor of the borehole. In the case of large diameter holes, these tools will be more influenced by the formation below

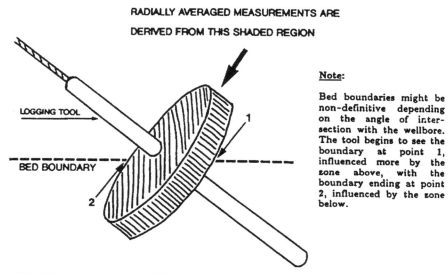

RADIALLY AVERAGED MEASUREMENTS ARE

DERIVED FROM THIS SHADED REGION

LOGGING TOOL

BED BOUNDARY

1

2

Note:

Bed boundaries might be non-definitive depending on the angle of intersection with the wellbore. The tool begins to see the boundary at point 1, influenced more by the zone above, with the boundary ending at point 2, influenced by the zone below.

FIGURE 3–97 Illustration of a radially averaged logging tool (after Struyk and Poon, 1990).

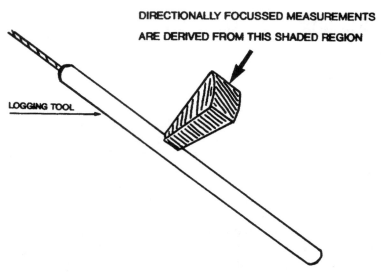

FIGURE 3–98   Illustration of a directionally focused logging tool (after Struyk and Poon, 1990)

the wellbore. It must also be kept in mind that laterolog readings are shifted to a higher value when run on the end of the drillpipe. This is caused by the drillpipe acting as a current return. Correction charts should be generated to account for this shift (Struik and Poon, 1990).

Examples of Applications

Rospo Mare Field, Italy.   An excellent example of enhanced reservoir description is provided by horizontal well Rospo Mare 6d in the Italian Adriatic sea (Montigny et al, 1988). The logging program of this well included the spherical focused, induction resistivity, borehole compensated sonic, cement bond, variable density, dual laterolog, and gamma ray logs. Production logs included the full bore spinner as well as sensors for pressure and temperature measurements. The comparison of all these logs with cores indicated that the fluid entry occurred at intervals characterized by natural fractures and karstic voids.

In other wells the logging programs included the sonic, natural gamma ray, formation microscanner, dual laterolog, microspherical focused, compensated neutron, and lithodensity logs, auxiliary measurement tool (AMT) for detection of theft zones by measuring temperature, the general purpose inclination tool (GPIT), which was used for pad orientation of tools in the lower part of the hole, and the measurement–while–drilling (MWD) gamma ray tool.

Figure 3–99 shows an interesting correlation of well logs and geology. The upper portion of the figure shows the percent flow from each one of the horizontal production intervals. Next is the geological interpretation. The well intercepted many vertical fractures and allowed a reasonable estimate of the distance between large natural fractures (approximately 100 ft.) This distance between macro-fractures in emphasized because there is a tendency among some professionals of the oil industry to believe that this kind of fracture spacing is unrealistically large. *It is not!*

Total loss of circulation occurred very soon after the well crossed the first fracture. The well came out of the reservoir twice: first, when it passed through a sinkhole, and second, when it went through a depression. It is interesting to note that nearly one third of the horizontal length, or approximately 620 ft, ended up being outside the reservoir.

These results indicated that a conventional seismic survey using a 1500 ft grid is not good enough for this kind of reservoir, as it was not able to pick up the depressions shown on Figure 3–99. Consequently, before further field development, a three-dimensional,

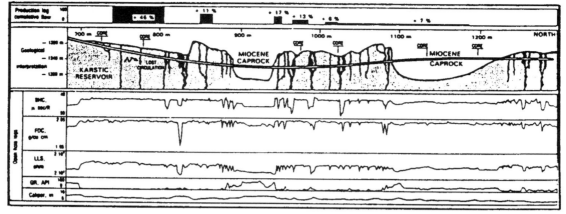

FIGURE 3–99 Correlation of geology and well logs, Rospo Mare 6d (after de Montigny and Combe, 1988)

high-resolution seismic survey was carried out using an 80 ft grid. This survey gave a good description of the irregularities at the top of the reservoir.

In the end, the most useful logs run in the horizontal wells to interpret the geology of Rospo Mare were as follows (de Montigny et al, 1988):

1. The formation microscanner (FMS) gave a good indication of natural fractures, vugs, and caves.
2. The compensated neutron and the lithodensity logs also pinpointed wide natural fractures and large caves.
3. The natural gamma ray, after calibration with cores, distinguished the natural fractures filled with clays from overlying shales containing thorium, uranium, and potassium from those fractures filled with weathered clays that contained only thorium.

Austin Chalk, Texas. Schlumberger's formation microscanner (FMS) has been reported to give good indications of natural fractures in the Austin chalk (Fett and Henderson, 1990; Stang, 1989). If there are open fractures and they are filled with salty water, they are obviously conductive and show up in the microscanner as a black feature. If they are mineralized and healed they show up in the microscanner as white. High inclination natural fractures show up as black sine waves. Multiple sine waves are indicative of multiple fractures.

Figure 3–100 shows an example of a formation microscanner in the Austin chalk in south central Texas. The log was made by attaching the logging tool to the bottom of the drillpipe and operating from a wet connector. High inclination natural fractures in the horizontal borehole appear as very dark, steeply inclined lines. A large natural fracture appears just left of X958 ft in Figure 3–100.

Another tool that has been reported to provide good results in horizontal wells drilled in the Austin chalk is the Western Atlas circumferential borehole imaging log (CBIL) (Western Atlas, 1990). The tool is pipe-conveyed and a combination of centralizers and knuckle joints prevent eccentric positioning of the CBIL instrument.

Figure 3–14 shows a combination of CBIL, gamma ray, and spectralog. The spectralog was recorded while going into the hole. The CBIL and gamma ray measurements were made coming out of the hole through the entire 1500 ft of horizontal section. Logging operations were carried out at a standard speed of 600 ft/hr.

The left side of Figure 3–14 shows a natural fracture intercepted by the horizontal wellbore. This is confirmed by the spectralog which displays an increase in uranium counts.

The right side of Figure 3–14 shows vee pattern in the CBIL image indicating that the wellbore exited the chalk and drifted into a marl. The spectralog verifies the presence of the chalk-marl interface.

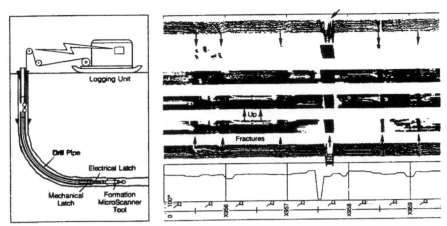

FIGURE 3–100 Formation microscanner of horizontal well in Austin chalk (after Bourke et al, 1989)

MEASURE OF UNCERTAINTY IN LOG CALCULATIONS

Calculating original oil-in-place (or gas-in-place) by volumetric means in naturally fractured reservoirs requires previous knowledge of area, net pay, oil (or gas) formation volume factor, matrix and fracture porosities, and matrix and fracture water saturations.

To establish exact values for the previous parameters is essentially impossible. This is especially true for fracture porosity and fracture water saturation. Fracture reservoirs may produce at initially high oil rates which in many cases lead to very optimistic forecasts. These initially high rates are due to the original oil-in-place within the fractures. Upon depletion of the fracture system, the oil production may decline drastically, with strong increases in gas-oil ratios. From there on, production depends on the facility with which the oil bleeds-off from the matrix into the fractures. Consequently, the importance of having reasonable estimates of original oil-in-place (or gas-in-place) within the fracture network cannot be overemphasized.

The Monte Carlo simulation method has been used by Aguilera (1978, 1979) to calculate ranges and probability distributions of total porosity (ϕ), double-porosity exponent (m), total water saturation (S_w), fracture porosity (ϕ_2), water saturation within the fractures (S_{wf}), original oil-in-place within the fractures (N_f), and total original oil-in-place (N) in a naturally fractured reservoir.

ORIGINAL OIL-IN-PLACE

Total original oil-in-place in a naturally fractured reservoir can be calculated volumetrically from the relationship:

$$N = \frac{7758\,Ah\phi(1-S_{wi})}{B_{oi}}$$
(3–140)

where:
    N = original oil-in-place, stbo
    A = area, acres
    h = net pay, ft
    ϕ = total porosity, fraction
    S_{wi} = initial average water saturation of composite system, fraction
    B_{oi} = initial oil formation volume factor, b/stbo

The original oil-in-place within the fractures can be approximated from:

$$N_f = \frac{7758\,A_f h_f \phi_2 (1-S_{wf})}{B_{oi}}$$
(3–141)

where: N_f = original oil-in-place within the fractures, stbo
A_f = fracture area, acres
h_f = fracture net pay, ft
ϕ_2 = fracture porosity, fraction
S_{wf} = initial water saturation within the fractures, fraction

The following sections present procedures for estimating the parameters that go into Equations 3–140 and 3–141.

Total Porosity. Total porosity can be calculated from logs, whole core analysis, and/or as a function of the matrix porosity and the partitioning coefficient from the equation (Pirson, 1962):

$$\phi = \frac{\phi_b}{1 - v(1 - \phi_b)} \qquad (3\text{–}142)$$

where: ϕ_b = matrix porosity attached to matrix bulk volume (or core porosity from unfractured plugs), fraction
v = partitioning coefficient, fraction

The partitioning coefficient represents the fraction of total porosity made up of fractures. Its value can be estimated from cores using the method of Locke and Bliss (1950), from well logs using Figures 3–48 to 3–52 or from pressure analysis. The matrix porosity can be obtained from intergranular unfractured core plugs in the laboratory. Experience indicates that matrix porosity in many cases has a normal distribution (Figure 3–101).

Double-porosity exponent. The double-porosity exponent (m) for naturally fractured reservoirs is smaller than the porosity exponent of intergranular unfractured systems. Sometimes, it is possible to estimate the value of m from logs alone when a statistically significant number of zones is available (Figure 3–75). If this is not the case, it is still possible to estimate the value of the double-porosity exponent (m) from the relationship:

$$m = -\log\left[\frac{1}{v\phi + (1-v)/\phi_b^{-mb}}\right]/\log\phi \qquad (3\text{–}143)$$

where: m = double-porosity exponent
m_b = matrix porosity exponent

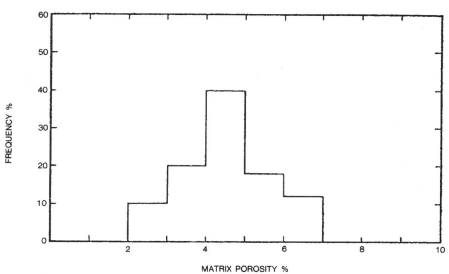

FIGURE 3–101 Distribution of matrix porosity for example problem.

The above equation allows a continuous calculation of m with depth and the generation of an "m" log. The smaller the value of m the larger the degree of fracturing. The matrix porosity exponent is obtained in the laboratory from the analysis of intergranular unfractured plugs. Other parameters that go into Equation 3–143 are obtained as described in previous sections.

Total water saturation. Total water saturation can be calculated using Pickett's cross-plots or the $P^{1/2}$ statistical analysis discussed previously.

Fracture porosity. An estimate of fracture porosity is necessary to calculate the original oil-in-place within the fracture system. The fracture porosity is equal to the total porosity times the partitioning coefficient. Estimates can also be obtained from cores, from the FMS/FMI/EMI, from Equations 3–8 or 3–9, from Equations 3–55 to 3–58, and from Equations 3–59 to 3–64.

Water saturation in the fractures. The possibility of estimating water saturation in fractures is usually overlooked in petroleum engineering literature; but it is possible to make such an estimate with the use of Equation 3–79.

Reservoir Area. Aerial photography, fractured indices, seismic data, radius of curvature, and remote sensing imagery can obtain estimates of the fractured reservoir area.

Net Pay. Conventional methods to determine net pays can be used if there is evidence that vertical fractures extend all over the section of interest. Other potential tools that may prove useful for determining fracture pay include variable intensity logs, dipmeter, Fracture identification log (FIL), dual induction-laterolog 8, sonic and neutron or density logs, core porosity and neutron logs, borehole televiewer, comparison of matrix porosity exponent (m_b) with the double-porosity exponent (m), cycle skipping, SP curves, correction curve on the compensated density log, comparison of shale volume to uranium index, lithoporosity crossplot, comparison of long and short normal curves, production index, temperature and sibilation logs, P_e curve on lithodensity log, dual laterolog-microspherical log combination, gamma ray with and without introducing a tracer, formation micro-scanner and micro-imager (FMS and FMI), electrical micro-imager (EMI), circumferential borehole imaging log (CBIL), and the fracture completion log (FCL).

MEASURE OF UNCERTAINTY

The log analyst is painfully aware that the previous equations or any other set of equations for evaluating naturally fractured reservoirs are not exact. These equations can provide meaningful information if the analyst places reasonable ranges on the parameters that go into the solution of each equation. The choice of range must reflect, as close as possible, the present knowledge of the basic data. The analyst must also choose the best probability distribution for data at hand. This information permits a probabilistic analysis with the use of the Monte Carlo simulation technique.

Essentially, a Monte Carlo simulation takes the probability distributions of the input data and generates random values within the preestablished minimum and maximum values of the data. By repeating a calculation many times (for instance 100, 200, or perhaps 1000 times) until the distribution of the answer does not change, the analyst obtains enough information to generate a plot of the various answers (for instance oil-in-place) vs. probability of occurrence. The generated data permit analysts and managers to decide whether or not they like the odds.

Several distributions are considered in the literature (McCray, 1975; Walstrom et al, 1967). This chapter considers histogram, triangular and rectangular distributions.

Histogram Distribution. Amyx et al (1960) give evidence that most intergranular porosity distributions are symmetrical. Consequently, the histogram appears to be a good choice for inputting data on primary porosity in naturally fractured reservoirs. Random values from a histogram are selected as shown in Figure 3–102.

Triangular Distribution.   This distribution can be used knowing the lower, most likely, and upper values of the data. The selection of random values from a triangular distribution is illustrated in Figure 3–103.

Rectangular Distribution.   It is also referred to as uniform distribution. This type of distribution is used when there is a high degree of uncertainty in the input data and only the lower and upper limits of the range are known. The selection of random values from a uniform distribution is depicted in Figure 3–104.

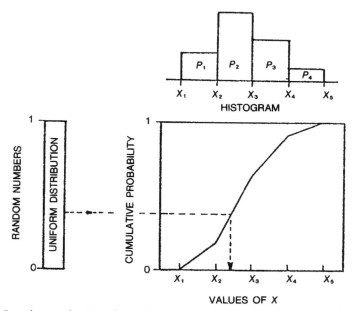

FIGURE 3–102   Random selection from histogram distribution (after McCray, 1975)

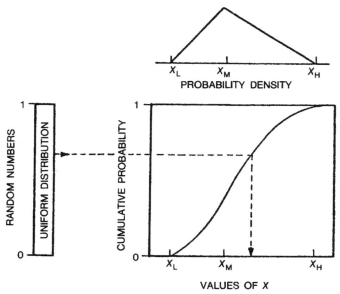

FIGURE 3–103   Random section from triangular distribution (after McCray, 1975)

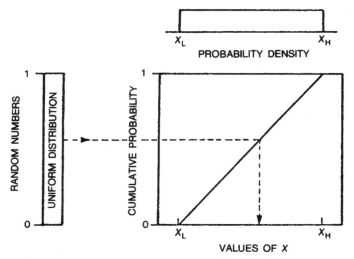

FIGURE 3–104 Random selection from uniform distribution (after McCray, 1975)

Example 3–15. This example is designed to provide a working knowledge of how to quantify uncertainty in the evaluation of naturally fractured reservoir. The data available for the analysis are presented below:

Parameter	Distribution	Lower	Most Likely	Higher	Source of Data
v	Triangular	0.10	0.20	0.40	Core, logs and pressure analysis
m_b	Rectangular	1.90	—	2.10	Analysis of inter-granular cores
R_t	Triangular	30.00	35.00	50.00	Logs
R_w	Rectangular	0.05	—	0.07	SP log, water analysis, R_w tables, Pickett plots
B_o	Rectangular	1.30	—	1.45	PVT analysis, empirical charts
h	Triangular	90.00	110.00	150.00	Logs, core analysis

In addition, a histogram of matrix porosity (ϕ_b) has been constructed (Figure 3–102) using unfractured core data with the following results:

Matrix Porosity Interval	Frequency
0.02 to 0.03	0.10
0.03 to 0.04	0.20
0.04 to 0.05	0.40
0.05 to 0.06	0.18
0.06 to 0.07	0.12
	1.00

Values of total porosity (ϕ) were calculated using Equation 3–142 and the Monte Carlo simulation method [Figure 3–105 (A)]. The calculated total porosities range between a minimum of 2.6% and a maximum of 9.9%. Figure 3–105(A) also indicates that there is a 50% probability that the total porosity will be 5.6% or greater, and an 80% probability that it will be greater than 4.3%.

The calculated values of the double porosity exponent (m) are presented in Figure 3–105(B). Note that the calculated values of m range between 1.321 and 1.678 as compared with the input m_b values of the matrix which range between 1.9 and 2.1.

Figure 3–105(C) shows that the total water saturation (S_w) varies between 9.7 and 57%. It also shows that there is a 50% chance that the water saturation is 22% or less and an 80% probability that it will be 27% or less.

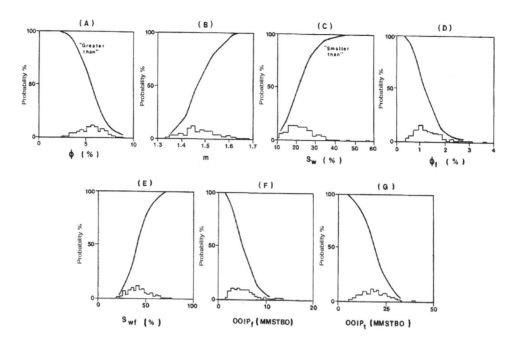

FIGURE 3–105 Probability of occurrence of calculated values in example problem (after Aguilera, 1978)

Notice that in Figure 3–105(A) the probability of occurrence is plotted to indicate values of porosity greater than a certain value. In Figure 3–105(C), the probability of occurrence is plotted in such a form as to indicate values of water saturation smaller than a certain value. It is plotted in this fashion because we are looking for the largest possible porosities and the smallest possible water saturations.

Fracture porosity ranges between 0.47% and 3.70% [Figure 3–105(D)] and there is a 50% probability that fracture porosity is equal to or greater than 1.8%.

Values of water saturation in the fractures are calculated to range between 21 and 77% with a most likely value of 41% [Figure 3–105(E)].

The main objective of this example, i.e., the quantification of oil-in-place in the fracture system, is illustrated in Figure 3–105(F). The variation is striking as the oil-in-place within the fractures ranges between 1.85 and 13.46 MMstbo. There is a 50% probability that the oil-in-place will be 5.1 MMstbo or greater. There is an 80% probability that the oil-in-place will be 3.5 MMstbo or greater.

Finally, the total oil-in-place (matrix plus fractures) ranges between 3.6 and 40.9 MMstbo with a 50% probability that it will be 18 MMstbo or greater [Figure 3–105(G)].

The Monte Carlo approach can be extended for calculating recoverable volumes (RV) with the use of equation:

$$RV = N_f \, (R_{ec})_f + N_m \, (R_{ec})_m \tag{3-144}$$

Where R_{ec} is fractional recovery, N is original oil-in-place, and the subscripts f and m stand for fractures and matrix, respectively. Inclusion of economic considerations in the above recoveries leads to estimates of reserves.

HYDROCARBON RECOVERIES FROM LOGS

Crossplots of economic recoverable oil as determined from decline curves vs. $\Sigma \, S_o \phi h$ using the ratio k/ϕ as a parameter have led to reasonable estimates of hydrocarbon recoveries from new wells. The approach requires good production history from older wells in such a way as to be able to determine recovery from decline curves.

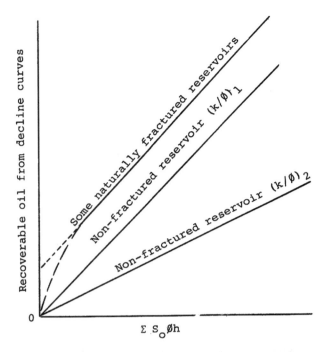

FIGURE 3–106 Schematic of crossplot for estimating recoveries from well logs. There are two different rock types in the non-fractured reservoir.

Figure 3–106 shows this type of crossplot. It should be possible to determine recoverable oil for a new well based on knowledge of $\Sigma S_o \phi h$ and the ratio k/ϕ. The approach has been used in Ordovician Carbonates of the Williston basin (Pickett and Artus, 1970), Wasatch fractured sandstones of the Uinta basin (Aguilera, 1973), and Minnelusa and Muddy Sands of the Powder River basin (Oduolowu, 1976).

GEOTHERMAL RESERVOIRS

This subject has been covered in detail in SPWLA reprint volume "Geothermal Log Interpretation Handbook" (1982). These reservoirs are characterized by a large degree of fracturing even when they occur in sedimentary rock with intergranular porosity.

Table 3–7 lists some of the high temperature equipment for geothermal wells available from Dresser Atlas and Schlumberger. Table 3–8 describes some of the essential parameters for log interpretation of Geothermal reservoirs.

MISCALIBRATION OF LOGS

One of the advantages of a Pickett plot is that under certain conditions it allows calibration of porosity and resistivity logs. Three types of calibration errors have been discussed by Pickett (1973):

1. Zero type error
2. Sensitivity type error
3. Combination type error

These errors are illustrated using the response of the sonic log. The same treatment, however, would apply to other logs.

Zero type error. In this case the log is miscalibrated and gives an apparent sonic transient time (Δt_a) in microseconds per foot:

$$\Delta t_a = C_1 + \Delta t \qquad (3\text{–}145)$$

where Δt is the correct transient time and C_1 is a constant.

TABLE 3–7 Dresser atlas logging services for geothermal wells (high-temperature equipment) (Source: Geothermal Log Interpretation Handbook, SPWLA, 1982)

	Maximum Pressure	Maximum Temperature
*High-Temperature Densilog	138 MPa (20,000 psi)	260°C (500°F)
*High-Temperature GR/Neutron Log (Thermal and Epithermal)	172 MPa (25,000 psi)	260°C (500°F)
**High-Temperature Caliper/Temperature Log	138 MPa (20,000 psi)	288°C (550°F)
*High-Temperature Spinner Flowmeter	117 MPa (17,000 psi)	260°C (500°C)
**High-Temperature Fluid Sampler	138 MPa (20,000 psi)	288°C (550°F)

*Tested at 288°C (550°F)
**Tested at 316°C (600°F)

Schlumberger Open Hole Hostile Environment Logging (HEL) (high temperature equipment for geothermal wells)

	Maximum Pressure		Maximum Temperature
*Induction Logging		25,000 psi	260°C (500°F)
*Dual Laterolog	172 MPa	25,000 psi	260°C (500°F)
*Sonic Log		25,000 psi	260°C (500°F)
*Gamma Ray-Neutron	138 MPa	20,000 psi	260°C (500°F)
*Gamma Ray		25,000 psi	260°C (500°F)
*Formation Density	172 MPa	25,000 psi	260°C (500°F)
*Compensated Neutron		25,000 psi	260°C (500°F)
*Flowmeter-Temperature	138 MPa	20,000 psi	260°C (500°F)

*Tools can be operated at 550°F for short periods of time.

In this case the Pickett plot is constructed using $\Delta t_a - \Delta t_{ma}$ or $\Delta t_a - A_a$ vs. R_t rather than $\Delta t - \Delta t_m$ or $\Delta t - A$. The subscript a represents apparent properties. We can *only* get a straight line if:

$$(C_1 + \Delta t - \Delta t_{ma}) = \Delta t - \Delta t_m \tag{3–146}$$

Consequently, we have found a Δt_m such that:

$$\Delta t_{ma} = C_1 + \Delta t_m \tag{3–147}$$

Through this analysis the error due to a zero type miscalibration has been compensated for, although Δt_{ma} is not the true Δt_m; rather, it is a parameter with little physical significance similar to A in Equation 3–51.

A problem for calculating porosity is the denominator B in Equation 3–51. B can be determined if water resistivity and constant a in the formation factor equation are known. One procedure is to enter Figure 3–62 with the product aR_w in the x-axis and read $\Delta t - A$ in the 100% water saturation line. At this point porosity is 100% and $\Delta t - A$ is equal to the desired denominator B, i.e., $\phi = 1 = (\Delta t - A)/B$. Another possibility is to calculate B from the equation:

TABLE 3–8 Essential parameters for log interpretation of geothermal reservoirs (Source: Geothermal Log Interpretation Handbook, 1982)

Formation Evaluation: Those measurements made in the borehole that are ultimately used to characterize the entire reservoir.	*Production Management:* Those measurements in or near the wellbore required in an engineering sense to keep the well producing over a number of years and to provide data for design and operation of surface facilities.

Formation Evaluation:

1. Time-lapse temperature profile measurements for true formation temperature profile.
2. Lithology, depth, and thickness of formations.
3. Permeability, both intergranular (matrix) and fracture.
4. Porosity, again both intergranular (matrix) and fracture.
5. Fracture system re-emphasized with regard to location in depth, orientation, permeability, and other characteristics.
6. Borehole geometry as an indicator of fractures, particularly with regard to size as an indication of quality and corrections for other logging data.
7. Fluid composition.
8. Thermoconductivity and heat capacity.
9. Several elastic moduli of rock that are useful in designing well stimulation.

Production Management:

1. Flow profile including flow rate.
2. Pressure profile.
3. Fluid Composition.
4. Hole and/or casing mechanical conditions such as:
   a. Scaling
   b. Corrosion
   c. Cement quality
   d. Mechanical properties of the borehole system itself

$$B = (\Delta t_a - A_a)(aR_w/R_o)^{-1/m} \qquad (3\text{–}148)$$

or using more conventional nomenclature have A_a replaced by the apparent Δt_{ma}.

Sensitivity Type Error. In this case the sonic log is miscalibrated and gives an apparent sonic transit time (Δt_a) equal to $C_2\Delta t$, where C_2 is a constant, or

$$\Delta t_a = C_2 \Delta t \qquad (3\text{–}149)$$

In the Pickett method we crossplot ($\Delta t_a - \Delta t_{ma}$) vs. R_t on log-log paper, or ($C_2\Delta t - \Delta t_{ma}$) vs. R_t. The only way we can obtain a straight line is if:

$$(C_2\Delta t - \Delta t_{ma}) = (\Delta t - \Delta t_m) C_2 \qquad (3\text{–}150)$$

Consequently,

$$\Delta t_{ma} = C_2 \Delta t_m \qquad (1\text{–}151)$$

Taking logarithm in both sides of Equation 3–150 leads to:

$$\log (C_2\Delta t - \Delta t_{ma}) = \log (\Delta t - \Delta t_m) + \log C_2 \qquad (3\text{–}152)$$

and,

$$\log (\Delta t_a - \Delta t_{ma}) = \log (\Delta t - \Delta t_m) + \log C_2 \qquad (3\text{–}153)$$

Equation 3–153 indicates that if we crossplot ($\Delta t_a - \Delta t_{ma}$) rather than ($\Delta t - \Delta t_m$), we have shifted the water-bearing line by a factor "$\log C_2$", i.e., by a constant amount. Therefore, all hydrocarbon points will be shifted by that same amount and the calculated resistivity index "I" and water saturation will be correct. Denominator B in Equation 3–51 is calculated as explained in the previous section, graphically or with the use of Equation 3–148.

Combination Type Error. In this case the sonic log is miscalibrated and gives an apparent sonic transit time:

$$\Delta t_a = C_1 + C_2 \, \Delta t \qquad (3\text{–}154)$$

where C_1 and C_2 are constants.

In the Pickett approach we crossplot $\Delta t_a - \Delta t_{ma}$ vs. R_t on log-log paper or $(C_1 + C_2\Delta t - \Delta t_{ma})$ vs. R_t. We can get a straight line *only* if:

$$(C_1 + C_2\Delta t - \Delta t_{ma}) = C_2 \, (\Delta t - \Delta t_m) \qquad (3\text{–}155)$$

Thus,

$$\Delta t_{ma} = C_1 + c_2 \, \Delta t_m \qquad (3\text{–}156)$$

Following the same reasoning presented in the previous two sections, the conclusion is readily reached that when we prepare a Pickett plot we compensate the "zero" part of the miscalibration, and we shift the water-bearing trend and hydrocarbon points by the constant amount "$\log C_2$" (sensitivity error). In this case the calculated values of I and S_w would be correct even if the porosity log is miscalibrated.

Summary

The same procedure presented above for calibrating the sonic log can be followed with other porosity and resistivity logs. Usually the more drastic corrections for porosity logs will occur in the low porosity regions. The biggest corrections in resistivity logs will occur in the low resistivity (or high porosity) regions. Practical applications of weighted least square methods to Pickett plots when there are intervals with 100% of water saturation have been presented by Etnyre (1984).

GENERAL REMARKS

Log analysts are aware that their results are not always correct. Moreover, they are aware that their results are not exact, as the log analysis science and art is based upon empirical relationships. Uncertainties arise especially when dealing with naturally fractured reservoirs. Uncertainty can be quantitatively measured using the Monte Carlo simulation approach. The set of equations presented in this chapter has provided valid results in the author's experience. However, this or any other set of equations is imperfect and must be treated as such. It is important to quantify the amount of oil-in-place (or gas-in-place) within the fracture system to make sound projections of reservoir life.

High initial oil rates in naturally fractured reservoirs are the result of oil stored within the fractures. Strong production declines may arise when the fracture network is depleted, depending on the rate with which oil bleeds-off from the matrix into the fractures. Better decisions can be made regarding naturally fractured reservoirs if the uncertainty of calculating oil-in-place is quantitatively measured.

REFERENCES

Aguilera, Roberto. "Evaluation of Fine-Grained Laminated Systems from Logs, Wasatch Formation, Utah." Ph.D. Thesis 1569, Colorado School of Mines (1973).

Aguilera, Roberto. "Analysis of Naturally Fractured Reservoirs from Sonic and Resistivity Logs." *Journal of Petroleum Technology* (November 1974), 1233–1238.

Aguilera, Roberto. "Analysis of Naturally Fractured Reservoirs from Conventional Well Logs." *Journal of Petroleum Technology* (July 1976), 764–772.

Aguilera, Roberto. "The Uncertainty of Evaluating Original Oil-in-place in Naturally Fractured Reservoirs." SPWLA (June 13–16, 1978), paper A.

Aguilera, Roberto. "Uncertainty in Log Calculations Can be Measured." *Oil and Gas Journal* (September 10, 1979), 126–128.

Aguilera, Roberto, and H.K. van Poollen. "Naturally Fractured Reservoirs—Porosity and Water Saturation Can be Estimated from Well Logs." *Oil and Gas Journal* (January 8, 1979), 101–108

Aguilera, Roberto. "A New Approach for Log Analysis of the Pulsed Neutron and Resistivity Log Combination." *Journal of Petroleum Technology* (April 1979), 415–418.

Aguilera, Roberto and Luis Acevedo. "FCL—Computerized Well Log Interpretation Process for Evaluation of Naturally Fractured Reservoirs." *Journal of Canadian Petroleum Technology* (January–February, 1982), 31–37.

Aguilera, R. "New Methods for Evaluation of Shaly Formations by Well Logs." SPE paper 17517 presented at the SPE Rocky Mountain Regional Meeting, held in Casper, Wyoming (May 11–13, 1988).

Aguilera, Roberto. "A New Approach for Analysis of the Nuclear Magnetic-Resistivity Log Combination." *Journal of Canadian Petroleum Technology* (January–February, 1990), v. 29, no. 1, 67–71.

Aguilera, R. "Extensions of Pickett Plots for the Analysis of Shaly Formations by Well Logs." *The Log Analyst* (September–October, 1990), 304–313.

Aguilera, R., J.S. Artindale, G.M. Cordell, M.C. Ng, G.W. Nicholl, and G. A. Runions. *Horizontal Wells*, Contributions in Petroleum Geology and Engineering No. 9, Gulf Publishing Company, Houston, Texas (1991), 401 p.

Alger, R.P. et al. "Formation Density Log Applications in Liquid Filled-Holes." *Journal of Petroleum Technology* (March, 1963).

Allaud, L.A., and J. Ringot. "The High Resolution Dipmeter Tool." *The Log Analyst* (May–June 1969), 3.

Anderson, J.A. et al.: "Determination of Fracture Height by Spectral Gamma Log Analysis." SPE paper 15439 presented at the 61st Annual Technical Conference and Exhibition of the Society of Petroleum Engineers held in New Orleans, Louisiana (October 5–8, 1986).

Amyx, J.W., D.M. Bass, and R.L. Whiting. *Petroleum Reservoir Engineering—Physical Properties*. McGraw-Hill Co. (1960), 112.

Archie, G.E. "The Electrical Resistivity Log as an Aid in Determining Some Reservoir Characteristics." *Transactions.* AIME 146 (1942), 54–67.

Bamber, C.L. and J.R. Evans, Jr. "ϕ-k Log-Permeability Definition from Acoustic Amplitude and Porosity Logs." SPE paper 1971 presented at the midway USA Oil and Gas Symposium held in Wichita, Kansas (November 9–10, 1967).

Beck, J., A. Shultz, and D. Fitzgerald. "Reservoir Evaluation of Fractured Cretaceous Carbonates in South Texas." SPWLA Logging Symposium Transactions (1977), paper M.

Beiers, R.J. "Vast Sedimentary Basin of Quebec Lowlands Major Interest to SOQUIP." *Oil and Gas Journal* (January 26, 1976).

Biot, M.A. "Theory of Propagation of Elastic Waves in a Fluid-Saturated Porous Solid." *Journal Acoustic Society American*, Vol 28 (1956) 168–191.

Birks, J. "Coning Theory and Its Use in Predicting Allowable Producing Rates of Wells in a Fissured Limestone Reservoir." *Iranian Petroleum Institute Bulletin* (December 1970) No. 12 and 13, 470–480.

Bishop, W.D., M.R. DeVries, and W.H. Fertl. "Well Log Analysis in the Austin Chalk Trend." SPWLA Eighteenth Annual Logging Symposium (June 5–8, 1977), paper W.

Bourke, L., et al., "Using Formation Microscanner Images," *The Schlumberger Technical Review* (January 1989) 16–40.

Boyeldieu, C. Private communication. Schlumberger-Surenco (1975).

Boyeldieu, C. and C. Martin. "Fracture Detection and Evaluation." Presented at the Ninth International Formation Evaluation Symposium, SAID, Paris (October 24–26, 1984).

Boyeldieu, C. and A. Winchester. "Use of the Dual Laterolog for the Evaluation of the Fracture Porosity in Hard Carbonate Formations." Presented at the Offshore Southeast Asia Conference, Singapore (February 9–12, 1982).

Burke, J.A., R.L. Campbell, and A.W. Schmidt. "The Litho-Porosity Crossplot." *Transactions.* 10th Annual Symposium of SPWLA (1969).

Burke, J.A., M.R. Curtis and J.P. Cox. "Computer processing of Log Data Enables Better Production in Chaveroo Field." *Journal of Petroleum Technology* (July, 1967).

Butler, M. et al. "Buchan Field: Evaluation of a Fractured Sandstone." *The Log Analyst* (March, April, 1977), 23–31.

Butsch, R.J. "Evaluation of Fractured Reservoirs Through Casing." *Schlumberger Technical Review* (September 1982).

Cannon, D.E. "Log Evaluation of a Fractured Reservoir-Monterey Shale." SPWLA 20th Annual Logging Symposium (June 3–6, 1979).

Chardac, J.L.M. "EPT Applications in the Middle East." SPE paper 13737 presented at Middle East Oil Technical Conference and Exhibition held in Bahrain (March 11–14, 1985).

Cheung, P.S.Y. and D. Heliot. "Workstation—Based Fracture Evaluation Using Borehole Images and Wireline Logs." SPE 20573 presented at the 65th Annual Technical Conference and Exhibition held in New Orleans, LA (Sept. 23–26, 1990).

Cox, J.W. "Long Axis Orientation in Elongated Boreholes and its Correlation with Rock Stress Data." SPWLA 24th Annual Logging Symposium (June 27–30, 1983).

Crampin, S., H.B. Lynn, and D.C. Booth. "Shear-Wave VSP's: A Powerful New Tool for Fracture and Reservoir Description." *Journal of Petroleum Technology* (March 1989), 283–288.

Crary, S., C. Boyeldieu, S. Brown, et al. "Fracture Detection With Logs." *The Technical Review*, 37 no. 1 (January 1987), 22–34.

Daniel, G.A. and Oscar Hvala. "Análisis de Rocas Tobáceas y Volcánicas a Partir de los Perfiles Sónicos y de Resistividad." Presented at the First Hydrocarbons National Congress, Buenos Aires, Argentina (December 3, 1982)

Delhomme, J.P. et al. "An Integrated Approach to Fracture Detection from Multipad Sensors." *Schlumberger and AGIP* publication (1982).

de Montigny, O., et al. "Horizontal-Well Drilling Data Enhance Reservoir Appraisal," *Oil and Gas Journal* (July 4, 1988) 40–48.

Dempsey, J.C. and J.R. Hickey. "Use of a Borehole Camera for Visual Inspection of Hydraulically-Induced Fractures." *Producers Monthly* (April, 1958), 14–21.

Dennis, B. and E. Standen. "Fracture Identification Techniques in Carbonates." CIM paper 88–39–110 presented at the 39th Annual Technical Meeting of the Petroleum Society of CIM held in Calgary, Alberta (June 12–16, 1988).

Dewan, J.T., "Thermal Neutron Decay Time Logging Using Dual Detection." SPWLA Fourteenth Logging Symposium (May 6–9, 1973).

Draxler, J.K. and D.P. Edwards. "Evaluation Procedures in the Carboniferous of Northern Europe." Presented at the Ninth International Formation Evaluation Symposium, SAID, Paris (October 24–26, 1984).

Edmunson, H.N. "Archie's Law: Electrical Conduction In Clean, Water-Bearing Rock." *The Technical Review*, Schlumberger, vol. 36, No. 3 (1988), 4–13.

Engelke, C.P. and D.W. Hilchie. "A New Qualitative Permeability Indicator." SPWLA Twelfth Annual Logging Symposium (May 2–5, 1971).

Etnyre, Lee. "Fracture Identification in the Panoma Field Council Grove Formation." *The Log Analyst* (November–December, 1981), 3–6.

Etnyre, L.M. "Practical Applications of Weighted Least Squares Methods to Formation Evaluation. Part I: The Logarithmic Transformation of Non-Linear Data and Selection of Dependant Variable." *The Log Analyst* (January–February, 1984), 11–21.

Eubanks, D.L. "Electric Micro Imaging Tool." *A Halliburton Publication* (1994).

Eubanks, D.L. "Use of Interactive Graphics Workstation Applications and Borehole Imaging Data in Geologic Studies." *A Halliburton Publication* (1994).

Fatt, Irwin: "The Network Model of Porous Media, II—Dynamic Properties of a Single Size Tube Network." *Transactions*. AIME 207 (1956), 160–163.

Ferrie, G.H., B.O. Pixier and S. Allen. "Well-site Formation Evaluation By Analysis of Hydrocarbon Ratios." Paper 81–32–20 presented at the 32 Annual Technical Meeting of the Petroleum Society of CIM held in Calgary, Alberta (May 3–6, 1981).

Fett, T. and J. Henderson. "Fracture Evaluation in Horizontal Wells Using the Formation Microscanner (FMS) Imaging Device." *Schlumberger* (1990).

Focke, J.W. and D. Munn. "Cementation Exponents in Middle Eastern Carbonate Reservoirs." *SPE Formation Evaluation* (June 1987), 155–167.

Frac-Finder/Micro-Seismogram Log, Basic Acoustics. *Welex* Publication.

Gardner, J.S. and J.L. Dumanoir. "Litho-Density Log Interpretation." Presented at the 21st Annual SPWLA Symposium held in Lafayette, Louisiana (July 8–11, 1980).

Geothermal Log Interpretation Handbook. SPWLA Reprint Volume, Tulsa, Oklahoma (1982).

Gianzero, S., R. Chemali, and S.M. Su, "Induction Resistivity, and MWD Tools in Horizontal Wells," *The Log Analyst* (May–June, 1990) 158–171.

Halliburton. "Tool Pusher Pipe-Conveyed Logging Systems," *Halliburton Horizontal Completions Seminar Manual* (1990).

Halliburton. "Measurement While Drilling," *Halliburton Horizontal Completions Seminar Manual* (1990).

Heflin, J.D., B.E. Neill, and M.R. Devries. "Log Evaluation in the California Miocene Formation." SPE 6160 presented at the 51st Annual Meeting of SPE of AIME, New Orleans (October 3–6, 1976).

Herrick, R.C. et al. "An Improved Nuclear Magnetism Logging System and Its Application to Formation Evaluation." SPE paper 8361 presented at the 54th Annual Fall Technical Conference and Exhibition held in Las Vegas, Nevada (September 23–26, 1979).

Hilchie, W.W., and S.J. Pirson. "Water Cut Determination from Well Logs in Fractured and Vuggy Formations." *Transactions*. SPWLA (Dallas, 1961).

Hilton, Jon. "Wireline Evaluation of the Devonian Shale: A Progress Report." First Eastern Shales Symposium, Morgantown Energy Research Center, ERDA (October 17–19, 1977).

Johnson, W.L. and W.A. Linke. "Some Practical Applications to Improve Formation Evaluation of Sandstones in the MacKenzie Delta." Paper C, Nineteenth Annual Logging Symposium Transaction, El Paso, Texas (June 13–16, 1978).

Kleinberg, R.L. et al. "Sensitivity and Reliability of Two Fracture Detection Techniques for Borehole Application." *Journal of Petroleum Technology* (April, 1984), 657–663.

Koerperich, E.A. "Evaluation of the Circumferential Microsonic Log, A Fracture Detection Device." SPWLA 16th Annual Logging Symposium, New Orleans, Louisiana (June 4–7, 1975).

Koerperich, E.A. "Application of Waxman-Smits and Archie Equations for Determination of Oil Saturation in Shaly Sand Reservoirs." SPE paper 5038 presented at the 49th Annual Fall Meeting of SPE of AIME held in Houston, Texas (October 6–9, 1974).

Ladeira, F.L. and N.J. Price. "Relationship Between Fracture Spacing and Bed Thickness." *Journal of Structural Geology*, (1981), v. 3, P. 179–183.

Leeth, R. and M. Holmes. "Log Interpretation of Shaly Formations Using the Velocity Ratio Plot." Presented at the SPWLA 19th Annual Symposium, El Paso, Texas (June 13–16, 1978).

Locke, L.C., and J.E. Bliss. "Core Analysis Technique for Limestone and Dolomite." *World Oil* (September 1950), 204.

Log Interpretation Principles. Schlumberger (1969), 105.

Luthi, S. and P. Souhaite. "Fracture Aperture from Electrical Borehole Scans." *Geophysics* (July, 1990), v.55, no. 7.

Luthi, S. "Borehole Imaging Devices." *Development Geology Reference Manual*, AAPG (1992).

Lyttle, W.J., and R.R. Ricke. "Well Logging in Spraberry." *Oil and Gas Journal* (December 13, 1951), 201.

McCray, A.W. *Petroleum Evaluations and Economic Decisions.* Prentice Hall, Inc. (1975), 201.

McQuillan, H., "Fracture-Controlled Production From The Oligo-Miocene Asmari Formation In Gach Saran and Bibi Hakimeh Fields, Southwest Iran." P.O. Roehl and P.W. Choquette, eds, *Carbonate Petroleum Reservoirs*, (1985), P. 511–524.

Mardock, E.S., and J.P. Myers. "Radioactivity Logs Define Lithology in the Spraberry Formation." *Oil and Gas Journal* (November 29, 1951), 90.

Mihram, G.R. "Flow to obtain More Fracturing Information by the Use of Radioisotopes." Halliburton Laboratories, Duncan, Oklahoma (March, 1964).

Morris, R.L., D.R. Grine, and T.E. Arkefeld. "The Use of Compressional and Shear Acoustic Amplitudes for the Location of Fractures." *Journal of Petroleum Technology* (June 1964).

Myung. J.T. "Fracture Investigation of the Devonian Shale Using Geophysical Well Logging Techniques." Proceedings of the Seventh Appalachian Petroleum Geology Symposium, Morgantown, West Virginia (March 1–4, 1976).

Neutron Lifetime Interpretation. *Dresser Atlas* publication (circa 1970).

O'Brien, W.J. "Porosity from Well Logs of the Kurparuk Reservoir-Procedures for using Core Data." SPE paper 12095 presented at the 58th Annual Technical Conference and Exhibition held in San Francisco, California (October 5–8, 1983).

Oduolowu, A. "A Case Study of the Recoverable Hydrocarbon Volumes for the Minnelusa and Muddy Sands, Powder River Basin, Wyoming." *The Log Analyst* (March–April 1976), 24–32.

Paauwe, E.F. "Quantitative Anaysis of Potential Flow in Fractured Reservoirs." A Schlumberger of Canada Publication (1994).

Paauwe, E.F. "Fracture Analysis." In *Geological Applications of Dipmeters and Borehole Imaging*, a Schlumberger of Canada Short Course (1994).

Patchett, J.G., and R.W. Rausch. "An Approach to Determining Water Saturation in Shaly Sands." *Journal of Petroleum Technology* (October, 1967).

Pickett, G.R. "Acoustic Character logs and Their Application in Formation Evaluation." *Transactions.* AIME 228 (1963), 659–667.

Pickett, G.R., and E.B. Reynolds. "Evaluation of Fractured Reservoirs." *Society of Petroleum Engineers Journal* (March 1969), 28–38.

Pickett, G.R. and S.D. Artus. "Prediction from logs of Recoverable Hydrocarbon Volume-Ordovician Carbonates: Williston Basin." *Geophysics* (1970), v.35, 113–123.

Pickett, G.R. *Formation Evaluation*, Colorado School of Mines class notes, copyright by G.R. Pickett (1972).

Pickett, G.R. "Pattern Recognition as a Means of Formation Evaluation." paper presented at the 14th Annual Logging Symposium of SPWLA (May 6–9, 1973).

Pirson, S.J. "Petrophysical Interpretation of Formation Tester Pressure Build-Up Records." Paper presented at the third Annual Logging symposium, SPWLA (May 17–18, 1962).

Pirson, S.J. "How to Map Fracture Development from Well Logs." *World Oil* (1967), 106–114.

Pirson, S.J. *Geologic Well Log Analysis.* Gulf Publishing Company (1970), 203.

Pirson, S.J. "Log Interpretation in Rocks with Multiple Porosity Types—Water or Oil Wet". *World Oil* (June 1975), 196–198.

Porter, C.R., G.R. Pickett, and W.W. Whitman. "A statistical Method for Determination of Water Saturation from Logs." *Transactions.* SPWLA, New Orleans (May 25–28, 1969).

Poupon, A., W.R. Hoyle, A.W. Schmidt. "Log Analysis in Formations with Complex Lithologies." SPE paper 2925 presented at 45th Annual Fall Meeting of SPE of AIME (October 1970).

Poupon, A., M.E. Loy and M.P. Tixier. "A Contribution to Electrical Log Interpretation in Shaly Sands." *Journal of Petroleum Technology* (June, 1954).

Poupon, A., I. Strecker, and L. Gartner. "A Review of Log Interpretation Methods Used in the Niger Delta." SPWLA Symposium (1967).

Rasmus, J.C. "Variable Cementation Exponent, m, for Fractured Carbonates." *The Log Analyst* (November–December, 1983), 13–23.

Rasmus, J.C. "Determining the Type of Fluid Contained in the Fractures of the Twin Creek Limestone by using the Dual Laterolog Tool." *Journal of Petroleum Technology* (November, 1982), 2673–2682.

Sanyal, S.K. and J.E. Ellithorpe. "A Generalized Resistivity-Porosity Crossplot Concept." Paper SPE 7145 Presented at the California Regional Meeting held in San Francisco (April 12–14, 1978).

Schafer, J.M. "A Practical Method of Well Evaluation and Acreage Development for the Naturally Fractured Austin Chalk Formation." *The Log Analyst* (January, February, 1980), 10–23.

Schlumberger. "Production Log Interpretation" (1973).

Schlumberger. "Log Interpretation Principles" (1979).

Schlumberger Well Evaluation Conference, Egypt (1984).

Schlumberger Product Development-CAU (April 23, 1986).

Schlumberger. "Evaluación de Formaciones en la Argentina." (1973), 94–95.

Schlumberger. "Current Affairs in Saturation." *Middle East Well Evaluation Review* (1987), 49–57.

Schmoker, J.W. "Density Variations in Quartz Diorite Determined from Borehole Gravity Measurements, San Benito County, California." *The Log Analyst* (March–April, 1977), 32–38.

Setser, G.G. "Fracture Detection by Circumferential Propagation of Acoustic Energy." SPE paper 10204 presented at the Annual Technical Conference and Exhibition held in San Antonio, Texas (October 4–7, 1981).

Shankland, T.J. and H.S. Waff. "Conductivity in Fluid-Bearing Rocks." *Journal of Geophysical Research* (November 10, 1974), 4863–4868.

Sherman, M.M. "The Determination of Cementation Exponents Using High Frequency Dielectric Measurements." SPWLA 24th Annual Logging Symposium (June 27–30, 1983).

Sibbit, A.M. and O. Faivre. "The Dual Laterolog Response in Fractured Rock." *Trans.* 26th Annual Logging Symposium, Paper T, Dallas, Texas (1985).

Simandoux, P. "Measures Dielectriques en Milieu Poreux, Application a Mesure des Saturations en Eau, Etude du Comportement des Massifs Argileux." (Dielectric Measurements in Porous Media and Application to Shaly Formations), *Revue de l'institut Francais du Petrole*, Supplementary Issue." (1963).

Souder, W.W., and G.R. Pickett. "A Computerized Method for the Zonation of Digitized Well Logs." Paper SPE 4019 presented at the 47th Annual Fall Meeting of SPE of AIME in San Antonio, Texas (October 8–11, 1972).

Standen, E. "Formation MicroScanner, MST/SHDT Imaging Dipmeter." *Schlumberger*, Calgary, Alberta (circa 1980).

Stang, C.W., "Alternative Electronic Logging Technique locates Fractures in Austin Chalk Horizontal Well," *Oil and Gas Journal*, (November 6, 1989) 42–45.

Struyk, C. and Poon, A., "Log Interpretation in Horizontal Wells," *Horizontal Wells, Evaluation and Completion*, A Schlumberger Well Services Publication (January 1990).

Suau, J., et al. "Evaluation of Very Low-Porosity Carbonates, Malossa, Italy." SPWLA (June 13–16, 1978). Paper W

Sutton, Todd. "Cased Reservoir Evaluation in Michigan with TDT and GST." Michigan SPE/SPWLA Symposium (March 21–22, 1984).

Taylor, D.B. *ANASEIS II Manual*, Applied Geophysical Software Inc., Houston, (1987).

Timko, D.T. "A Case Against Oil Muds." *The Log Analyst* (November 1966), 4.

Tixier, M.P. et al. "Estimation of Formation Strength from the Mechanical Properties Log." paper SPE 4532 presented at the 48th Annual Fall Meeting held in Las Vegas, Nevada (September 30–October 3, 1973).

Towle, Guy. "An Analysis of the Formation Resistivity Factor-Porosity Relationship of Some Assumed Pore Geometries." Presented at the 3rd SPWLA Annual Meeting, Houston, Texas (May 17–18, 1962).

Vogel, C.B. and R.A. Herolz. "The CAD—A Circumferential Acoustical Device for Well Logging." *Journal of Petroleum Technology* (October, 1981), 1985–1987.

Walker, Terry. "Progress Report on Acoustic Amplitude Logging for Formation Evaluation." paper SPE 451 presented at the 37th Annual Fall Meeting of SPE of AIME, Los Angeles (October 7–10, 1962).

Walstron, J.E., T.E. Mueller, and R.C. McFarland. "Evaluating Uncertainty in Engineering Calculations." *Journal of Petroleum Technology* (December 1967), 1595.

Waxman, M.H. and L.J.M. Smits. "Electrical Conductivities in Oil-Bearing Shaly Sands." SPE of AIME, *Transactions*. Vol 243 (1968).

Waxman, M.H. and Thomas, E.C. "Electrical Conductivities in Shaly Sands-I. The Relation Between Hydrocarbon Saturation and Resistivity Index; II. The Temperature Coefficient of Electrical Conductivity." *Journal of Petroleum Technology* (February, 1974), v. 14, 213–225.

Western Atlas: "Extended-Reach and Horizontal Well Services." Houston, Texas (1990).

Wharton, R.P. et al. "Electromagnetic Propagation Logging: Advances in Technique and Interpretation." paper SPE 9267 presented at the 55th Annual Fall Technical Conference and Exhibition held in Dallas, Texas (September 21–24, 1980).

Wyllie, M.R.J., A.R. Gregory, and L.W. Gardner. "Elastic Wave Velocities in Heterogeneous and Porous Media." *Geophysics* 21, (January 1956), 1, 41.

Yale, D.P. and E.S. Sprunt. "Prediction of Fracture Direction Using Shear Acoustic Anisotropy." *The Log Analyst* (March–April, 1989), 65–70.

Zemanek, J., et al. "The Borehole Televiewer: A New Logging Concept for Fracture Location and Other Types of Borehole Inspection." *Journal of Petroleum Technology* (June 1969), 762–774.

Chapter 4

Tight Gas Reservoirs

Production from tight gas reservoirs is possible in many cases thanks to the presence of natural fractures. Many of these reservoirs are also multi-layered adding a great deal of complexity to the evaluation. The problem we face from a practical point of view is that productivity is usually low and as a consequence it is non-economic to spend significant amounts of money in all wells to rigorously characterize the reservoir.

Fortunately funding provided by the U.S. Department of Energy (DOE) and the Gas Research Institute (GRI) has led to research and improved techniques to evaluate and forecast performance of tight gas reservoirs in the United States. These techniques should also prove valuable in other countries.

Tight reservoirs can be grouped into (1) Devonian shales and (2) tight gas sandstones. Because of the low productivity of tight reservoirs I place emphasis on techniques that in my experience have proven cost-efficient although they are not necessarily highly rigorous.

FRACTURED SHALES

Fractured shales have produced gas since the early 1900s along the western margin of the Appalachian Basin. The amount of gas-in-place in this basin was originally estimated at 460 quadrillion scf (Schumaker, 1976). Subsequent estimates indicates values of gas-in-situ ranging between 585 and 2500 trillion scf (USGS report 82–474, 1982). More recent estimates range between 200 and 1860 trillion scf (de Witt, 1986). The point is that there is a large uncertainty with respect to the original volumes of gas-in-place and even more uncertainty with respect to the recoverable volumes. An extensive area of fractured shales is also present in the lowlands of Quebec which appears to have reasonable potential.

Since gas production of fractured shales is usually low, the economics of producing these reservoirs is in many cases regarded as marginal. This creates problems in gathering the necessary data for rigorous reservoir characterization. Water saturation in most Devonian shales appear to be at irreducible conditions.

Oil production is also obtained from some fractured shales. For example, the Woodford shale of Upper Devonian age in Oklahoma is oil productive in certain areas. The same holds true for the Eagle Ford shale in Texas. The Eagle Ford is of Cretaceous age and it separates the more prolific Austin Chalk and Buda limestone. Oil production is also obtained from the Mancos Shale, Rangely field, Colorado, black shales in the Bakken formation, Williston Basin, North Dakota, and the Monterey "Shale" in California. Actually the lithology in the Monterey is very complex. Some geologic aspects of this reservoir are discussed in Chapter 1 and the history is presented in Chapter 5. Fox and Martiniuk (1994) have reported on reservoir characteristics and petroleum potential of the Bakken formation in south-western Manitoba.

WORLDWIDE DISTRIBUTION OF BLACK SHALE

Devonian-Mississippian organic rich black shales are present in 26 states of the United States and six provinces and territories of Canada. They are also found in northern

Mexico, along the USA-Mexican border, and in South America, Africa, and Europe (Provo, 1976).

Many stratigraphic names have been assigned to the Devonian-Mississippian black shales. The most widely used are Chattanooga and Ohio shales in the Appalachian basin. Black shales of Devonian-Mississippian age are also found in the Exshaw formation of southern Alberta and the Rapids formation of the Moose River Basin.

Thickness of black shales varies widely. For example, the Double Horn shale of central Texas and the New Albany shale of Kentucky average less than 20 ft. On the other hand, the shale has a thickness of over 2000 ft in West Virginia.

Black shales of North America are characterized by two tectonic settings. They may occur as distal facies of turbidity associated with the great Catskill Delta complex, and they may occur in the craton far from marginal geosynclines.

Black shales of Africa and South America are similar to those of North America in their association with sandstone and siltstones. Black shales of central Europe and western Soviet Union are found as basinal facies near growing carbonate reefs which stood high above the flow of the basin. This third tectonic setting that associates reefs and black shales is not known in North America.

Figure 4–1 is a highly generalized map showing major tectonic elements of middle and upper Palaeozoic age in Eastern USA.

A computerized gas resource information system (GRIS) was established by the Gas Research Institute (GRI) to assist companies working in the eastern U.S. gas shales (Zielinski et al, 1982).

OIL AND GAS OCCURRENCE IN FRACTURED SHALES

Oil and gas production is obtained in northern West Virginia, primarily from the Cottageville field (Martin & Nuckols, 1976). Production is attributed mainly to fractured porosity and permeability. Shale sections of interest are highly radioactive and can be easily identified in gamma ray logs (Figure 4–2), as indicated by zones, I, II, and III.

The main production at Cottageville comes from Zone II, which is characterized by brown shale. An isopach map of this zone is shown in Figure 4–3. The stratigraphic unit thickens toward the east and reaches a limit of about 500 ft in the eastern limit of mapping.

Gas production from shales in northern West Virginia is shown in Figure 4–4. The largest field is the Cottageville located in Jackson and Mason Counties.

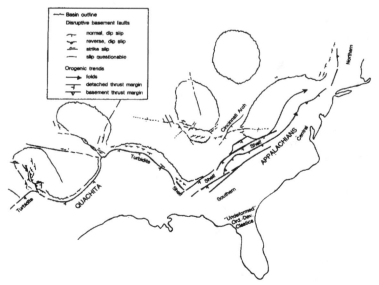

FIGURE 4–1 Schematic tectonic elements of middle and upper Palaeozoic age (after Schumaker, 1976)

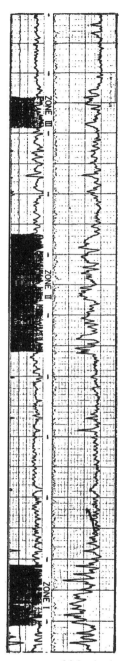

FIGURE 4–2 Gamma ray indicates presence of black shales (after Martin & Nuckols, 1976)

Shale fracturing is possibly related to basement faulting along the west side of the Rome Trough. Production from this field has been estimated at over 15 Bscf during the last 25 years. A total of about 90 wells were drilled originally.

More than 3000 Devonian fractured shale wells have been drilled in southwestern West Virginia. Estimated ultimate recovery from these wells amounts to 1 Tscf. As more wells are completed open hole, it is possible that some gas attributed to Devonian shale has actually come from shallower Mississippian beds.

Bagnal and Ryan (1976) have reported that better wells are associated with fractured brown, kerogen-rich shales which forms between 10% and 60% of the total shale thickness.

Figure 4–5 shows a typical driller log of a Devonian shale well in southwestern West Virginia.

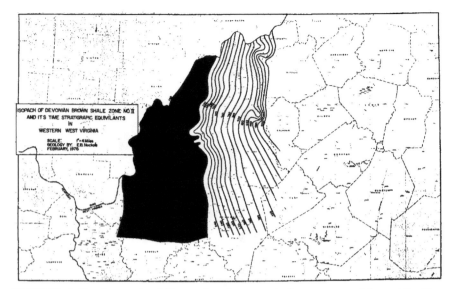

FIGURE 4–3 Isopach of Devonian brown shale zone no. 11 and its time equivalents in West Virginia (after Martin and Nuckols, 1976)

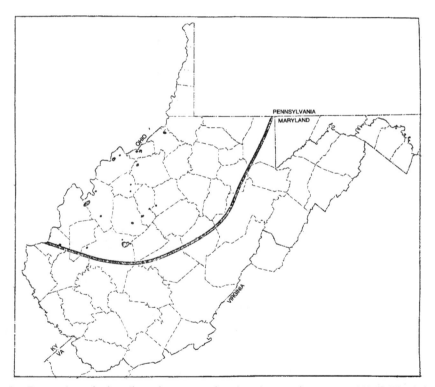

FIGURE 4–4 Devonian shale oil and gas production in northwestern West Virginia (after Martin & Nuckols, 1976)

Neal (1979) and Schumaker (1982) have indicated that basement growth faulting enhances local fracture density generating areas of greater production. Since wells located in the flanks of certain structures have the higher initial open flows, Schumaker considers as prospective for future exploration the flanks of flexures over normal faults, particularly at intersecting basement fault trends.

Some advantages of drilling in the flanks when exploring for naturally fractured reservoirs were highlighted in Chapter 1.

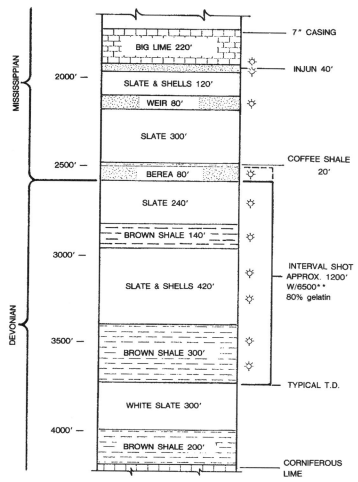

FIGURE 4–5 Typical drillers log of a Devonian shale well, S.W. West Virginia (after Bagnal & Ryan, 1976)

In Quebec, black shales are found mostly in the external zone of the St. Lawrence Lowlands forming part of the Utica Group (Ordovician Age).

The sedimentary basin of the lowlands of Quebec is located between Montreal and Quebec City. It has an approximate length of 200 mi., an approximate width of 50 mi., and an area of about 10000 sq. mi.

Figure 4–6 shows a generalized stratigraphic section of the area. The Utica Group is characterized lithologically by black shales. These shales have a thickness of up to 1200 ft and they are characterized for their odour of petroleum when freshly broken.

Figures 4–7, and 4–8 show the tectonic zones of the Quebec Lowlands, i.e., the pre-Cambrian shield, platform, external zone, internal zone, and nappe zone.

The platform is characterized by normal faulting. In some cases throws up to 3000 ft are known.

The external zone is characterized by thrusting in the Utica and Lorraine groups. The lower parts of the lithological column have not been disturbed. The black shales are located in this zone (Figure 4–9) and have been reached in wells Villeroy 1, Villeroy 2, and Ste-Francoise Romaine 1. The area of the external zone is about 900 sq. mi.

The internal zone is characterized by thrusting in all the sedimentary sequence, including the Ordovician carbonates and Cambrian sands.

The nappe zone is characterized by imbrications of carbonates and sands. In this zone, due to upheaval and compression of the platform associated subduction zone, gravity sliding and thrusting at the beginning of the Tacony orogeny have carried deep basinal sediments from the southeast over the previously described zones (Beiers, 1976).

SYSTEM	SERIES		ST. LAWRENCE LOWLANDS GROUPS/FORMATIONS		GENERALIZED LITHOLOGICAL CHARACTER
Ordovician	U	Cincinnatian	Richmond		Red sh and ss
			Lorraine		Interbedded sh and siltstone
			Utica		Black sh
	M	Champlanian	Trenton		Thin-bedded ls
			Black River		Ls
			Chazy		Ls and dolo
	L	Canadian	Beekmantown	Beauharnois fm	Dolo
				Theresa fm	
Cambrian	U	Croixan	Potsdam	Chateauguay fm	Quartz sands
				Covey Hill fm	Feldspathic sands

FIGURE 4–6   Stratigraphic chart, Saint Lawrence Lowlands of Quebec (after Beiers, 1976)

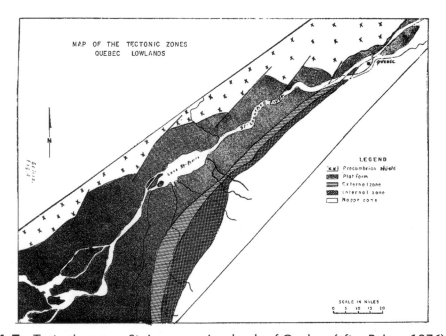

FIGURE 4–7   Tectonic zones, St. Lawrence Lowlands of Quebec (after Beiers, 1976)

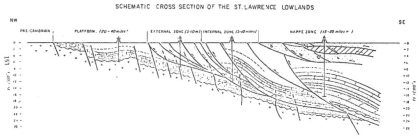

FIGURE 4–8   Cross section of St. Lawrence Lowlands

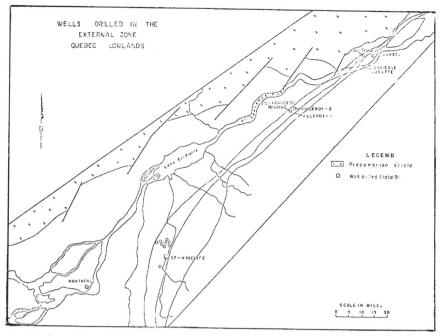

FIGURE 4–9 Wells drilled in external zone, St. Lawrence Lowlands

High grade oil was produced from the Mancos shale, Rangely field, Rio Blanco County, Colorado, for many years. The fractures were related to axial bending and arching in the Rangely anticline (Peterson, 1955).

Regan (1953) has reported on oil production from fractured chert and silicious shales in the Santa Maria coastal district and the southwestern part of the San Joaquin Valley, California. Initial production from individual wells ranged from 200 to 1000 bo/d. The reservoirs were characterized by high permeability due to fracturing and low porosity.

REMOTE SENSING SURVEYS

Delineation of fractured trends is very important for proper exploitation of fractured reservoirs. Aerophotography and fracture indices can be used to map fracture trends. Other powerful tools aimed at the same objective are remote sensing surveys. Ryan (1976) has used the following sensing surveys in conjunction with seismic and geologic maps to delineate fracture traces in the Haysi area of Western Virginia and southeastern Kentucky; side looking radar imagery (SLAR), color aerial photography, black and white infrared photography, color infrared photography, and thermal infrared imagery.

From the analysis it was found that better wells were associated with natural fractured zones. As an example, Berea wells hydraulically fractured within 1500 ft from a major lineation yielded an average open flow of 1287 Mscfd, while those wells more than 1500 ft from a major lineation yielded only an average of 637 Mscfd.

The conclusion was reached that from the type of geologic and topography terrain under consideration, the radar imagery was a useful tool in delineating natural fractured zones and black and white infrared photography was useful in adding detail. More information on the optical processing of remote sensor imagery are provided in a paper by Rabshevsky (1976).

McIver et al (1982) have used aerial photography to locate potential drilling sites in the Appalachian basin by choosing those areas that appear to have a high density of natural fracturing in the surface.

Empirically, 220 locations were selected where three or more intersecting fracture traces fell within an area circumscribed by a circle 1600 feet in diameter.

GEOPHYSICAL INVESTIGATION

Tegland (1976) has suggested a program leading to sound evaluation of fractured shales from geophysics. The program encompasses four main phases; geologic framework analysis that must be regional in nature; seismic research project whose objective is to evaluate the seismic response of the geologic section at each of the key well locations; seismic reconnaissance programs which allow a sound comparison of two types of explosive source techniques and vibrator; and the use of 3-dimensional techniques to provide detail control. This allows one to improve the signal-to-noise ratio of the data by means of 3-D migration stack, 3-D dip filtering, or both.

Ruotsala and Williams (1982) observed that seismic reflection data collected in the Cottageville field in 1977 showed a normal basement fault extending up into the Ordovician. Gas production along this fault occurs in an area several miles long and about 1 mile wide.

Because of the presence of gas in the shale and the large degree of fracturing, the seismic velocities were reduced. Similarly greater attenuation of the high frequency components of the seismic wave were observed.

This observation led to the conclusion that reflection seismology can be used to target specific exploration wells following preliminary reconnaissance with other methods.

PHYSICAL CHARACTERIZATION

Physical characterization has been carried out as part of the Eastern Gas Shales Project sponsored by the Department of Energy (DOE) by measuring physical properties such as bulk density, porosity, permeability, tensile strength, compressive strength, and breaking strength in all cores obtained. Figure 4–10 shows a map of the core locations.

Total porosities have been reported to range between 1.63 and 15.8%. Open porosity has ranged between 0.59 and 12.6%. Permeability parallel to bedding planes ranges from 180 microdarcies to 0.63 millidarcies in rocks which according to X-ray analysis were not fractured.

Diameters of pore spaces in the shales are usually less than one micrometer. Orientation of induced and natural fractures has been determined. Stress analysis has been helpful in evaluating hydraulic fracturing possibilities and the effect of tectonic transport of the shales. An interesting finding was presented by Yost et al (1982) indicating that in well Baler 11940, Jackson County, W. Va., natural fractures had a preferential orientation of approximately N 45° E in most intervals. However, in a zone at 3725–3775 ft a series of multiple fracture orientations was noted in core. This corresponded to an unusual large rate of 1.05 MMscfd.

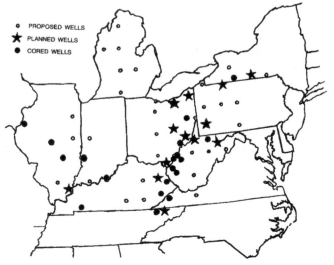

FIGURE 4–10 Map of Easter Gas Project core locations (after Overbey, 1978)

More recently Watson and Mudra (1994) reported on the use of computed tomography (CT) for characterizing Devonian shales. They were able to detect fractures that were considered smaller than the resolution of the instrument. They further determined distribution and magnitude of storage that were consistent with a medium of small permeability and large adsorption.

GEOCHEMICAL CHARACTERIZATION

Many of the cores of the Eastern Gas Shales Project were sampled every 10 ft to carry out geochemical tests. These tests included 1) elemental and trace elemental analysis, 2) organic carbon content, 3) total hydrocarbon content, 4) material balance Fisher assay to determine gas, oil, and water content, 5) bitumen content, 6) kerogen content and kerogen type, 7) vitrinite reflectance of kerogen material, and 8) gas concentration in scf of gas per cu ft of shale.

The gas concentration in shales was reduced from an average of 0.82 scf/cu ft of shale in 1977 to 0.52 scf/cu ft of shale in 1978. This reduced the in-place estimate of gas from 1500 Tscf to 800 Tscf.

In some cases there is a good correlation between carbon content, gamma radiation, and bulk density.

Marine, restricted marine, and nonmarine facies have been identified. Laboratory analysis indicates that higher gas yields are associated with the marine and restricted marine facies.

Key minerals have been identified as quartz, illite clay, chlorite clay, and mixed layer clays. Accessory minerals are pyrite, calcite, and dolomite.

Based on geochemical data Zielinski and McIver (1982) have provided another estimate of gas-in-place in the Appalachian basin at approximately 2500 trillion scf. Furthermore, the geochemical data have allowed pinpointing those areas where the greatest volumes of gas are contained in the matrix giving direction to future selection of locations for artificial fracturing research.

STIMULATION

Funding provided by the U.S. Department of Energy (DOE) and the Gas Research Institute (GRI) has lead to important advances in hydraulic fracturing. The gas industry, however, has been plagued by very optimistic forecasts following hydraulic fracturing treatments due to the use of single-layer models that do not represent accurately the multi-layer characteristics of these reservoirs. For example Shaw et al (1989) have reported on stimulation treatments conducted in Mason County, W. Va. The jobs were successful in initiating gas production, but the rates expected from the stimulation jobs were not achieved. Part of the problem might be the difficulty in properly evaluating pre-stimulation test data due to long afterflow periods (Holgate et al, 1988). Under these conditions conventional semilog and type curve methods might yield questionable results that in turn lead to very optimistic post-stimulation forecasts. Also in many cases pre-stimulation well testing data are simply not available.

The following methods have been used in the past to stimulate shales with varied degree of success (Norton, 1976):

Shooting.   It was used mostly in the old days. It was restricted to open hole completions. The main disadvantage of this type of treatment was very small horizontal penetration into the zone. Hunter and Young (1953) have reported on gas wells in the Big Sandy field shot with 80% gelatinated nitroglycerine throughout sections of 350 to 750 ft. Such shooting required 3000-7000 pounds of explosives.

Water fracturing.   This method has been used effectively for many years. By continuously improving additives and techniques, the results should also improve.

Gas fracturing. It has proven to be an efficient stimulation method, without the disadvantage of having a water base fluid. The economics of the technique are a problem.

Methanol fracturing. In this technique, 30% methanol is added to the frac fluid to improve water recovery.

Foam fracturing. In this method, about 75% of the water in the fracture fluid is replaced with nitrogen and a surfactant that creates a foam. One advantage of this method is that it reduces the amount of water to be used in the frac.

Acid-methanol fracturing. This treatment has provided excellent results in stimulating calcareous shales in the lowlands of Quebec (Sanchez, 1977). A typical frac consists of 25,000 gal 28% HCL mixed with 12,500 gal methanol. Injection pressures have averaged 5500 psi at a rate of 6 bbl/min. About 90% of treating fluids have been flowed back following the treatment.

In general, it is recognized that shales can be effectively stimulated, especially with large volumes of frac fluids that provide deeper penetration and a more extensive fracture network. In hydraulic fracturing, the fluid loss of frac fluid must be controlled and the size of sand must be large enough to bridge small fractures to reduce the movement of fluid to them. Good clay control agents are available for use in water base fluids. However, it is experience that indicates the best treatment in each particular area. Some of the stimulations conducted under the Eastern Gas Shales Project (EGSP) are summarized in Table 4–1.

The highlights of this stimulation program are summarized by Overbey (1978):

- Massive hydraulic fracture treatments using gelled water and sand is not a good technique to use in low pressure shale reservoirs due to problems associated with cleanup time (Columbia Gas 20401).
- Low residual fluid (LRF) stimulations cleanup faster in large volumes (MHFF). Fastest cleanup may be CO_2-H_2O fracture at moderate depth (to 5000 ft).
- Low residual fluid stimulations (Foam fracture) may have a practical depth limitation of 5000 ft to 6000 ft because of excessive N_2 requirements.
- When stimulating the Rhinestreet-Marcellus zone, consideration should be given to the density of the fracturing fluid and the possibility of downward growth into an aquifer.
- Screen outs may occur in the formation even with cross-linked polymer gels of high viscosity (low leakoff rate) carrying 50% of rated capacity perhaps due to spalling or other mechanisms in the formation.
- Hydraulic fracturing of Devonian shales has proved to result in at least 60% more production over two 5-year period above borehole shooting in eastern Kentucky.
- LRF-Foam and LRF-CO_2-H_2O stimulations give indications of consistently higher production rates over gelled water stimulations. Additional testing and production is needed to confirm.
- Out-of-borehole chemical explosive fracturing has been successful in three Lincoln, West Virginia tests. This technique may be economical in areas of isotropic stresses.

Opposite highlights presented by Shea and Bucher (1982) include:

- Total fracturing and cleanup costs of foam fractured wells are approximately 17.5% higher than water fracturing methods.
- Although open flows from foam fractured wells are significantly higher than open flows from water fractured wells, it seems to be that actual in-line production of the foam-fractured wells is not higher than production from water fractured wells.

Columbia Gas System's stimulation studies indicate that commercial shale wells can be obtained by modern fracturing methods (Smith, 1978). From the well tests, they observed that gas production increases after the frac jobs, and that induced fracture lengths are shorter than those calculated.

They consider that only production history provides dependable data for evaluating shale wells. Figure 4–11 shows a correlation of 20 years' cumulative production vs. initial open flow for 35 wells in West Virginia. The correlation is not very good, but the best curve

TABLE 4–1 EGSP Stimulation Results (after Overbey, 1978)

			Appalachian Basin		
Well-location	**Interval**	**Type treatment**	**Treatment vol.-gal**	**Flow after stimulation**	**Approx. clean-up time, days**
Columbia 20401	1	MHF	517,000	110	200
Lincoln Co. W. Va.	2	MMHF	104,160	111	30
	3	MMHF	100,600	80 Total	30
	4	MMHF	115,000	21 (322)	30
Columbia 11236	1	CO_2-H_2O	117,180	36	
	2	CO_2-H_2O	113,400	NA	
Columbia 20403	1	MHFF	250,000	110	30
Lincoln Co. W. Va.	2	MHFF	320,000	200	30
	3	MHFF	346,000	107	30
	4	MHFF	347,000	NT (381)	30
Consolidated 12041	1 ?	MHFF	163,000	173	30
Jackson Co. W. Va.					30
Ky-WVa 7239	1	Foam	46,000	0	30
Perry Co. Ky.	2	Foam	48,000	42	30
	3	Foam	44,000	20	30
	4	Foam	45,000	0 (62)	30
Ky-WVa 7246	1	Foam	50,000	337	30
Leslie Co. Ky.					
Ky-WVa 1627	1	Foam	50,000	103	
Leslie Co. Ky.					
PTC-686-1	1	CEF	3,000	260	20
Lincoln Co. W. Va.					
Columbia 20338-T	1	CO_2-H_2O	98,000	54-107	30
Wise Co. Va.	2	CO_2-H_2O	112,500	40 (150)	30
PTC-686-2	1	CEF	3,000	300	15
Lincoln Co. W. Va.					
Columbia 20337	1	CO_2-H_2O	3,000		
Martin Co. Ky	2	CO_2-H_2O	3,000		
PTC-686-3	1	CEF	3,000	310	15
Lincoln Co. W. Va.					
Columbia 20336	1	CO_2-H_2O	61,300	370	30
Martin Co. Ky.	2	CO_2-H_2O	64,500	250 (620)	30
People NG Co.	1	MHFF	175,000	50-100*	Mechanical failure - plugged
Mercer Co. Pa.					
Fleet Energy	1	CO_2-H_2O	50,000	100	Could not get rid of
Mason Co. W. Va.	2	Foam	50,000	600**	H_2O-TP
Columbia 20402	1	MHFF	150,000	145	30
Lincoln Co. W. Va.	2	MHFF	150,000	NA	30
PTC-685-2	1	CEF	3,000	NA	NA
Pike Co. Ky.					
DOE-MERC #1	1	CO_2-H_2O	105,000	0 ?	30
Mon. Co. W. Va.	2	CO_2-H_2O	50,000	NA	30
			Illinois basin		
Ray Clark #1	1	Foam	50,800	2	30
Christian Co. Ky.					

TABLE 4–1 Continued

			Michigan Basin		
Well-location	**Interval**	**Type treatment**	**Treatment vol.-gal**	**Flow after stimulation**	**Approx. clean-up time, days**
Welch 1-15 Otsego Co. Mi.	1	Foam	46,000	550	30

MHF	=	Massive hydraulic fracture
MMHF	=	Modified massive hydraulic fracture (water followed by gas)
MHFF	=	Massive hydraulic foam fracture
CEF		Chemical explosive fracture
CO_2-H_2O	=	20% CO_2 and 80% gelled water (also referred to as cryogenic frac)

* Mechanical Failure

** Production lost to H_2O from Onadaga-Oriskany

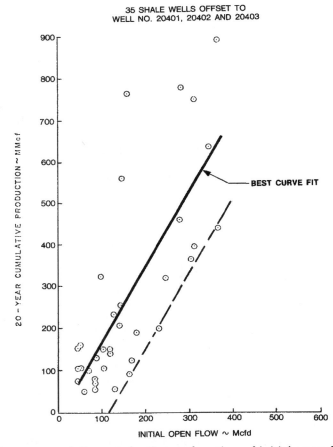

FIGURE 4–11 20-year cumulative production as function of initial open flow (after Smith, 1978)

fit allows order-of-magnitude estimates with regard to recoverable reserves as a function of initial open flow.

INTERACTION BETWEEN HYDRAULICALLY INDUCED AND NATURAL FRACTURES

In many instances hydraulic fracturing jobs are performed in naturally fractured reservoirs in efforts to intercept the system of natural fractures. Blanton (1982) has carried out laboratory studies on naturally fractured blocks of Devonian shales, fracing hydraulically the pre-fractured samples under triaxial states of stress.

Based on this work the conclusion has been reached that hydraulic fractures tend to cross pre-existing (natural) fractures only under high differential stresses and high angles of approach. In most instances the hydraulic fractures were either diverted or arrested by the pre-existing fractures.

Under this laboratory scenario Blanton believes that for hydraulic fractures in the field; symmetrical, double-winged, vertical fractures are probably very rare. His work indicates that it would be most likely to have fractures with wings diverted at different angles or with truncated wings of different lengths.

The interaction between hydraulically induced and natural fractures has been studied by Yost et al (1982) using oriented cores, induced fracture orientation, lineaments and a Seisviewer in Devonian shale wells.

Their work led to the conclusion that primary major surface lineament direction is similar to the natural fractures direction observed in cores taken from the Cottageville field. In the Mt. Vernon area the secondary major surface lineament is similar to the orientation of both natural and hydraulically induced fractures. This type of behavior has been observed in many places.

Yost et al (1982) indicated that there was an adequate correlation between the fracture identification log (FIL) and oriented core vertical fractures. Low angle fractures along bedding planes were not defined properly.

METAMORPHISM

Metamorphism can occur in sedimentary rocks or other metamorphic rocks. Metamorphism is the result of high levels of heat and pressure. For example various studies of the St. Lawrence Lowlands of Quebec have indicated that the black shales of the external zone are mature to gas.

Figure 4–12 shows plots of crystallinity of the illite and reflectometry of bitumen vs. depth for well Villeroy 1.

The crystallinity of the illite reflects the level of metamorphism of a rock. Empirically, there are not possibilities of hydrocarbon presence when the rock is in the anchizone. Figure 4–12 shows that the crystallinity of illite in the Utica group does not reach the anchizone and places the group in a zone potentially mature to gas.

Studies of reflectometry power of bitumen and/or vitrinite indicate the following maturity levels for hydrocarbon presence: immature 0.1–0.5%, mature to oil 0.5–1.35%, mature to gas 1.35–3%. Figure 4–12 shows that the Utica Group falls in a reflectometry power between about 1% and 2% and consequently places this group in a zone mature to gas.

The various studies of metamorphism in the St. Lawrence Lowlands of Quebec correlate with studies in eastern United States (Appalachian) which indicate an increase in metamorphism to the east in the direction of more intense diastrophism (Landes, 1959).

In conclusion, metamorphism studies indicate that shales in the St. Lawrence Lowlands are located in a zone mature to gas.

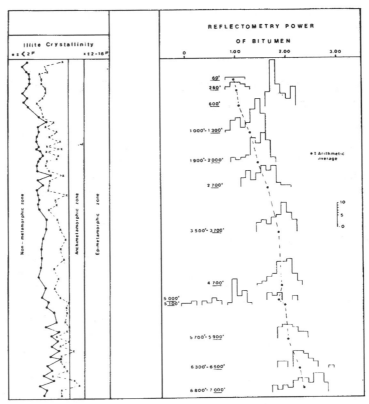

FIGURE 4–12 Illite crystallinity and reflectometry power of bitumen indicate Utica Group is mature to gas

LOG ANALYSIS

Smith (1978) indicates that the gas content of shales increases with thermal maturity, rock pressure, and organic carbon; radiation increases with organic carbon (Figure 4–13).

Figure 4–14 shows that the amount of off-gas (scf of gas/cu ft of shale) increases with gamma radiation. The term "off-gas" refers to "those tighter hydrocarbons which can escape from a rock volume under ambient conditions given sufficient time without any heat stimulation and without grinding the rock" (Smith, 1978).

Figure 4–15 shows a crossplot of density (g/cc) vs. organic carbon. There is a rough correlation which indicates a formation density decrease with increase in organic carbon.

Figure 4–16 shows a correlation between off-gas and bulk density. The smaller the formation density, the larger the amount of off-gas.

The correlations presented in Figures 4–14 to 4–16 suggest that gamma ray and density logs are valuable tools for evaluating Devonian shales. Smith (1978) has indicated that these two logs have correlated better with gas-bearing zones than the synergetic, computer-generated logs.

Temperature and sibilation surveys discussed in Chapter 3 have reportedly yielded successful results in localizing fractured zones in shales of West Virginia and Ohio.

Hilton (1977) has described the use of Kerogen Analysis for evaluating Devonian shales. This method was presented in Chapter 3. King and Fertl (1979) presented methods involving density-acoustic and density-neutron overlays together with gamma ray spectral logging for evaluating shales. Truman and Campbell (1987) have also developed a technique for well log interpretation of Devonian shales. Luffel et al (1992) presented the interesting finding that all porosity exceeding about 2.5% in Devonian shales is occupied by free hydrocarbons, mostly gas. Based on the analysis of 519 ft of core and well log interpretation they concluded that the average reservoir porosity was approximately 5% and the free-gas content averaged about 2% by bulk volume.

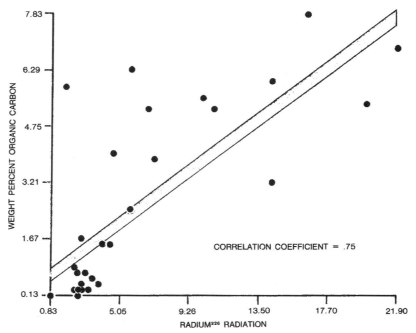

FIGURE 4–13   Radium[226] Gamma radiation vs. total organic carbon (after Smith, 1978)

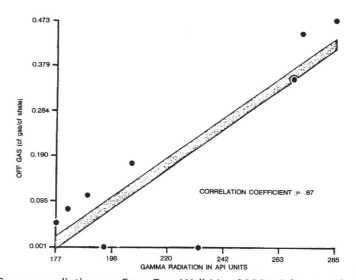

FIGURE 4–14   Gamma radiation vs. Free Gas, Well No. 20336 (after Smith, 1978)

A quick sonic shear wave-resistivity overlay allows identification of potential pay zones in certain Devonian shale wells (Flower, 1981). In this method the induction log is flipped creating a condition whereby resistivity increases towards the depth track rather than away from it. The flipped induction log is placed on top of the wave train making sure that both logs are in depth.

Since in some cases the shear wave is attenuated by fractures, the flipped resistivity and shear wave curves will move in opposite directions when a fracture is present. For unfractured intervals the two curves stack together on the overlay. An additional verification with respect to the presence of a fracture might be provided by a bulk density log when it shows some porosity at the same level.

Experience with Devonian shales has shown that his overlay gives reasonable results using scales of 10 ohm-m and 50 microsecond/ft per chart division. Flower (1981) has indicated that the rate of success using this approach is high.

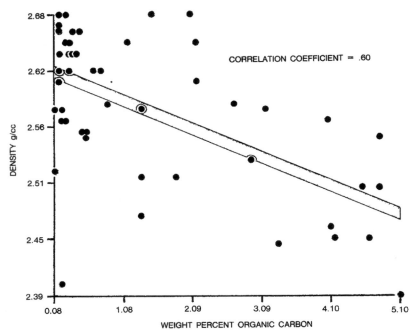

FIGURE 4-15 Percent organic carbon vs. density (after Smith, 1978)

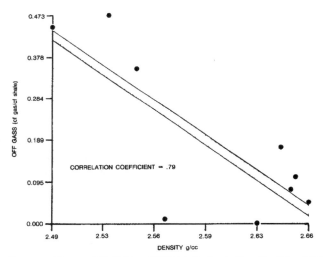

FIGURE 4-16 Density vs. off gas, Well No. 20336 Data (after Smith, 1978)

In wells drilled with air the method still finds application. In this case a Differential Temperature and/or Sibilation log can be run together with the density-sidewall neutron-gamma ray-caliper combination since the sonic log cannot be run in empty holes. The induction log is run on the second trip into the hole.

A seismic spectrum can be obtained later inside the cased hole and the overlay is prepared as discussed above. According to Flower's experience this procedure also yields good results. There are occasions in which the compressional wave works better than the shear wave. This is probably dependent on the inclination of the fractures, since the shear wave tends to be attenuated mostly by low angle and horizontal fractures, while the compressional wave is attenuated mainly by high angle and vertical fractures (see Chapter 3).

Hashmy et al (1982) have presented an approach to evaluate quantitatively Devonian shales. The response equations for density and neutron logs are expressed conventionally as follows:

$$\rho_B = (1 - \phi - V_{sh})\rho_{ma} + V\rho_{sh} + \phi\rho_f \qquad (4\text{--}1)$$

$$\phi_N = \phi\phi_{N,f} + V_{sh}\phi_{N,sh} \qquad (4\text{--}2)$$

where
  ρ = density, gr/cc
  ϕ = porosity, fraction
  V_{sh} = bulk volume fraction of shale in formation

and subscript B stands for bulk, sh is shale, ma is matrix, f is fluid, and N is Neutron. Symbol ϕ without subscripts stands for fractional total porosity.

Effective porosity will decrease as the shale volume increases. In the limiting case, as $V_{sh} \to 1$, $\phi = 0$ and Equations 4–1 and 4–2 become:

$$\rho_B = \rho_{sh} \qquad (4\text{--}3)$$

$$\phi_N = \phi_{N,sh} \qquad (4\text{--}4)$$

The assumption is made that the shale is pure, i.e., it is not contaminated with calcite, silica or any other mineral. This leads to:

$$\rho_B = (1 - \phi)\,\rho_{ma,sh} + \phi\,\rho_f \qquad (4\text{--}5)$$

$$\phi_N = (1 - \phi)\,\phi_{N,ma} + \phi\phi_{N,f} \qquad (4\text{--}6)$$

where $\rho_{ma,sh}$ and $\phi_{N,ma}$ refer to matrix density and apparent neutron porosity of shale matrix, respectively. These would correspond to dry clay parameters (Poupon et al, 1979). Note that the term $\phi_{N,ma}$ includes the apparent neutron porosity due to the chemically bound water in the shales. The term $\phi\phi_{N,f}$ accounts for the hydrogen index and the corresponding apparent neutron porosity attributable to the intrinsic porosity of shale and the fracture porosity, when present.

Due to the presence of gas, hydrocarbon corrections (Poupon et al, 1971; and Gaymard and Poupon, 1968) are introduced into Equations 4–5 and 4–6 as follows:

$$\rho = \rho_{ma,sh}\,(1 - \phi) + \phi\rho_{mf} - \{(1.19 - 0.16P_{mf})\,\rho_{mf} - 1.33\,\rho_h\}\,\phi S_{hr} \qquad (4\text{--}7)$$

$$\phi_N = (1 - \phi)\,\phi_{N,ma} + \phi\phi_{N,f} - \left\{1 - \frac{2.2\rho_h}{\rho_{mf}(1 - P_{mf})}\right\}\phi S_{hr} \qquad (4\text{--}8)$$

where
  ρ_{mf} = mud filtrate density, gr/cc
  P_{mf} = salinity of mud filtrates
  S_{hr} = hydrocarbon saturation of flushed zone, fraction

Using the value of the intrinsic porosity of the shale, ϕ_i, in conjunction with bulk density $\rho_{B,w}$ and the neutron porosity $\phi_{N,w}$ of the "wet" or organic-poor shales in Equations 4–5 and 4–6 , and substituting in Equations 4–7 and 4–8, we obtain:

$$\phi_D = \phi + \phi S_{hr}\,\{(1.19 - 0.16P_{mf})\,\rho_{mf} - 1.33\rho_h\}\,(1 - \phi_i)/\{\rho_{B,w} - \phi_i - \rho_{mf}\,(1 - \phi_i)\} \qquad (4\text{--}9)$$

$$\phi_N = (1 - \phi)\,(\phi_{N,w} - \phi_i)\,/\,(1 - \phi_i) + \phi\left\{\phi_{N,f} - S_{hr}\left[1 - \frac{2.2\rho_h}{\rho_{mf}(1 - P_{mf})}\right]\right\} \qquad (4\text{--}10)$$

Equations 4–9 and 4–10 represent the response of the density and neutron logs in gas-bearing shales. If residual hydrocarbon saturation (S_{hr}) is known, the above equations can be solved for total (intrinsic plus fracture) porosity and hydrocarbon density. If we have a good estimate of gas density, the total porosity and the residual hydrocarbon saturation (S_{hr}) can be calculated.

In the case of wells drilled with air, invasion effects are ignored and ρ_{mf}, P_{mf} and S_{hr} can be replaced by water density (ρ_w), water salinity (P_w), and hydrocarbon saturation of non-invaded formation (S_h).

Water saturation is calculated from:

$$S_w = \left(\frac{R_t}{R_{sh}}\right)^{-1/n} \qquad (4\text{--}11)$$

where R_t is true resistivity (ohm-m) and the exponent n ranges from 1 to 1.2 in Ohio shales and is approximately equal to 1.5 in the Woodford Shales of Oklahoma (Hashmy et al, 1982). Note that Equation 4–11 is the same as Archie's (1942) equation having R_o replaced by R_{sh}. Small values of n for naturally fractured reservoirs are reasonable as discussed in Chapter 3.

An example of analysis using the above procedure is depicted in Figures 4–17 and 4–18; and Table 4–2. A typical set of logs for Devonian shales including Gamma ray, caliper, resistivity, density, and neutron logs are presented in Figure 4–17. Computed results are presented on Table 4-2 and Figure 4–18. The last column on Table 4–2 (OC) represents organic content. It is estimated empirically as a function of gamma ray response. The perforated zones are pinpointed by arrows. The intervals were fractured and the well came in with 700 Mscfd of gas.

A computerized shale evaluation technique (COMSET) using the above method has been developed by Hashmy et al (1982).

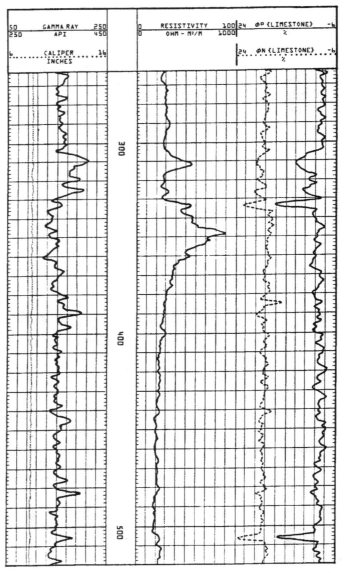

FIGURE 4–17   Field logs for well B (after Hashmy et al, 1982)

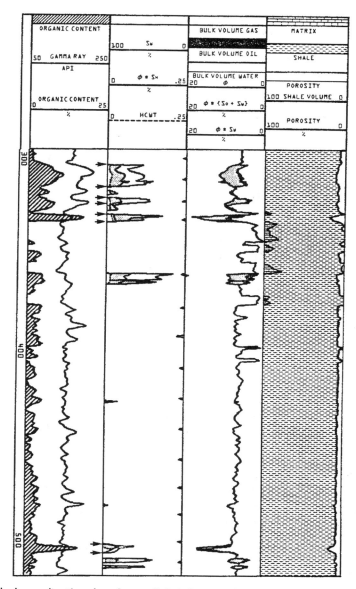

FIGURE 4-18   Shale evaluation log for well B (after Hashmy et al, 1982)

Aguilera (1978) discussed methods for log analysis of gas-bearing fractured shales in the St. Lawrence Lowlands of Quebec. The methods consider the following situations:

1. Determination of water saturation
2. Evaluation of porosity and shale volume
3. Identification of fractured zones
4. Quantitative estimates of fracture porosity
5. Quantitative estimates of water saturation in matrix and fractures
6. Differentiation of zones which are gas-bearing but have amounts of gas which are commercial or noncommercial
7. Estimates of formation strength parameters

It is impossible to warrant exactitude when trying to calculate the previous parameters from well logs. However, the results correlate reasonably well with laboratory data, pressure surveys, and actual stimulation and production information.

TABLE 4–2 Computed Results from Shale Evaluation Technique for Well "B" (after Hashmy et al., 1982)

DEPTH	GR	R_t	ρ_b	ϕ_N	V_{SH}	ϕ	ρ_h	S_w	S_o	S_g	OC
304.5	148	31	2.71	14.5	100	6.6	1.00	100	0	0	3.9
305.0	156	32	2.69	15.1	100	7.4	1.00	100	0	0	4.8
305.5	170	33	2.68	15.5	100	8.4	0.20	56	4	40	6.0
306.0	177	33	2.66	16.0	100	9.0	0.20	59	3	39	6.7
306.5	186	34	2.66	16.4	100	9.4	0.20	62	1	37	7.4
307.0	193	37	2.65	16.9	100	9.6	0.20	69	0	31	7.4
307.5	199	39	2.65	17.2	100	9.5	0.20	75	0	25	7.4
308.0	202	41	2.65	17.4	100	10.0	0.18	74	0	26	7.4
308.5	209	42	2.65	17.6	100	10.2	0.22	71	0	29	7.8
309.0	212	44	2.62	17.8	100	10.9	0.20	67	0	33	9.4
309.5	212	50	2.61	17.8	100	11.5	0.21	60	2	38	10.4
310.0	210	51	2.60	17.7	100	11.8	0.20	56	4	40	10.8
310.5	204	55	2.60	17.9	100	11.7	0.23	55	5	40	10.9
311.0	197	55	2.60	18.2	100	11.7	0.26	55	5	40	10.9
311.5	183	55	2.61	18.2	100	11.5	0.26	55	5	40	10.5
312.0	181	56	2.62	17.4	100	10.8	0.24	54	6	41	9.6
312.5	182	50	2.63	17.0	100	10.7	0.20	56	4	40	9.2

Total Water Saturation

The suite of logs presented in Figure 4–19 have been used to obtain estimates of total water saturation in the black fractured shales of Quebec.

Archie's equations can be combined into a single equation to yield:

$$\log R_t = -m\log \phi + \log (a R_w) + \log I \tag{4–12}$$

The above equation indicates that a log-log plot of R_t vs. ϕ must result in a straight line with a slope of $-m$ for zones of constant aR_w and constant I as discussed in Chapter 3. To insert a value of ϕ in the previous equation is rather difficult for Villeroy since the formation is a fractured shale and the constants in the porosity tool response are not known with a good degree of confidence.

Wyllie et al (1956) indicated in their original study that the sonic log which fits the equation:

$$\phi = \frac{\Delta t - \Delta t_m}{\Delta t_f - \Delta t_m} \tag{4–13}$$

sees only matrix porosity and bypasses vuggy and fracture porosities. Experience indicates that the total ϕ of a double-porosity system can be calculated from the relationship (Schlumberger, 1972):

$$\phi = \frac{\Delta t - A}{B} \tag{4–14}$$

where A and B no longer correspond to well known matrix and fluid transit times. While the values of A and B do not have much physical meaning, they do provide reliable estimates of total porosity in certain fractured reservoirs (see also Chapter 3).

Inserting Equation 4–14 in 4–12 results in:

$$\log R_t = -m \log (\Delta t - A) + m\log (B) + \log (aR_w) + \log I \tag{4–15}$$

Equation 4–15 indicates that a log-log plot of R_t vs. $(\Delta t - A)$ should result in a straight line with a slope of $-m$ for zones with constant A, B, aR_w, and I. The previous equation has evaluated the Utica shale of Villeroy.

Figure 4–20 shows a log-log crossplot of R_{LL8} vs. $(\Delta t - A)$. The value of A for Equation 4–15 was found to 52 on a trial and error basis. In fact, a log-log plot of R_{LL8} vs. $(\Delta t - 52)$ resulted

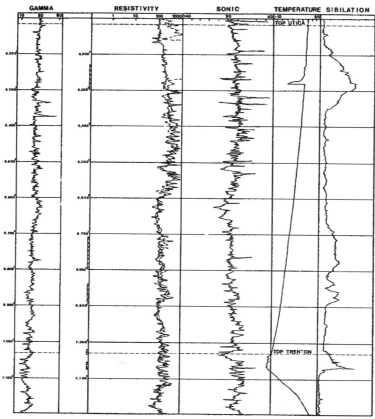

FIGURE 4–19   Logs of Villeroy no. 2 (after Aguilera, 1978)

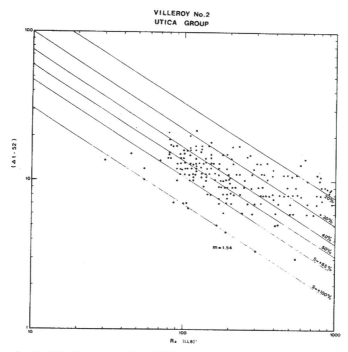

FIGURE 4–20   Sonic-Resistivity cross plot, Villeroy 2 (after Aguilera)

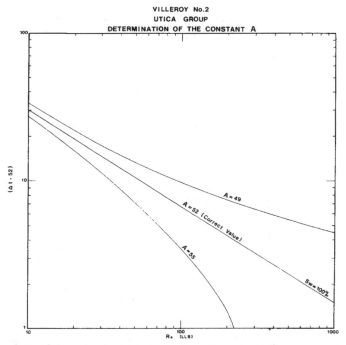

FIGURE 4–21 Sonic-Resistivity cross plot permits estimation of correct value of A (after Aguilera, 1978)

in a straight line for zones of constant I (for example I = 1, S_w = 100%) as shown in Figure 4–20. Using a value of A different from 52 generated curves (Figure 4–21).

The laterolog-8 was selected for resistivity because it has better vertical definition than the deep induction log in laminated systems. The slope of the water-bearing trend (S_w = 100%) was 1.54. This low value of m indicated the possible presence of fractures in the Utica Group since the porosity exponent of the matrix alone (m_b) was probably 2 or higher.

One of the main advantages of this crossplotting technique is that the water saturation can be calculated without having previous knowledge of porosity and water resistivity, using the equation (Archie, 1942):

$$S_w = I^{-1/n} = (R_t/R_o)^{-1/n} \tag{4–16}$$

For this evaluation this was very important as there were no estimates of R_w in the Utica Group.

The values of R_o are the resistivities read all through the straight line that represents the 100% water saturation trend. The values of R_t are resistivities of each zone in the crossplot. The water saturation exponent (n) is assumed to be equal to the porosity exponent (m). This assumption, based on my experience, provides reasonable results. This would mean a value of n equal to 1.54. Note that this is similar to the 1.5 value used by Hashmy et al (1982) for analyzing the Woodford shale of Oklahoma.

The value of R_o in Equation 4–16 can be correlated with the value of R_{sh} used in Equation 4–11 since all the resistivities belong to the shale formation.

Based on Equation 4–16, straight lines that represented various water saturations were generated (Figure 4–20) without having previous knowledge of porosity. These were total values that included water saturation in both matrix and fractures. The resulting values of total S_w for perforated interval 6,232–6,300 ft are presented in Table 4–3.

Porosity and Shale Volume

Figure 4–19 shows gamma ray, resistivity, sonic, temperature, and sibilation logs for the Utica Group of Villeroy 2. In addition, a CNL log was run. Figure 4–22 shows a crossplot of Δt vs ϕ_{CNL} in Cartesian coordinates for various perforated and nonperforated zones.

TABLE 4-3 Log Analysis of Perforated Interval 6232'-6300'

Zone	Depth (ft)	R_{LL8} (Ω-m)	Δt (μsec/ft)	ϕ_n (%)	Δt-52 (μsec/ft)	S_w	$\phi_{n,s}$ (%)	V_{sh} (%)	ϕ_s	ϕ_b	ϕ_2 (fraction)	V	S_{wm}
50	6,232-38	175	62.0	11.5	10.0	.47	6.4	37	.073	.057	.016	.23	.61
51	6,238-40	190	61.0	10.5	9.0	.49	6.0	34	.066	.052	.014	.22	.63
52	6,240-41	160	62.0	11.5	10.0	.49	6.4	37	.073	.057	.016	.23	.64
53	6,241-42	200	61.0	10.8	9.0	.48	5.8	36	.066	.052	.014	.22	.62
54	6,242-44	160	62.0	11.0	10.0	.49	6.8	34	.073	.057	.016	.23	.64
55	6,244-46	170	59.5	11.0	7.5	.64	4.0	43	.055	.044	.011	.21	.81
56	6,248-50	170	63.0	12.0	11.0	.43	7.0	38	.080	.062	.018	.24	.57
57	6,250-54	200	59.0	10.0	7.0	.61	4.3	37	.051	.042	.009	.18	.75
58	6,254-56	230	59.0	10.0	7.0	.56	4.3	37	.051	.042	.009	.18	.69
59	6,256-60	200	62.0	11.5	10.0	.43	6.4	37	.073	.057	.016	.23	.56
60	6,263-64	220	63.0	10.0	11.0	.37	8.7	22	.080	.062	.018	.24	.49
61	6,264-66	200	61.5	10.0	9.5	.45	7.1	28	.069	.053	.016	.24	.60
62	6,266-68	180	73.0	10.0	21.0	.22	15.0	0	.150	.108	.045	.33	.40
63	6,271-72	180	63.0	10.0	11.0	.42	8.7	23	.080	.062	.018	.24	.55
64	6,272-76	190	62.0	9.5	10.0	.44	8.0	22	.073	.057	.016	.23	.57
65	6,276-78	130	64.0	11.5	12.0	.47	8.5	31	.088	.068	.020	.24	.62
66	6,278-80	145	60.0	8.0	8.0	.66	7.0	18	.058	.047	.011	.20	.82
67	6,280-82	185	61.0	8.0	9.0	.50	8.0	14	.066	.052	.014	.22	.64
68	6,282-85	175	61.0	9.0	9.0	.52	7.2	22	.066	.052	.014	.22	.67
69	6,285-90	140	62.5	10.0	10.5	.51	8.0	24	.077	.060	.017	.23	.67
70	6,290-92	190	59.5	8.0	7.5	.59	6.5	20	.055	.044	.011	.21	.75
71	6,292-94	170	64.0	9.0	12.0	.40	10.4	12	.088	.068	.020	.24	.53
72	6,294-96	230	60.0	8.5	8.0	.49	6.8	22	.058	.047	.011	.20	.61
73	6,296-98	220	59.0	8.0	7.0	.58	6.0	21	.051	.042	.009	.18	.71
74	6,298-6,300	200	59.0	7.5	7.0	.58	6.4	18	.051	.042	.009	.18	.71

Subscripts: LL8 = laterolog 8, n = neutron, w = water, s = sonic, sh = shale, b = matrix block, 2 = fracture, m = matrix

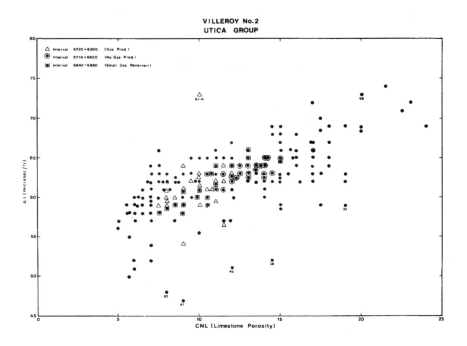

FIGURE 4–22 Sonic-Neutron cross plot, Villeroy 2 (after Aguilera, 1978)

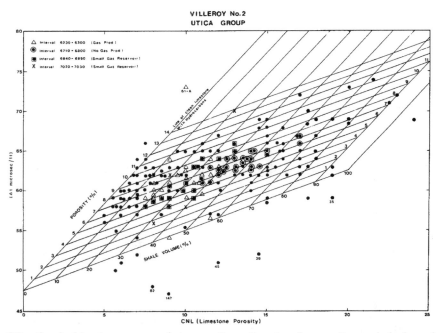

FIGURE 4–23 Sonic-Neutron cross plot permits estimate of porosity and shale volume in consolidated gas-bearing formations (after Aguilera, 1978)

Figure 4–23 shows estimated porosity and shale volumes based on the data points of Figure 4–22. Experience indicates that the sonic-neutron combination is not very useful for shale volume analysis because both logs are affected in a similar fashion by shales. However, Figure 4–23 shows clearly that for the consolidated gas-bearing fractured shales of Quebec, the sonic-neutron combination provided a valuable formation evaluation tool for estimating porosity and shale volumes.

Table 4–3 shows Δt and ϕ_{CNL} data of a perforated interval and the calculated porosities and shale volumes. Shale volumes in this well do not approach 100% as occurs in most black shales in the Appalachian basin.

Another set of porosity values was calculated from the sonic log alone with the use of Equation 4–14 and the basic parameters A = 52 and B = 137. Agreement was reasonable as shown in Table 4–3.

Identification of Fractured Zones

Temperature and sibilation logs were run in empty hole to locate gas entry into the well bore. From the logs of Figure 4–19, the temperature log gave a kick toward lower values of temperature at 6270–80 ft.

The sibilation survey detected noise from 6220 to 6320 ft. In addition, the sibilation log indicated the presence of noise from 6700 to 6780 ft and from 6800 to 6890 ft. These indications of gas entry were indicative of the possible presence of fractures.

Figure 4–24 shows sibilation and dipmeter logs for the Utica Group. The correlation is excellent and suggests that the dipmeter log is an additional valuable tool for detecting potential fractured zones.

Estimates of Matrix and Fracture Porosities

Matrix porosities were found from Figure 4–25 as a function of m = 1.54, total porosities, and the matrix porosity exponent (m_b) which was assumed equal to 2.2. Fracture porosities were taken as the difference between total and matrix porosities. The calculated values of matrix and fracture porosities are tabulated in Table 4–3 for an interval perforated in the Utica Group. Although the fracture porosities around the wellbore are shown to be relatively high, deliverabilities are low due to localized mineralization within the fractures which is a common constrictive effect.

Estimates of Water Saturation in Matrix and Fractures

Production of gas from the fracture system of Villeroy 2 was water-free, indicating that the water saturation within the fracture system was nearly zero. Because of capillarity, water tends to occupy the finer spaces (intergranular porosity) of the rock, leaving the coarse porosity (fractures) to hydrocarbons.

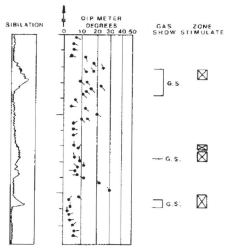

FIGURE 4–24 Sibilation and dipmeter logs reveal gas-bearing fractured zones (after Beiers, 1976)

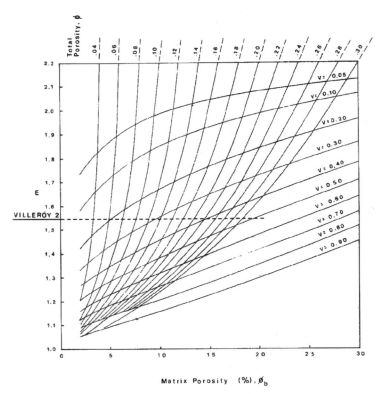

FIGURE 4–25 Chart used to evaluate double-porosity systems (after Aguilera, 1978)

The water saturation in the matrix (S_{wm}) was estimated from the equation (Hilchie & Pirson, 1961):

$$S_{wm} = \frac{S_w - vS_{wf}}{(1 - v)} \tag{4–17}$$

where:

v = partitioning coefficient, fraction
S_{wf} = water saturation in the fractures, fraction

The values of v and S_{wm} are shown in Table 4–3 for a perforated interval.

Distinguishing Productive Intervals

Villeroy 2 was drilled with air from 3900 to 7200 ft. No significant has shows were reported while drilling the Lorraine and Utica Groups (Refer to Stratigraphic chart on Figure 4–6).

Interval 6232–6300. This interval was perforated on February 11, 1976, with one shot per foot. The interval was acidized with 25,000 gal of 28% HCL mixed with 50% methanol. Injection pressures remained at 5500 psi while injecting acid at an average rate of 6 bbl/min. This high volume acid treatment was successful, as 90% of the treating fluids were flowed back to the surface and a steady production of 300 Mscfd was obtained through a ¼ in. choke.

A qualitative indication of the presence of gas had been given by the temperature and sibilation logs (Figure 4–19). The combination sonic-resistivity resulted in an order-of-magnitude quantitative evaluation of S_w (Table 4–3). Perforated interval 6232–6300 ft is represented by zones 50 to 74 (Table 4–3).

Gas saturation within the fracture system was estimated at 100% since no water production was obtained from this interval.

In conclusion, the combination temperature-sibilation and sonic-resistivity provided means to evaluate qualitatively and quantitatively this gas-bearing interval.

Interval 6710–6800. This interval was perforated on February 1, 1976, with one shot per foot for a zero initial gas production. The sibilation survey detected some noise indicating that the interval was probably gas-bearing. The temperature log did not give any indication of gas. The combination sonic-resistivity indicated attractive values of porosity and water saturation. However, an attempt to acid-frac the interval failed, as the formation did not take any acid at surface pressures as high as 5000 psi. This result was difficult to explain at first glance. However, a log-log crossplot of $(\Delta t$–A) vs. R_a (Ll–8) for intervals 6230–6300 and 6710–6800 ft showed distinct patterns for each interval (Figure 4–26), separated by a cutoff resistivity in the order of 145 Ω – m. This indicated that resistivity could be used as a cut-off criterion for selecting potential gas-bearing intervals.

Interval 6840–6890. This interval was perforated on January 10, 1976, with 2 shots per foot. The interval was acidized with 1500 gal of 15% HCL and produced at an average rate of 476 Mscfd. Pressure surveys indicated that the reservoir was under strong depletion, thus suggesting a small reservoir.

To corroborate this finding, 2⅜-in tubing and packer were set in the 5½-in. casing at 6651 ft. A sliding sleeve (open) was placed in the tubing at 6302 ft. Next, continuous flow meter, temperature, and caliper logs were run past the sliding sleeve (Figure 4–27). The bottom hole pressure on the top zone was 3000 psi and on the bottom zone 409 psi. Several weeks before, the wellhead presure was 500 psi, and at the time of logging the wellhead pressure was 1145 psi. This indicated that the gas from the top zone was "charging up" the bottom zone. The temperature log also verified this finding by the cooling effect above the sliding sleeve. The amount of flow from the top to the bottom zone as indicated by the flow meter was 398 Mscfd or 29,800 cu ft at downhole conditions.

The sibilation survey (Figure 4–19) indicated this interval was gas-bearing, even though the temperature log did not show any gas. This suggested that gas was not entering the hole, but that is was moving near the well bore producing some noise.

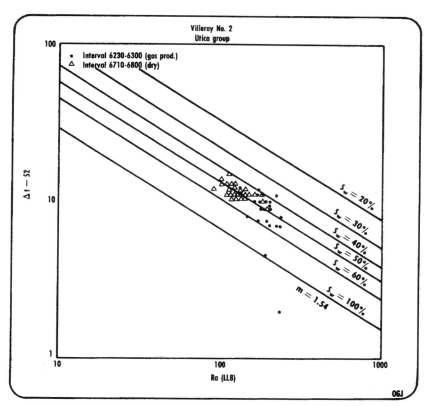

FIGURE 4–26 Sonic-resistivity cross plot for dry and gas-bearing intervals (after Aguilera, 1978)

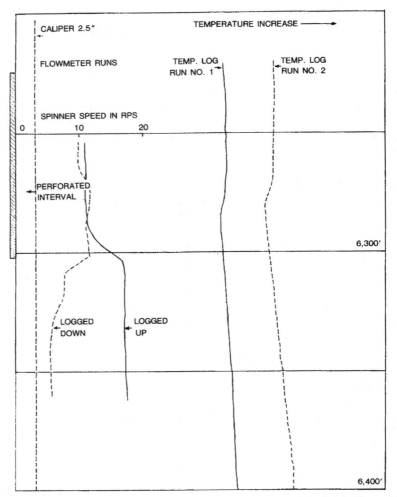

FIGURE 4–27 Temperature and flowmeter surveys, Villeroy 2 (after Aguilera, 1978)

The combination sonic-resistivity indicated attractive values of water saturation ranging from 37% to 68%. Fractured porosities ranged between 0.8% and 2.6%.

The problem here was to determine from well logs when a zone would deplete very quickly due to the small size of the reservoir.

It was found that small reservoirs are associated with abnormal, over-pressured zones in the Utica Group of the Villeroy area. Figure 4–28 shows resistivity and sonic logs of Villeroy 2. A clear trend of increasing resistivity is shown from the top of the Utica Group down to about 6530 ft, where a drastic reduction of resistivity occurs indicating the top of a probable over-pressured zone.

Interval 6840–6890 ft falls in this region. A linear trend is also delineated by the sonic log from the top of the Utica Group down to about 6530 ft where a drastic increase of Δt occurs. This also tends to corroborate the presence of an over-pressured zone below 6570 ft.

Comparing the performance of perforated intervals seems to indicate that small reservoirs are associated with abnormal over-pressures in the fractured shales of Quebec.

FORMATION STRENGTH PARAMETERS

Hubert and Willis (1957) have indicated that in regions where there is normal faulting, the greatest stress should be approximately vertical and equal to the effective pressure of the overburden. The least stress should be horizontal, ranging between one half and one third of the overburden.

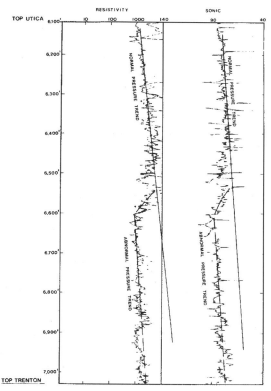

FIGURE 4–28 Resistivity and sonic logs indicate the presence of overpressed zones below 6550 ft (after Aguilera, 1978)

The equation relating stresses and Poisson's ratio for rocks in compression can be written as:

$$\sigma_x = \sigma_y = \frac{\mu}{1 - \mu}\,\sigma_z \qquad (4\text{–}18)$$

where σ's are stresses in the x, y, and z axes, and μ is Poisson's ratio. For a more detailed discussion on stresses see Chapter 1.

Solving the previous equation indicates that for normal faulting, the value of μ should be between 0.25 and 0.33. This range has proven to be very realistic in the Gulf Coast.

Hubbert and Willis (1957) further indicated that in regions of thrust faulting, the least stress should be vertical and equal to the net overburden pressure while the horizontal stress should be horizontal and approximately between two and three times the net overburden pressure. Solving Equation 4–18 indicates that Poisson's ratio (μ) could be approximately between 0.67 and 0.75 for regions where thrust faulting occurs.

These experimental findings seem to hold in some parts of the St. Lawrence Lowlands of Quebec, which are characterized by thrusting tectonics in the Utica Group. In some instances it was possible to carry out stimulations with pressure gradients of about 1 psi/foot, while in other cases it was impossible to stimulate some zones with pressure gradients greater than 1.5 psi/foot.

ESTIMATES OF GAS-IN-PLACE

Calculating gas-in-place in fractured shales is a complex problem. The presence of absorbed gas, trapped gas, and free gas in fractures contributes to the complexity of the problem.

Most estimates of gas-in-place in the Appalachian Basin are based on measured or assumed concentrations of gas expressed in scf gas/cu ft shale.

Smith (1978) made four estimates of gas-in-place in Devonian shales. These estimates are reproduced below and include gas-in-place in West Virginia, Kentucky, Ohio, Pennsylvania, and Virginia.

The estimates are based on the general equation:

$$G = 27,878,400 \, A \, h \, (G.C.) \tag{4–19}$$

where:
G = gas in place, scf
A = area, sq mi.
h = average net thickness, ft
G.C. = gas concentration, scf of gas/cu ft shale
1 sq mile = 27,878,400 sq ft

First Estimate. Assume that the average gas concentration is 0.6 scf of gas/cu ft shale. The area is estimated at 90,000 sq mi. and the shale thickness at 576 ft.

From Equation 4–19 the gas-in-place for the five states mentioned above is calculated to be:

$$G = (27,878,400) \, (90,000) \, (576) \, (0.6) = 867 \times 10^{12} \text{ scf}$$

or 867 trillion cu ft. This figure compares well with the estimate of 800 trillion cu ft presented by Overbey (1978).

Second estimate. This is based on a gas concentration of 0.6 scf gas/cu ft shale. The average net pay is 137 ft based on a gamma ray cutoff of 200 API units.

From Equation 4–19 the gas-in-place is calculated to be:

$$G = (27,878,400) \, (90,000) \, (137) \, (0.6) = 206 \times 10^{12} \text{ scf}$$

or a total of 206 trillion cu ft for the five-state area.

Third estimate. Assume a gas concentration of 0.2 scf of gas/cu ft shale. This is an overall concentration which includes the average gray, organic-poor shale. Average thickness of both black and gray shales is estimated at 1800 ft. The gas-in-place is calculated from Equation 4–19 as follows:

$$G = (27,878,400) \, (90,000) \, (1,800) \, (0.2) = 903 \times 10^{12}$$

or a total of 903 trillion cu ft gas.

Fourth estimate. It is based on individual areas (Figure 4–29) of the Appalachian Basin. Thickness of organic shale facies for numbered areas and the gas-in-place estimate for each area are presented in Table 4–4. Total gas-in-place is calculated to be 247.2 trillion cu ft.

In conclusion, Smith's (1978) estimates vary between 200 and 900 trillion cu ft. of gas. This wide variation will be reduced with further research in Devonian shales.

Based on limited information in the St. Lawrence Lowlands of Quebec, it appears that free gas-in-place within the fracture network can be calculated volumetrically from the equation:

$$G = \frac{43,560 \, A \, h \, \phi_2 \, (1 - S_{wf})}{B_{gi}} \tag{4–20}$$

where:
G = gas-in-place associated with the naturally fractured system, scf
A = area, acres
h = net pay, ft
ϕ_2 = fracture porosity, fraction
S_{wf} = water saturation in the fractures, fraction
B_{gi} = initial gas formation volume factor, cu ft/scf

Gas-in-place has been calculated to be 5.9 Bscf per section (1 section equals 640 acres) using the following average data:

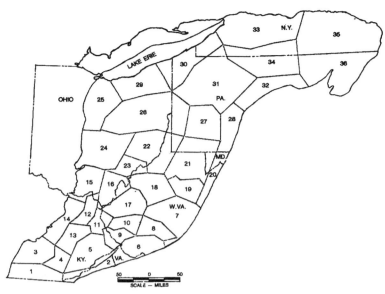

FIGURE 4–29 Appalachian Basin area underlain by Devonian shale (after Smith, 1978)

$$
\begin{aligned}
h &= 70 \text{ ft} \\
\phi_2 &= 0.014 \\
S_{wf} &= 0 \\
B_{gi} &= 0.00461 \text{ cu ft/scf (at bottom hole conditions, } T = 575° \text{ R, } p = 3{,}000 \text{ psi,} \\
&\quad \text{and } z = 0.85).
\end{aligned}
$$

Using Equation 4–19 or 4–20 depends on whether the gas is thought to be sorbed (Equation 4–19) or free in open pores and fractures (Equation 4–20).

RECOVERABLE GAS

Long production histories in some areas of the Appalachian Basin make it possible to predict recoveries with some level of confidence. The following methods will be discussed in this section.

1. Comparison of cumulative gas production with initial open flow potential
2. Comparison of cumulative gas production with drainage area by well
3. Comparison of degassing studies with production curves
4. Material balance and sorption studies
5. Linear flow analysis
6. Decline curve analysis
7. Higher order polynomial models
8. Hydrodynamic analogy
9. Empirical performance equations

Comparison of Cumulative Gas Production with Initial Open Flow Potentials

Figure 4–11 shows a graphic of 20-year cumulative production as a function of initial open flow for 35 shale wells of West Virginia. The correlation is not very good, but the apparent trend of better recoveries for wells with better open flow potentials is evident.

The best curve fit can be represented by the equation:

$$ G_p = 1.76 \times \text{IOF} \tag{4–21} $$

where: G_p = cumulative recovery in a 20-year period, MMscf
          IOF = initial open flow potential, Mscfd

TABLE 4–4 Estimates of Gas-in-place By Area (after Smith, 1978)

Area #	Average organic (radioactive) shale thickness (ft)	Gas content (trillions of cu ft)
1	78	2.8
2	267	15.4
3	87	3.5
4	83	1.5
5	199	10.7
6	206	28.9
7	45	2.7
8	106	6.8
9	347	8.4
10	305	5.8
11	219	3.9
12	106	2.1
13	146	4.1
14	107	3.7
15	170	6.8
16	135	4.7
17	88	3.5
18	100	4.6
19	56	1.7
20	63	5.2
21	56	4.4
22	69	5.2
23	210	4.6
24	206	15.7
25	230	14.2
26	115	15.0
27	51	2.6
28	77	2.8
29	83	4.6
30	120	7.5
31	99	13.4
32	100	5.0
33	41	4.5
34	69	6.5
35	60	8.4
36	60	6.0

The dashed line in fig. 4–11 shows the most conservative possible recovery as a function of initial flow. The most conservative recoveries (dashed line) for initial open flows greater than 111 Mscfd can be obtained from the equation:

$$G_p = 1.76 \times IOF - 195 \tag{4-22}$$

The method is empirical and very elementary. However, for the same area it is not unreasonable to anticipate recoveries as calculated from Equations 4–21 and 4–22.

Comparison of Cumulative Gas Production with Drainage Area by Well

This is another elementary approach based on production history. Smith (1978) calculated the gas concentration of Devonian shales using Columbia Gas production data.

He considered that the average 10-year cumulative production of a well with 500 ft of pay draining 40 acres was 198 MMscf. He calculated the gas concentration to be 0.23 scf gas/cu ft shale from the equation:

$$G.C. = \frac{average\ 10\ year\ production}{43,560\ A\ h} \qquad (4-23)$$

Recovery from any well in the area for a 10-year period can be calculated with the use of the equation:

$$G_p = 27,878,400\ A\ h\ (G.C.) \qquad (4-24)$$

Comparison of Degassing Studies with Production Curves

Schettler et al (1977) have presented equations which describe gas production from wells in Devonian shales. The amount of gas M_t which diffuses out of shale into a low pressure fracture is given by:

$$M_\tau = 2\ (C_2 - C_o)\ A\ (Dt/\pi)^{1/2} \qquad (4-25)$$

$$\frac{dM_\tau}{dt} = (C_2 - C_o)\ A\ (D/\pi t) \qquad (4-26)$$

where:
A =   area of rock degassing
C_2 =   initial gas concentration in the rock
C_o =   gas concentration in the rock as $t \to \infty$
D =   diffusion constant
t =   time

The idea that the rock is the source of gas can be better visualized by placing reasonable Devonian shale values into Equation 4–25. For example one square centimetre of rock with diffusion constant (D) equal to 5×10^{-7} cm²/sec. and $(C_2 - C_o)$ equal to 3 cm³ of gas per cm³ of rock would produce 42 cm³ of gas over a 10–year period. This gas volume is larger than the average fracture volume. Consequently, it is not unreasonable to state that the matrix rock is the source of gas.

The upper curve of Figure 4–30 shows a fit to actual data points using Equations 4–25, 26. The match is reasonably good except at late times. It is likely that these equations may find application in wells with high initial deliverabilities which drop to lower rates in rather short times and are followed by long tails of nearly constant production.

In some cases the high initial rates are missing. This might be the result of some constriction on the fractures. Schettler et al (1977) indicated the presence of slickensides and mineralization within the fracture is a common constrictive effect, which can be taken into account with the equations:

$$\frac{dM_\tau}{dt} = \frac{D(C_2 - C_o)}{g} \exp\ (Z^2)\ erfc\ (Z) \qquad (4-27)$$

$$Z = \frac{D^{1/2}t^{1/2}}{g} \qquad (4-28)$$

where g is a constrictive parameter which corresponds to some kind of impermeable coating on the fracture surface. The more impermeable the coating, the larger g is. As the value of g approaches zero, Equation 4–27 approaches Equation 4–25.

Figure 4–30 shows matches of actual production data obtained by Schettler et al (1977) for various values of the constrictive parameter, g.

Material Balance and Sorption Studies

Fucuk et al (1978) have developed an interesting model for reservoir analysis of gas-bearing fractured shales. Typical behavior of this conceptual model is presented as a plot of p/z vs. cumulative gas production (Figure 4–31).

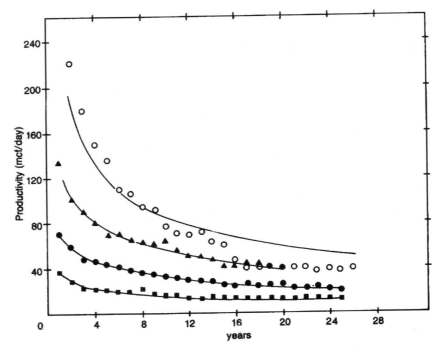

FIGURE 4–30 Flow rate (MMcfd) plotted against time. Values of parameters from upper to lower curves:

Area (meters²) × 10¹¹	g (cm)
1.210	0
.845	4.79
.506	6.02
.254	6.32

(after Schettler et al., 1977)

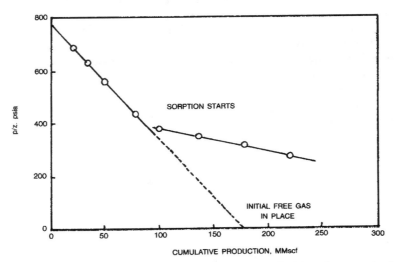

FIGURE 4–31 Theoretical values of p/Z vs. cumulative production from a dual porosity gas reservoir (after Fucuk et al 1978)

The first part is a straight line and represents a conventional material balance for gas reservoirs. Extrapolation of this straight line to p/Z = 0 permits an estimate of initial free gas-in-place. This gas most likely corresponds to free gas within the fractured network.

Deviation from the straight line to the right corresponds to water influx in conventional reservoirs. In Devonian shales, this deviation is believed to correspond to the initiation of the sorption process.

Figures 4–32 and 4–33 present plots of p/Z vs. cumulative gas production for Wells 6630 and 6654 in lincoln County, West Virginia. As in the case of the conceptual model, there are early straight lines which, when extrapolated to p/Z = 0, reveal values of initial gas-in-place most likely within the fractures.

The same type of behavior has been observed in other parts of the basin. Figures 4–34 and 4–35 present the material balance crossplot for two wells in Meigs County, Ohio. Although there is scattering these plots compare well with the conceptual model discussed above.

Fucuk et al (1978) have used a two-porosity model to demonstrate that the mechanism of sorption-diffusion can explain the deviation from volumetric reservoir behavior when no water influx is present.

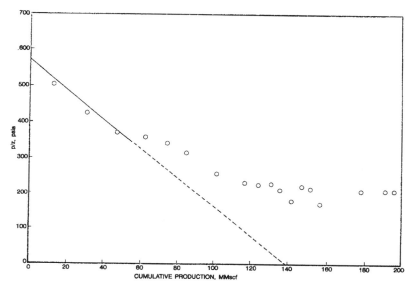

FIGURE 4–32 Values of p/Z vs. cumulative production for Well No. 6630, Lincoln County, WV (after Fucuk et al, 1978)

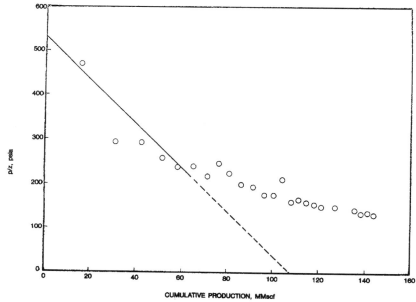

FIGURE 4–33 Values of p/Z vs. cumulative production for Well No. 6654, Lincoln County, WV (after Fucuk et al, 1978)

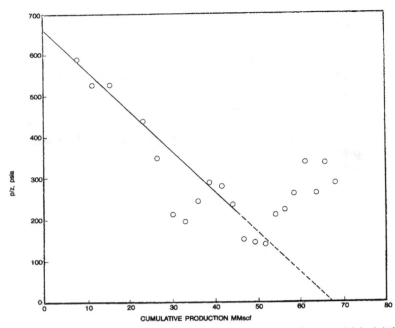

FIGURE 4–34 Values of p/Z vs. cumulative production for Well No. 4121, Meigs County, OH (after Fucuk et al, 1978)

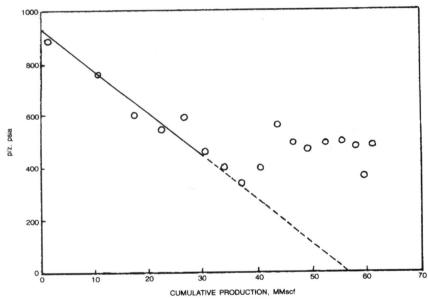

FIGURE 4–35 Values of p/Z vs cumulative production for Well No. 9553, Meigs County, OH (after Fucuk et al, 1978)

The two-porosity model was developed by Chase (1980), who considered the following flow equations for gas and water in a fractured reservoir:

$$\nabla \bullet \left[\frac{\rho_w k k_{rw}}{\mu_w} \left(\nabla p_w - \rho_w g \nabla h \right) \right] - q_{wv} = \frac{\partial}{\partial t} \left(\phi \rho_w S_w \right) \tag{4-29}$$

$$\nabla \bullet \left[\frac{\rho_w k k_{rg}}{\mu_g} \left(\nabla p_g - \rho_g g \nabla h \right) \right] + N_v - q_{gv} = \frac{\partial}{\partial t} \left(\phi \rho_g S_g \right) \tag{4-30}$$

where N_v is the rate of methane desorption per unit matrix volume, in g/cm³ sec, k is permeability, p is pressure, g is gravity, h is thickness, q is rate, t is time, S is saturation, μ is viscosity, ρ is density, ϕ is porosity, and the subscripts w = water and g = gas.

Since there are only water and gas in the system:

$$S_g + S_w = 1.0 \qquad (4\text{--}31)$$

The capillary pressure at any point in the system is given by:

$$p_c = p_g - p_w \qquad (4\text{--}32)$$

The gas dispersion term (N_v) in Equation 4–30 is found using the solution of the diffusion equation for the boundary conditions given by Equations 4–34 and 4–35:

$$\frac{1}{r^2}\frac{\partial}{\partial r}\left(r^2\frac{\partial C}{\partial r}\right) = \frac{1}{D}\frac{\partial C}{\partial t} \qquad (4\text{--}33)$$

$$\frac{\partial C}{\partial r} = 0 \ @ \ r = 0 \qquad (4\text{--}34)$$

$$C = f(p_g) \ @ \ r = a \qquad (4\text{--}35)$$

Where C is gas concentration in g mole/cm³ and r is radius.

Equations 4–33 to 4–35 describe the diffusion of gas in a solid spherical particle where the concentration of gas at the surface of the sphere is a function of the gas pressure in the fractures.

Equation 4–33 can be solved for the concentration distribution of gas in a sphere, which in turn is used in the computation of the sorption rate given by:

$$N_v = \frac{3\,(M.W.)\,D}{r}\frac{\partial C}{\partial r}\Big|_a \qquad (4\text{--}36)$$

where M.W. is molecular weight and D is the diffusion coefficient.

The previous equations were solved using an alternating direction implicit procedure (ADIP) and the data presented in Table 4–5 for a single porosity reservoir without sorption and a dual porosity reservoir with sorption included.

Figure 4–36 shows a crossplot of p_{avg}/Z_{avg} vs. cumulative gas production for the two cases under consideration. The single porosity reservoir without sorption reveals a conventional straight line which is characteristic of volumetric reservoirs.

The dual porosity reservoir with sorption results in a curve which deviates form the volumetric straight line. This behavior indicates conventional reservoir engineering methods

TABLE 4–5   Input Data for the Model (after Fucuk et al, 1978)

Wellbore radius	0.292 ft
Drainage radius	1600 ft
Reservoir temperature	575°R
Initial reservoir pressure	284 psi
Specific gravity of gas	0.62
Average compressibility factor	0.96
Shale bulk density	2.70 g/cm³
Initial gas concentration	2.84 cm³/g
Gas viscosity	0.0125 cp
Formation thickness	622 ft
Formation depth	3,720 ft
Porosity	0.02
Permeability	0.10 md
Gas saturation	0.458
Water saturation	0.542
Diffusion coefficient	0.1×10^{-7} cm²/s
Particle radius	10 cm
Skin factor	-3.5
Total simulation time	15 years

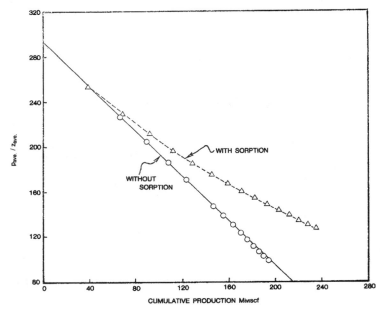

FIGURE 4–36 Simulated values of p/Z vs. cumulative production for a dual porosity gas reservoir (after Fucuk et al, 1978)

may result in very conservative estimates of gas-in-place in Devonian shales and other tight gas reservoirs.

Figure 4–37 presents simulated results of production rate vs. time for a dual-porosity reservoir with sorption and a single porosity reservoir without sorption. The importance of sorption is evident in the later production stages of the well. Figure 4–38 shows the effect of sorption on cumulative gas production.

Linear Flow Analysis

The analysis is carried out by assuming that the well produces from connected fractures in a reservoir where performance is dominated by linear flow. For conditions of constant pressure at the reservoir face, a crossplot of $1/q$ vs $\sqrt{t}$ should result in a straight line in Cartesian coordinates, or:

$$1/q = m_q \sqrt{t} + 1/q_o \qquad (4\text{--}37)$$

where
q = gas rate at times greater than zero, Mscfd
q_o = gas rate at time zero, Mscfd
m_q = slope of straight line in Cartesian coordinates, $(\text{day/scf} \times 10^6)/\sqrt{years}$
t = time, years

Table 4–6 shows data read from production curves published by Bagnal and Ryan (1976). Also shown in the Table are the square roots of time and the inverse of the production rates.

These data were crossplotted (Figure 4–39) and resulted in approximate straight lines for group of wells 2, 3, and 4. The data for group 1 (AOF > 300 Mscfd) are somewhat more scattered. Fortunately, a hyperbolic decline seems to fit the data for group 1.

The importance of this finding cannot be overemphasized. In fact, it means that flow during the 20–year period has been linear and, consequently, production has probably come from a network of fractures dominated by linear flow.

In cases where some history is available, the linear flow characterization might yield reasonable production forecasts using Equation 4–37.

In addition it may be possible to calculate formation permeability from the equation (Timmerman, 1978):

$$m = \frac{136.7}{m(p_i) - m(p_{wf})} \frac{T}{A} \sqrt{\frac{1}{k\phi\mu c_t}} \qquad (4\text{--}38)$$

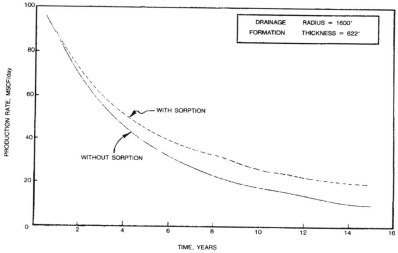

FIGURE 4–37 Simulated values of production rate vs. time for dual porosity gas reservoir (after Fucuk et al, 1978)

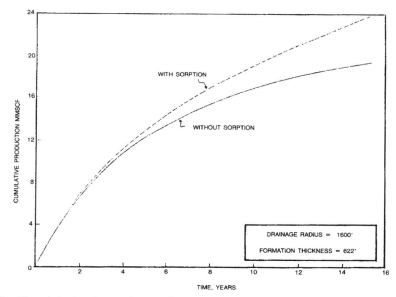

FIGURE 4–38 Simulated values of cumulative production vs. time for dual porosity gas reservoir (after Fucuk et al, 1978)

where: $m(p_i)$ = real gas pseudo-pressure at initial conditions, psi²/cp
$m(p_{wf})$ = real gas pseudo-pressure at flow conditions, psi²/cp
T = temperature, °R
A = area = net pay × half fracture length, sq ft
ϕ = porosity, fraction
k = permeability, md
c_t = total compressibility, psi⁻¹
μ = viscosity, cp
m = slope of straight lines in a plot of 1/q vs. $\sqrt{t}$ where t is in hours and q is in Mscfd.

Probably the largest uncertainty in the calculation of k from Equation 4–38 stems from not having a good knowledge of A.

TABLE 4–6 Production Data of Devonian Shales from Mingo, Lincoln, and Wayne Counties, West Virginia

| t | | Mscfd | | | | day/(scf x 10⁶) | | | |
Yrs	√t	q₁	q₂	q₃	q₄	1/q₁	1/q₂	1/q₃	1/q₄
1	1.00		135	70			7.41	14.3	
2	1.41	220	100	59	30.0	4.5	10.0	16.9	33.3
3	1.73	180	90	45		5.6	11.1	22.2	
4	2.00	150	80	45	22.0	6.7	12.5	22.2	45.5
5	2.24	135	70	43		7.4	14.3	23.3	
6	2.45	110	70	40	21.0	9.1	14.3	25.0	47.6
7	2.65	105	65	38		9.5	15.4	26.3	
8	2.83	95	64	36	20.0	10.5	15.6	27.8	50.0
9	3.00	90	60	35		11.1	16.7	28.6	
10	3.16	78	64	33	19.0	12.8	15.6	30.3	52.6
11	3.32	70	55	31		14.3	18.2	32.3	
12	3.46	68	52	30	18.0	14.7	19.2	33.3	55.6
13	3.61	70	50	29		14.3	20.0	34.5	
14	3.74	60	48	29	17.0	16.7	20.8	34.5	58.8
15	3.87	59	42	26		16.9	23.8	38.5	
16	4.00	45	42	24	16.0	22.2	23.8	41.7	62.5
17	4.12	40	43	25		25.0	23.3	40.0	
18	4.24	40	41	24	16.0	25.0	24.4	41.7	62.5
19	4.36	40		24		25.0		41.7	
20	4.47	40		24	15.0	25.0		41.7	66.7
21	4.58	40		22		25.0		45.5	
22	4.69	40		22		25.0		45.5	
23	4.80	39		22		25.6		45.5	
24	4.90	39		21		25.6		47.6	
25	5.00	39		20		25.6		50.0	
26	5.10	39				25.6			

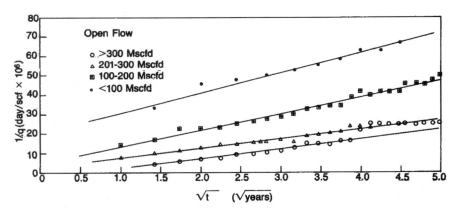

FIGURE 4–39 Drawdown linear analysis for well producing from connected fractures in an infinite reservoir

Decline Curve Analysis

Decline curve analysis has been useful to the petroleum industry in all phases of oil production. This section shows that gas production decline in Devonian shales might be analyzed with the use of hyperbolic decline curves. The equation governing hyperbolic decline can be written as follows:

$$q_g = (q_g)_i \, (i + nD_i t)^{-1/n} \tag{4–39}$$

where: q_g = flow rate at time t

$(q_g)_i$ = flow rate at time zero

n = exponent governing character of resulting curve

D_i = initial rate of decline, month^{-1}

The exponent, n, ranges between 0 and 1.0. An increase in the value of n generates a more pronounced curvature in a Cartesian plot of production rate vs. time.

Bass (1973) has constructed hyperbolic graph paper for values of n equal to 0.25, 0.50, 0.75, and 1.0 (Figures 4–40 to 4–43). Production data can be plotted in this kind of paper until a straight line is obtained. Once the appropriate value of n has been determined, the cumulative production at any future date can be calculated as follows:

1. Calculate the ratio $(q_g)_i/q_g$ for the period of interest.

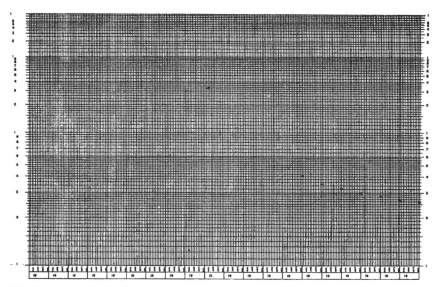

FIGURE 4–40 Hyperbolic decline, n = 0.25 (after Bass, 1973)

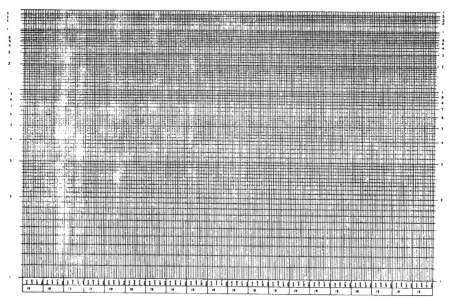

FIGURE 4–41 Hyperbolic decline, n = 0.50 (after Bass, 1973)

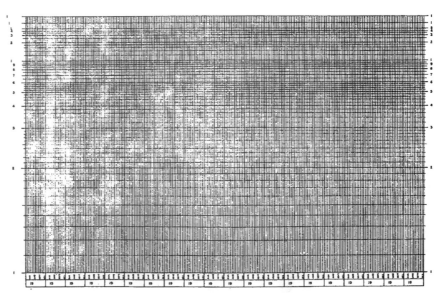

FIGURE 4–42 Hyperbolic decline, n = 0.75 (after Bass, 1973)

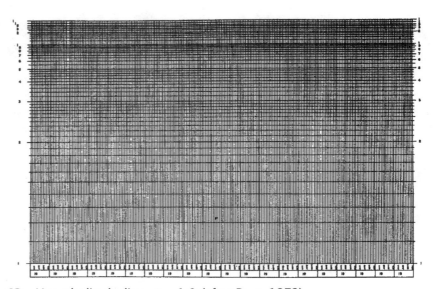

FIGURE 4–43 Hyperbolic decline, n = 1.0 (after Bass, 1973)

2. Calculate the initial monthly rate of decline from the equation:

$$D_i = \left\{ \left[\frac{(q_g)_i}{q_g} \right]^n - 1 \right\} / (nt) \qquad (4\text{–}40)$$

where t is in months.

3. Calculate the cumulative gas production using the equation:

$$G_p = \frac{(q_g)_i}{(n-1)\,D_i} \{ [(q_g)_i / q_g]^{n-1} - 1 \} \qquad (4\text{–}41)$$

for the case in which n is less than one, and from the equation:

$$G_p = \frac{(q_g)_i}{D_i} \ln \frac{(q_g)_i}{q_g} \qquad (4\text{–}42)$$

for the special case when n = 1.0. Flow rates in Equations 4–41 and 4–42 are in Mscf/month.

The use of the approach will be illustrated with data presented by Bagnal and Ryan (1976).

Figure 4–44 shows typical plots of logarithm of production rate vs. time. The same data is plotted in Cartesian paper on Figure 4–30 and was used by Schettler (1977) to illustrate his matching procedure. The semilog plot (Figure 4–44) shows upward curves indicating hyperbolic declines, except for the lower set of data points which form a straight line (exponential decline).

The data of the three upper curves is plotted again on hyperbolic paper with n = 1.0 (Figure 4–45). Approximate straight lines are obtained from which it is possible to extrapolate to any desired limit as shown by the dashed lines.

As an example, the well represented by the lower curve will be producing 14.5 Mscfd after 40 years. The cumulative production at that moment can be calculated as follows:

1. Calculate the ratio $(q_g)_i/q_g$ for the period under consideration:

$$(q_g)_i/q_g = 20 \; Mscfd/14.5 \; Mscfd = 1.38$$

2. Calculate the initial monthly rate of decline from Equation 4–40 as follows:

$$D_i = [(20/14.5)^1 - 1] \; / \; (1 \times 15 \times 12) = 0.0021 \; month^{-1}$$

the product, 15×12, represents the months required for the production to decrease from 20 Mscfd to 14.5 Mscfd.

3. Calculate the cumulative gas production from Equation 4–42 as follows:

$$G_p = \{ \; (20) \; (30.4) \; / \; .0021 \; \} \; \ln \; (20 \; / \; 14.5) = 93,106 \; Mscf$$

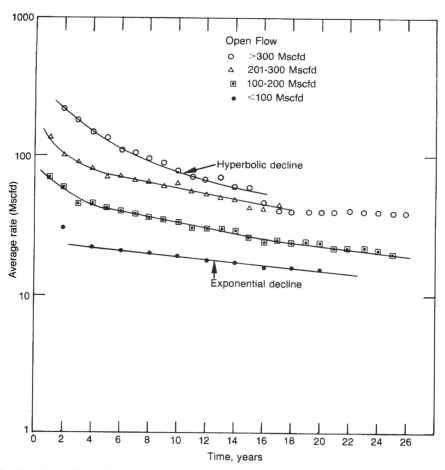

FIGURE 4–44 Log of production rate vs. time

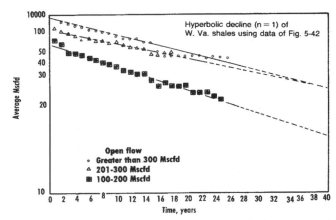

FIGURE 4–45   Hyperbolic decline (n = 1) of West Virginia shales, using data points of Figure 4–44

In general, decline curves are not used for evaluating gas reservoirs. However, the hyperbolic decline with n = 1.0 seems especially useful for evaluation of Devonian shales.

A decline analysis approach using production type curves has been developed by Hazzlett et al (1986) and its application has been demonstrated by Gatens et al (1989). The type curve is based on the observation that key reservoir characteristics including the storativity ratio, ω, and the interflow parameter, λ, correlate with observed well performance. Individual type curves have to be developed for each case depending on the values of ω and λ.

Higher-Ordered Polynomial Models

A higher-ordered polynomial model is a linear regression model of the form:

$$Y = B_o + B_1 X + B_2 X^2 + B_3 X^3 \ldots \tag{4–43}$$

where Y is production and X is time. De Wys and Schumaker (1978) have used Equation 4–43 in the Cottageville field of West Virginia and have found good fits with actual production data (Figure 4–46).

This method does not explain the data and their characteristics; rather, it matches the production data during an interval of time and forecasts production performance during short time spans.

Based on production data of Devonian shales, De Wys and Schumaker (1978) have presented the schematic of natural fractures shown on Figure 4–47. Initially there is production of free gas, i.e., gas stored within the fractured network at original conditions. Since the gas storage capacity within the fractures is small there is drastic production decline.

This is followed by production of residual (or the remaining) gas within the fractures and the beginning of production of adsorbed gas.

The last part of the decline curve corresponds to production of remaining adsorbed gas and absorbed gas. In general, over a 20–25 year production life, the tail of the curve comprises up to 90% of the production curve in terms of time.

Hydrodynamic Analogy of Production Decline

An approach to forecasting production decline using a simple hydrodynamic analogy has been presented by Pullé (1982).

Figure 4–48 presents a schematic of the idealized reservoir. It presents two broad compartments; the first one called R_1, represents the low porosity and low permeability gas-generating shale, the second one (R_2) represents a part of the reservoir with better porosity and better permeability. This second domain is connected to the wellbore; thus flow occurs from domain R_1 to R_2 and from R_2 to the wellbore.

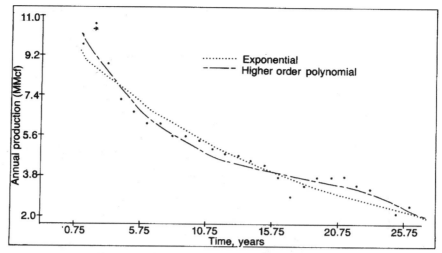

FIGURE 4–46 Best fit exponential and higher order polynomial curves to annual production data for Well no. 715 over 26 years (after De Wys and Schumaker, 1978)

Figure 4–49 presents the hydrodynamic analogy. It consists of two circular cylinders containing liquid, which are interconnected with inlets (sources) and outlets (sinks).

The rate Q_1 is outflow (sink) from R_1 but inflow (source) to R_2; C_1 (capacitance) is uniform cross-sectional area; h_1 (t) is pressure at time t at the boundary, which is proportional to the height of liquid; Ω_1 is resistance to flow at the boundary separating R_1 from R_2, and is a measure of the permeability in the region of the boundary. Similar interpretations are applied to the R_2 reservoir.

A first-order equivalence relationship between pressure-volume (PV) initial conditions in the actual reservoirs with the analog cylinders "reservoir" has been established by Pullé (1982):

For R₁:

$$(\pi r_1^2)\, d_1 = V_1 \tag{4–44}$$

ie.

$$C_1 d_1 = V_1 \tag{4–45}$$

$$d_1 \sim p_1 \tag{4–46}$$

where d_1 is liquid height in compartment 1 and p_1 is pressure in compartment 1.

For R₂:

$$(\pi r_2^2)\, d_2 = V_2 \tag{4–47}$$

$$C_2\, d_2 = V_2 \tag{4–48}$$

$$d_2 \sim p_2 \tag{4–49}$$

where d_2 is liquid height in compartment 2 and p_2 is pressure in compartment 2.

Based on the above equations, C_1 and C_2 can be uniquely determined from initial conditions, since $d_1 = d_2 \simeq p$, the equilibrium pressure at the boundary at time t = 0. The pressures in the hydrodynamic system are a function of depth and would vary with depth even at time t = 0.

The following two assumptions regarding boundary flow are made:

1. Ω_1 and Ω_2, the resistances to boundary flow are independent of position and time and are constants, i.e., Ω (x, y, z, t) = Ω, at the boundary. This is equivalent to the permeability k(x,y,z) = k being a constant on the boundary. Referring to Figure 4–49, the resistances are at the intersections of R_1 and R_2, and R_2 and the wellbore.

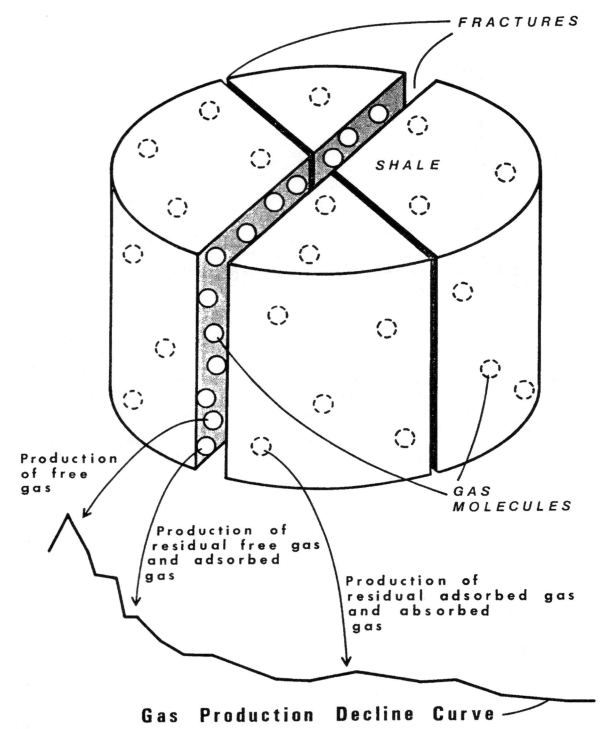

FIGURE 4–47 Schematic sketch of fractures in Devonian shale and possible relationships of gas production (after de Wys and Schumaker, 1978)

2. There are no independent sources and sinks within R_1 and R_2, apart from mass transfer resulting from flow in the system. For example, kerogen sources are ignored on the assumption that the gas released by these sources over geologic time is almost complete, and further gas release and rates of release are negligible in comparison to the total volume of gas in matrix and fracture pores and interstices.

Table 4–7 summarizes the equivalence between the hydrodynamic and reservoir systems.

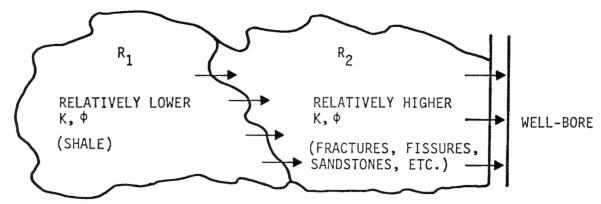

FIGURE 4–48   Reservoir partition (after Pullé, 1982)

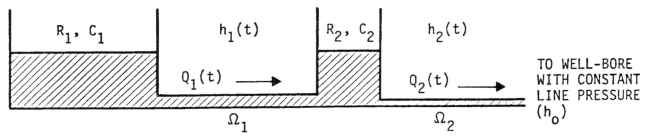

FIGURE 4–49   Hydrodynamic system (after Pullé, 1982)

TABLE 4–7   Factor Equivalence (after Pullé, 1982)

Hydrodynamic System	Shale Reservoir
Capacitance, C	Porosity, ϕ; $C = C(\phi)$
Pressure h(x,y,z,t) = h (t) at the boundary	Pressure, p(x,y,z,t)
Resistance, Ω(x,y,z,t) = Ω, at the boundary	Permeability k(x,y,z) = k(1/Ω)
Bulk Flow Rate, Q(t,Ω,∇h)	Darcy Flow Rate, Q(x,y,z,t,k,∇p,v,.....,)

Derivation of Production Decline

Using as a base the nomenclature presented in Table 4–7 and Figure 4–49 the following equations are obtained (Pullé, 1982):
   For "Reservoir" R_1:
      (Continuity Equation)

$$C_1 \frac{dh_1(t)}{dt} = Q_1(t) \qquad (4\text{--}50)$$

      (Flow Equation)

$$Q_1(t) = \frac{h_1(t) - h_2(t)}{\Omega_1} \qquad (4\text{--}51)$$

   For "Reservoir" R_2
      (Continuity Equation)

$$C_2 \frac{dh_2(t)}{dt} = -Q_2(t) + Q_1(t) \qquad (4\text{--}52)$$

(Flow Equation)

$$Q_2(t) = \frac{h_2(t) - h_o}{\Omega_2} \tag{4-53}$$

where h_o is constant line pressure.

substituting for $h_2(t)$ in Equation 4–52 with Equation 4–53 gives:

$$C_2\Omega_2 \frac{dQ_2(t)}{dt} + Q_2(t) = Q_1(t) \tag{4-54}$$

$$\frac{dQ_2(t)}{dt} + \lambda Q_2(t) = \lambda Q_1(t) \tag{4-55}$$

where lambda is a production decline rate given by:

$$\lambda = 1 / C_2\Omega_2 \tag{4-56}$$

and,

$$Q_2(t) = e^{-\lambda t} [C + \lambda \int e^{\lambda t} Q_1(t) \, dt] \tag{4-57}$$

where C is a constant of integration.

At initial conditions, t = 0 and $Q_1(t) = Q_1(0) = 0$, i.e. , when the height of liquid initially in both cylinders is the same, there exists not pressure gradient and no flow is possible from R_1 to R_2, and

$$C = Q_2(0) \text{ (initial open flow)} \tag{4-58}$$

Consequently, Equation 4–57 becomes:

$$Q_2(t) = Q_2(0)e^{-\lambda t} + \lambda e^{-\lambda t} \int_0^t e^{\lambda \theta} Q_1(\theta) \, d\theta \tag{4-59}$$

where θ is integration parameter.

If the rate is a constant, $Q_1(\theta) = Q_1$, the rate of change of velocity at the boundary of R_1 and R_2 is a constant, or

$$\frac{dQ_1(t)}{dt} = 0 \tag{4-60}$$

For the case of shales the rate of movement of gas molecules is considered infinitesimal at and near the fracture faces, so as to be considered constant. For this case $Q_1(\theta) = Q_1$ and Equation 4–59 becomes:

$$Q_2(t) = Q_2(0)e^{-\lambda t} + Q_1\lambda e^{-\lambda t} \int_0^t e^{\lambda \theta} \, d\theta \tag{4-61}$$

$$Q_2(t) = Q_2(0) e^{-\lambda t} + Q_1(1 - e^{-\lambda t}) \tag{4-62}$$

For large values of t, $e^{-\lambda t}$ approaches zero and $(1 - e^{-\lambda t})$ approaches 1.0. Thus, for large times:

$$Q_2(t) \rightarrow Q_1 \tag{4-63}$$

Figure 4–50 shows a theoretical decline curve generated with the above equation. The constant flow rate $Q_1 = Q_2(\infty)$ will be maintained as long as the flow from the shale matrix into the fractures is steadily sustained.

If Q_1 approaches zero or is very small, an exponential curve dominated by the first term on the right hand side of Equation 4–62 will provide the best fit. If Q_1 is large a hyperbolic decline will fit better the data.

When Q_1 approaches zero and λ is small the reservoir can be considered to have single porosity. When Q_1 is large the reservoir is considered to have dual-porosity.

A practical application using Equation 5–62 is illustrated in Figures 4–51 and 4–52 for well 9417 of Jackson County, West Virginia. The plus signs represent production history. The circles represent calculated values. The match is good.

Production rate, $Q_2(t)$, is maximized when (1) the initial open flow, $Q_2(0)$ is large, (2) the flow of gas from the matrix shale, Q_1, is large, and (3) the production decline rate, λ, is small.

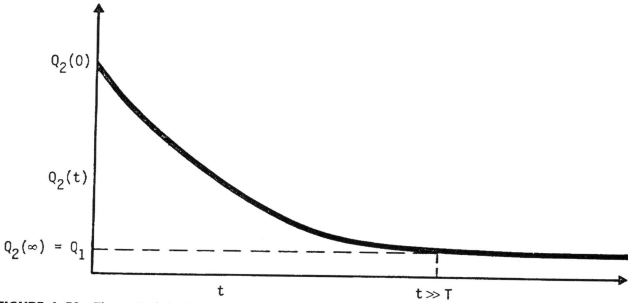

FIGURE 4–50 Theoretical decline curve (after Pullé, 1982)

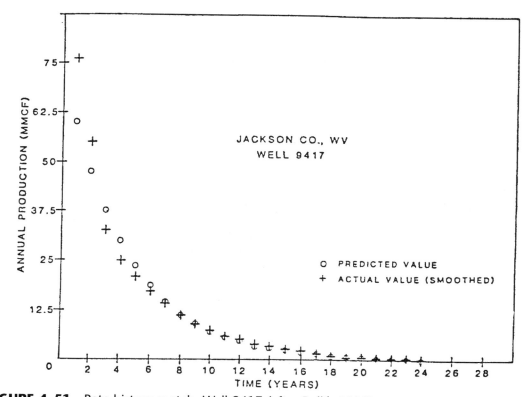

FIGURE 4–51 Rate history match, Well 9417 (after Pullé, 1982)

Empirical Performance Equations

Gatens et al (1989) have presented empirical performance equations based on the analysis of production data from more than 800 Devonian shale wells in Kentucky, Ohio and West Virginia. Figure 4–53 shows the four areas where the wells are located. Area 1 is usually referred to as Big Sandy area. It has many wells with long production histories. Area 2 in central West Virginia produces dry gas from average quality wells. Area 3 in southeastern Ohio has poorer-quality wells that produce mainly dry gas. Area 4, usually referred to as

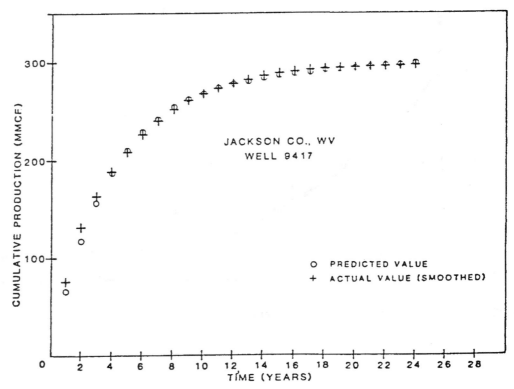

FIGURE 4-52 Cumulative history match, Well 9417 (after Pullé, 1982)

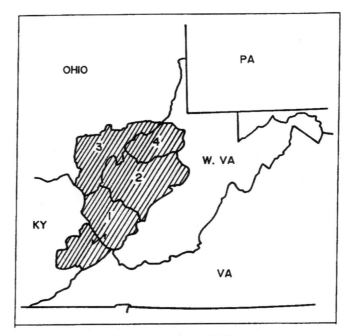

FIGURE 4-53 Map of study areas (after Gatens, 1989)

Burning Springs area, produces gas and liquids from a large number of wells completed recently.

Equation Based on Reservoir Properties. This empirical equation is based on reservoir properties and can be written as follows:

$$Q_t = a \, (k_{eff}h)^b \, (G)^c \qquad (4\text{-}64)$$

where

Q_t	=	cumulative production at time t, Mscf
a,b,c	=	reservoir coefficients
k_{eff}	=	effective permeability, md
h	=	net pay, ft
G	=	initial gas-in-place, Mscf

The regression coefficients required to solve Equation 4–64 are presented in Table 4–8. Included are correlation coefficients (R^2).

The regression coefficients apply to areas A, B, C and D located in the following counties:

Area	Counties
A	All not included in Areas B through D
B	Putnam and Jackson counties, WV
C	Lincoln and Kanawha counties, WV; Floyd County, KY
D	Boone, Logan, and Mingo counties, WV; Pike, Martin, and Johnson counties, Ky

Equation 4–64 indicates that at early times, Devonian shale wells produce as if they were in an infinite-acting reservoir. At this stage the capacity, $k_{eff}h$, is the main factor controlling production. At late stages the volume of original gas-in-place, G, becomes very important.

As an example a well located in Lincoln County, West Virginia (Area C), with $K_{eff}h = 10$ md–ft, and an original gas-in-place, $G = 1.56 \times 10^9$ scf, will have accumulated after 10 years of production:

$$Q_{10} = 1763 \times (10)^{0.5996} (1.56 \times 10^6)^{0.2148}$$

$$Q_{10} = 120,382 \ Mscf$$

It is interesting to note that this represents only 7.72% of the estimated original gas-in-place.

Equation Based on Past Well Performance. This empirical equation is based on past well performance rather than on estimates of reservoir properties. The equation can be written as follows:

$$Q_{tj} = a(Q_{tk})^{bj} (Q_{tl})^{cj} \tag{4–65}$$

where the subscripts j, k, and l refer to production time, and t is total time. In the above equation $t_j > t_k > t_l$. Values of j, k and l analyzed were 1, 3, 5, 10, 20, 30 and 40 years. Table 4–9 displays the regression coefficients required to use Equation 4–65. Also included are the correlation coefficients (R^2).

Assume for example that five years of production history are available in a given well. The appropriate equations to forecast performance are:

$$Q_{10} = a_{10}(Q_5)^{b10} (Q_3)^{c10}$$

$$Q_{20} = a_{20}(Q_5)^{b20}(Q_3)^{c20}$$

$$Q_{30} = a_{30}(Q_5)^{b30} (Q_3)^{c30}$$

$$Q_{40} = a_{40}(Q_5)^{b40}(Q_3)^{c40}$$

Good forecasting results have been published by Gatens et al (1989) using Equations 4–64 and 4–65.

WELL TESTING

In some instances, analytical transient pressure methods for dual-porosity systems are useful for evaluating gas-bearing fractured shales. When these methods fail, it is better to resort to simulation techniques which might provide an insight into fracture length (XF), reservoir permeability (k), and porosity (PHI) by a history-matching process.

TABLE 4-8 Empirical Coefficients for Predicting Performance Using $k_{eff}h$ and G (after Gatens et al., 1989)

Cum. prod (Mscf)	Area A				Area B				Area C				Area D			
	a	b	c	R²	a	b	c	R²	a	b	c	R²	a	b	c	R²
Q_1	672	0.4961	0.1605	0.739	*	*	*	*	280	0.7059	0.1871	0.803	78	0.3885	0.3389	0.759
Q_3	2,984	0.4861	0.1229	0.759	*	*	*	*	1,102	0.6933	0.1623	0.851	104	0.3303	0.3963	0.771
Q_5	3,223	0.5457	0.1516	0.791	*	*	*	*	1,493	0.6609	0.1772	0.850	80	0.2829	0.4530	0.768
Q_{10}	2,084	0.3628	0.2384	0.832	4,683	0.4113	0.1902	0.749	1,763	0.5996	0.2148	0.822	73	0.2310	0.5075	0.775
Q_{20}	395	0.2541	0.4146	0.870	1,214	0.3015	0.3429	0.861	1,451	0.5154	0.2792	0.776	57	0.1856	0.5671	0.793
Q_{30}	487	0.1726	0.4375	0.822	-	-	-	-	1,725	0.5104	0.2862	0.765	37	0.1456	0.6193	0.704
Q_{40}	191	0.1156	0.5257	0.738	-	-	-	-	1,320	0.4549	0.3280	0.841	-	-	-	-

366

TABLE 4–9 Empirical Coefficients for Predicting Performance Using Past Production (after Gatens et al., 1989)

Time (years)	a	b	c	R^2
Less Than 3 Years of Production, $Q_t = a(Q_1)^b$				
3	4.866	0.9346	-	0.972
5	11.22	0.8925	-	0.944
10	35.0	0.8318	-	0.898
20	103.9	0.7736	-	0.843
30	303.3	0.6991	-	0.746
40	903.5	0.6130	-	0.752
Less Than 5 Years of Production, $Q_t = a(Q_3)^b(Q_1)^c$				
5	1.340	1.283	-0.2982	0.995
10	2.847	1.468	-0.5234	0.975
20	7.098	1.539	-0.6437	0.939
30	12.58	1.493	-0.6240	0.870
40	132.3	0.7294	0	0.839
Less Than 10 Years of Production, $Q_t = a(Q_5)^b(Q_3)^c$				
10	1.509	1.844	-0.8657	0.991
20	2.493	2.372	-1.430	0.965
30	4.036	2.377	-1.441	0.908
50	22.54	1.850	-1.034	0.879
Less Than 20 Years of Production, $Q_t = a(Q_{10})^b(Q_5)^c$				
20	1.285	1.739	-0.7528	0.993
30	1.566	1.899	-1.019	0.974
40	3.410	2.084	-1.156	0.937
Less Than 30 Years of Production, $Q_t = a(Q_{20})^b(Q_{10})^c$				
30	1.136	1.476	-0.4838	0.997
40	1.225	1.812	-0.8247	0.989
Less Than 40 Years of Production, $Q_t = a(Q_{30})^b(Q_{20})^c$				
40	1.129	1.550	-0.5572	0.999

Sawyer et al (1976) provided a numerical simulation case history for a Devonian shale. Using a general purpose diffusivity model and, as basic data, a pay of 30 ft, porosity of 0.0025, permeability of 0.121 md, and fracture length of 628 ft, they were able to match drawdown, buildup, and isochronal data of a stimulated Devonian shale well (Figure 4–54). Fracture length and permeability were consistent with core analysis and sand volume injected during stimulation. The low value of effective gas porosity, however, suggested either a low porosity or a very high water saturation system.

Use of this method requires a computer. It has the advantages that (1) a constant rate test is not required; (2) virtually all types of well performance data can be analyzed; and (3) production anomalies resulting from stratification or heterogeneity may be explained by careful analysis of how fracture length and permeability affect well pressure behavior.

Various interference tests have been conducted in Devonian shales (Lee et al, 1982, Frohne and Mercer, 1984). In work done by Lee et al they calculated wellbore storage, fracture permeability, fracture porosity and skin in the production well.

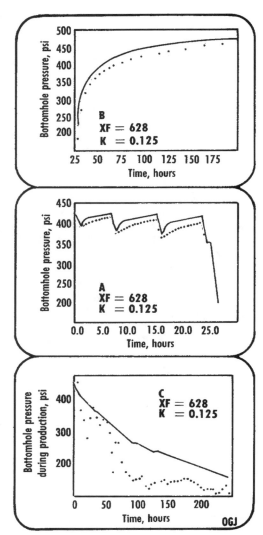

FIGURE 4–54 Best curve match for well test and production data, PHI = 0.25% (after Sawyer et al, 1976)

The interference test was analyzed using Warren and Root's (1963) model. Results were good leading to the conclusion that Devonian shales must be handled, at least in some cases, with dual-porosity models. Pulse test analysis showed inconsistent results. However, pulse testing might be of little value in naturally fractured reservoirs.

Frohne and Mercer (1984) and Lee et al. (1982) used SUGAR, a numerical simulator for unconventional gas resources, to simulate the well testing data. SUGAR was developed by the Department of Energy's (DOE) Eastern Gas Shales Project (EGSP), National Energy Software Center, Argonne National Laboratory, Argonne, Illinois. Results were satisfactory and led to the conclusion that anisotropy is large in Devonian shales, and that the reservoirs are drained in highly elliptical patterns.

This suggests that overall gas recovery from shales could be increased if development wells are drilled in such a way as to take advantage of the elliptical drainage pattern. The same holds true for other naturally fractured reservoirs as discussed by Aguilera (1983) who suggested rectangular rather than square pattern configuration for certain naturally fractured reservoirs.

Another interesting finding was that reservoir pressures close to existing wells were relatively high even after producing for 22 years. This suggests that infill drilling could improve overall recoveries.

TIGHT GAS RESERVOIRS

Most tight gas reservoirs are sandstones and siltstones of marine and fluvial origin. There are also occurrences of low-permeability marine carbonate reservoirs. Spencer and Mast (1986) have indicated that tight sandstone gas reservoirs have recoverable resources in the order of 100 to 400 trillion scf. Their work is an excellent source of information which has been used extensively in this section on tight gas reservoirs.

As in the case of Devonian shales a significant amount of research if funded by the U.S. Department of Energy (DOE) and the Gas Research Institute (GRI).

PENNSYLVANIA

Laughrey and Harper (1986) have discussed the geological and engineering characteristics of Upper Devonian and Lower Silurian formations in Pennsylvania. Permeabilities and porosities, most of which are secondary tend to be low. Although initial rates are low, modern stimulation techniques increase the rates substantially to the point that many wells in Indiana county test from 700 to 1000 Mscfd.

Reserves are difficult to establish but Laughrey and Harper (1986) indicate that as a rule a well is economical if a simple initial volumetric calculation of recoverable gas does not fall below 200 MMscf. Figure 4–55 shows a subsurface correlation diagram of Upper and Middle Devonian rocks in western Pennsylvania including shales and tight sandstones. Black shales were discussed previously in this chapter.

The Venango, Bradford and Elk Groups are composed of interbedded sandstone, conglomerates and mud rocks with minor contributions of carbonates. Deposition was mostly marine to transitional marine.

The Venango Group outcrops extensively in northwestern Pennsylvania and is known to have the coarser clastics. As a result of the outcrop the Venango Group has been studied more intensively than the Bradford and Elk Groups.

Natural fracturing is present in the Bradford Group including the Kane formation (drillers' nomenclature) and is responsible for enhanced productivity in some wells. This is illustrated with log interpretation and well testing data of well Vernon Henry 3, Indiana county.

Log Interpretation. Figure 4–56 shows Servipetrol's Fracture Completion Log (FCL) of well Henry 3, Indiana County. The principles behind the FCL interpretation are presented in detail in Chapter 3. Figure 4–56 shows in the first track the partitioning coefficient and the gamma ray curve. The next two tracks show total and fracture porosity, and total and matrix water saturation. The last track shows the estimated lithological components. In this analysis the FCL shows via the partitioning coefficient (v) that the Lower Bradford and the Kane have some natural fracturing.

Table 4–10 shows a listing of FCL results for Kane interval 3760–3774 ft. The first column shows depth in feet. The second column shows the statistical parameter $P^{1/2}$. This parameter allows to distinguish hydrocarbon from water-bearing intervals. $P^{1/2}$ is expressed as a function of resistivity, porosity log response, and the dual-porosity exponent (m) as discussed in Chapter 3.

The third column shows total porosity; i.e., it includes matrix and fracture porosity. Total porosity ranges between 0.6% and 10.5%. The fourth column presents fracture porosities ranging between 0.1% and 0.2%. Both total and fracture porosities are attached to total bulk volumes (matrix plus fracture bulk volumes).

The fifth column shows matrix porosities. These volumes are expressed as a function of only the matrix bulk volume (excluding fracture volume), and as such they can be compared directly with core porosities from unfractured plugs. Matrix porosities range between 0.6% and 10.3%

The sixth column shows the partitioning coefficient, i.e., the fraction of total porosity that is made out of fractures. Partitioning coefficients range between zero and 1.7%.

The next two columns show total and matrix water saturations. Total water saturations are calculated as functions of their $P^{1/2}$ values and the $P^{1/2}$ for water bearing intervals. Matrix

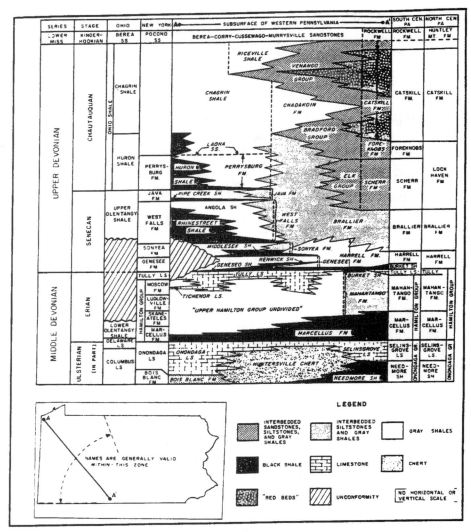

FIGURE 4–55 Subsurface correlation diagram of Upper and Middle Devonian rocks in Western Pennsylvania (after Laughrey and Harper, 1986)

water saturations are calculated as a function of total water saturation, fracture water saturations and the partitioning coefficient as explained in Chapter 3. Water saturations in this well are relatively low compared to other wells in Indiana County.

The ninth column shows cumulative pore volume, i.e., the product of total porosity and net pay.

The tenth, eleventh and twelfth columns show total, fracture and matrix cumulative hydrocarbons, i.e., the product of porosity, hydrocarbon saturation and net pay.

Most conventional well log interpretations introduce porosity, shale volume, water saturation and permeability cutoffs in order to calculate net pay. Recent work with naturally fractured reservoirs, however, indicate that the porosity and permeability cutoff approach might lead to pessimistic values of net pay. In fact, it has been found that for the same physical environment the degree of fracturing increases as porosity decreases (Nelson, 1985; Aguilera, 1992; Stearns et al, 1994). As a consequence, I do not recommend (in general) the use of porosity cutoffs for estimating net pay in naturally fractured reservoirs. This is specially important when the intervals to test are being selected. Introducing a cutoff might leave untested a low porosity interval with good fracture development.

It has also been shown with numerical simulators and analytical expressions that matrix permeabilities smaller than 0.01 md can contribute efficiently to production (Kazemi,

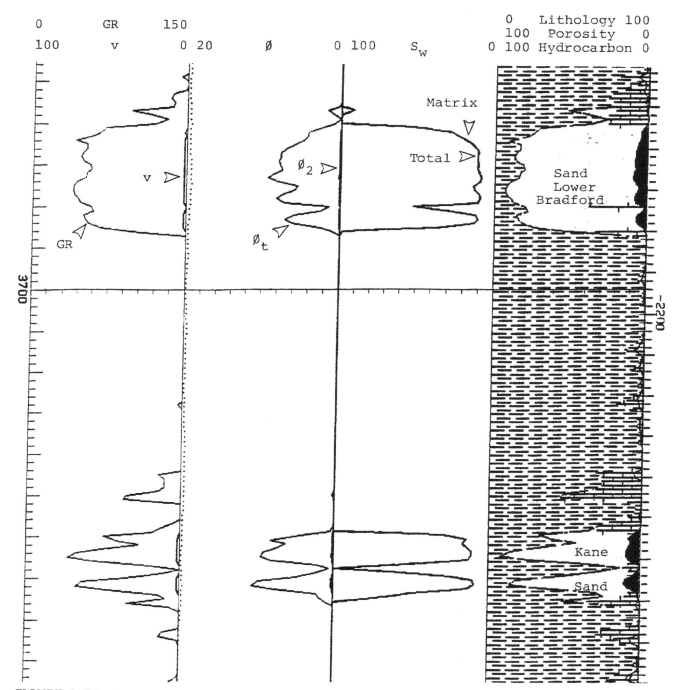

FIGURE 4–56   Fracture Completion Log (FCL), well Henry #3

1969; Aguilera, 1987) in naturally fractured reservoirs. As a consequence, the use of permeability cutoffs must be used with significant care.

My experience is that water saturation provides the most reliable cutoff in naturally fractured reservoirs. For example core data in most naturally fractured reservoirs I am familiar with show that S_w increases as permeability decreases. Consequently using S_w as a cutoff takes care of eliminating the lower permeabilities in the net pay analysis.

Table 4–11 shows the cutoffs used for estimating net pay in the Kane and a summary for interval 3759–3774 ft. The average properties presented in Table 4–11 apply only to the net pay of 10 ft. The "free" gas in place within the fractures amounts to 29.45 MMscf assuming a reservoir area of 640 acres. The bulk of the gas volume, however, is in the tight

TABLE 4–10 Fracture Completion Log Well: Henry No. 3, January 1, 1993

Depth ft	$P^{1/2}$	Total Porosity %	Fracture Porosity %	Matrix Porosity %	Partitioning Coefficient %	Total Water Sat %	Matrix Water Sat %	Cum. Porosity Feet	Cum. Total Hydrocarbon Feet	Cum. Fracture Hydrocarbon Feet	Cum. Matrix Hydrocarbon Feet
3760.00	0.58	4.4	0.0	4.4	1.1	29.0	29.4	5.645	4.699	0.064	4.635
3761.00	1.02	8.1	0.1	8.0	1.5	14.1	14.4	5.727	4.769	0.065	4.704
3762.00	0.88	6.3	0.1	6.2	1.3	17.0	17.2				
3763.00	1.06	7.7	0.1	7.6	1.4	13.5	13.7	5.803	4.835	0.066	4.769
3764.00	1.22	9.2	0.1	9.0	1.5	11.2	11.4	5.895	4.916	0.067	4.849
3765.00	1.25	9.4	0.2	9.3	1.6	10.9	11.1	5.989	5.001	0.069	4.932
3766.00	1.09	8.1	0.1	8.0	1.4	13.0	13.2	6.070	5.071	0.070	5.001
3767.00	0.38	2.2	0.0	2.2	0.0	50.6	50.6				
3768.00	0.12	0.6	0.0	0.6	0.0	100.0	100.0				
3769.00	0.32	2.1	0.0	2.1	0.0	63.4	63.4				
3770.00	0.67	5.0	0.1	4.9	1.3	24.4	24.7	6.119	5.108	0.070	5.038
3771.00	1.28	9.8	0.2	9.6	1.6	10.5	10.6	6.217	5.196	0.072	5.124
3772.00	1.41	10.5	0.2	10.3	1.7	9.3	9.4	6.322	5.291	0.073	5.218
3773.00	1.08	7.6	0.1	7.5	1.4	13.1	13.3	6.398	5.357	0.074	5.283
3774.00	0.39	2.2	0.0	2.2	0.0	48.8	48.8				

TABLE 4–11 Fracture Completion Log. Well: Henry No. 3 January 1, 1993

Cutoffs Used	
Porosity >=	0.00
Water saturation <=	50.00
Permeability >=	0.00
Shale volume <=	50.00
Type of reservoir: Gas	
Hydrocarbon form volume factor	0.0098
Interval Summary	
Top of Interval	3759.00
Bottom of Interval	3774.00
Gross feet	15.00
Net feet	10.00
Average Total Porosity	7.97
Average Fracture Porosity	0.12
Average Matrix Porosity	7.85
Average Total Water Saturation	13.51
Average Fracture Water Saturation	0.00
Average Matrix Water Saturation	13.52
Cumulative Total Hydrocarbon feet	0.69
Cumulative Fracture Hydrocarbon feet	0.01
Cumulative Matrix Hydrocarbon feet	0.68
Total Hydrocarbons in place-scf/ac	3063989.25
Fracture Hydrocarbons in place-scf/ac	46008.88
Matrix Hydrocarbons in place-scf/ac	3017980.37

matrix. The average matrix water saturation (13.52%) is low compared with other wells I have analyzed in Pennsylvania.

Well Test Analysis. Figure 4–57 shows a Horner plot from the same Kane interval discussed above. Pressures were recorded at the wellhead and were converted to bottomhole conditions using standard correlations available in the literature. The pressure behavior is typical of dual-porosity systems. The early portion of the test is dominated by wellbore storage. This is followed by a flow period dominated by natural fractures and a transition period initiated when the matrix starts to feed the fractures. The next flow period corresponds to radial flow dominated by the composite system of matrix and fractures, and finally the transient analysis gives indications of a possible sealing fault (it could be any other type of linear discontinuity).

Flow rate data used in the analysis is presented in Table 4–12. During the test the well flowed at a maximum rate of 275,000 scfd. Table 4–13 shows other input data. Notice that the net pay, porosity and water saturation values are the same ones calculated in Table 4–11. Table 4–14 presents results of the buildup analysis. The effective permeability to gas is 0.05 md. A negative skin of –4.49 indicates improved conditions around the wellbore as a result of a mild acid wash and the presence of natural fractures. The production history of well Henry 3 is presented in Figure 4–58.

The average porosity from well logs (7.97%) and the effective permeability to gas from the buildup analysis (0.05 md) compare well with ranges published by Glohi (1984) who

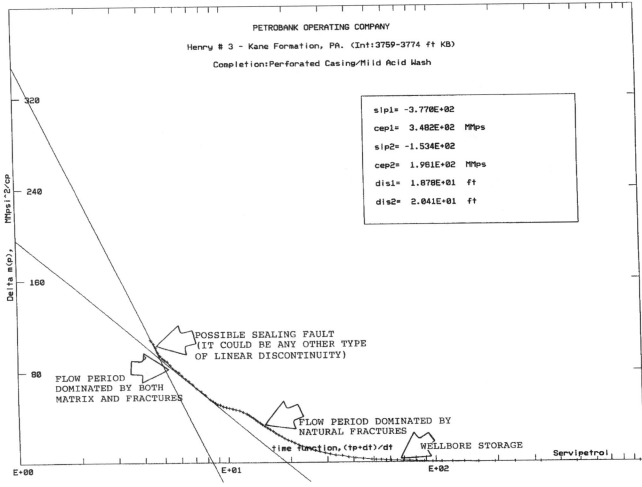

FIGURE 4–57 Horner crossplot, well Henry #3

TABLE 4–12 Petrobank Operating Company Henry #3–Kane Formation, PA. (Int: 3759–3774 ft KB) Completion: Perforated Casing/Mild Acid Wash Press. Buildup: Jan 22-25, 93. Surf. Pres. corrected to downhole

			Flowrate Data		
No.	**Elapsed time Hr**	**Total Res bbl/D**	**Oil STB/D**	**Gas MSCF/D**	**Water STB/D**
1	8.000E+00	1.062E+03	0.000E–01	2.750E+02	0.000E–01
2	2.300E+01	8.300E+02	0.000E–01	2.150E+02	0.000E–01
3	3.200E+01	3.667E+02	0.000E–01	9.500E+01	0.000E–01
4	4.700E+01	3.165E+02	0.000E–01	8.200E+01	0.000E–01
5	5.300E+01	2.779E+02	0.000E–01	7.200E+01	0.000E–01
6	5.500E+01	0.000E–01	0.000E–01	0.000E–01	0.000E–01
7	6.200E+01	3.860E+02	0.000E–01	1.000E+02	0.000E–01
8	7.100E+01	2.895E+02	0.000E–01	7.500E+01	0.000E–01
9	7.750E+01	2.934E+02	0.000E–01	7.600E+01	0.000E–01
10	1.740E+02	0.000E–01	0.000E–01	0.000E–01	0.000E–01
11	1.950E+02	5.018E+02	0.000E–01	1.300E+02	0.000E–01
12	2.190E+02	4.131E+02	0.000E–01	1.070E+02	0.000E–01
13	2.390E+02	3.281E+02	0.000E–01	8.500E+01	0.000E–01

TABLE 4–13 Basic Data, Well Henry 3, Indiana Co., PA

Pressure Run Depth	3.766E+03 ft
Fluid Analyzed	Gas
Production Time	2.758E+02 hr
Total Flow Rate	2.353E+02 Res bbl/D
Oil Flow Rate	0.000E–01 STB/D
Gas Flow Rate (Free Gas)	6.095E+01 MSCF/D
Water Flow Rate	0.000E–01 STB/D
Bottomhole Flowing Pressure	2.154E+01 psia
Specific Gravity of Gas	5.800E–01
Gas Deviation Factor	9.152E–01
Oil Formation Volume Factor	1.000E+00 Res bbl/STB
Gas Formation Volume Factor	3.860E+00 Res bbl/MSCF
Water Formation Volume Factor	1.000E+00 Res bbl/STB
Solution Gas-Oil Ratio	0.000E–01 SCF/STB
Solution Gas-Water Ratio	0.000E–01 SCF/STB
Oil Viscosity	1.000E+00 cp
Gas Viscosity	1.243E–02 cp
Water Viscosity	1.000E+00 cp
Reservoir Temperature	9.200E+01 Degree F
Pay Thickness	1.000E+01 ft
Porosity	7.970E–02
Oil Saturation	0.000E–01
Gas Saturation	8.649E–01
Water Saturation	1.351E–01
Total Compressibility	1.339E–03 1/psia
Wellbore Radius	1.875E–01 ft

found that for Upper Devonian sandstones porosities ranged between 0.5 and 12%, and permeabilities ranged between 0.01 and 2 md.

I have used the same type of evaluation with reasonable success in other tight gas reservoirs including the Morrow sandstones of the Anadarko Basin in Oklahoma, the lower Tertiary Wasatch formation and Upper Cretaceous Mesaverde Group of the Piceance Basin in Colorado, and the Niobrara Chalk in the eastern part of the Denver Basin. The advantage of this type of evaluation is that it provides reasonable results and is cost efficient. There are more rigorous approaches to reservoir characterization of tight sands but usually they are very expensive relative to the low production rates that can be obtained from tight formations.

OKLAHOMA

The Morrow sandstone's porosity is secondary and is provided by dissolution of clayey matrix, carbonate fragments, cement, glauconite and quartz grain (Al-Shaieb and Walker, 1986). Natural fractures do not seem to play an important role in the Morrow sandstone.

Secondary porosity by dissolution varies between 2 and 25%. The maximum porosity development occurs in the quartz-dominated sandstones. The development of secondary porosity is related to (1) Source of H^+ ions in the bleaching fluids. The sources are organic acids, carbonic acid and dissolved H_2S. (2) Composition and texture of the various sandstone lithologies (Al-Shaieb, 1986).

TEXAS

Low permeability gas sandstones with a blanket geometry are found in the Cotton Valley, Travis Peak, Cleveland and Olmos formations of Texas (Finley, 1986).

Hydraulic fracturing has contributed to improved gas production from the Cotton Valley sandstone (Upper Jurassic). Circulation losses likely associated with the presence of

TABLE 4–14 Well Test Analysis Results, Well Henry 3

Slope	1.534E+02
Intercept	1.961E+02 MMpsi^2/cp
Dp or Dm(p) at 1 hour	–1.689E+02 MMpsi^2/cp
Extended Pressure	2.369E+03 psia
Total Transmissibility	4.027E+01 mD-ft/cp
Oil Transmissibility	0.000E–01 mD-ft/cp
Gas Transmissibility	4.027E+01 mD-ft/cp
Water Transmissibility	0.000E–01 mD-ft/cp
Oil Reservoir Capacity	0.000E–01 mD-ft
Gas Reservoir Capacity	5.005E–01 mD-ft
Water Reservoir Capacity	0.000E–01 mD-ft
Total Mobility	4.027E+00 mD/cp
Oil Mobility	0.000E–01 mD/cp
Gas Mobility	4.027E+00 mD/cp
Water Mobility	0.000E–01 mD/cp
Effective Permeability to Oil	0.000E–01 mD
Effective Permeability to Gas	5.005E–02 mD
Effective Permeability to Water	0.000E–01 mD
Skin Factor	–4.491E+00
Pressure Drop due to skin	–2.604E+03 psia
Total Ideal Productivity Index	6.627E–02 Res bbl/D/psia
Total Actual Productivity Index	1.398E–01 Res bbl/D/psia
Total Flow Efficiency	2.109E+00
Oil Ideal Productivity Index	0.000E–01 STB/D/psia
Oil Actual Productivity Index	0.000E–01 STB/D/psia
Oil Flow Efficiency	0.000E–01
Gas Ideal Productivity Index	1.717E–02 MSCF/D/psia
Gas Actual Productivity Index	3.621E–02 MSCF/D/psia
Gas Flow Efficiency	2.109E+00
Radius of Investigation	9.730E+01 ft
Hydrocarbons-in-Place based on Radius of investigation (Meaningful only during the transient flow period)	
Oil-in-Place (OOIP)	0.000E–01 STB
Gas-in-Place(OGIP)	9.498E+02 MSCF
Late time slope	0.377E+03
Late time intercept	0.348E+03 MMps
Early time slope	0.153E+03
Early time intercept	0.196E+03 MMps
Fault distance by Davis and Hawkins	0.188E+02 ft
Fault distance by Standing	0.204E+02 ft

natural fractures occur in some wells. Contribution of natural fractures in the Travis Peak formation (Lower Cretaceous) is not well known but according to Finley (1986) is considered to be minimal.

The Cleveland sandstone (Pennsylvanian) of the Anadarko basin produces gas from thin, distal, deltaic facies. There is no clear evidence of natural fracturing. The Olmos formation (Upper Cretaceous) contains fine-grained to very fine-grained silty sandstones within massive shales. The extent of natural fracturing is unknown.

ROCKY MOUNTAIN REGION

Niobrara Formation. The Upper Cretaceous Niobrara formation in the Denver Basin produces biogenic gas from thermally immature, organic-rich chalk beds.

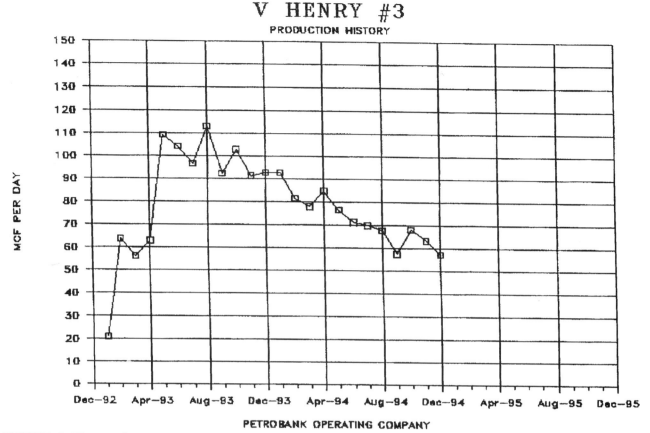

FIGURE 4–58 Production history, well Henry #3.

At shallow depths, fault-related natural fractures are probably essential for commercial gas production. Chalk Porosities are high (up to 50%) but permeabilities are low (< 0.5 md) and hydraulic fracturing is normally required to initiate commercial gas production (Pollastro and Scholle, 1986).

More to the west of the Denver Basin, oil (rather than gas) is produced from much tighter but more fractured chalks. It is estimated that Niobrara chalk matrix porosities are less than 10% and effective permeabilities to gas are less than 0.01 md at depths greater than 6000 ft. Natural fractures at Niobrara not only allow oil production from reservoirs that otherwise would be non-commercial but also provide some storage capacity.

Initial Niobrara wells were drilled with air in 1972 at the Beecher Island field, Colorado. Following open-hole completions without stimulation the wells produced 20 to 60 Mscfd and cumulative gas productions were in the order of 50 MMscf. Later on, foam fracturing led to improved recoveries with initial rates ranging from 100 to 1200 Mscfd and cumulative productions in excess of 200 MMscf (Pollastro and Scholle, 1986).

J (Muddy) and Codell Sandstones. The J (Muddy) sandstone of the Cretaceous period is the main oil producing reservoir in the Denver Basin (Weimer et al., 1986). The porosity is mostly intergranular and there are minor amounts of microporosity. Some secondary porosity is present due to dissolution of feldspars and lithic rock fragments. Furthermore there is some natural fracturing developed late in the diagenetic process.

At the Wattenberg field (Colorado) the J sandstone extends over an area of approximately 600,000 acres. Production depths range between 7600 and 8400 ft. The average water saturation is 44%. Porosities range from 8% to 12% and permeabilities range from 0.05 to 0.005 md. Cumulative production from Wattenberg wells developed in 320-acre spacing is usually less than 400 MMscf per well. The typical production decline is 50% for the first one or two years and subsequently 10% to 20%/yr through out the remaining life of the well.

The Codell (Cretaceous) is a bioturbated, silty, shaly and very fine-grained sandstone. Structural movements have led to tensional natural fracturing which significantly enhances reservoir performance (Weimer et al, 1986).

The Federal Energy Regulatory Commission has designated the Codell sandstone as a tight gas reservoir. Early production from the Codell was obtained through natural fractures at the Boulder field. Usual spacing for oil wells is 80 acres and for gas wells 160 acres. Porosities range between 12% and 20%.

Both the J and the Codell sandstones are stratigraphic traps. They are similar to the deep basin traps in Alberta described by Masters (1979). In these traps there is no downdip water. However, water might be found updip due to small permeability and capillary pressure variations caused by diagenesis. Figure 4–59 presents a schematic published by Brown et al (1986) illustrating the capillary pressure gas trap model.

Piceance Basin. Studies of tight formations in the Piceance Basin have been conducted by Brown et al (1986) and Pitman and Sprunt (1986). The most productive is the Piceance Creek field located on the Piceance Creek anticline. The best production is obtained from naturally fractured rocks within structurally closed areas. Without natural fractures the very low matrix permeability would not allow commercial production.

Pitman and Sprunt (1986) studied the distribution of natural fractures in the Wasatch formation and Mesaverde Group and found that the largest degree of natural fracturing occurs in gas fields along the eastern margin of the basin. This pervasive fracturing might be the result of close proximity to the White River uplift. Lithologically, they found that fracturing is more intensive in sandstone dominated sequences. Typically the fractures are vertical or subvertical and terminate at sand-shale interfaces. Once in a while they terminate within the same sandstone bed.

Fractures in shales tend to be non-mineralized and closed. Usually they are polished and slickensided. It is likely that under net overburden conditions the permeability of these fractures is close to zero.

UTAH

Pitman et al (1986) have studied the hydrocarbon potential of Non-marine Upper Cretaceous and Lower Tertiary rocks in the eastern Uinta Basin of Utah. Cores of the Tuscher formation (Cretaceous) at the Natural Buttes field contain incipient hairline fractures,

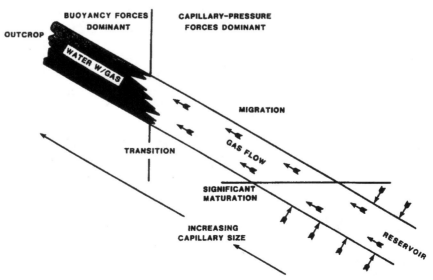

FIGURE 4–59 Capillary pressure gas trap model (after Brown et al, 1986)

while the Wasatch formation did not show any fracturing. It is interesting that at the Tuscher formation fractures are best developed in shale and are rarely observed in sandstone.

Fractures in shale are parallel or at low angles to bedding. It is speculated that they were formed during burial as a result of high pore-fluid pressures or during a period of regional uplift and erosional unloading. The gas is stratigraphically trapped in the lenticular, low-permeability, diagenetically modified sandstones.

SUMMARY

Problems associated with characterization of tight reservoirs have been discussed in an excellent paper by Lee and Hopkins (1994). One of the main problems is that more often than not, economic considerations do not allow sufficient data collection for rigorous reservoir characterization. As a consequence geologists and engineers must find a balance between the amount of data they can collect to properly describe the reservoir and the economics of the play.

Basically all wells in tight formations require hydraulic stimulation to establish commercial production. It has been found historically, however, that the usual reservoir characterization lumping all parameters into a single layer lead to very optimistic forecasts of reservoir performance. This is due to the complexity of tight reservoirs which are often composed of multiple layers with varying permeabilities and sometimes natural fractures. Lee and Hopkins (1994) have shown that when all layers are taken into account the performance forecasts are closer to the actual production results.

REFERENCES

Aguilera, Roberto. "Analysis of Naturally Fractured Reservoirs from Conventional Well Logs." *Journal of Petroleum Technology* (July 1976), 764–772.

Aguilera, Roberto. "Log Analysis of Gas-Bearing Fracture Shales in the St. Lawrence Lowlands of Quebec." Paper SPE 7445 presented at the 53rd Annual Fall Meeting of SPE of AIME, Houston (October 1–4, 1978).

Aguilera, R. and H.K. van Poollen. "Fractured Shales Present an Attractive Gas Potential." *Oil and Gas Journal* (March 5, 1979) 167–178.

Aguilera, Roberto. "Exploring for Naturally Fractured Reservoirs." SPWLA Twenty-Fourth Annual Logging Symposium (June 27–30, 1983).

Aguilera, Roberto. "Well Test Analysis of Naturally Fractured Reservoirs." *SPE Formation Evaluation* (September 1987), 239–252.

Aguilera, Roberto. "Undiscovered Naturally Fractured Reservoirs, Why and How?" *The Petroleum Explorer* (December 1992) 8–11.

Al-Shaieb, Z. and P. Walker. "Evolution of Secondary Porosity in Pennsylvanian Morrow Sandstones, Anadarko Basin, Oklahoma." in *Geology of Tight Gas Reservoirs*, edited by C.W. Spencer and R.F. Mast, AAPG Studies in Geology #24 (1986), 45–67.

Archie, G.E. "The Electrical Resistivity log as an Aid in Determining Some Reservoir Characteristics." *Transactions*. AIME 146 (1942), 54.

Arps, J.J. "Analysis of Decline Curves." *Transactions*. AIME 160 (1945), 228–247.

Bagnal, W.D., and W.M. Ryan. "The Geology, Reserves, and Production Characteristics of the Devonian Shales in Southwestern West Virginia." Proceeding of the Seventh Appalachian Petroleum Geology Symposium, Morgantown, West Virginia (March 1–4, 1976), 41.

Bass, D.M. "Decline Curve Evaluation." Petroleum Engineering Department, Colorado School of Mines (circa 1973).

Beirs, R.J. "Quebec Lowlands: Overview and Hydrocarbon Potential." Proceeding of the Seventh Appalachian Petroleum Geology Symposium, Morgantown, West Virginia (March 1–4, 1976), 142.

Blanton, T.L. "An Experimental Study of Interaction Between Hydraulically Induced and Pre-Existing Fractures." SPE/DOE paper 10847 presented at the Unconventional Gas Recovery Symposium held in Pittsburgh, PA (May 16–18, 1982).

Brown, C.A. et al. "Southern Piceance Basin Model-Cozette, Corcoran and Rollins Sandstones." in *Geology of Tight Reservoirs*, edited by C.W. Spencer and R.F. Mast, AAPG Studies in Geology #24 (1986), 207–219.

Carlson, E.S. and J.C. Mercer. "Devonian Shale Gas Production: Mechanisms and Simple Models." *Journal of Petroleum Technology* (April 1991), 476–482.

Chase, R.W. "Degassification of Coal Seams Via Vertical Bore Hole: A Field and Computer Simulation Study." Ph.D. Dissertation. Penn. State University (circa 1980).

Clark, J.A. "Glacial Loading: A Cause of Natural Fracturing and a Control of the Present Stress State in Regions of High Devonian Shale Gas Production." SPE/DOE paper 10798 presented at the Unconventional Gas Recovery Symposium held in Pittsburgh, PA (May 16–18, 1982).

Cook, T.L. et al. "Tracer Experiments in Eastern Devonian Shale." SPE/DOE paper 10797 presented at the Unconventional Gas Recovery Symposium held in Pittsburgh, PA (May 16–18, 1982).

Coulter, G.R. "Hydraulic Fracturing—A Summary." Proceeding of the Seventh Appalachian Petroleum Geology Symposium, Morgantown, West Virginia (March 1–4, 1976), 115.

Craft, B.C., W.R. Holden, and E.D. Graves. *Well Design: Drilling and Production.* Prentice Hall, Inc. (1962), 485.

De Witt, W. "Devonian Gas Bearing Shales in the Appalachian Basin." *Geology of Tight Gas Reservoirs,* edited by C.W. Spencer and R.F. Mast, AAPG Studies in Geology #24, (1986), 1–8.

De Wys, J.N., and R.C. Schumaker. "Pilot Study of Gas Production Analysis Methods Applied to Cottageville Field." Morgantown Energy Research Center (July 1978).

Finley, R.J. "An Overview of Selected Blanket-Geometry, Low-Permeability Gas Sandstones in Texas." In *Geology of Tight Gas Reservoirs,* edited by C.W. Spencer and R.F. Mast, AAPG Studies in Geology #24, (1986), 69–85.

Flower, J.G. "Use of Sonic Shear Wave–Resistivity Overlay as a Quick Look Method for Identifying Potential Pay Zones in the Ohio (Devonian) Shale." SPE paper 10368 presented at the Eastern Regional Meeting held in Columbus, Ohio (November 4–6, 1981).

Fox, J.N. and C.D. Martiniuk. "Reservoir Characteristics and Petroleum Potential of the Bakken Formation, Southwestern Manitoba." *The Journal of Canadian Petroleum Technology* (October 1994), vol. 37, No. 8, 19–27.

Frantz, J.H. "Reservoir and Stimulation Evaluation of the Berea Sandstone Formation in Pike County, Kentucky." SPE paper 25896 presented at the SPE Rocky Mountain Regional/Low Permeability Reservoirs Symposium held in Denver, Co. (April 12–14, 1993).

Frohne, K.H., and J.C. Mercer. "Fracture Shale Gas Reservoir Performance Study. An Offset Well Interference Field Test." *Journal of Petroleum Technology* (February 1984), 291–300.

Fucuk, F., J. Alam, and D.L. Streib. "Reservoir Engineering Aspects and Resource Assessment Methodology of Eastern Devonian Gas Shales." Science Applications, Inc. (October 5, 1978).

Gatens, J.M. et al. "Analysis of Eastern Devonian Gas Shales Production Data." *Journal of Petroleum Technology* (May 1989), 519–525.

Gaymard, R. and A. Poupon. "Response of Neutron and Formation Density Logs in Hydrocarbon-Bearing Formations." *The log Analyst* (Sept–Oct., 1968) 9, No. 5.

Glohi, B.V. "Petrography of Upper Devonian Gas Bearing Sandstones in the Indiana 7 1/2-minute Quadrangle, Well #11237, Ind 25084, Indiana County, Pennsylvania." University of Pittsburgh, Unpublished Master's Thesis (1984), 143 p.

Hashmy, K.H. et al. "Quantitative Log Evaluation in the Devonian Shales of the Northeast United States." SPE/DOE paper 10795 presented at the Unconventional Gas Recovery Symposium held in Pittsburgh, PA (May 16–18, 1982).

Hashmy, K.H. et al. "Computerized Shale Evaluation Technique—COMSET." SPWLA Twenty Third Annual Logging Symposium (July 6–9, 1982).

Hazlett, W.G. et al. "Production Data Analysis Type Curves for the Devonian Shales." Paper SPE 15934 presented at the SPE Eastern Regional Meeting held in Columbus, Ohio (November 12–14, 1986).

Hilchie, D.W., and S.J. Pirson. "Water Cut Determination from Well Logs in Fractured and Vuggy Formations." *Transactions.* SPWLA, Dallas (1961).

Hilton, Jon. "Wireline Evaluation of the Devonian Shale: A Progress Report." Presented at First Eastern Gas Shales Symposium, Morgantown Energy Research Center, ERDA (October 17–19, 1977).

Holgate, K.E. et al. "Analysis of Prestimulation Test Data in Devonian Shale Reservoirs." *SPE Formation Evaluation* (June 1988), 364–370.

Hubbert, M.K., and D.G. Willis. "Mechanics of Hydraulic Fracturing." *Transactions.* AIME 210 (1957), 153–166.

Hunter, C.D. and D.M. Young. "Relationships of Natural Gas Occurrence and Production in Eastern Kentucky (Big Sandy Gas Field) to Joints and Fractures in Devonian Bituminous Shale." *AAPG Bulletin* (February 1953), vol. 37, No. 2, 282–299.

Jochen, J.E. et al. "Quantifying Layered Reservoir Properties with a Novel Permeability Test." paper SPE 25864 presented at the SPE Rocky Mountain Regional/Low Permeability Reservoir Symposium, Denver, Co (April 26–28, 1993).

Kazemi, H. "Pressure Transient Analysis of Naturally Fractured Reservoirs with Uniform Fractured Distribution." *Society of Petroleum Engineers Journal* (December 1969), 461–462.

King, E.E. and Fertl, W.H. "Evaluating Shale Reservoir Logs." *The Oil and Gas Journal* (March 26, 1979), 166–168.

Kubik, W. and P. Lowry. "Fracture Identification and Characterization Using Cores, FMS, CAST, and Borehole Camera: Devonian Shale, Pike County, Kentucky." Paper SPE 25897 presented at the SPE Rocky Mountain Regional/Low Permeability Reservoir Symposium, Denver, Co (April 26–28, 1993).

Lancaster, D.E. et al. "A Case Study of the Evaluation, Completion, and Testing of a Devonian Shale Gas Well." *Journal of Petroleum Technology* (May 1989), 509–518.

Landes, K.K. *Petroleum Geology.* 2nd ed. John Wiley and Sons, Inc. (1959).

Laughrey, C.C. and J.A. Harper. "Comparisons of Upper Devonian and Lower Silurian Tight Formations in Pennsylvania—Geological and Engineering Characteristics." In *Geology of Tight Gas Reservoirs,* edited by C.W. Spencer and R.F. Mast, AAPG Studies in Geology #24 (1986), 1–8.

Lee, B.O. et al. "Evaluation of Devonian Shale Reservoir Using Multi-Well Pressure Transient Testing Data." SPE/DOE paper 10838 presented at the Unconventional Gas Recovery Symposium held in Pittsburgh, PA (May 16–18, 1982).

Lee, W.J. and C.W. Hopkins. "Characterization of Tight Reservoirs." *Journal of Petroleum Technology* (November 1994), 956.

Luffel, D.L. et al. "Evaluation of Devonian Shale with New Core and Log Analysis Methods." *Journal of Petroleum Technology* (November 1992), 1192–1197.

Martin, P., and E.B. Nuckols. "Geology and Oil and Gas Occurrence in the Devonian Shales: Northern West Virginia." Proceeding of the Seventh Appalachian Petroleum Geology Symposium, Morgantown, West Virginia (March 1–4, 1976), 20.

Masters J.A. *Elworth—Case Study of a Deep Basin Gas Field*, AAPG Memoir 38 (1984), 315 p.

McIver, R.D. et al. "Location of Drilling Sites in the Devonian Shales by Aerial Photography." SPE/DOE paper 10794 presented at the Unconventional Gas Recovery Symposium held in Pittsburgh, PA (May 16–18, 1982).

Myung, J.I. "Fracture Investigation of the Devonian Shale Using Geophysical Well Logging Techniques." Proceeding of the Seventh Appalachian Petroleum Geology Symposium, Morgantown, West Virginia (March 1–4, 1976), 212.

Neal, D.W. "Subsurface Stratigraphy of the Middle and Upper Devonian Classic Sequence in Southern West Virginia and Its Relation to Gas Production." West Virginia University, Department of Geology and Geography, Ph.D. Dissertation, Morgantown (1979) 144 p.

Nelson, R.A. *Geologic Analysis of Naturally Fractured Reservoirs*. Gulf Publishing Company, Houston, Texas (1985).

Ning, X. et al. "The Measurements of Matrix and Fracture Properties in Naturally Fractured Cores." Paper SPE 25898 presented at the SPE Rocky Mountain Regional/Low Permeability Reservoirs Symposium, Denver, Co. (April 26–28, 1993).

Norton, J.L. "Stimulation of the Devonian Shale." Proceeding of the Seventh Appalachian Petroleum Geology Symposium, Morgantown, West Virginia (March 1–4, 1976), 173.

Overbey, W.K. "Present Status on the Eastern Gas Shales Project." Second Eastern Gas Shales Symposium, Morgantown, West Virginia (October 1978) 1.

Peterson, B.E. "Fracture Production from Mancos Shale, Rangely Field, Rio Blanco County, Colorado." *Bulletin of American Association of Petroleum Geologists* (April 1955), 39, 532.

Pickett, G.R. "Pattern Recognition as a Means of Formation Evaluation." *Transactions*. SPWLA, 14th Annual Logging Symposium (May 6–9, 1973).

Pitman, J.K. and E.S. Sprunt. "Origin and Distribution of Fractures in Lower Tertiary and Upper Cretaceous Rocks, Piceance Basin, Colorado, and their Relation to the Occurrence of Hydrocarbons." In *Geology of Tight Gas Reservoirs*, edited by C.W. Spencer and R.F. Mast, AAPG Studies in Geology #24 (1986), 221–233.

Pitman, J.K. et al. "Hydrocarbon Potential of Non-Marine Upper Cretaceous and Lower Tertiary Rocks, Eastern Uinta Basin, Utah." In *Geology of Tight Gas Reservoirs*, edited by C.W. Spencer and R.F. Mast, AAPG Studies in Geology #24 (1986), 235–251.

Pollastro, R.M. and P.A. Scholle. "Exploration and Development of Hydrocarbons from Low-Permeability Chalks—An Example from the Upper Cretaceous Niobrara Formation, Rocky Mountain Region." In *Geology of Tight Gas Reservoirs*, edited by C.W. Spencer and R.F. Mast, AAPG Studies in Geology #24 (1986), 129–141.

Poupon, A. et al. "Log Analysis of Sand-Shale Sequences—A Systematic Approach." *Journal of Petroleum Technology* (July 1970).

Poupon, A. et al. "Log Analysis in Formations with Complex Lithologies." *Journal of Petroleum Technology* (August 1971).

Provo, L.J. "Upper Devonian Black Shale—Worldwide Distribution and What It Means." Proceeding of the Seventh Appalachian Petroleum Geology Symposium, Morgantown, West Virginia (March 1–4, 1976), 1.

Pullé, C.V. "A Hydrodynamic Analogy of Production Decline for Devonian Shale Wells." SPE/DOE paper 10837 presented at the Unconventional Gas Recovery Symposium held in Pittsburgh, PA (May 16–18, 1982).

Rabshevsky, G.A. "Optical Processing of Remote Sensor Imagery." Proceeding of the Seventh Appalachian Petroleum Geology Symposium, Morgantown, West Virginia (March 1–4, 1976), 100.

Regan, L.J. "Fracture Shale Reservoirs of California." *Bulletin of American Association of Petroleum Geologists* (February 1953), 37, 201–216.

Roth, E.E. "Natural Gases of the Appalachian Basin." *Natural Gases of North America*, Beebe, B.W., Ed., AAPG (1968).

Ruhovets, N. and W.H. Fertl. "Volumes, Types, and Distribution of Clay Minerals in Reservoir Rocks Based on Well Logs." SPE/DOE paper 10796 presented at the Unconventional Gas Recovery Symposium held in Pittsburgh, PA (May 16–18, 1982).

Ruotsala J.E. and R.T. Williams. "Reflection Seismology as an Exploration Tool for Fractured Zones in Gas-Bearing Shale—Cottageville, West Virginia: A Case Study." SPE/DOE paper 10792 presented at the Unconventional Gas Recovery Symposium held in Pittsburgh, PA (May 16–18, 1982).

Ryan, W.M. "Remote Sensing Fracture Study—Western Virginia and Southeastern Kentucky." Proceeding of the Seventh Appalachian Petroleum Geology Symposium, Morgantown, West Virginia (March 1–4, 1976), 94.

Sanchez, Mario. Personal communication. SOQUIP (October 1977).

Sawyer, W.K., and C.D. Locke. "General Purpose Diffusivity Model for Fluid Flow and Heat Conduction in Porous Media." Morgantown Energy Research Center (January 1976).

Sawyer, W.K., J.C. Mercer, and K.H. Frohne. "Prediction of Fracture Extent by Simulation of Gas Well Pressure and Production Behavior." Morgantown Energy Research Center, ERDA (October 1976).

Schettler, P.D., et al. "The Relationship of Thermodynamic and Kinetic Parameters to Well Production in Devonian Shale." First Eastern Gas Shales Symposium, Morgantown, West Virginia (October 17–19, 1977), 450–461.

Schlumberger. *Log Interpretation Principles*. (1972), 39.

Shaw, J.S. et al. "Reservoir and Stimulation Analysis of a Devonian Shale Gas Field." *SPE Production Engineering* (November 1989), 450–458.

Shea, C. and D.H. Bucher. "Foam Fracturing of the Upper Devonian Benson Formation in Central West Virginia." SPE/DOE paper 10826 presented at the Unconventional Gas Recovery Symposium held in Pittsburgh, PA (May 16–18, 1982).

Schumaker, R.C. "A Detailed Geologic Study of Three Fractured Devonian Shale Gas Fields in the Appalachian Basin." SPE/DOE paper 10791 presented at the Unconventional Gas Recovery Symposium held in Pittsburgh, Pa (May 16–18, 1982).

Shumaker, R.C. "A Digest of Appalachian Structural Geology." Proceeding of the Seventh Appalachian Petroleum Geology Symposium, Morgantown, West Virginia (March 1–4, 1976), 75.

Smith, E.C. "A Practical Approach to Evaluating Shale Hydrocarbon Potential." Second Eastern Gas Shales Symposium, Morgantown, West Virginia (October 1978), 73.

Sterns, D.W., M. Friedman, R. Nelson, and R. Aguilera. *Fractured Reservoir Analysis.* AAPG School Notes, Great Falls, Montana (1994).

Swartz, G. and U. Ahmed. "Optimization of Hydraulic Fracturing Techniques for Eastern Devonian Shales." SPE/DOE 10848 presented at the Unconventional Gas Recovery Symposium held in Pittsburgh, PA (May 16–18, 1982).

USGS. "Resource Appraisal for the Devonian Shale and the Appalachian Basin." Open file report No. 82–474, USGS, Reston, Va (May 1982) 35.

Tegland, E.R. "Geophysical Investigations Related to the Devonian Shales of the Eastern U.S." Proceeding of the Seventh Appalachian Petroleum Geology Symposium, Morgantown, West Virginia (March 1–4, 1976), 104.

Timmerman, E.H. "Analysis of a Tight Gas Well Flow Test." *Petroleum Engineer* (December 1977–January 1978).

Truman, R.B. and R.L. Campbell. "Devonian Shale Well Log Interpretation." ResTech Houston final Report to GRI, Contract 5083–213–1390 (April 1987).

Venorsdale, C.R. "Evaluation of Devonian Shale Gas Reservoirs." *SPE Reservoir Engineering* (May 1987), 209–216.

Voneiff, G.W. and J.M. Gatens. "The Benefits of Applying Technology to Devonian Shale Wells." Paper SPE 26890 presented at the SPE Eastern Regional Meeting, Pittsburgh, PA (November 2–4, 1993)

Warren, J.E. and P.J. Root. "The Behavior of Naturally Fractured Reservoirs." *Society of Petroleum Engineers Journal* (September 1963), 245–255.

Watson, A.T. and J. Mudra. "Characterization of Devonian Shales with X-Ray Computed Tomography." *SPE Formation Evaluation* (September 1994), 209–212.

Weimer, R.J. et al. "Wattenberg Field, Denver Basin, Colorado." in *Geology of Tight Gas Reservoirs*, edited by C.W. Spencer and R.F. Mast, AAPG Studies in Geology #24 (1986), 143–164.

Wyllie, M.R.J., A.R. Gregory, and J.W. Gardner. "Elastic Wave Velocities in Heterogeneous Porous Media." *Geophysics* 21 (January 1956), 1, 41.

Yost, A.B. et al. "Techniques to Determine Natural and Induced Fracture Relationships in Devonian Shale." *Journal of Petroleum Technology* (June 1982), 1371–1377.

Zielinski, R.E. and R.D. McIver. "Synthesis of Organic Geochemical Data From the Eastern Gas Shales." SPE/DOE paper 10793 presented at the Unconventional Gas Recovery Symposium held in Pittsburgh, PA (May 16–18, 1982).

Zielinski, R.E. et al. "The GRI/Industry Eastern Gas Shale Data Base." SPE/DOE paper 10823 presented at the Unconventional Gas Recovery Symposium held in Pittsburgh, PA (May 16–18, 1982).

Chapter 5

Case Histories

The geologic characteristics of naturally fractured reservoirs found in sandstones, carbonates, shales, cherts, basement rocks, and coal seams were discussed in Chapter 1. This chapter will consider the case histories of fractured reservoirs in some of these lithologies.

FRACTURED SANDSTONES

Spraberry Field, USA (Oil, Underpressured)

The Spraberry field is located in the Midland Basin, Texas. It is cut by regional fractures and is considered to be a reservoir of Type A. Very low oil recoveries (< 10% OOIP) are attributed to low reservoir energy as the reservoir is underpressured (0.33 psi/ft) and capillary end effects.

The geologic aspects were discussed originally by Wilkinson (1953) who indicated that the main producing structure was a fracture permeability trap on a homoclinal fold made out of alternate layers of sand, siltstone, shales, and limestone. Oil was stored primarily in the sandstone matrix, and paper-thin fractures provided channels for conducting the oil to the wellbore.

More recently, Guevara (1988) and Tyler and Gholston (1988) presented a detailed geological characterization of the Permian submarine fan reservoirs of the Driver Waterflood Unit in the Spraberry Field. Guevara (1988) examined more than 350 well logs and substantial amounts of core and production data to characterize the reservoir and to determine the relationship between oil recovery and reservoir stratigraphy. From his study, Guevara (1988) concluded that it was possible to increase oil recovery by the proper selection of completion intervals and strategic infill drilling. The importance of natural fractures on oil production was stressed throughout the study. The possibility that oil migration occurred through vertical fractures from the underlying Wolfcamp source rocks was indicated. Another possibility is that there are oil-generating intervals within the Spraberry.

Mardock & Myers (1951) and Lyttle & Ricke (1951) discussed well logging activities. The authors considered formation evaluation by means of radioactive and induction logs, techniques which provided the means for distinguishing lithologies. However, quantitative analysis was not possible.

Dyes and Johnstone (1953) presented buildup curve analysis for this reservoir in 1953. They found that the effective permeability of the Spraberry field determined from buildup tests was larger than the matrix permeability determined from cores. They concluded that a large part of the fractured systems noted in the cores was native to the formation. Dyes and Johnstone (1953) noticed only one straight line in a conventional semilog plot of buildup pressure vs. time. This is opposed to some situations in naturally fractured reservoirs where an "S" shape buildup or two parallel straight lines are found.

Elkins and Skov (1963) analyzed pressure interference of the Spraberry in 1960. They assumed anisotropic permeability and considered that the pressure reduction due to production expanded in an elliptical form with the length-width ratio varying as the square root of the ratio of permeability along and at right angles to the fracture trend.

The technique consisted of solving the interference equation on a trail and error basis by assuming effective compressibility of rock and fluids and permeabilities in the X and Y directions until good match between calculated and measured pressures was obtained.

The fracture orientation for the Spraberry as determined by pressure analysis and fluid injection was good, leading to the conclusion that anisotropic permeability is real, and that it must be included in analysis of pressure transients in naturally fractured reservoirs.

Christie and Blackword (1952) published detailed history of the Spraberry field, which included production data, reservoir fluid properties, bottom hole pressures, flow test data, and production methods. They indicated that fractures served as the principal flow channels, and that the matrix contained the principal oil storage volume (naturally fractured reservoir of Type A). They referred to lifting problems associated with small oil production and unusually high volumes of gas.

Elkins (1952) has presented an excellent study on reservoir performance and well spacing in the Spraberry. The paper discussed history, geology, drilling, completion practices, sand properties, vertical fractures analysis, oil properties at reservoir conditions, initial oil-in-place calculations, measurement and interpretation of initial pressures in wells, interference tests, gas-oil ratios, productivity indices, general reservoir performance, production history, decline in well productivity, pressures and recoveries.

From the study it was concluded that the Spraberry oil was contained primarily in the matrix. Flow channels were provided by paper thin fractures. It was further concluded that a spacing of 160 acres was efficient enough to deplete the reservoir. Capillary "end effects" in the small fracture blocks limited the recovery to only a small percentage of the initial oil-in-place.

Brownscombe and Dyes (1952) researched the possibilities of successfully using imbibition displacement in the Spraberry field. During imbibition part of the injected water soaks into the matrix and releases oil that becomes part of the flow string.

Elkins and Skov (1963) described a cyclic operation carried out in Spraberry. The process consisted of a period of water injection followed by a period of production without water injection. Then the cycle was repeated. They indicated that this technique had provided for oil recoveries at least 50% faster and with lower water cuts than with the imbibition method.

An important part of the Spraberry history has been the hydraulic fracturing required for obtaining commercial production rates. Hoel (1988) has indicated that cumulative oil production data demonstrate that crosslinked gelled water provides the optimum fracture fluid for the Spraberry. He indicated that large volume treatments provide the best results and that large proppant concentrations are not required. His conclusions were based on the study of 100 wells drilled on 80 and 160-acre spacing. The wells were distributed in four areas.

Barba (1989) indicated that 3,000 new wells were drilled in the Spraberry between 1982 and 1986. All of these wells were hydraulically fractured. Barba used a group of wells drilled in Midland County and data from geologic mapping, well logs, cores, post-hydraulic fracture pressure transient data, production declines, and well and completion-cost data to determine the optimum fracture length that maximizes the internal rate of return.

White (1989) described the drilling and completion of a horizontal well in the Spraberry field. The well was drilled in the direction of the least principal stress in such a way that the multiple fractures would radiate at right angles from the horizontal well. Multiple hydraulic fracture treatments were carried out. Isolation for the fracturing jobs was achieved by a cemented liner, tubing conveyed perforating and bridge plugs. Another horizontal well experience was published by Edlund (1987) who indicated that the well was designed to intersect as many vertical natural fractures as possible.

Blauer et al (1992) evaluated the production performance of 120 Spraberry wells hydraulically fractured with different quantities of sand. Estimates of formation permeability, hydraulic fracture length and future recoveries were obtained with the use of a production decline type curve developed by Locke and Sawyer (1975). From the study it was concluded that wells fractured with 400,000 to 450,000 pounds of sand produced a cumulative average of 76,200 stbo and 190.6 MMscf of gas. On the other hand fracturing jobs carried out with 300,000 to 320,000 pounds of sand resulted in wells producing an average of 49,000 stbo and 129.1 MMscf of gas. Furthermore it was concluded that each Spraberry well had a unique production decline curve which was strongly influenced by the hydraulic fracture.

Uinta Basin, USA (Oil, Overpressured)

The Uinta Basin is located in the State of Utah. It is limited to the north by the Uinta Mountains, to the east by the Douglas Creek Arch, to the southeast by the Uncompahgre Uplift, to the southwest by the San Rafael Swell, and to the west by the Wasatch Mountains and Plateau (Figure 5–1). Production is obtained from reservoirs of Type A cut by regional fractures. Thus from this point of view the reservoirs are similar to the Spraberry. Recoveries, however, are larger (>30%) than in the Spraberry due to a large amount of energy as the reservoirs are overpressured (0.7 to 0.9 psi/ft).

The Uinta Basin trends east-west with an axis approximately parallel to the Uinta Mountains. Sediments are of Tertiary age and consist of fluvial and lacustrine deposits that contain abundant clay minerals.

Drilling problems in the Altamont trend and Bluebell field arose from lost circulation and abnormal pressures coupled with difficult weather, rugged terrain, and remoteness (Findley, 1972). Figure 1–37 illustrated the complex geology of the Tertiary formations. Pressure gradients ranged between 0.7–0.9 psi/ft in the Altamont trend.

The large amount of drilling problems led to the design of a complete system of methods and equipment which greatly improved drilling efficiency. Findley (1972) has summarized the methods and equipment as follows:

1. A carefully-designed casing program
2. Degasser
3. Mud gas separator
4. Pump stroke counter
5. Pit volume totalizer and warning device
6. Flow-line sensor and warning device
7. Upper and lower Kelly cocks
8. Inside BOPs, well design BOP stack and minifold, and regular testing of BOP equipment
9. In and out mud weight recording

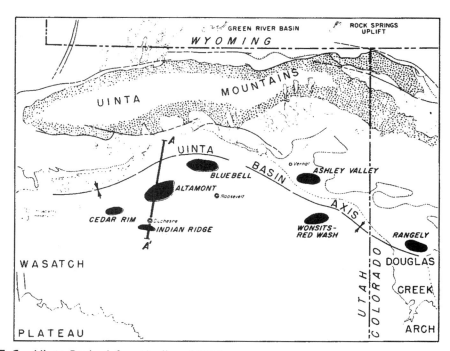

FIGURE 5–1 Uinta Basin (after Findley, 1972)

10. Hydraulic choke
11. Penetration rate recorder
12. Crew education in blowout detection and control

Baker and Lucas (1972) indicate that over 90% of the fractures are approximately vertical. Many of these fractures are partially to completely filled with calcite. Fracture vertical length averages less than one foot.

Flow rates of over 1,000 bo/d on ½ in. choke indicate that performance is a function of fracture permeability. Conventional core analysis has indicated matrix permeabilities of a fraction of millidarcy. However, buildup analysis has resulted in effective permeabilities between 14 and 18 md (Aguilera, 1973), indicating that part of the fractures observed in cores are native to the formation.

Narr and Currie (1982) have presented an excellent study dealing with the origin of fracture porosity at the Altamont trend. Initially, they computed a stress history for the Altamont strata. The calculations indicated that the fractures probably formed in extension. The well-cemented brittle rocks are the most likely to fracture. Fractures probably started to develop only after the strata was buried to great depth. They continued developing when the strata was uplifted as the overburden was eroded.

Figure 5–2 presents stress history curves for Altamont reservoir strata. Data used for generating these curves are presented in Table 5–1. This presentation dealing with the origin of fracture porosity in the Altamont trend follows very closely the original work of Narr and Currey (1982).

General trends in stress history are demonstrated by Figure 5–2a. Geostatic load or pressure, the vertical total stress (S_z), increases continuously as the rock is buried. The S_z curve for erosional unloading is different from the path followed during burial due to the density increases of rocks as they are buried to greater depths. Erosion strips off the lightest strata first, which results in an increase of the average density of the remaining rock. Consequently, the total weight of overburden on a rock will be larger when it arrives at any given depth during uplift and erosion than at the same depth during progressive burial.

The maximum horizontal principal effective stress, σ_x, increases during burial, then decreases rapidly during uplift. The minimum horizontal effective stress, σ_y, also increases with depth to a point, then decreases in value with further burial. As the rock approaches its maximum burial depth, σ_y enters the regime of negative effective stress (extension). During uplift and erosional unloading the horizontal effective stresses become increasingly negative.

Although the model does not account for the stress relief or changes in pore pressure that could be associated with fracturing, it does demonstrate that uplift and erosional unloading are accompanied by increasing horizontal effective tension. Under these conditions, it is likely that a system of extension fractures will develop during uplift and erosional unloading.

The stress history model can be used to compare the effects of differences among input parameters. The value of Young's modulus used in constructing the curves of Figure 5–2a is 70,000 MPa and for Figure 5–2b it is 40,000 MPa; otherwise, all input parameters are equal. These ultimate values of Young's modulus represent approximately the maximum and minimum values measured in core from the Altamont reservoir. Comparing Figure 5–2a and b, one notes that rock having a smaller modulus does not develop large compressive stress; however, during unloading the stress on rock with a larger modulus will decrease much more rapidly. It is reasonable to infer from Figure 5–2a that rock of lower Young's modulus might fail in extension earlier in its cycle of loading and unloading, but during uplift and denudation those rocks have a greater elastic modulus (Figure 5–2a) and would develop a more extensive system of extension fractures.

Figure 5–2c shows the stress history of a rock in which fluid pressure remains hydrostatic (0.45 psi/ft) throughout its burial history; otherwise, all model input parameters are equal to those of Figure 5–2a where fluids are overpressured (0.90 psi/ft). In Figure 5–2c the values of σ_x and σ_y do not become negative until substantial uplift and unloading have occurred. These results indicate that high fluid pressure can play a significant role in development of fractures at depth in reservoirs such as Altamont.

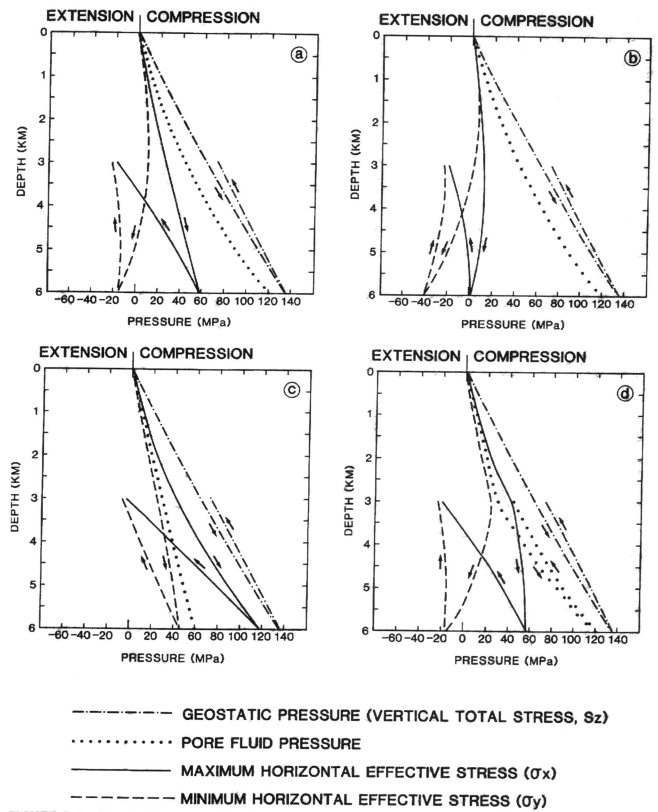

FIGURE 5–2 Stress history curves for Altamont reservoir strata. Input parameters are given in Table 5–1 (after Narr and Currie, 1982)

TABLE 5–1 Parameters Input for Computation of Stress History Curves Shown in Figure 5–2

Burial History Curve (Fig. 5–2)	a	b	c	d
Maximum burial depth (meters)	6,000	6,000	6,000	6,000
Initial density (g/cm³)	2.0	2.0	2.0	2.0
Ultimate density (g/cm³)	2.6	2.6	2.6	2.6
Initial thermal expansion coefficient ($\times 10^{-6}$)	3.0	3.0	3.0	3.0
Ultimate thermal expansion coefficient ($\times 10^{-6}$)	5.4	5.4	5.4	5.4
Initial temperature (°C)	10	10	10	10
Ultimate temperature (°C)	160	160	160	160
Initial Young's modulus (MPa)	100	100	100	100
Ultimate Young's modulus (MPa)	70,000	40,000	70,000	70,000
Initial Poisson's ratio	0.45	0.45	0.45	0.45
Ultimate Poisson's ratio	0.25	0.25	0.25	0.25
Horizontal strain gradient, ε_x ($\mu\varepsilon$/m)	0.4	0.4	0.4	0.4
Horizontal strain gradient, ε_y ($\mu\varepsilon$/m)	0.0	0.0	0.0	0.0
Initial fluid pressure gradient (MPa/m)	0.010	0.010	0.010	0.010
Ultimate fluid pressure gradient (MPa/m)	0.020	0.020	0.010	0.020

Narr and Currie's (1982) final application of the model is to examine how stress history is influenced by differences in the manner in which input geologic and physical parameters may change with depth of burial. The stress history curves of Figure 5–2d were generated using the same initial and ultimate input parameters as those employed for Figure 5–2a. However, for the model depicted by Figure 5–2a, fluid overpressure increases with depth from P = 0.010 MPa/m (0.44/psi/ft) at deposition to P = 0.020 MPa/m (0.90 psi/ft) at maximum depth of 19,700 ft (6,000 m). On the other hand in the model shown on Figure 5–2d, the fluid pressure remains hydrostatic (P = 0.010 MPa/m or 0.44 psi/ft) from deposition to a depth of 9,800 ft (3,000 m). At this depth overpressure starts to develop and reaches 0.02 MPa/m (0.90 psi/ft) at 19,700 ft (6,000 m).

This fluid-pressure path might occur if overpressuring were to result from temperature-dependent processes such as clay dewatering or hydrocarbon generation. Conditions favorable for creation of extension fractures will occur at slightly greater depth in the stress history modeled in Figure 5–2d than in that of Figure 5–2a, but one notes that stress conditions leading to development of a system of extension fractures during uplift are similar in both cases.

In order to evaluate model predictions regarding fracture genesis, it was necessary to study the geologic information associated with the fracturing history of Altamont reservoir rocks. This geologic assessment required an examination of the control that sediment diagenesis exerted over the occurrence of natural fractures. It also required an analysis of temperatures, pressures, and depths at which fractures opened, determined from observation of fluid inclusions in secondary mineralization on fracture surfaces. Fracture orientation was studied in joint patters observed at outcrops and in oriented cores from the reservoir. Some of the pertinent physical properties of the reservoir rock were determined in the laboratory.

The discovery well in the Altamont trend (Shell, Miles 1) was completed in May 1970 and produced an average rate of 1,150 bo/d. Seventy-nine feet of this well were perforated in three intervals between 12,341 and 12,942 ft. However, petrophysical evaluation suggested that only 26 ft (less than 35% of the perforated interval) were contributing to the net pay (Baker & Lucas, 1972).

The Bluebell is the largest field in the Uinta Basin having produced over 118 million barrels of oil from 213 active wells. However, close examination of openhole logs, cores, mud logs and reservoir data indicate that, in general, less than 25% of the perforated intervals contribute to production. This observation is not unusual in many naturally fractured reservoirs.

In the Altamont trend, a perforated completion was preferred to open hole completions in order to have the possibility of selective stimulation.

The difficulty in clearly determining the net pay prevented volumetric calculations of oil-in-place with a good degree of certainty. Baker and Lucas (1972) indicate the reserves estimates in the area depended mainly on conventional crossplots of p vs. cumulative oil production. Care was suggested when using this approach because variations in fracture communication as pressure declined could produce changes in the slope of the p vs. N_p curve.

Initial economic spacing was 640 acres for the Altamont trend. Closer spacing in the nearby Bluebell field proved uneconomic, according to Baker and Lucas (1972). More recently the well spacing in many areas has been reduced to 320 acres.

Total cumulative production of over 378 million barrels have been obtained from naturally fractured sandstones deposited in fluvial dominated deltas in the Palaeocene/Eocene Green River and Wasatch formations. Approximately 25% of the wells have been abandoned and the U.S. Department of Energy (DOE) estimates that at least 1 billion bbls of oil is at risk of being left behind due to poor completion practices and portions of this heterogeneous reservoir that remain undrained. This has led to a study of the Bluebell field, sponsored by the U.S. Department of Energy which is currently underway.

Palm Valley Field, Australia (Gas, Normally Pressured)

The Palm Valley gas field of Australia provides a very interesting case history that has been discussed by various authors including McNaughton and Garb (1975); Strobel, Gulati and Ramey (1976); Sabet and Franks (1985): Do Rozario and Baird (1987); Milne and Barr (1990); Au, Franks and Aguilera (1991); Aguilera, Franks and Au (1992); Aguilera, Au and Franks (1993).

The Palm Valley Gas Field is situated in the central-northern Amadeus Basin, Northern Territory, Australia (Figure 5–3), approximately 120 km. southwest of Alice Springs. The structure is an arcuate anticline mapped from surface expression and seismic data (Figure 5–4). The amount of seismic data, however, is very limited due to very rough surface topography. The western and eastern plunges are poorly defined. However, the anticline axis can be traced for over 40 km.

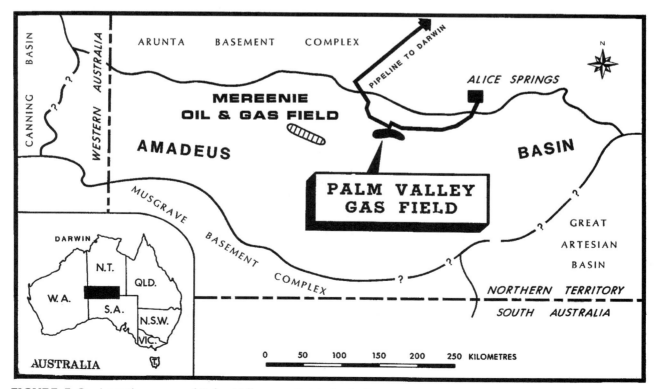

FIGURE 5–3 Location map of Palm Valley gas field, Australia

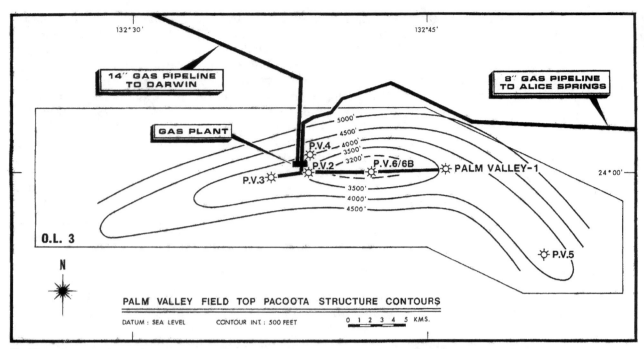

FIGURE 5-4 Palm Valley gas field—top Pacoota structure showing location of gas pipelines

A stratigraphic column of the Amadeus Basin including distribution of gas reservoirs in Palm Valley Field is presented on Figure 5–5. Gas has been found in the lower Stairway sandstone, the basal Horn Valley siltstone, and the Pacoota sandstone, all within the Larapinta Group of Ordovician Age. These reservoir units can be summarized as follows:

The Stairway Sandstone varies in thickness between 295 and 310 m. (969 and 1018 ft.). It is a shallow, intertidal, marine sequence of sandstones, siltstones and shales. A basal quartizitic unit of approximately 40 m. (130 ft.) is occasionally pyritised and fractured, and has associated good gas production.

The sandstones, also termed orthoquartzites, are generally very fine to fine, and once in a while, medium to coarse grained. The siltstones are partially sandy, dark grey to black, micromicaceous, hard and generally silicified. The shales are dark, brittle and hard.

The Horn Valley Siltstone consists of siltstone and shale with thin interbeds of limestone and dolomite. The formation was deposited on a shallow marine shelf and ranges in thickness between 100 and 114 m. (330 and 375 ft.) A limestone and dolomite layer 3 m. (10.15 ft.) thick at the base of the formation is gas productive in Palm Valley 1.

The Pacoota Sandstone, the main gas bearing reservoir, lies conformably below the Horn Valley Siltstone. It consists of nearshore, marine, interbedded sandstones, siltstones, and shales. The formation is subdivided into four distinct units—P1, P2, P3, and P4—on the basis of lithology and log character. No wells drilled in the Palm Valley field have penetrated the entire Pacoota section which is estimated to be approximately 518 m. (1700 ft.) thick from seismic and regional data. Gas flows have so far been encountered in the upper three Pacoota Units. Reserves are estimated at 19.25×10^9 m³ or 680 Bscf (Aguilera et al, 1991).

Production from the field commenced in August 1983 with the completion of an eight-inch pipeline to Alice Springs. Natural gas has been used as a replacement for liquid fuels in electricity generation. Gas production from the field has increased steadily currently averaging 141,000 standard m³/d (5 MMscfd) to Alice Springs.

In September 1986 a 14-inch trunk pipeline was completed connecting the field to the city of Darwin, 1300 km. to the north, and to several major towns en-route. Production for this pipeline has reached 622,000 standard m³/d (19 MMscfd) and again has been used as a liquid fuel replacement in electric power generation.

Total production from the field amounts to 763,000 standard m³/d (24 MMscfd). A crossplot of pool rate vs. time is presented in Figure 5–6.

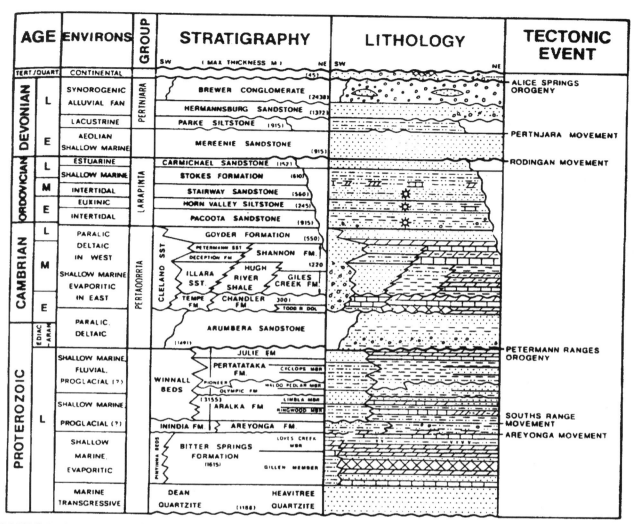

FIGURE 5–5 Stratigraphic column of the Amadeus Basin including distribution of gas reservoirs in the Palm Valley field

Buildup data display the "S" shape typical of dual-porosity behaviour in a Horner cross-plot as shown in Figure 5–7. Evaluation of the Horner plot led to a gas transmissibility equal to 35,020 md-m/mPa.s (114,894 md-ft/cp) and a skin of –3.18. Basic input data used for the analysis is presented in Table 5–2. Figure 5–8 shows a type curve match developed with the use of WELLTEST-NFR, a software package for well test analysis of naturally fractured reservoirs (Aguilera and Song, 1988). The derivative shows the typical signature of dual-porosity behavior. From this match a transmissibility equal to 33,470 md-m/mPa.s (109,808 md-ft/cp) and a skin equal to –3.26 were calculated.

Validation of results was accomplished by developing a single pressure type curve and single derivative that were compared against the real pressure data (represented by crosses) as shown in Figure 5–9. The comparison is considered reasonable. It must be noted that the pressure data were neither smoothed nor edited.

A numerical simulator (TETRAD) with dual-porosity capabilities was also used to validate the analysis. In this case, results from the build-up including fracture permeability, fracture porosity, fracture spacing, wellbore storage and skin, were input in the simulator to generate a theoretical build-up.

Simulation results are presented in the form of a Horner cross-plot in Figure 5–10. The "S" shape typical of dual porosity behaviour is clearly displayed. From the straight line, a transmissibility equal to 33,630 md-m/mPa.s (110,333 md-ft/cp) and a skin equal to –2.98 were calculated.

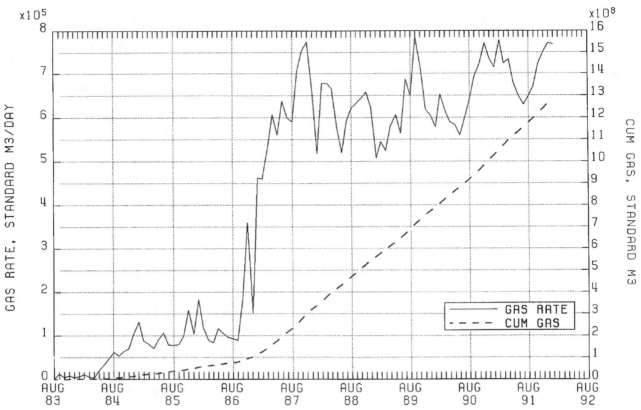

FIGURE 5–6 Production history of Palm Valley gas field, Australia (after Aguilera et al., 1993)

Simulation results were also evaluated using WELLTEST-NFR as shown in Figure 5–11. The match is excellent and led to a transmissibility of 33,930 md-m/mPa.s (111,318 md-ft/cp) and a skin equal to –2.95 were calculated. These results ere validated as shown in Figure 5–12, where the pressures from WELLTEST-NFR's analytical solutions (continuous solid lines) and TETRAD's numerical schemes (crosses) compare very well, mutually validating both interpretation tools.

It is worth examining the similarity between the Horner crossplots presented in Figure 5–7 (real pressure data) and Figure 5–10 (simulated data). The similarity is excellent. The same holds true for the type curve matchings shown in Figures 5–9 and 5–12.

Table 5–3 displays a summary of buildup results. The comparison between real and simulated data are excellent as in the comparison of results obtained from Horner and type curve analysis.

Different sources of information have led to the conclusion that the tight low permeability matrix contributes to production in the Palm Valley gas field of Australia including:

1. The well test analysis discussed previously that shows very clearly pressure derivatives that are typical of naturally fractured reservoirs with matrix contribution.
2. The numerical simulation work using a single well, radial dual-porosity model. When the model is run using matrix permeabilities of less than 0.01 md and other average properties of the reservoir, the pressures in the matrix start to decline, clearly indicating matrix contribution. The same has been observed in computer runs with a full three-dimensional model (Au et al, 1991).
3. The simulated data of a single well model can be matched properly with the dual-porosity type curve that takes into account contribution from the matrix.
4. Observation well PV5 in the south-east of the field which is shut-in and has shown continuous pressure decline due to production in wells PV1, 2, 3 and 6. Well PV5 was

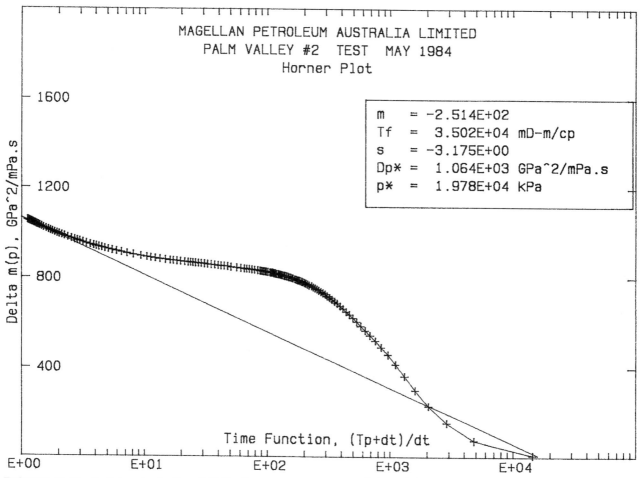

FIGURE 5–7 Buildup test of well PV2. Crosses represent real data (after Aguilera, Franks and Au, 1992)

TABLE 5–2 Input Data for Drawdown and Build-up Analysis, Palm Valley Well No. 2 (after Aguilera, Franks and Au, 1992)

Pressure Run Depth	1.859E+03 m
Fluid Analyzed	Gas
Production Time	8.000+00 hr
Total Flow Rate	1.679E+03 Res m³/D
Gas Flow Rate (Free Gas)	3.115E+05 Standard m³/D
Water Flow Rate	0.000E-01 Res m³/D
Bottomhole Flowing Pressure	1.934E+04 kPa
Specific Gravity of Gas	6.210E-01
Gas Deviation Factor	8.699E-01
Gas Formation Volume Factor	5.389E-03 Res m³/m³
Gas Viscosity	1.837E-02 mPa.s
Reservoir Temperature	7.140E+01 degree C
Pay Thickness	1.603E+01 m
Porosity	4.800E-02
Gas Saturation	4.200E-01
Water Saturation	5.800E-01
Total Compressibility	5.134E-05 kPa^{-1}
Wellbore Radius	7.619E-02 m

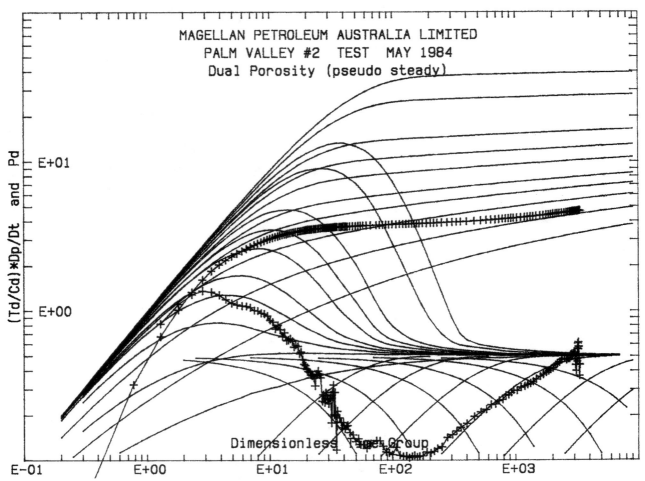

FIGURE 5–8 Buildup test of well PV2, type curve analysis. Crosses represent real data (after Aguilera, Franks and Au, 1992)

drilled mainly through the matrix system and did not intercept any significant fractures. A match of PV5 pressure history using TETRAD has been presented by Au et al (1991).

5. Recent capillary pressure work (Fig. 5–13) demonstrating that threshold pressures are very low (in the order of 5 psi) in clean unfractured sandstone samples suggesting that gas moves easily from the low permeability matrix into the fractures (Berry, 1991). Also note that the irreducible water saturation is only 22% in spite of the low porosity and permeability.

6. A detailed well log interpretation using the Fracture Completion Log (FCL) (Aguilera and Acevedo, 1982).

7. One more important source of information is Duncan McNaughton (1989) from whom comes the following quote:

"Core No. 17 from PV No. 1 well (1796.5–1805.5 m or 5894 ft–5924 ft) bubbled gas uniformly from all over the outside surface of the core when the core barrel was broken. At that time, I initially suspected gas seepages along micro-fractures. However, closer core examination revealed random rather than aligned gas seepages on the outer surface of the core, so I concluded that we were seeing gas emanations from very low permeability material.

"Thus I am one of the few people, still on the scene, who has seen gas flowing from tight Pacoota matrix."

It must be noted that well PV1 was completed as a shutin gas well on May 20, 1965, after reaching a total depth of 2,029.4 m (6,658 ft).

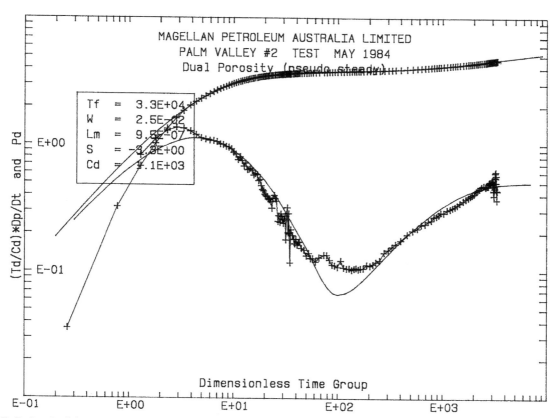

FIGURE 5–9 Buildup test of well PV2, validation of type curve analysis. Crosses represent real data (after Aguilera, Franks and Au, 1992)

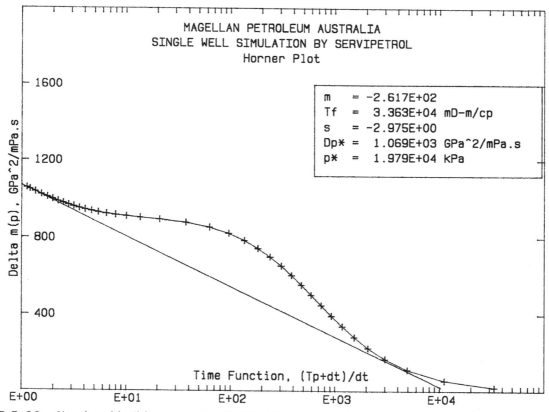

FIGURE 5–10 Simulated buildup test of well PV2, Horner analysis. Crosses represent TETRAD data (after Aguilera, Franks, and Au, 1992)

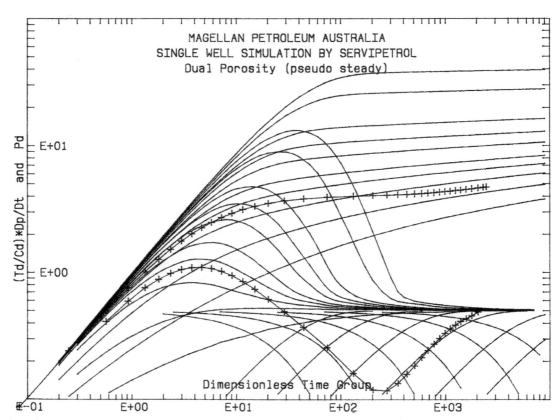

FIGURE 5–11 Simulated buildup test of well PV2, type curve analysis. Crosses represent TETRAD data (after Aguilera, Franks and Au, 1992)

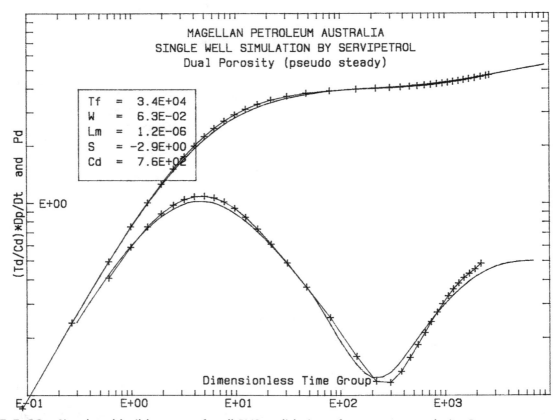

FIGURE 5–12 Simulated buildup test of well PV2, validation of type curve analysis. Crosses represent TETRAD data (after Aguilera, Franks and Au, 1992)

TABLE 5–3 Summary of results, buildup, Palm Valley well No. 2 (after Aguilera, Franks and Au, 1992)

	Real Data		Simulated Data	
	Horner	Type Curve	Horner	Type Curve
Transmissibility, md-m/mPa.s	35,020	33,470	33,630	33,930
Permeability, md	40.13	38.35	38.53	38.88
Skin factor	–3.18	–3.26	–2.98	–2.95
Dimensionless wellbore storage	—	1,114	—	758
Omega	—	0.03	—	0.03
Lambda	—	0.95×10^{-6}	—	1.1×10^{-6}

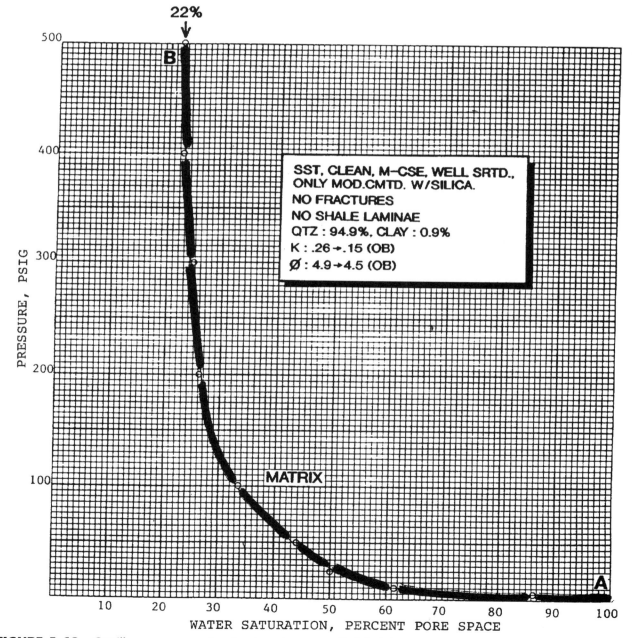

FIGURE 5–13 Capillary pressure of unfractured core, well PV7 (after Aguilera, Franks and Au, 1992)

The presence of natural fractures at Palm Valley has been corroborated from direct and indirect sources of information.

Cores are the direct source of information. Drilling records, well logs, transient pressure data and production history represent the indirect source of information.

Average values of fracture permeability attached to bulk properties (k_2), fracture spacing (h_m), and the storativity ratio, omega, were determined from the well testing procedure presented previously to be as follows:

WELL	k_2(md)	h_m (ft)	ω
PV1	7.4	35.0	0.074
PV2	40.0	41.3	0.030
PV3	4.6	25.7	0.072
PV4	0.3	26.8	0.110
PV5	—	—	—
PV6	0.4	73.9	0.087
PV7	8.1	54.3	0.050

A dramatic example of natural fracturing is provided by well PV2, which was drilled with air and tested over 70 MMscfd after penetrating only 1 or 2 feet of Pacoota reservoir. This in spite that geometric mean permeabilities from unfractured Pacoota plugs most of the times yielded values smaller than 0.1 md.

A detailed log evaluation using a dual-porosity model and a statistical procedure published by Aguilera and Acevedo (1982) also indicated the presence of natural fractures.

Well PV6 was drilled vertically but did not encounter commercial amounts of gas. Well PV6 was subsequently deviated and intersected high angle natural fractures which yielded over 130 MMscfd.

A reservoir simulation of the Palm Valley gas field was conducted using a multi-purpose reservoir simulator, TETRAD. The program is capable of modelling naturally fractured reservoirs with dual porosity and dual permeability behaviour.

The reservoir grid model for Palm Valley gas field was designed to satisfy three basic conditions. First, the model should adequately represent the structure of the reservoir. Second, the model should accommodate the well locations, and provide sufficient grid nodes between wells to show the pressure and fluid saturation gradients. Finally, the number of grid nodes in the model should be minimized to provide optimum computing time.

The top Pacoota structure contour is shown on Figure 5–4. Given the shape of the structure, a radial grid was selected to represent the reservoir (Figure 5–14). The centre of the radial grid was determined in such a way that wells PV3, PV2, PV6 and PV5 lie along the same grid row. The grid block sizes in the radial direction are given a smaller value for well blocks and larger values away from the well blocks. The grid block sizes in the angular direction are specified in degrees and a two degree angle provides a reasonable length-to-width ratio for well blocks.

Gas at Palm Valley is very dry with no liquid hydrocarbon at reservoir conditions. The gas formation volume factor was calculated to 0.005193 cf/scf based on an initial pressure of 2,875 psia at –3,145 ft, a reservoir temperature of 148°F, and a deviation factor (z) of 0.8678.

From the model initialization, the total gas-in-place was determined to be 844.8 Bscf. Gas-in-place in the fractures was 10.79 Bscf and gas-in-place in the matrix 834.01 Bscf.

It must be noted that well PV5 was drilled through the matrix but did not intercept any significant natural fractures. As a consequence, this well was never capable of commercial gas production, in spite of a massive hydraulic stimulation job. The well has been used successfully, however, as a pressure observation well. The fact that the pressure drops continuously while the other wells have been in production gives a clear indication that the matrix system contributes to production. Figure 5–15 shows the history match of well PV5.

The simulation study suggests that well PV5 provides a good approximation to the average reservoir pressure. This was corroborated by shutting-in all the wells in the simulator on December 1990, as shown on Figure 5–16.

Notice that the pressure of observation well PV5 remains approximately constant once the reservoir is shut-in. On the other hand, the other observation well (PV4) shows an

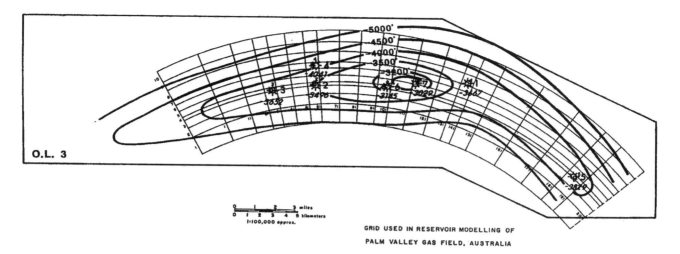

GRID USED IN RESERVOIR MODELLING OF
PALM VALLEY GAS FIELD, AUSTRALIA

MAGELLAN PETROLEUM (N.T.) PTY. LTD.

PALM VALLEY FIELD

TOP PACOOTA DEPTH STRUCTURE CONTOURS

JANUARY 1990

FIGURE 5–14 Radial grid used for numerical simulation of Palm Valley (after Au, Franks and Aguilera, 1991)

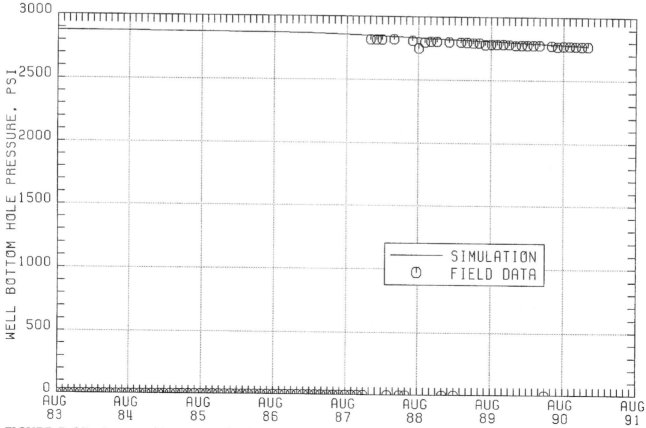

FIGURE 5–15 Pressure history match of well PV4 (after Au, Franks and Aguilera, 1991)

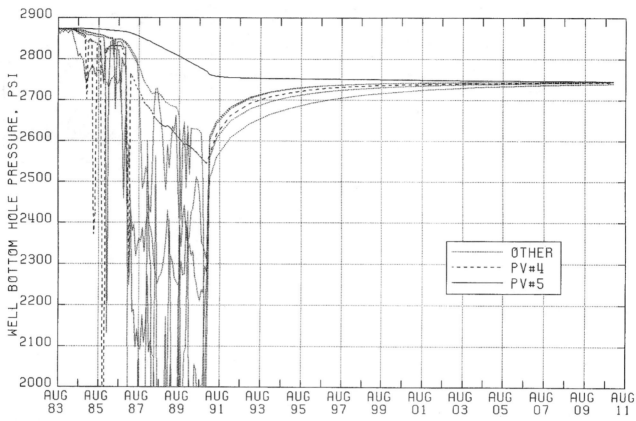

FIGURE 5–16 Simulated shut-in of all wells in Palm Valley (after Au, Franks and Aguilera, 1991)

increase in pressure towards PV5 once the reservoir is shutin. This suggests that the pressure of observation well PV5 can be used for estimating volumes of gas-in-place from a conventional crossplot of p/z vs cumulative gas production. This plot is shown on Figure 5–17, using data from the simulation work.

The linear trend in Figure 5–17 points towards a value of original gas-in-place equal to 844.8 Bscf. Recoverable gas reserves at an assumed abandonment pressure of 600 psi amount to 680 Bscf.

The communication between well PV5 and the other wells is corroborated with the use of Figure 5–18. In fact, decreases in PV5 pressures correspond to increases in gas rates in well PV2 and the rest of the field, with a time lag of approximately two–three months.

PV2 is a very important well that has accumulated approximately 50% of the total gas production of Palm Valley. This well was drilled with air and flowed gas at a rate of about 70 MMscfd from only one or two feet of penetration into the Pacoota P1 gas reservoir.

FRACTURED CARBONATES

Beaver River Field, Canada (Gas Reservoir with Infinite Aquifer)

The Beaver River field is located on the border of British Columbia and the Yukon Territory (Fig. 5–19). The reservoir is characterized by a high relief middle Devonian carbonate (Davidson and Snowdon, 1978) crossed by regional fractures.

The first well, the Beaver River A-1 (Figure 5–19), was drilled in 1958. The well indicated the presence of potential gas zones but was abandoned due to mechanical problems.

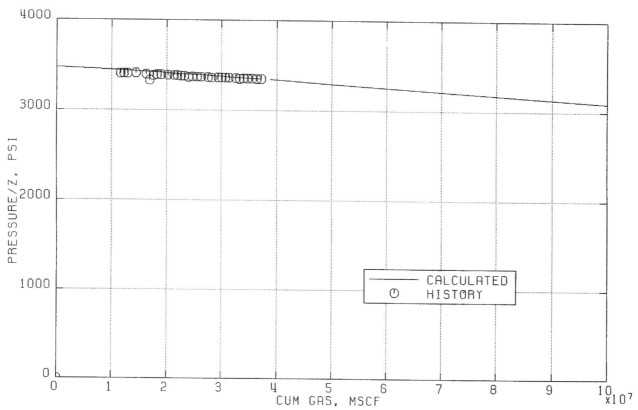

FIGURE 5–17 Material balance crossplot (after Au, Franks and Aguilera, 1991)

Well A-2 was drilled in 1961 to a total depth of 13,500 ft. The well penetrated 1,200 ft. of Devonian carbonate without reaching the gas-water contact. Absolute open flow potential was 85 MMscfd.

Four additional successful wells were completed from 1967 to 1970, leading to the idea that the Beaver River was the largest field in British Columbia.

Core analysis indicated that the reservoir could be described as a two-porosity system made out of fracture vugs and matrix porosities. The average matrix porosity was estimated to be 2% or less, and the fracture vug porosity was estimated to range between 0 and 6%. Further, the fracture porosity alone was estimated to vary between 0.05 and 5%. Table 5–4 shows porosity results from core analysis in eight wells of the area.

The fracture system contained an important part of the storage capacity of the reservoir. This is emphasized because there is a general tendency among reservoir engineers to neglect in all cases the storage capacity of the fractures compared with the storage capacity of the matrix.

Water saturations were estimated by conducting mercury injection capillary pressure tests in cores with only matrix porosity and cores with matrix, vugs and fracture porosity (Figure 5–20).

The results indicated that water saturation ranged between about 50% and 90% for the matrix system. Water saturation for the two-porosity system (matrix and vugs-fractures) ranged between 10% and 40% from the capillary pressure curves. Finally, it was estimated that the fracture system contained 0% water saturation. This is not unreasonable, as capillarity makes water occupy the finer pores (matrix) and leaves the coarser pores (fractures and vugs) to the hydrocarbons.

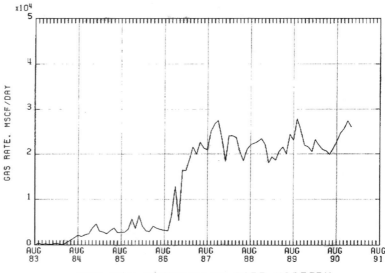

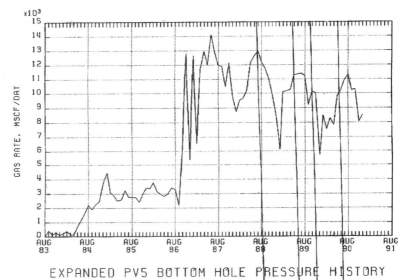

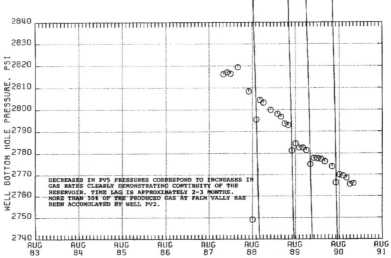

FIGURE 5–18 Plots showing communication between well PV5 and the rest of the reservoir (after Au, Franks, and Aguilera, 1991)

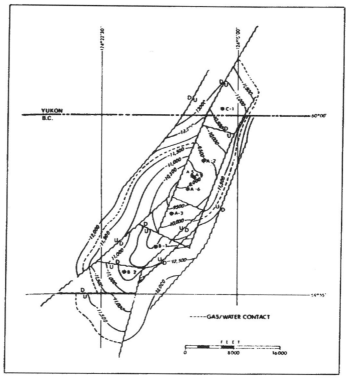

FIGURE 5–19 Schematic of Beaver River Gas Field showing top of Middle Devonian carbonate (after Davidson & Snowdon, 1978)

TABLE 5–4 Beaver River Core Analysis (after Davidson & Snowdon, 1978)

Well	Upper zone (top 350 ft)		Lower zone		Total reservoir	
	Feet Analyzed	Weighted average ϕ (%)	Feet Analyzed	Weighted average ϕ (%)	Feet Analyzed	Weighted average ϕ (%)
B-2	13.5	3.15	260.0	1.93	273.5	1.98
B-1	30.0	3.59	281.8	2.09	311.8	2.27
A-3	59.4	2.41	470.3	1.41	529.7	1.52
A-6	127.5	2.77	192.5	1.81	320.0	2.19
A-5	37.6	6.60	97.5	1.80	135.1	3.16
A-1	248.5	3.64	—	—	248.5	3.64
A-2	35.2	4.43	13.0	0.65	48.2	3.41
C-1*	190.7	0.79	139.0	1.51	329.7	1.10
	724.4 Total	3.54 Average*	1,454.1 Total	1.72 Average	2,196.5 Total	2.09 Average

*Upper zone in Well C-1 is not porous because of a facies change; therefore, it was not included in the upper-zone average

Average horizontal matrix permeabilities for the best zones ranged between 20 and 200 md, and vertical permeabilities varied between 2 and 25 md.

Initial estimates of recoverable reserves were encouraging, as they indicated an ultimate recovery of 1.47 Tscf (Table 5–5).

Production started in October 1971 at average rates above 200 MMscfd (Figure 5–21). After six months of production, a decline in gas rates was noticed in Wells A-5 and C-1 at the same time that the water-gas ratio increased to about 200 bbl/MMscf. Water coning

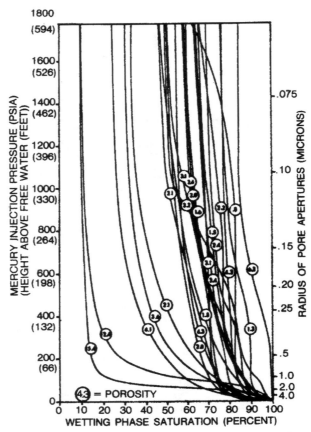

FIGURE 5–20 Mercury-injection capillary pressure tests (after Davidson & Snowdon, 1978)

TABLE 5–5 Gas Reserves and Reservoir Data (after Davidson & Snowdon, 1978)

Reservoir

Area at gas-water contact, acres	10,700
Reservoir volume (gross), million acre-ft	10.5
Initial temperature, °F	353
Initial pressure, psig	5,836
Gas gravity	0.653
Gas composition, %	6.0 CO_2
	0.5 H_2S
	92.5 CH_4

Reservoir Parameters

Porosity cut-off, %	2
Porosity average, %	2.7
S_w: Matrix, %	25
Fractures-vugs, %	0
Average, %	20
Net pay, ft	888
Volume (net), acre-ft	7,210,664
Recovery factor (with volumetric depletion), %	90
Gas deviation factor	1.10
Recoverable reserves (raw), Bcf	1,470

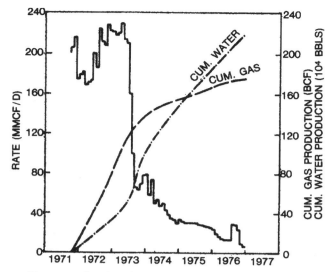

FIGURE 5–21 Beaver River production history (after Davidson & Snowdon, 1978)

was suspected, and Well C-1 was temporarily suspended, while Well A-5 was perforated in a higher interval. In June 1973, however, all wells started producing water and gas rates declined drastically (Figure 5–21).

Detailed computer analysis with a two-phase coning simulator indicated that coning was not the cause of the water production problems. In fact, perforations were generally 1,000 to 1,500 ft above the gas-water contact; to obtain water coning under these conditions it would be necessary to have a vertical permeability of about 40,000 md and a horizontal permeability of less than one md. This, of course, did not represent the actual permeability characteristics of the reservoir.

Further modelling with actual petrophysical data allowed a good match of production history and indicated that the water level was raising uniformly throughout the field.

These results were further corroborated by actual field observation. In fact, the water-gas ratios continued to increase even after imposing production restrictions. Recompletion of well C-1 in a high interval resulted in 100% water production. The field was closed in October 1978 after it had recovered 178 bscf of raw gas.

The forecasted recoverable reserves were 1,470 Bscf as shown on Table 8–5. So the actual recovery was only 12.1% of the forecasted gas recovery.

Three schools of thought have evolved to try to explain the problem. The first one says that because of the strong water influx, most of the gas was trapped in the fractures resulting in a extremely low recovery.

Davidson and Snowdon (1978) have explained the strong water influx at Beaver River by considering the field's relationship with the regional aquifer (Figure 5–22). They indicated that there are various pressure discontinuities in areas adjacent to Beaver River field. In each of these pressure systems, it is possible to correlate the pressure-depth relationship with the outcrop elevation.

Figure 5–23 shows a hydrodynamics plot for Beaver River and nearby Pointed Mountain field. The potentimetric surface elevation is about 1,500 ft above sea level. This corresponds to the ground surface elevation of the middle Devonian carbonate outcrops in the canyons of South Nahanni River about 85 miles north.

These findings can be interpreted by saying that the aquifer elevation increased about 13,000 ft over a distance of 85 miles, or nearly 150 ft/mile (Figure 5–22).

The relief allows quick water influx as the reservoir is being depleted. Also, the relatively short distance (85 miles) to the outcrop produces a low pressure decrease in the aquifer, which consequently maintains a pressure at the gas-water contact consistent with the hydrodynamic pressure-depth relationship.

Based on their experience, Davidson and Snowdon (1978) recommended careful examination of the "nature of the aquifer regarding permeability, hydrodynamic potential

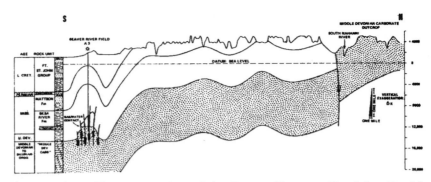

FIGURE 5–22 North-south cross section of the Beaver River aquifer (after Davidson & Snowdon, 1978)

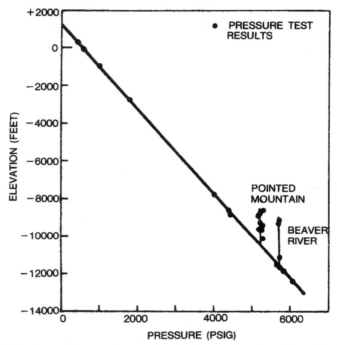

FIGURE 5–23 Middle Devonian carbonate hydrodynamics plot (after Davidson and Snowdon, 1978)

(distance and elevations change to outcrop) and the reservoir pore system" when evaluating naturally fractured reservoirs with possible water influx.

The second school of thought is advanced by this author and indicates that a large portion of the gas was recovered. In this case the original gas-in-place was overestimated, probably there was no gas or very little gas in the matrix and the fracture porosity was smaller than the one used in the original study. The idea that very little gas was in the matrix is not unreasonable. In fact, it is difficult to visualize a matrix water saturation of only 25% when the matrix is so tight, porosity is 2.7% (Table 5–5) and the matrix capillary pressures in Figure 5–20 show wetting phase saturations ranging between 50% and 90%.

The third school of thought has been advanced by Gray et al (Circa 1992) who conducted a feasibility study that they say confirmed that an active water drive was not present, and that unexploited gas volumes exceeding 1.0 Tcf are present in the reservoir out of which up to 500 Bscf or more are potential recoverable reserves. Their study indicated that the rate of gas production from the total reservoir as a unit must be balanced against water influx into the highly permeable, low porosity fracture system. Excessive water to this equation must be removed by a high capacity submersible pump(s). Mining companies routinely conduct dewatering to permit removal of an ore body.

Reservoir simulation calculations by Gray et al (circa 1992) have indicated that the BHP had risen by 500 PSI as of 1988 with a drop in the water table of 1500 ft. Simulation studies also indicate a production rate of 50 million scf/day for 30 years is feasible, accompanied by water production of between 5,000 and 20,000 bbls/day, or 25 million scf/day for five years water free.

Corroboration of this third school of thought awaits the acid test of dewatering.

Weyburn Unit, Canada (Water Injection and Polymers)

The Weyburn Unit is located in southeastern Saskatchewan approximately 130 Km southeast of the city of Regina. The field was discovered in 1955 and was developed by drilling 675 wells on an 80-acre spacing (Mazzocchi and Carter, 1974).

The reservoir is found in a stratigraphic trap of Mississippian age. It is divided into two lithological units: the Marly zone (higher structurally) and the Vuggy zone (lower).

The Marly is a chalky intertidal dolostone with occasional limy or limestone interbeds. Porosity in the dolostones ranges from 16% to 38%, with an average of about 26%. Matrix permeability within the Marly ranges from 1 to > 100 md, with an average of < 10 md. Although both lithological units are fractured, the Marly zone is less intensely fractured than the Vuggy zone.

The Vuggy zone is a heterogeneous, subtidal limestone. Varied diagenetic and depositional environments result in a wide range of porosities (3% to 18%) and matrix permeabilities (< 0.01 to > 500 md) for this reservoir. Natural fractures within the Vuggy zone control the magnitude and direction of the permeability anisotropy.

Oil is produced from depths of 1310 to 1500 m. The original oil in place (OOIP) was 178 $\times 10^6$ stock-tank m^3 (1.1×10^9 stbo).

A water flood was initiated in June 1964. Figures 5–24 to 5-26 show the importance of fracture orientation for secondary recovery operations.

Figure 5–24 shows the primary and waterflood performance of on-trend well (10) 31 represented by a double circle. The fracture system is oriented northeast-southwest (Figure 5–25). Since the fracture system was nearly connecting the injection and production well, water break-through occurred rapidly and water production increased drastically. Mazzocchi and Carter (1974) estimated that oil recovery for this category of wells was between 8% and 12% of the oil-in-place.

Figure 5–26 shows primary and waterflood performance of off-trend well (12) 32, represented by a double circle. Since waterflood direction was perpendicular to the direction of the fracture system, there was a gradual increase in oil production for about one year,

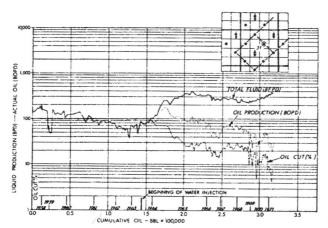

FIGURE 5–24 Primary and waterflood performance of on-trend Well (10) 31 (after Mazzocchi & Carter, 1974)

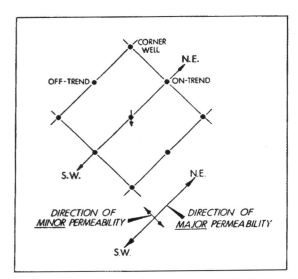

FIGURE 5–25 Inverted nine-spot pattern (after Mazzocchi & Carter, 1974)

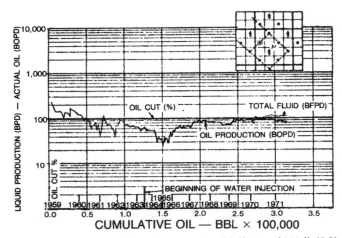

FIGURE 5–26 Primary and waterflood performance of off-trend Well (12) 31 (after Mazzocchi & Carter, 1974)

followed by a sustained stabilization. By early 1971, water breakthrough had not yet occurred.

The importance of the performances presented in Figures 5–24 and 26 cannot be overemphasized and is directly related to sweep efficiencies. The biggest sweep efficiencies occur when the direction of flood goes perpendicular to the fractures.

In 1971 a gelatinous blocking agent was tried in an effort to reduce channelling caused by the fracture system and streaks of very high permeability. A pilot was carried out along the east edge of the Weyburn unit.

This area presented favorable conditions for a pilot test due to minimum interference from injection wells surrounding the area. In addition, all pilot water injection wells were connected to the same injection satellite, and the available pressure (1,500 psi) was suitable for the expected average injection rates of 2,600 bw/d at 650 psi.

Figure 5–27 shows the performance of the entire pilot area. Actual results compared with extrapolated regression lines suggest that the blocking-agent treatment was successful. Oil production increased substantially, while water production followed the same trend projected by the extrapolated regression line.

After 28 months of regular water injection, no traces of the gel component were found in the water samples obtained from production wells, indicating the successful application of the gelatinous blocking agent on the edge of this fractured limestone reservoir.

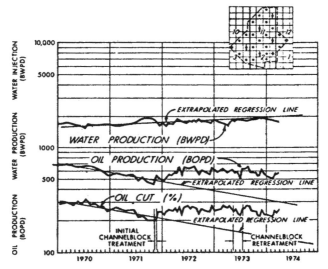

FIGURE 5–27 Treatment performance of all wells in the pilot area (after Mazzocchi & Carter, 1974)

More recently a multidisciplinary reservoir characterization and simulation study of the Weyburn Unit was carried out by Elsayed et al (1993). The reservoir model was developed by integrating geologic, petrophysical, and engineering data. The authors indicate that the reservoir characterization was so effective that, 70% of the wells were history-matched on the first simulation runs without modification to the original geological/petrophysical data.

The field presently has 627 producing wells and 162 water injection wells on 24-hectare ($\simeq$ 60 acres) spacing.

To conduct the reservoir simulation the Marly and Vuggy zones were split into flow units using petrophysical and petrographic data. The Marly was subdivided into three flow units (M1 to M3), and the Vuggy was split into six flow units (V1 to V6). These flow units represent alternating layers of high and low permeability. Fracture spacings or distance between fractures were estimated to be 3 m. in the Marly dolostones, 0.3 m. in the Vuggy intershoal limestones, and 2.5 m. in the Vuggy shoal limestones. Fracture apertures were estimated from FMS data to range between 50 and 100 μm. Average fracture permeability was estimated to be 40 md.

Elsayed et al (1993) compared single vs. dual-porosity simulation models and concluded that a single-porosity model was adequate to simulate the Weyburn unit. Figure 5–28 shows a good history match for oil and water cumulatives starting in 1956. Although the match is good, I do not recommend the use of single-porosity simulators to study naturally fractured reservoirs. It is feasible to match the history of a naturally fractured reservoir with a single-porosity model as shown by Elsayed et al's study. However, it is extremely dangerous to carry out forecasting with a single-porosity model because it does not take into account imbibition and other interactions between matrix and fractures. The danger is increased if the field operation mode is changed for example from waterflooding to CO_2 flooding.

Midale Field, Canada (Water Injection and CO₂ Injection)

The Midale field is located in the same general area as the Weyburn Unit discussed in the previous case history. The original oil-in-place is 500 MMstbo. The field was discovered in 1953 and developed on 80-acre spacing. Midale was unitized in 1962 with eighty-three 320-acre inverted nine-spot patterns. The geology and reservoir characteristics are very similar to those discussed in the case of the Weyburn Unit. Cumulative oil to September 1993 was in the order of 100 MMstbo ($\simeq$ 20% of OOIP). The water cut was in the order of 80%.

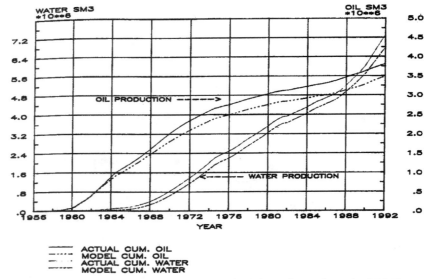

FIGURE 5–28 Weyburn waterflood history match (after Elsayed et al., 1993)

Beliveau et al (1993) conducted a study to evaluate the feasibility of carrying a CO_2 flooding in the Marly and Vuggy zones. The study involved a multidisciplinary team that analyzed core data, residual oil measurements, time lapse logging, tracer studies, pressure transient testing, phase-behaviour studies, and geotomographic measurements.

Profile logging indicated that the water was injected in the more fractured underlying Vuggy zone where oil is displaced efficiently by both viscous drive effects and capillary imbibition. However, analysis of the waterflood performance suggests that there are opportunities to increase sweep efficiency through infill drilling and horizontal wells. So far, 15 horizontal wells have been drilled which have produced approximately six times more than vertical wells.

Sweep of the Marly zone, however, is slower and not as efficient. This leaves the Marly zone as a potential enhanced oil recovery (EOR) target, given the tendency for CO_2 to rise naturally by gravity forces. In order to investigate this possibility a CO_2 pilot was conducted. It included the following phases: (1) planning and design, (2) Pre-CO_2 brine flood, (3) CO_2 flood and (4) post CO_2 flood operations.

In the planning and design stage (Phase 1), studies with a slim-tube indicated that the CO_2 minimum miscibility pressure was about 2,250 psi (average reservoir pressure is 2,600 psi). The crude oil has a gravity of 29° API and a 3-cp viscosity. The design included four injectors, three producers, two logging observation wells, and a downhole fluid sampling well.

In the pre-CO_2 brine flood stage (Phase 2), an average of 140 b/d of brine was injected into each of the four injection wells for a period of nine months with a view to ensure that the pilot area was at waterflood residual oil conditions. Profile logs showed that 100% of the water was injected into the Vuggy zone, while at the producers up to 25% of the inflow came from the Marly zone. The average water cut at the end of the brine flood was 95%.

In the CO_2 flood stage (Phase 3) there were some pleasant surprises. For example, no CO_2 was produced until almost one month after CO_2 injection was started. This is a clear indication that gravity was playing an important role and that CO_2 was penetrating large amounts of the tight matrix porosity. The first tertiary oil was produced after five weeks of injection, just a few days after CO_2 breakthrough in each producer. A total of 14,600 barrels of tertiary oil was produced at a net CO_2 utilization factor of about 3 Mscf/stb. This suggests a 14% OOIP recovery from the pilot. Beliveau et al (1993), however, estimate that the CO_2 displacement efficiency will amount to 27% OOIP. The difference between the 14% and 27% indicated previously is assigned to the unconfined leaky nature of the pilot.

In the post-CO_2 flood operations stage (Phase 4) and to avoid a quick water breakthrough and the loss of some tertiary oil the CO_2 flood was followed by two periods of controlled production without injection support. This production was followed by a traced (salt and alcohol) brine flood. The controlled production proved successful in recovering additional tertiary oil, while the brine flood caused the producers to water out rapidly.

Reservoir simulation was used to model the water flood and CO_2 behaviour. A dual-porosity model taking into account capillary imbibition, local gravity drainage, pressure equilibration and viscous displacement was used for the reservoir simulation. The Midale waterflood history match is presented in Figure 5–29. The CO_2-flood pilot history match is presented in Figure 5–30. The matchings are reasonable. An effective vertical/horizontal

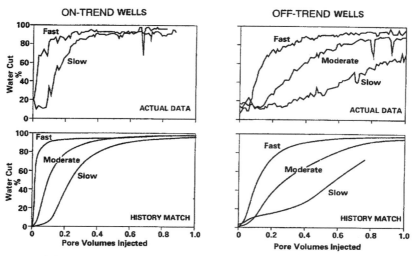

FIGURE 5–29 Midale waterflood history match (after Beliveau et al., 1993)

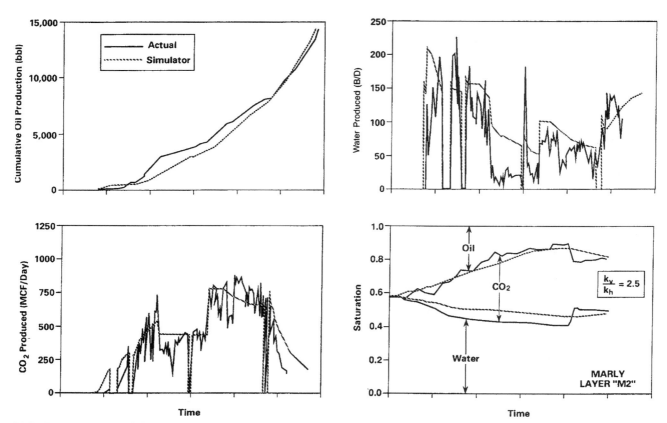

FIGURE 5–30 Midale CO_2 flood pilot history match (after Beliveau et al., 1993)

permeability ratio of 2.5 was required to match the saturation histories. The good results of the simulation provided a high degree of confidence in the model and matched reservoir parameters.

Based on the successful pilot and successful simulation of the pilot, field-scale CO_2 flood predictions were carried out. From here the field-scale tertiary recovery is forecasted to be 20% OOIP. Beliveau et al (1993) assign this excellent recovery to the unique interaction of the different geological layers combined with favorable gravity effects.

From the simulation results it is forecasted that on-trend wells will show peak tertiary oil rates two to three years after CO_2 injection begins. On-trend wells will be converted to CO_2 injections when it is judged economical to do so (probably four to five years after injection begins).

Based on the previous results Midale owners began a commercial-scale CO_2 flood covering approximately 10% of the unit in February 1992.

Cottonwood Creek Field, USA (Water Injection in Karst Reservoir)

The Cottonwood Creek field is located approximately 11 miles west of Ardmore, Oklahoma. The reservoir was discovered with the drilling of well No. 32-1 in 1987. Production is obtained from the Brown zone, a highly fractured, vuggy to cavernous dolomite, that occurs within a 480 ft interval of the lower West Spring Creek/Upper Kindblade formations.

Production rates of > 4,000 bo/d and 3.0 MMscfd were obtained from the Arbuckle Brown zone in November 1987 at CNG Producing Co.'s Cottonwood Creek 32-1. Data from 21 offset wells, including four deep penetrations in the field, indicate that production is from a northwest-southeast-oriented anticline which is over turned to the northeast and cut by a southwest-directed backthrust (Figure 5–31). Fourteen successful Brown zone wells define a productive area measuring 2.5 by 0.75 mi, covering 1,200 acres, and containing an oil column of more than 900 ft (Read and Richmond, 1993).

The Brown zone is composed of at least four types of dolomite. The earliest form is a mudstone-replacive dolomicrite/dolomicrospar. The second type is fine to coarsely-crystalline, xenotropic dolomite that replaced the original lime mud and dolomicrite constituents. The third type is fine to medium-crystalline sucrosic dolomite. The fourth type, which fills or partially fills vugs and fractures, is subhedral to euhedral, saddle or baroque dolomite. Data from wells spaced as closely as one well to 40 acres indicate that the distribution of dolomite in the field is very erratic.

Dolomite matrix porosity and permeability in the Brown zone are very low, typically < 2% and 0.01 md, respectively.

Early to Middle Pennsylvanian erosion removed as much as 800 ft of West Spring Creek from the anticline and karsted the Brown zone and the Kindblade to a depth of at least 1,500 ft below the pre-Pennsylvanian unconformity. Zones of cavernous porosity as thick as 24 ft were encountered in the Brown zone during drilling in the field (Read and Richmond, 1993).

The intensive brecciation and fracturing interconnect much of the dissolution porosity in the reservoir. The origin of fractures in the field is not fully understood. Probably much of the fracturing is due to Pennsylvanian tectonism. However, core studies of the Brown zone at Healdton field and in the Texaco-Mobil no. 1 Criner (sec.1, T.5S., R.1W.) indicated that a significant part of the brecciation observed at these two localities may be related to paleokarstification and solution collapse (Lynch and Al-Shaieb, 1991). The same interpretation may be valid for Cottonwood Creek.

Reservoir performance data helps to understand this naturally fractured reservoir. The varied nature of dolomitization, karst development and natural fracturing precludes accurate determination of reservoir characteristics through petrophysical evaluation alone. Initial production from the 32-1 discovery well, in which 350,000 bbl of oil were produced in five months, suggested that the cavernous Karst reservoir was prolific. Further data shows that all wells produce from a common reservoir, and that there is good interconnection of the vugular and fractured, high productivity zones to less permeable types of porosity. The

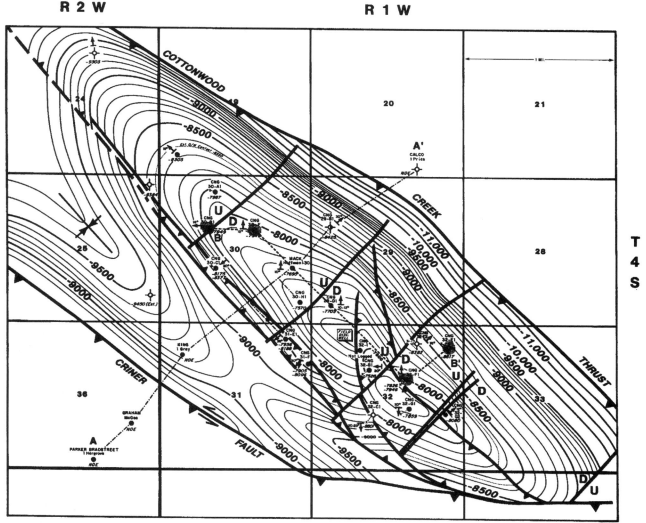

FIGURE 5–31 Cottonwood Creek structural map. Structural datum, Brown zone marker. Depths subsea. Contour intervals is 100 ft (after Read and Richmond, 1993)

low-porosity tight matrix is believed to contain most of the hydrocarbon-in-place and to be a major contributor to the ultimate oil recovery.

Most of the wells exhibit similar pressures throughout the field's productive history, suggesting a common reservoir in direct communication. The southeast end of the field (wells to the southeast of the 32-B-1) has maintained pressures slightly above the rest of the field, although these values declined before any wells in the area had produced. This suggests that there is communication between these two portions of the field through either a low permeability zone or across the northeast-southwest-trending normal fault mapped through the center of sec. 32 (Fig. 5–31).

The decline in pressure (Figure 5–32) provides data for a material balance calculation of the oil-in-place. Values from both volumetric and material balance methods were subsequently compared until consistent interpretations could be made. Based on these calculations, the Cottonwood Creek OOIP was over 40 million stbo with an assumed average effective Brown zone porosity of approximately 3%.

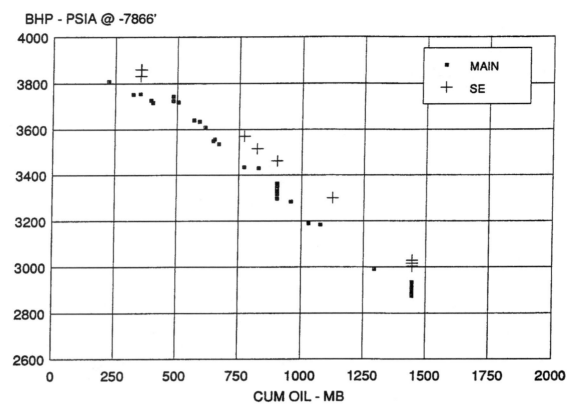

FIGURE 5–32 Reservoir pressure vs. field cumulative oil production prior to pressure mainte-nance. Data points denoted with a (+) are from the southeast end of the field which has maintained pressure readings slightly higher than the main portion of the field (after Read and Richmond, 1993)

Pressure buildup and drawdown tests suggest contributions from both the fractures and the tight matrix porosity. A plot of bottomhole pressure vs. time in hours (Fig. 5–33) uses data taken from a nonproducing well while (A) the field is producing at a rate of 4,000 bopd, (B) the field is shutin with only one well producing at approximately 150 bopd, and (C) a second well is brought on at an additional rate of 600 bopd.

The immediate response to each of these actions (in a well more than 2,000 ft from the changes) shows the high permeability of the fractures. Each change in the producing rate yields a nearly linear pressure vs. time curve that suggests a closed depletion system with very large transmissibility. The buildup portion of the graph, with its slope of 15 psi in 100 hours, indicates flow from the tight matrix into the fractures. The rate of flow was nearly constant over the test period but has changed over the life of the field. Evaluation of the variations in slope for various producing rates indicates that the tight matrix probably holds most of the oil-in-place.

Buildup plot of four wells across the field (Fig. 5–34) indicates that all wells are experi-encing the same rate of influx. The top pressure curve is from well 32F-1 (SW¼ NE¼ sec. 32) located in the southeastern part of the field. Although the early-time rise shows a greater pressure drawdown was necessary for flow, the late-time data increases at a rate similar to other wells in the field. The other wells in Figure 5–34 are the 31E-1 (NE¼ NE¼ sec. 31) and the 29D-1 (SW¼ SW¼ sec. 29), situated in the center of the field, as well as the 30A-1 well (NE¼ NW¼ sec. 30), located in the northwest corner of the field.

Pressure maintenance by water injection into the Arbuckle aquifer was started in June 1990. Pressure response to the water injection was noticed within 30 minutes. Oil produc-tion rates increased to approximately 5,500 bo/d by September 1990 with injection vol-umes increased to 8,000 bw/d to balance reservoir withdrawal. Cumulative oil production to October 1992 was 6.1 million bbl (less than 15.3% OOIP).

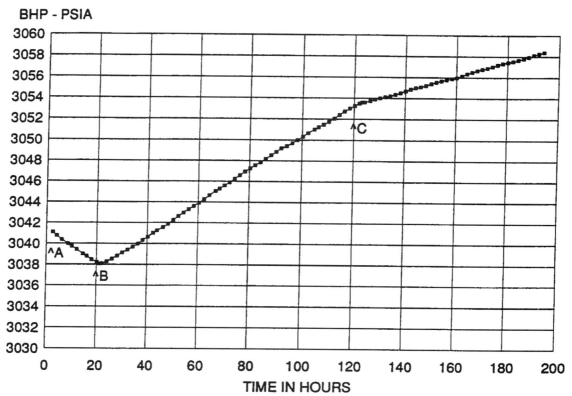

FIGURE 5–33 Pressure buildup and drawdown data from nonproducing well (29B-1, SW¼ NW¼ sec. 29) with field producing at (A) 4,000 BOPD, (B) 150 BOPD, and (C) 750 BOPD (after Read and Richmond, 1993)

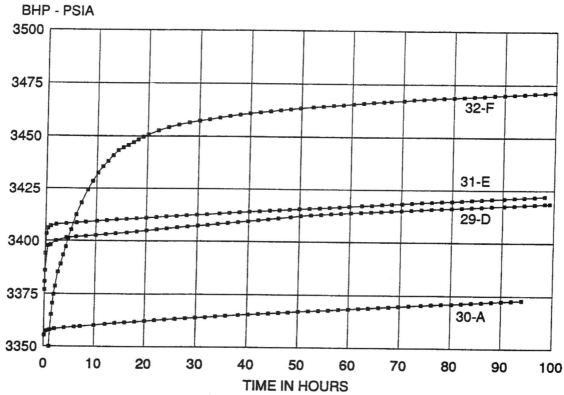

FIGURE 5–34 Simultaneous pressure buildup plots for widely spaced field wells. Wells no. 32F and no. 30A are 1.8 mi apart (after Read and Richmond, 1993)

Cottonwood Creek field, with oil-in-place estimates of over 40 million bbl, has been heralded as the largest southern Oklahoma oil find since the initial Brown zone discovery at Healdton field in 1960. Cumulative Arbuckle production from these two fields may ultimately reach 35–40 million bbl of oil.

The Arbuckle carbonates provide very promising potential for large hydrocarbon discoveries in southern Oklahoma. Detailed understanding of both the Healdton and Cottonwood Creek fields are believed to be necessary to develop exploration models which might lead to the discovery of additional Arbuckle/Ellenburger accumulations in the southern Midcontinent (Read and Richmond, 1993).

Ekofisk Field, Norway (Overpressured, Gas Injection, Water Injection, Subsidence)

An excellent review of the production history of Ekofisk field during its first 20 years has been presented by Sulak (1991). This case history follows very closely Sulak's original work.

Ekofisk accumulated 894×10^6 bbl of oil and 2.8×10^{12} scf of gas (net of injection) during its first 20 years. The dominant natural production mechanisms have been solution-gas drive and compaction drive. Oil expansion and water influx were significant contributions to recovery during the first few years. Natural drive mechanisms have been enhanced by gas and water injection.

The Ekofisk discovery well tested 1,070 bo/d on a 34/64-in. choke on December 31, 1969. This test was limited by equipment capacity. The well subsequently was tested at sustained rates exceeding 10,000 bo/d. Initially, the center of the structure was interpreted, on the basis of seismic data, to be depressed and consequently was avoided by the discovery and appraisal wells. Subsequently, well 2/4 C-8 drilled in 1974 showed that seismic data were affected by gas over the center of the structure and that a normal anticline existed.

The anticline, based on current interpretations, is elongated and covers some 12,000 acres. The major north/south axis is about 6 miles long, while the east/west axis extends approximately 4 miles (Fig. 5–35). The origin of the structure is still open to interpretation.

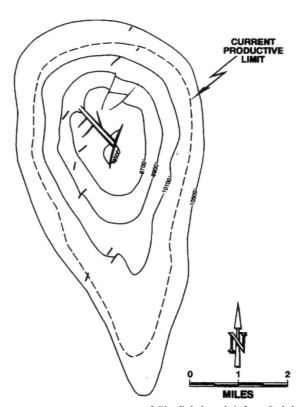

FIGURE 5–35 Current structure map at top of Ekofisk level (after Sulak, 1991)

Some geologists think that the structure was created by an underlying salt dome. However, no well has penetrated salt. Another interpretation is that compressional forces leading in a deep flower fault could be responsible for the anticline (Fig. 5–36).

Production is obtained from the Ekofisk and Tor formations. Both formations are chalks (fine-grained limestones), composed of the skeletal debris of pelagic unicellular algae known as coccolithophorides.

The Ekofisk (Palaeocene) formation is located at a depth of about 9,600 ft. The Tor (Cretaceous) formation underlies the Ekofisk formation. The Tight Zone, where it exists, forms an impermeable barrier between the Ekofisk and Tor formations. The chalks have high porosity (25% to 48%) and reasonable matrix permeability (1 to 5 md). It must be noted that most naturally fractured reservoirs have matrix permeabilities that are a fraction of millidarcy. Natural fractures increase the effective permeability to up to 100 md. Overall pay thicknesses of more than 1,000 ft has been penetrated.

The overburden consists of about 9300 ft of shales and clays. It is overpressured below about 4500 ft.

The reservoir was initially undersaturated at 7135 psia and 268°F at datum (–10,400 ft). The volatile 36° API-gravity crude oil had a solution GOR of 1530 scf/bbl. The bubblepoint pressure was 5560 psia.

Initially it was thought that the fractures could tend to close as the reservoir was being depleted. To investigate this possibility, an extended flow test was conducted. Within 18 months of the discovery, Well 2/4-1AX and three appraisal wells were tied into the converted jackup platform Gulftide. Approximately 28×10^6 bbl of oil were produced from July 1971 to May 1974. No closing of the fractures was evident.

Field development of this offshore reservoir required three drilling/production platforms, a field terminal platform, and living quarters. Also, a one million-bbl concrete storage tank was installed for use when sea conditions prevented offshore loading. The permanent facilities became operational in May 1974.

Following the installation of 9000 psi design compressors in Feb. 1975 and until the gas pipeline to Emden, Germany became operational, all produced gas was injected. After that, only gas in excess of sales was injected. An average of 324 MMscf/d was injected during this period.

Subsidence was discovered in November 1984. The seabed under the Ekofisk complex had subsided some 10 ft by the end of 1984. The approximately 15 ft of reservoir compaction required for this amount of subsidence had not caused any detectable change in productivity.

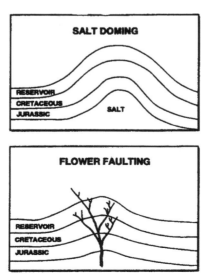

FIGURE 5–36 Possible mechanisms for the formation of the Ekofisk structure (after Sulak, 1991)

The main concern at this stage was protection of the steel platforms and the concrete storage tank. To buy time while trying to find a solution, gas was diverted from sales to injection. The final solution, reached on January 1986, was to elevate the steel platforms and to construct a concrete protective barrier around the tank. Gas sales were started again while design work for the elevation and the protective barrier began.

Approximately $1 billion was spent on the deck elevation project and the protective barrier to remedy the subsidence problem.

To this date no reduction in productivity has been observed. Analysis of well tests under two-phase conditions is cumbersome and interpretation could mask some reductions in absolute permeability. However, oil production from Ekofisk is almost twice what it was three years ago.

The effective permeability of the Ekofisk is mostly controlled by the fracture system. Since the matrix permeability ranges between 1 md and 5 md, the reservoir can be considered as one where the fracture permeability assists an already producible reservoir. The rates from the matrix alone, however, would be non-economic.

The reservoir volume represented by the fracture system is estimated to be less than 0.5%. The relatively tight matrix contains 99.5% of the reservoir volume.

The matrix permeability declines with matrix compaction, but because matrix permeability is not within the same order of magnitude as effective permeability, this effect is inconsequential according to Sulak (1991). If reservoir compaction creates new fractures, there is a possibility that reservoir productivity can increase.

Almost 1 Tcf of dry has been injected into the Upper Ekofisk formation since February 1975. Five crestal injection wells provide up to 500 MMscf/d injection capacity. Natural gas injection at Ekofisk has always been for operational reasons—i.e., subsidence mitigation or "swing gas" injection. Nevertheless, it has improved recovery.

Water injection was initiated at Ekofisk in November 1987. Following a successful waterflood pilot in the Tor formation, the decision was made in 1983 to waterflood the Tor formation in the northern two-thirds of the field at a cost of approximately $1.5 billion.

Platform 2/4K was designed to inject up to 375,000 bw/d and included 30 injection well slots. Only 20 slots were designated for the original waterflood project.

Laboratory experiments indicated that imbibition was good in the Tor formation but poor in the Ekofisk formation. However, a pilot project indicated that, contrary to laboratory results, water injection could be very effective in the Lower Ekofisk formation. Two potential reasons exist for the difference between field and laboratory results: (1) imbibition test results obtained at room temperature were assumed applicable under reservoir temperature and (2) viscous displacement may be playing a more important role than previously thought. The importance of temperature on imbibition is highlighted in Figure 5–37.

This pilot resulted in a $320-million waterflood expansion project to include waterflooding the Tor and Lower Ekofisk formations fieldwide.

Shortly after the waterflood expansion project was approved in 1988, drilling technology improved that permitted high-angle wells (more than 60°) to be drilled. This, combined with a relatively low-cost method for increasing the capacity of Platform 2/4K from 375,000 to 500,000 bw/d, led to a waterflood extension project approved in 1989. It is anticipated that the project, costing $110 million, will increase recovery as a result of extending the waterflood to the southernmost tip of the field, increasing the overall water injection rate, and extending gas lift to the southern portion of the field.

Upper Ekofisk pressure support, provided by water injection into the Lower Ekofisk and Tor formations, has started reducing the free-gas saturation, allowing low-GOR production from the Upper Ekofisk formation, especially in the flanks.

Production from Ekofisk field has increased from 70,000 bo/d in 1987 to 140,000 bo/d today. Two-thirds of that increase are attributed to the waterflood. The other one-third is the result of a remedial work program implemented over the past few years and improved communications among disciplines. Until the last three years, for every well completion added to Ekofisk, at least one was lost due to failure. More recently, however, failures decreased drastically while completion additions improved slightly (Sulak, 1991).

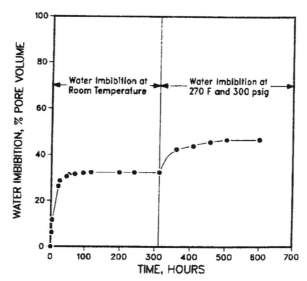

FIGURE 5–37 Temperature effect on water/oil imbibition data (after Thomas et al., 1987)

Alamein Field, Egypt (Oil Reservoir With Water Drive)

Alamein field is located about 80 miles southwest of Alexandria, Egypt. Production comes from a fractured dolomite. As of 1974, The reservoir had already produced over 30% of the original oil-in-place under the influence of an active water drive mechanism (El Banbi, 1974).

The reservoir is in an anticline of lower Cretaceous (Apian) age. Table 5–6 shows Alamein reservoir data. The water-oil contact has an area of about 3100 acres (Figure 5–38). Net pay thickness ranges between 70 and 200 ft.

The excellent recovery efficiency has been attributed to two main reasons: completion techniques and withdrawal rate policies.

Perforating was done with four ⅜ in. holes/ft using through-tubing jet perforators. The distance between the bottom perforation and the water-oil contact was usually 60 ft. The perforated intervals were generally located in the upper part of the pay zone.

All wells were acidized using 15% HCL at 25 to 100 gal/ft of perforated intervals. For example, the well whose logs are shown in Figure 5–39 was perforated initially from 8235 to 8255 ft. Distance to the water oil contact was about 100 ft. Following a 500 gal, 15% HCL acid treatment, the well produced at a rate of 5200 bo/d with a drawdown of only 90 psi. This well produced 3 MMstbo before the water cut reached 0.5%.

Later, these perforations were squeezed and interval 8214 to 8227 ft was open to production. Following an acid job, the well produced 3600 bo/d with a drawdown of only 46 psi. This well had accumulated 5.9 MMstbo by 1974.

Figure 5–39 shows the advance of the water table which was monitored with TDT logs. This information was useful in deciding on recompletions and drilling of additional infill wells.

Conventional buildup interpretation was not possible due to the immediate pressure stabilization obtained in all surveys. Therefore, a multirate test prior to placing the well in continuous production was carried out. This allowed estimates of some reservoir characteristics.

Maximum pressure drawdowns to avoid water coning were calculated using Muskat's equation. Whenever fractures or high porosity intervals were present below the perforations, the actual drawdown was kept well below the calculated maximum. This approach avoided coning problems.

Figure 5–40 shows the rate sensitivity of the reservoir. As the water cut increases, the well is choked back and there is an immediate decrease in fractional water cut. This minimized the upward movement of water.

The well of Figure 5–40 produced about 6 MMbo with less than 0.5% water cut. This is remarkable for a naturally fractured reservoir with water influx.

TABLE 5-6 Alamein reservoir data (after El Banbi, 1974)

Production area, acres	3,100
Original oil-water contact, feet	8,100
Average porosity, %	10.8
Average connate water saturation, %	23.7
Average residual oil saturation, %	38.5
Average permeability from Pls, md	2,000
Initial reservoir pressure, psi	3,613
Saturation pressure, psi	1,032
Solution gas, cu ft/bbl	205
Reservoir temperature, °F	191
Oil volume factor @ reservoir pressure	1.15
Oil viscosity @ reservoir conditions	1.42
Reservoir pore volume, acre-ft	23,800
Original oil-in-place, million STB	120

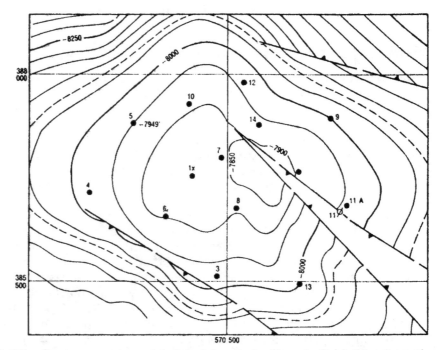

FIGURE 5-38 Contours on top of Apian dolomite in Alamein field, 80 mi. southwest of Alexandria, Egypt. Reservoir has strong natural water drive that will boost recoveries to over 37% without artificial lift or assisted recovery (after El Banbi, 1974)

Figure 5-41 shows a crossplot of bottom hole pressure vs. cumulative production. The maximum allowed pressure drop was 108 psi. As oil rate was reduced, the reservoir pressure increased significantly. Eight years after discovery of the field, the pressure was only 32 psi below the original reservoir pressure.

Reforma Area Fields, Mexico (Rapid Decline of Pressures)

Discoveries of the Reforma (Chiapas) and Samaria (Tabasco) fields transformed Mexico from an importer in 1973 to a net oil exporter in less than two years (Scott, 1975).

Exploratory wells Sitio Grande 1 and Cactus 1 led to the discovery of the Sitio Grande and Cactus fields in May 1972.

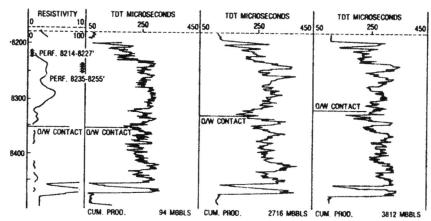

FIGURE 5–39 Example well completion illustrates method of perforating the top of the pay zone, well above the oil-water contact (after El Banbi, 1974)

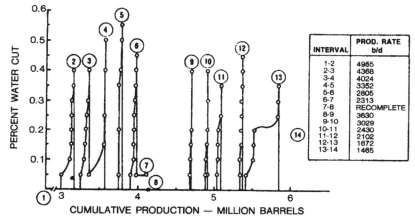

FIGURE 5–40 Rate sensitivity of the Alamein wells is demonstrated by history of the example (after El Banbi, 1974)

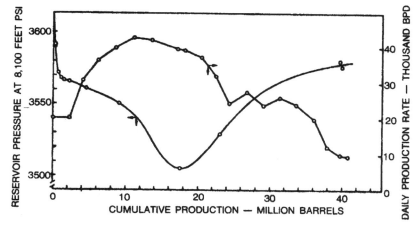

FIGURE 5–41 BHP vs. cumulative production performance for the entire Alamein field (after El Banbi, 1974)

Production in Sitio Grande field came from a Cretaceous dolomitic limestone of low matrix porosity. Excellent permeabilities were provided by caverns and microfractures. The structure is an anticline with an area of 12 square miles and a structural relief of 1650 ft.

Production in the Cactus field was tested in three Cretaceous beds, namely a calcarenite in the Upper Cretaceous and dolomites in the Middle and Lower Cretaceous.

One of the discovery wells, the Sitio Grande 1, was drilled in 1972 to a TD of 13,850 ft in 270 days. It was completed in a Middle Cretaceous dolomite at 13,652-685 and 13,596-629 ft. Figure 5–42 shows the drilling time, well program and geologic column of well Sitio Grande No. 1.

The other discovery well, the Cactus No. 1, was drilled in 1972 to a TD of 12,408 ft in 192 days. The well was completed in the Upper Cretaceous at 12,242-375 ft. Initial production was 1694 bo/d flowing through an 8 mm choke with a tubing pressure of 2659 psi. Gas-oil ratio was 2158 scf/STB.

Usual problems encountered in the Reforma area are summarized by Del Orbe-Valdiviezo (1975):

1. Poor rate of penetration in several zones
2. Lost circulation
3. Gasification
4. Frequent stuck casing
5. Frequent fishing operations
6. Salt water flows
7. Crumbling of hydrophillic shale in Oligocene, Eocene, and Palaeocene intervals
8. Overpressured zones

Problems were handled with efficiency. A typical casing program for some of the deepest wells at about 17,000 ft consisted of the following:

1. 30-in. OD conductor pipe
2. 20-in. OD surface casing

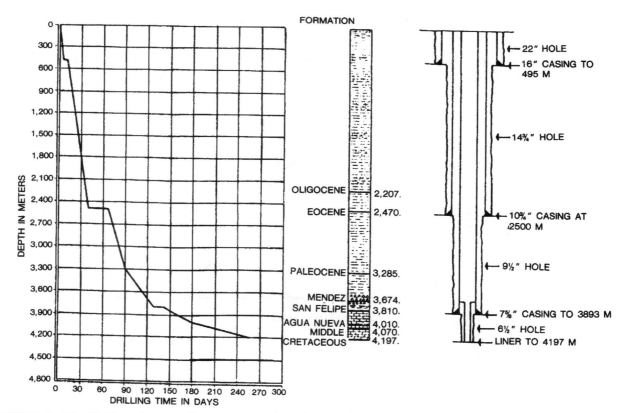

FIGURE 5–42 Drilling time, well program and geologic column of Reforma area discovery well, Sitio Grande No. 1, completed in May 1972 to 13,780 ft (after Del Orbe-Valdiviezo, 1975)

3. First intermediate 13⅜-in. OD casing
4. Second intermediate 9⅝-in. OD casing
5. Third intermediate 7-in. OD casing or first tubing string
6. First or second 4¼-in. OD tubing string

Average porosities range between 6% and 11%. Water saturations range between 13% and 20%. Several interference tests have allowed accurate determinations of permeability. Average values are 135 md for Sitio Grande and 16 md for Cactus.

Efforts to optimize ultimate recoveries led to a spacing of 3000 ft between wells in an hexagonal pattern. It was estimated that 19% of the original oil-in-place would be recovered in Sitio Grande at an assumed abandonment pressure of 1707 psi. Rapid decline of static pressures was characteristic of the Reforma area (Delgado & Loreto, 1975).

Consequently, plans were initiated to carry out a peripheral water injection program. Initial plans called for an injection of 150,000 bw/d through 10 wells at an average pressure of 3000 psi.

Handling of fluids at the surface was carried out in large installations where gas and oil were separated. Wells were kept under control and production in most cases was below capacity. Gas-oil ratios were carefully measured, bottom hole pressures were taken regularly, and attempts were made not to produce the wells below saturation pressures.

Asmari Reservoirs, Iran (Strong Segregation with Counterflow)

Andresen et al (1963) have indicated Asmari fields, although connected by a common aquifer, do not usually interact effectively during production life. The matrix is very tight and contains most of the hydrocarbons, while fractures are the channels that allow the fluids to flow.

The Asmari limestone of Lower Miocene to Oligocene age is a hard, compact rock, fine to coarse grained, and slightly marly anhydritic or sandy. It exhibits various degrees of recrystallization and dolomitization. The reservoir is folded in complex elongated anticlines.

Figure 5–43 shows regional Asmari aquifer conditions and the estimated original pressures. The coherence shown in this map suggests a continuous common aquifer. The isobaric lines suggest a pressure gradient of 10 psi/mile decreasing toward the southwest. This indicates water movement toward the southwest from charging outcrops located at the northeast.

Most outcrops have been eroded to elevations between 1500 and 2500 ft, which are too high to allow escape from the aquifer and to low to provide a charging head of surface water. Other local sinks are produced by oil and gas seepages.

Although the aquifer is common to various reservoirs it seems that communication is extremely poor, and no effective interference takes place between the various fields.

The presence of fractures at Asmari reservoirs is evidenced by mud losses, high productivities not correlatable to matrix permeabilities, pressure buildup characteristics, flowmeter surveys, and core analysis.

Table 5–7 shows the results of flow meter surveys in the Agha Jari field. It can be seen that oil enters through several fractured zones. Pressure communication within the reservoir for long distances suggests an intensive network of natural fractures. Pressure buildups have been analyzed using Pollard's (1959) method to obtain estimates of pressure differences between matrix and fractures.

Asmari reservoirs are highly fractured, as shown in the schematic illustration of Figure 5–44. The high degree of fracturing has suggested the numerical modelling of the Asmari reservoirs by identical matrix blocks which are separated by capillary discontinuities. The height is defined as the vertical extent of permeable matrix interrupted by capillary discontinuities.

During production from Asmari reservoirs, there are various active mechanisms, i.e., gas expansion, undersaturated oil expansion, solution gas drive, gravity drainage, and displacement by water imbibition. Figure 5–45 shows a schematic of fluids distribution in Asmari reservoir during production. There is an original gas cap. The original gas oil limit (GOL) has moved downwards to the current position generating a secondary gas cap.

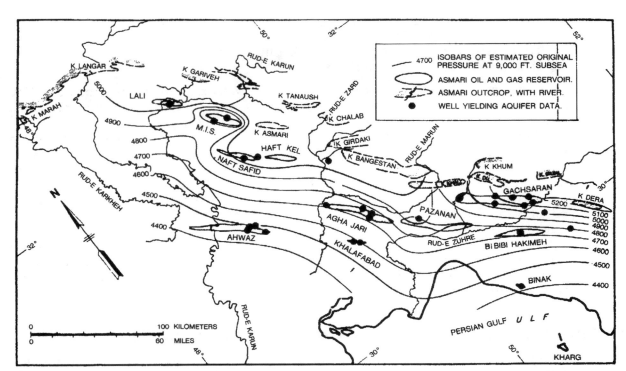

FIGURE 5–43 Regional Asmari aquifer conditions showing estimated original pressures (after Andresen et al., 1963)

TABLE 5–7 Agha Jari Results of Flowmeter Surveys (after Andresen et al, 1963)

Well number	Top of Asmari, ft	Total depth, ft	Inflow interval, ft	Percent of total flow
60	8126	8647	8405-19	16
			8456-61	17
			8469-74	8
			8501-08	19
			8519-23	14
			8540-46	26
54	6110	6933	6483-87	7
			6549-89	23
			6746-62	5
			6762-69	13
			6769-6835	18
			6835-43	23
			6843-TD	12
49	6378	7657	6393-97	5
			6508-11	35
			6515-21	9
			6549-61	8
			6573-TD	43
33	7072	7333	7072-78	18
			7079-83	32
			7103-12	3
			7127-46	23
			7148-7250	23
			7320-TD	1

FIGURE 5–44 Schematic of Asmari fissure development (after Andresen et al., 1963)

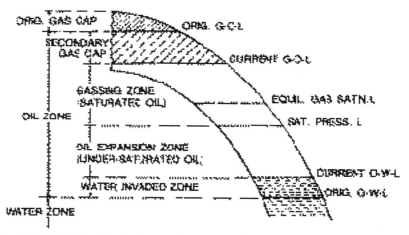

FIGURE 5–45 Fluid distribution in Asmari reservoirs during production (after Andresen et al, 1963)

There is an aquifer and the original oil water limit (OWL) has moved up to the current position generating a water-invaded zone where water imbibition takes place. There is a region of saturated oil where gas is coming out of solution forming a gassing zone.

In the early '60s mathematical treatment of the producing process included a material balance and water influx analysis which were carried out conventionally, although individual calculations were performed for each of the subdivisions of the reservoir.

The solution gas drive analysis was figured by accounting for the pressure difference between fractures and matrix.

The gravity drainage in the gas-invaded zone was calculated by considering blocks of identical dimensions whose vertical boundaries were gas-invaded fractures and whose horizontal separation was formed by capillary discontinuities. This analysis was carried out by two different methods: the relative permeability method in which drainage was assumed to start only when capillary isolation of each block was affected by the lowering of the gas-oil level in the fractures; and the capillary equilibrium method in which drainage was assumed to start from the moment in which the gas-oil level fell below the top of each block (Birks, 1955).

Displacement of oil by water was analyzed with a model developed by Aronofsky et al (1958) which made it possible to estimate recovery from the matrix due to water imbibition.

Andresen et al. (1963) indicated that for Asmari reservoirs with long production histories and good data, these methods worked very well and provided reliable forecasts. For newer reservoirs, projections were less reliable because many of the controlling reservoir parameters had to be estimated or derived by some indirect approach.

Although the above calculation techniques have been replaced by more modern simulation methods (Novinpour et al, 1994) the physical principles discussed above remain the same.

La Paz and Mara Fields, Venezuela (Fold and Fault-Related Fractures)

The Cretaceous limestone at La Paz field in western Venezuela was discovered in 1944. The matrix was very dense and had very low porosities and permeabilities (<0.1 md). Satisfactory flow rates were associated with extensive fracture systems (Smith, 1951).

The presence of fractures was usually associated with circulation losses. Fractures were located throughout the limestone section independent of stratigraphic position. Surveys revealed that, in general, production came from only a few zones of entry.

Core analysis indicated that most fractures were nearly vertical and occurred in closely spaced parallel planes. Fracturing intensity was directly associated with folding along a main NNE-SSW axial trend. Well productivity was better in La Paz where the upfolding was more intense.

The water table was suspected to be about 11,000 ft below sea level in both La Paz and Mara fields. This provided an extraordinary oil column of about 6000 ft in Mara and 7000 ft in La Paz.

The high degree of shattering in the crest of the structure was probably responsible for lack of seismic reflection. La Paz wells, with small penetration in the limestone, suggested the high degree of shattering through productions of over 20,000 bo/d. However, it was impossible to establish a relationship between productivity and depth. Figure 1–42 (Chapter 1) presents a structural map on La Paz and Mara fields (Smith, 1951).

Well P-83 was one of the upper wells in the structure and produced at very low rates. This indicated that the well was not connected to the network of fractures. Well P-111, on the other hand, was located in the flank and produced over 16,000 bo/d.

Figure 5–46 shows a schematic of jointing pattern for La Paz-Mara fields. A probable cause of fracturing was uplifting of the crestal area.

After development and expansion of the gas cap were noticeable, wells were located downflank along fault zones to avoid high gas oil ratios. There was some evidence that wells located in the upthrown were better producers than wells located downthrown. This was probably due to tension that provided good fracture permeability in the upthrown. By contrast, the downthrown was subjected to compression that reduced the fracture permeability drastically.

Wells drilled in areas of intense fracturing, but associated with reverse faulting, resulted in dry or poor producers. This was probably due to secondary infilling and cementing of the fractured system.

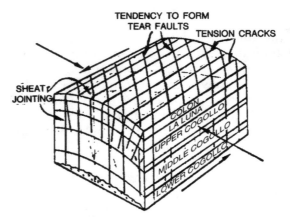

FIGURE 5–46 Theoretical fracturing pattern for La Paz-Mara fields (after Smith, 1951)

Maximum width of fractures observed in cores was ¼ in. It is important to emphasize this width as it is commonly stated in the literature that all natural fractures are "paper thin" or hairline fractures with negligible hydrocarbon storage.

Drilling difficulties arose from lost circulation in the limestone. A typical drilling program consisted of 17½ in. hole drilled to about 1000 ft and 13⅜ in. casing cemented to surface. Afterwards, the hole was drilled with either 12¼ in. or 8½ in. bits depending on the desired TD.

Caving problems were greatly reduced with emulsion muds. Circulation losses varied from a few barrels to thousands of b/d. To avoid possible damage, successful efforts were made to reduce the hydrostatic head as much as possible within safety limits. When mud losses were excessive, batches of limestone chips were introduced in the mud system and pulverized by the bit. This pulverized limestone acted as a filling and plugging agent. If necessary, this plugging agent was easily removed later with acid jobs.

The typical behavior of naturally fractured reservoirs in the Cretaceous of Venezuela was characterized by a drastic increase in gas-oil ratios once the bubble point was reached. This GOR increase is typical of at least 90% of the naturally fractured reservoirs I am familiar with.

Planning and Cooperation in Early Stages of Development Pays Off (Vertical Communication Through Fractures)

Freeman and Natanson (1963) have presented an excellent description of how close cooperation between reservoir and field engineers led to the evaluation of the basic characteristics of a naturally fractured carbonate in early stages of development.

Figure 5–47 shows a structural map on top of the upper pay, a massive tight limestone at a depth of about 6000 ft. Comparison of Figure 5–47 with Figure 1-39 leads to the conclusion that the reservoir described by Freeman and Natanson (1963) is the Ain Zalah field in Iraq. Wells in the crest and toward the crest of the structure produced between 1,000 and 20,000 bo/d from the upper pay. Wells downstructure did not prove any oil, and there was no evidence of edge water. The oil accumulation thus seemed to be isolated and restricted only to the fracture network.

As development continued, a new reservoir was discovered about 2000 ft deeper. This new reservoir was better than the one in the upper pay. It was also naturally fractured, and it seemed to be connected to an aquifer. Oil from the lower reservoir was heavier and more undersaturated.

There was concern about the commerciality of the reservoir in the upper pay. It was felt that the reservoir could deplete very quickly due to its fracture nature. At this stage, there was only one production well in the most promising lower pay. However, careful planning helped to clarify the situation.

In a preparatory period, it was decided to measure any pressure changes at the surface with dead weight testers. From January to nearly the end of May, everything was normal and pressure variations were not larger than 0.2 psi. By the end of May, well G was acidized and produced at about 3000 bo/d. Surprisingly, all other wells showed a pressure drop ranging from 1 to 2½ psi (Figure 5–48). When well G was shutin, all pressures recovered slowly but steadily. This behavior raised the question about the source of pressure recovery in the upper pay.

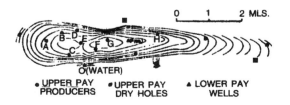

FIGURE 5–47 The upper pay (after Freeman & Natanson, 1963)

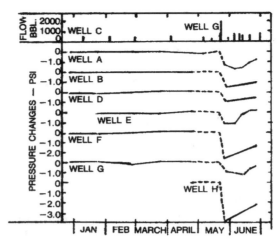

FIGURE 5–48 Upper pay pressure changes-preliminary period (after Freeman & Natanson, 1963)

The various logical possibilities were not reasonable in this case. For example, quick replenishment from the matrix was impossible because the matrix was very tight. There was no indication of water edge in the upper pay. And there was evidence that there were about 2000 ft of tight, unfractured limestone separating the upper and lower reservoirs.

Without having an answer to this question, a three-phase program was designed to bring the reservoir on to production. The first phase consisted of producing well P of the lower pay at about 23,000 bo/d during five days to fill one of the storage tanks. Pressure drops were recorded very carefully in wells A to H of the upper pay and well Q (water-bearing) of the lower pay (Figure 5–49). The results indicated that pressures in all wells were falling while well P (lower pay) was being produced. Pressure drops in the upper pay ranged from about 1 psi in H to about three-psi in wells A and B. This gave a clear indication there was a certain degree of communication between the upper and lower pay.

The second phase was carried out following a three-day general shutin. The upper pay of well C produced at a rate of 20,000 bo/d during five days. After five days of production, pressure in the nearest wells had dropped almost 100 psi (Figure 5–50). As soon as well C was shutin, pressure started to build in the observation wells. Water well Q, whose pressure was still rising when the second phase was begun, levelled off and only started to build up again some days later when pressures in the upper pay had substantially recovered.

In the third phase well P was open to production again for five days at a rate of about 20,000 bo/d. Findings with regard to communication were again very noticeable. Analysis of the upper pay pressure distribution provided the approximate location where the upper and lower pays were communicated.

Figure 5–51 shows upper pay pressure distribution following the first flow of the upper pay in well C. The wells are projected along the axis of wells A and G. The pressure drops are shown as a function of time for each well. At zero days there is a sink in well C, resulting from the production period. Pressures continue to level off with time, and at 25 days there is a west-east gradient.

The gradient at 25 days is more noticeable in Figure 5–52, where the pressure drop scale has been expanded. After two months, pressure drops are essentially the same at all wells, i.e., the gradient has virtually disappeared. This indicates that during this period oil was flowing from the lower to the upper pay until an equilibrium was reached.

As production continued from the lower pay with its corresponding pressure drop, oil started to flow from the upper to the lower pay, and an east-west gradient was generated (dashed lines in Figure 5–52). The west-east gradient during inflow and east-west gradient during outflow clearly indicated that the connecting channels between upper and lower pay were located in the region of wells A and B.

Freeman and Natanson (1963) calculated that approximately 40% of the total pore volume was made from fractures in the upper pay. They considered that the fractured system

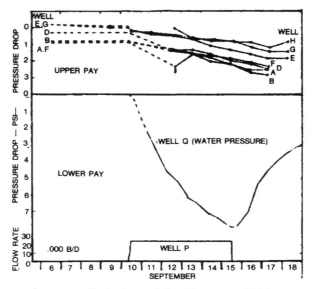

FIGURE 5–49 Pressure changes—first phase (after Freeman & Natanson, 1963)

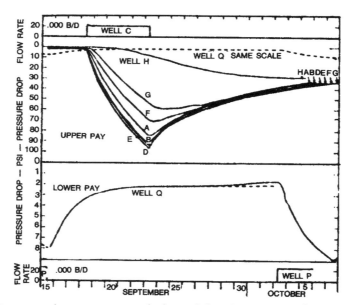

FIGURE 5–50 Pressure changes—second phase (after Freeman and Natanson, 1963)

in the upper pay included the oil. The tight matrix contained mainly water. So the upper pay was essentially a type C reservoir.

The information obtained from this early analysis made it possible to consider the upper pay as the production target. In this sense, the connecting channels and the upper pay itself were considered simply as some kind of pipeline that allowed production from the more prolific lower pay.

Bagre Field, Mexico (Strong Water Drive)

This field is located offshore in the Gulf of Mexico. Production is obtained from the El Abra carbonate Formation which contains matrix porosity, fractures and caverns with tight intercalations (Martinez, 1975).

Bagre is composed by three structures which appear to form independent reservoirs. The north and south portion of the field are denominated B and A as shown on Figure 5–53.

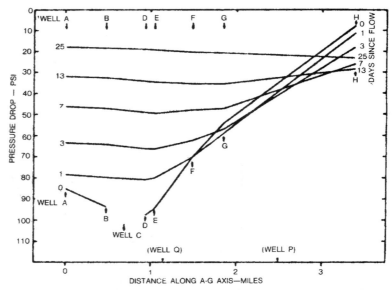

FIGURE 5–51 Upper pay pressure distribution -1 (after Freeman & Natanson, 1963)

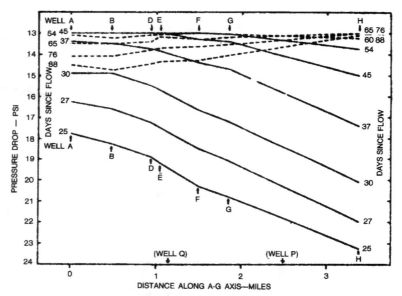

FIGURE 5–52 Upper pay pressure distribution -2 (after Freeman & Natanson, 1963)

An interesting feature is a very active water drive without a decline in pressure.

Average values of porosity and water saturation are as follows:

	Bagre A	**Bagre B**
ϕ	0.18	0.19
S_w	0.21	0.16

Since porosity is large this naturally fractured reservoir is probably of type A (refer to Figure 1–6). Residual oil saturation is high in the order of 50%.

Oil-in-place was calculated to be 46.5 MMstbo for Bagre A and 65.4 MMstbo for Bagre B by volumetric means using isopachous and isohydrocarbon maps. Oil recoveries for Bagre A and B were determined to be 36% and 40%, respectively, using a tank material balance for undersaturated reservoirs, which for the case in which there is no pressure drop ($\Delta p = 0$) becomes:

$$W_e = N_p B_o + W_p B_w \qquad (5–1)$$

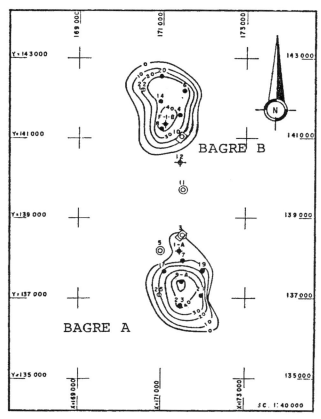

FIGURE 5–53 Isopachous Map, Bagre Field, Gulf of Mexico (after Martinez, 1975)

where W_e is water entrance, N_p is cumulative oil production, W_p is cumulative water production, B is the formation volume factor; and the subscripts o and w correspond to oil and water respectively.

Volumes of invaded rock (ΔV_{ri}) were calculated from:

$$\Delta V_{ri} = \frac{W_e - W_p B_w}{\phi\,(1 - S_w - S_{or})\,E_b} \qquad (5\text{--}2)$$

where ϕ is fractional porosity, S_w is fractional water saturation, S_{or} is fractional residual oil saturation and, E_b is the fractional sweep efficiency which was assumed to be equal to 1.0 due to the very active water drive. The position of the water oil contact during the life of the reservoir was calculated based on the remaining volume of rock (V_r), and the original rock volume (V_{ro}) from:

$$V_r = V_{ro} - \Delta V_{ri} \qquad (5\text{--}3)$$

For the case of Bagre A the following polynomial was found to represent the position of the water-oil contact (WOC):

$$WOC = 48.542 - 4.564 V_r + 0.345 V_r^2 - 0.0146 V_r^3$$
$$+\, 0.287 \times 10^{-3} V_r^4 - 0.207 \times 10^{-5} V_r^5 \qquad (5\text{--}4)$$

Equation 5–4 was obtained from the plot of net pay vs. rock volume presented in Figure 5–54.

Since the pressure is maintained practically constant, the gas oil ratio was maintained also constant at 1,324 scf/STB during forecasts of reservoir performance.

Crossplots of net present value vs. number of wells allowed determination of the optimum economic spacing. This type of plot is presented in Figure 5–55 using as a parameter different values of oil price. From this graph the optimum number of wells is 5.

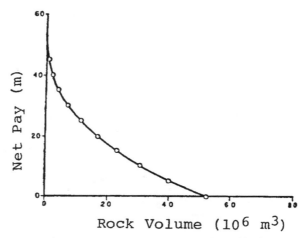

FIGURE 5–54 Net pay vs. volume of rock, Bagre A Field, Gulf of Mexico (after Martinez, 1975)

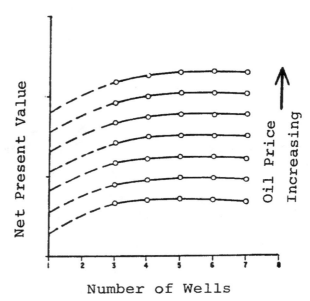

FIGURE 5–55 Crossplot used for choosing optimum number of wells in Bagre Field, Gulf of Mexico (after Martinez, 1975)

Norman Wells, Canada (Waterflood with Slanted Wells)

The Norman Wells oil reservoir is located about 900 miles north of Edmonton, Alberta in the Northwest Territories of Canada. The discovery well was drilled in 1920 by Imperial Oil. The remote location of the field delayed development until 1940.

Cumulative production to the end of 1980 was less than 4% of the original oil-in-place estimated at 630 MMstbo. The production mechanism had been by solution gas drive (Kempthorne and Irish, 1981).

Geologically, the Norman Wells are located in the Devonian Kee Scarp formation limestone reef buildup, 15 miles long and 5 miles wide. The reef is relatively shallow at depths ranging between 1000 and 1600 ft. Reservoir thickness is 360 ft. Regional structural dip is about 4.5° to the southwest. The bulk of the reservoir is under the Mackenzie River.

Oil is trapped stratigraphically in the extreme updip end of the reef buildup. There is no gas cap, and a limited water leg provides no pressure support.

Permeability is usually restricted to 10 md or less. Porosity can reach up to 24% and thus the reservoir is most likely of type A (see Figure 1–6).

The presence of natural fractures has been verified by direct observations in cores. There are numerous open and cemented vertical or nearly-vertical natural fractures. Widths can be as large as 0.4 in and lengths can reach 16 ft. Sonic log cycle skipping can be matched to the higher concentrations of fractures in cores. Zones of cemented fractures correspond to fast travel times, while intervals of open fractures can be matched to slow travel times. Figure 5–56 shows a correlation between fractures in cores and sonic log response. Table 5–8 shows some reservoir characteristics of the Norman Wells.

Oriented cores have indicated a preferential orientation N 26° E with a standard deviation of ± 21°. The same general direction has been obtained from a photo lineaments study. It seems to be that width of the fractures decreases as depth increases from 1150 to 1640 ft. This is due to a smaller overburden in the shallower areas.

A very interesting aspect of the Norman wells is that the reservoir is being waterflooded with directionally drilled wells. The need for directional wells is due to the bulk of the reservoir being under the MacKenzie River.

The fractures along the mean orientation are probably better developed and longer than those in other directions. Thus, on-trend fractures are expected to contribute more to flow in the reservoir.

Figure 5–57 shows a schematic of elongated target areas with similar oil recovery, R. Figure 5–58 presents the drilling platform; the wells enter the reservoir at one edge of the target area, pass through the idealized well location midway through the reservoir, and exit at the other end of the target area.

Notice that in this reservoir improved recoveries are expected by using a rectangular rather than a square well pattern configuration. There is a trade off in recovery efficiency for wells that can be made more vertical or more on-trend with the major fracture orientation.

Major factors influencing the effect of wells drilled directionally on pattern waterflood oil recovery are (1) average fracture orientation which at Norman wells is N 26°E, (2) average fracture length which is 164 ft, and (3) effective vertical permeability. Optimum spacing is estimated at 16 acre/well.

Figure 5–59 shows a schematic of surface installations. Located in the mainland, the complex includes a 30,000 b/d crude stabilizing unit, a 40,000 b/d water injection plant, and a 20,000 hp power plant (*Oil Week*, August 9, 1982).

The original budget was $800 million out of which 25% was planned to be spent in new plant construction. Waterflooding was expected to increase oil rate from 2,800 to 25,000 bo/d, and recovery from a primary of 17% to a secondary recovery of 42% of the original oil-in-place. This is significant for a waterflood in a naturally fractured reservoir which presents many indications of being preferentially oil-wet.

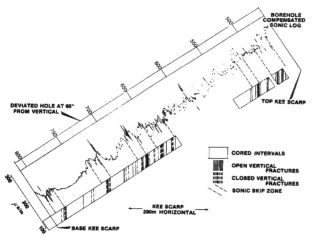

FIGURE 5–56 Fracture frequency and sonic log response from deviated Well 39X (after Kempthorne and Irish, 1981)

TABLE 5–8 Norman Wells Reservoir Characteristics (after Kempthorne and Irish, 1981)

Average reservoir depth, m	400
Average reservoir net pay, m	110
Average porosity, %	9.8
Reservoir temperature, °C	21
Initial reservoir pressure (-345 m subsea), kPa	4900
Bubble-point pressure, kPa	4700
Pour point, °C	–40
Crude gravity, °API	38.5
Viscosity of crude (21°C, 4900 kPa), kPa.s	1.32
Average horizontal permeability from core, md	4
Average vertical permeability from core, md	1
Connate water saturation, %	10
Residual oil saturation, %	31
End point water relative permeability	0.4
Critical gas saturation, %	2

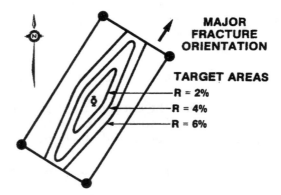

FIGURE 5–57 Norman Wells elongated target areas with similar oil recovery (after Bacon and Kempthorne, 1984)

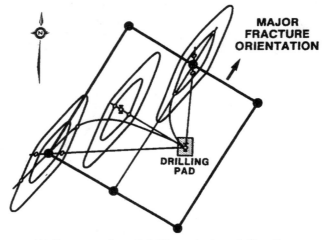

FIGURE 5–58 Norman Wells examples of drilling choices (after Bacon and Kempthorne, 1984)

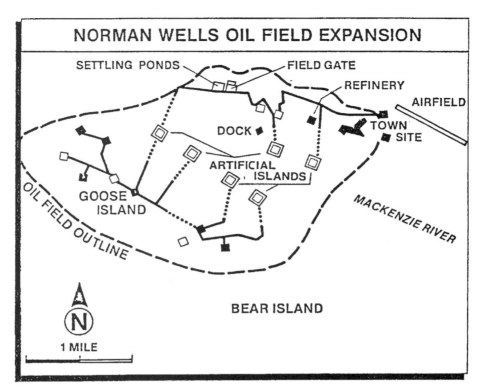

FIGURE 5–59 Norman well surface installations (from *Oil Week*, 1982)

Waterton Reservoir, Canada (Retrograde Condensate)

A model developed by Castelijns and Hogoort (1984) has been used for calculating recovery in the Waterton gas condensate reservoir of Alberta, Canada. In this reservoir, condensate may have accumulated at a down dip location by gravity drainage, as indicated by some production tests.

The reservoir is a heavily fractured dolomite and has been producing since 1962. Reservoir pressure had declined from 4600 psi initially to some 1600 psi by 1982. Typical rock and fluid data are given in Table 5–9. Matrix permeability is in the range of 0.1-1 md. Effective porosity is 0.025. Dip angle of reservoir layers is 30°, and thickness of permeable dolomite beds ranges from 0.3 to 5 m (1 to 16.4 ft).

Pressure history is presented in Figure 5–60. The condensate/gas interfacial tension is extremely low ($\cong 0.2$ dyne/cm) near the initial or dew point pressure (4600 psi). However, it may increase drastically upon pressure decline as shown on Figure 5–61.

Table 5–10 shows measured and calculated fluid compositions for $C_1 - C_6$ using Peng-Robinson (1976) equation of state and Kao's (1978) phase-equilibrium algorithm. The matched composition does not necessarily represent the actual gas composition; rather it gives a good description of its retrograde behavior as illustrated in Fig. 5–62.

Condensate and gas relative permeability and capillary-pressure curves were represented by the equations:

$$k^\star_{ro} = S^{\star n_o}_o \qquad (5\text{–}5)$$

$$k^\star_{rg} = S^{\star n_g}_g \qquad (5\text{–}6)$$

$$\frac{d}{d\,S^\star_o}\,P_{c,D} = -S^{\star n_c}_o \qquad (5\text{–}7)$$

where $n_o = 4$, $n_g = 2$, and $n_c = -1.6$. The first two values are considered typical of unfractured low permeability carbonate rock, such as that found in the Asmari limestone. The last value was estimated from actual Waterton drainage capillary pressure as shown on Figure 5–63.

TABLE 5–9 Typical Waterton Rock and Fluid Properties (after Castelijns and Hagoort, 1985)

Matrix permeability	k	$= 0.1 - 1. \times 10^{-15}$ m²
		(0.1 – 1 mD)
Effective matrix porosity	$\emptyset_m$	= 0.025
Dip angle	α	= 30°
Layer thickness	h	= 0.3 – 5 m
Dew-point pressure	p_{dp}	$= 31.7 \times 10^6$ Pa
		(4600 psia)
Reservoir temperature	T_R	= 343° K
		(156°F)
Condensate viscosity at dew point	μ_o°	$= 0.49 \times 10^{-3}$ Pa.s
		(0.49 cP)
Gas viscosity at dew point	μ_g°	$= 0.051 * 10^{-3}$ Pa.s
		(0.051 cP)
Condensate-vapour density difference at dew point	$\Delta\rho_{og}^\circ$	= 397 Kg/m³
		(24.77 lb/cf)
Condensate-vapour interfacial tension at dew point	σ_{og}°	$= 0.214 \times 10^{-3}$ N/m
		(0.214 dyne/cm)

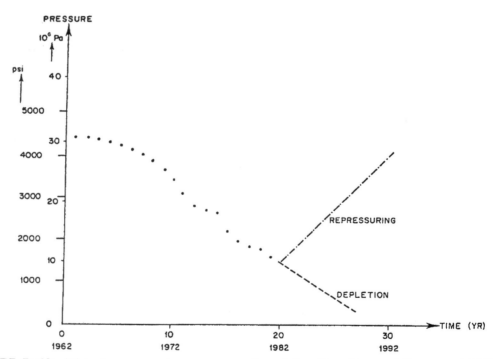

FIGURE 5–60 Waterton reservoir pressure vs. time (after Castelijns and Hagoort, 1985)

Using the analytical models of Castelijns and Hagoort (1984), the conclusion was reached that restoring the reservoir pressure would significantly increase the amount of condensate entering the fractures, specially due to reduced capillarity. Furthermore, the condensate still draining through the matrix rock when repressuring starts would evaporate and be produced with the gas stream.

Figure 5–60 shows pressure vs. time under natural depletion and with repressuring. Figure 5–64 presents the effect of pressure increase on the mean saturation of the condensate which, in the interior of the matrix block, is migrating towards the bottom layer. With methane (CH$_4$) injection, most of this condensate revaporizes and can be recovered via the vapor stream. Nitrogen (N$_2$) also causes a significant reduction in retrograde liquid loss, and consequently an improved recovery.

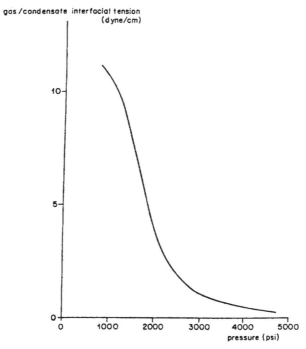

FIGURE 5–61 Waterton reservoir fluid: Condensate/Gas Interfacial Tension (after Castelijns and Hagoort, 1985)

TABLE 5–10 Comparison Between Measured and Matched Data For Waterton Reservoir Fluid (after Castelijns and Hagoort, 1985

	Measured		**Matched**
1. dew-point pressure (psia)	4595		4596
2. molecular weight	29.36		28.37
3. (composition) (mole %)			
methane	65.49		64.49
ethane	3.93		3.93
propane	1.53		1.53
isobutane	0.32		0.32
n-butane	0.92		0.92
isopentane	0.52		0.52
n-pentane	0.50		0.50
n-hexane	1.12		1.12
		n-heptane	2.61
		napthalene	1.75
		n-decane	0.35
C7+	5.19	biphenyl	0.16
		n-tetradecane	0.11
		terphenyl	0.20
		n-C24	0.01
nitrogen	0.97		0.97
CO_2	3.48		3.48
H_2S	16.03		16.03

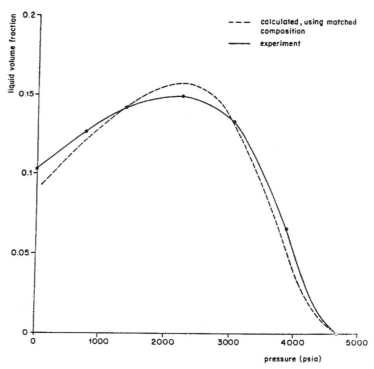

FIGURE 5–62 Waterton reservoir fluid: constant volume depletion test ($T_{Reservoir}$ = 156°F) (after Castelijns and Hagoort, 1985)

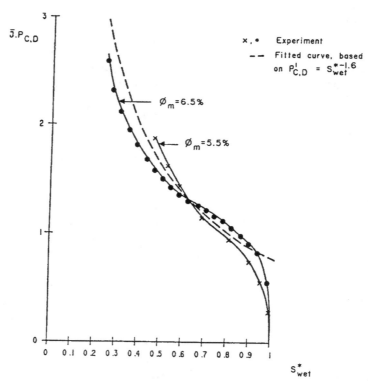

FIGURE 5–63 Waterton drainage capillary pressure curve $\bar{J}$ is Leverett's function (after Castelijns and Hagoort, 1985)

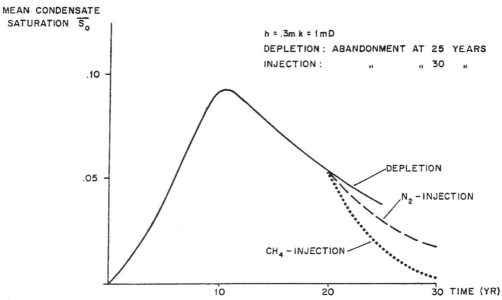

FIGURE 5–64 Mean condensate saturation S_o, in matrix rock; effect of repressuring (after Castelijns and Hagoort, 1985)

Lacq Superieur Field, France (Steam injection)

This reservoir is located in southwest France near the Pyrenees Mountains. Production began in 1950, and 80 wells were drilled for its development (Sahuquet and Ferrier, 1982). The reservoir of interest is the lower Senonian limestone of Upper Cretaceous age and is heavily fractured.

A summary of oil characteristics under field conditions is presented in Table 5–11. Reservoir data are displayed in Table 5–12. The fracture permeabilities are attached to single point properties, the fracture porosities are attached to bulk properties.

A pilot steam drive was initiated in 1977. Figure 5–65 shows a map of the pilot area. It is located in the central portion of the fracture limestone zone, near the top of the structure. The injection well (La 87) was drilled and completed thermally. Other wells were old and were not completed for steam injection. The pattern area is about 35 acres. Distances between injection and production wells range between 492 and 1312 ft.

Figure 5–66 shows the production history of various wells in the pilot. Steam injection was started in early October 1977. Only three months after injection began there was a noticeable increase in oil rates in wells La 85, La 2, and La 4 located as far as 1310 ft from the steam injection well. A production peak was achieved by mid-1979.

Figure 5–67 shows a plot of GOR vs. time. Well La 2 shows a GOR increase from 8 to 32 m^3/m^3. Analysis of associated gas indicated that this increase resulted from a dilution by CO_2. Based on various experiments the CO_2 generation can be explained by decomposition of the calcareous rock by steam.

By the end of June 1980 incremental oil production obtained from wells La 2, La 4, and La 85 was in the order of 176,000 stbo. At the same time 162,000 tons (147,000 Mg) of steam were injected.

A thermal, bidimensional, three-phase (oil, water and steam) numerical model using a steam-tube configuration was used to interpret the pilot. Comparison of actual and simulated results was good.

Results of this pilot led to the conclusion that a naturally fractured reservoir can be treated efficiently and oil recoveries can be significantly improved by a properly designed steam-drive process. Heat transfer was found to be efficient without early heat or steam breakthrough. From a thermal point of view the reservoir was found to have a homogeneous behaviour, as heat conduction smooths the temperature profile between the fracture and the matrix. Soon after the beginning of the steam injection, oil is expelled from the matrix into the fractures.

TABLE 5–11 Lacq Superieur Oil Characteristics Under Field Conditions (after Sahuquet and Ferrier, 1982)

Density, lbm/cu ft (kg/m³)	55.7 (893)
Viscosity, cp (Pa.s)	17.5 (17.5×10^{-3})
Saturation pressure, psi (MPa)	116 (0.8 MPa)
GOR, vol/vol	11
Formation volume factor	1.04

TABLE 5–12 Lacq Superieur Field-Reservoir Data (after Sahuquet and Ferrier, 1982)

Depth, ft (m)	1,970 to 2,300 (600 to 700)
Average thickness, ft (m)	400 (120)
Oil gravity, °API (g/cm³)	21.5 (0.925)
Oil viscosity at 140°F (60°C), cp (Pa.s)	17.5 (0.0175)
Reservoir temperature, °F (°C)	140 (60)
Reservoir pressure, psi (MPa)	870 (6)
OOIP, MMbbl (10^6 m³)	125 (20)

Characteristics of Fractured Carbonated Zone

Matrix blocks	
average permeability, md	1
average porosity, %	12
water saturation (at pilot start),	
% of matrix pore volume	60
Fracture Network	
permeability, md	5,000 to 10,000
average porosity, %	0.5
water saturation (at pilot start),	
% of fissure volume	100

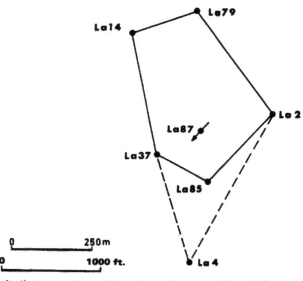

FIGURE 5–65 Map of pilot area, Lacq Superieur, France (after Sahuquet and Ferrier, 1982)

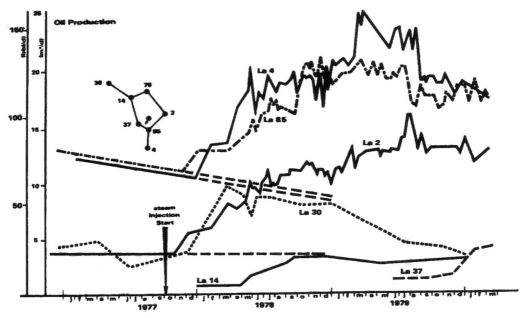

FIGURE 5–66 Production history, steam pilot, Lacq Superieur, France (after Sahuquet and Ferrier, 1982)

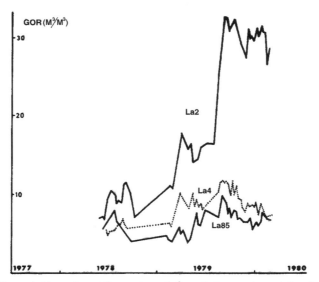

FIGURE 5–67 GOR evolution, Lacq Superieur (after Sahuquet and Ferrier, 1982)

Austin Chalk, Texas (Strongly Hydrocarbon-Price Dependent)

The Austin chalk of Cretaceous age is naturally fractured and generally parallels the Gulf Coast from Mexico to Alabama in a trend more than 50 miles wide (Holifield, 1982). It is a white to gray limestone intermixed with shale, and was formed during the late Cretaceous period some 60 to 90 million years ago. Migration of oil into the fractures was probably contemporaneous with the faulting that occurred shortly after burial. In some areas the Austin chalk is a naturally fractured reservoir of Type A with most of the hydrocarbon storage in the matrix. In other areas it is a reservoir of Type C with all the hydrocarbon storage in the fractures and the tight matrix acting as a seal that does not allow oil to escape via the "spill point" (see Figure 1–52). Some times the tight Austin chalk matrix acts as an impermeable barrier between wells.

The Eagleford shale, which underlies the Austin Chalk, is the source bed. Some good wells are found in areas where the Eagleford is thick (more than 200 ft). When the Eagleford is thin, wells often have a tendency to produce water. The Austin chalk is overlain by the Taylor formation (Figure 5–68). Some geologic aspects of this reservoir were presented in Chapter 1.

Natural fractures usually trend from the southwest towards the northeast parallel to the regional strike. Natural fractures at the Austin chalk can be classified as regional. Formation thicknesses vary greatly from 400–500 ft near Pearsall, Frio County, to approximately 1500 ft in the west, and 150 ft in the northeast over the San Marcos arch in Atascosa County.

Historically, one of the hottest areas in the Austin chalk has been around Giddings (Lee County) northwest of Houston. From the discovery date of the Giddings field in October 29, 1960 through December 31, 1981, operators completed about 2000 wells. More recently the Austin chalk has been the object of many successful horizontal well completions.

Most of the production in this field comes from the lower 200 feet. Matrix permeability is usually less than 0.1 md. Matrix porosity ranges between 0% and 9%.

In the deeper chalk, fractures provide both the necessary porosity and permeability (Type C reservoir), as it appears that significant matrix porosity is non-existent at greater depths (Holifield, 1982). In shallower reservoirs some matrix porosity is typically present (Type A reservoir).

In general, full potential of the well remains unknown until the well is cased, perforated and stimulated. A typical hydraulic fracturing job as discussed by Parker et al (1982) was presented in Chapter 2 (refer to Figure 2–55). One idea has been to stimulate the wells with large frac treatments and high pumping rates when the natural fractures are not intercepted in the hope of establishing communication with the network of natural fractures. When the well intercepts natural fractures, the treatment is carried out at low pumping rates, using low flow rates and small sand volumes in the hope of propping the natural fractures. In this author's experience efforts to prop natural fractures in the Austin chalk and many other naturally fractured reservoirs worldwide have not been very successful.

Routine and specialized core analysis are valuable in the Austin Chalk (Craft, 1982). Full core analysis is desirable. Unfortunately, many of the cores part along fracture planes when they are removed from the core barrel.

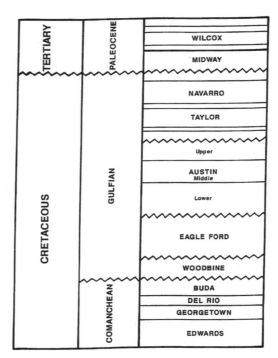

FIGURE 5–68 Stratigraphic section showing position of Austin chalk (after Fritz et al., 1991)

Many of the logs discussed in Chapter 3 have been found to give reasonable results (Fertl and Hotz, 1982), including the caliper, density, gamma ray, spectralog, sonic, resistivity, dipmeter, mud filtrate resistivity; and various interpreted logs such as secondary porosity index, pulse neutron vs. gamma ray, epilog, and fracture completion log. All of these devices and interpretation methods are presented in Chapter 3. A method of well evaluation and acreage development using a four-arm dipmeter has been published by Schafer (1980). Detection of fractures using the above suite of logs has been discussed by Julian (1982). More recently imaging devices such as Schlumberger's formation micro-scanner (FMS), formation micro-imaging (FMI), and Western Atlas circumferential borehole imaging log (CBIL) have been reported to provide good results in both vertical and horizontal wells in the Austin chalk.

Well testing data have been generally analyzed with log-log plots of Δp vs. time and conventional Horner plots (Claycomb, 1982). Much information can be gained from qualitative inspection of the early and late regions in these types of graphs. Quantitatively, it is sometimes possible to calculate average reservoir pressure, in-situ permeability, skin and fracture length. Under favorable conditions it is also possible to make estimates of the storativity ratio, ω, and the interporosity flow parameter, λ.

Forecasting of reservoir performance has sometimes been conducted using specialized type curves developed exclusively for Austin chalk wells (Poston et al, 1991; Chen et al, 1991).

The major factor affecting the economics of the Austin chalk play has been the decline in oil price (Holditch and Lancaster, 1982) followed by a considerable increase in operating expenses. Technology, even though important cannot reverse the trend by itself. For example, in spite of tremendous advances in hydraulic fracturing technology during the last four decades, the production from the average Austin Chalk vertical well has not really changed. Consequently, oil price was the unique parameter that made the difference and led to the Austin Chalk boom in 1980–1981.

The question is, from a technological point of view, have we really advanced in properly increasing recoveries with vertical wells, or as pointed out by van Everdingen and Kriss (1978, 1980) are oil recoveries today lower than in the past?

Something that has definitively improved the economics of the Austin chalk has been horizontal wells drilled perpendicular to the orientation of the fractures. Figure 5–69 shows initial rates of 48 horizontal wells drilled in the chalk. Two wells (4.2% of the wells) started with rates over 2500 bopd. Three wells (6.2%) started producing between 1000 and 2500 bopd. Eleven wells (22.9%) produced between 500 and 999 bopd. Fifteen wells (31.2%) started producing between 10 and 499 bopd, and seventeen wells (35.4%) started producing between 0 and 99 bopd.

FRACTURED CHERTS

Monterey Formation, California (Super-complex lithology)

The Miocene Monterey formation is a naturally fractured reservoir composed of rocks with very large proportions of silica. The lithology is extremely complex; it is a combination of siliceous shale, chert, phosphatic intervals, opaline limestone, quartz, dolomite, clay, diatomite, and other minerals (Grove and Whittaker, 1985). Because of this lithologic complexity the formation is interchangeably called Monterey "shale" or Monterey "chert." The Monterey thus could also be discussed under the heading of "Fractured Shales." The Monterey is both source rock and reservoir rock, and is considered to be preferentially oil-wet.

The formation comprises an enormous volume of sedimentary rock in the western half of California. Calculations by Pisciotto and Garrison (1981) suggest the onshore and offshore portions of the Monterey occupy as much as 100,000 cubic kilometres.

Figure 5–70 is a map of Neogene marine basins and extent of Monterey deposition. The strata are widespread in coastal California, and extends as far south as Newport Harbour and as far north as the Eel River basin.

Some geologic aspects of this reservoir were presented in Chapter 1. According to Redwine (1981), naturally fractured reservoirs in the Monterey area can be explained by a

AUSTIN CHALK HORIZONTAL WELLS
INITIAL PRODUCTION (B/D)

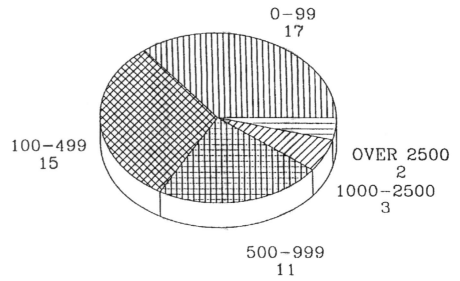

0-99
17

100-499
15

OVER 2500
2

1000-2500
3

500-999
11

FIGURE 5–69 Initial production of horizontal wells in the Austin chalk in bo/d (after Fritz et al., 1991)

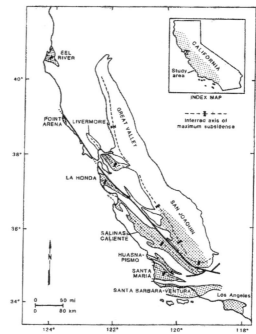

FIGURE 5–70 Neogene marine basins and extent of Monterey deposition (dotted) (after Isaacs, 1984)

hypotheses involving repeated episodes of rock dilation followed by natural hydraulic fracturing, all produced by episodic but continued tectonic compression of the region, where the principle maximum compressive stress is oriented approximately northeast–southwest. Dilation was explained in Chapter 1 with the use of Figure 1-19, and was indicated to be responsible for migration and accumulation of hydrocarbons in some naturally fractured reservoirs.

In the Mussel area the fracturing and oil accumulation occurred down the steep flanks of the Pezzoni anticline, approximately one half mile north of the crest line. Some of the theories presented by Redwine (1981) are illustrated in Figures 5–71 to 73.

Figure 5–71 shows a diagram illustrating proposed oil migration by dilation and natural hydraulic fracturing along planes of weakness due to bed or lamina partings in horizontal or low-dipping, dolomitic Monterey Shale. Both fracturing and migration are produced during one episode of tectonic compression. In Figure 5–71, σ_1 = direction of principal maximum compressive stress, σ_2 = direction of principal intermediate compressive stress, and σ_3 = direction of principal minimum compressive stress.

Figure 5–72 shows a diagram illustrating proposed horizontal extension fracturing, brecciation, and oil migration by dilation and natural hydraulic fracturing of moderate to steep-dipping, dolomitic Monterey Shale. Fracturing, brecciation, and migration are produced during one episode of tectonic compression. Stress system is the same one shown in Figure 5–71.

Figure 5–73 presents a diagram illustrating proposed horizontal extension fracturing, brecciation, and oil migration by dilation and natural hydraulic fracturing of slump-contorted, dolomitic Monterey Shale. Fracturing, brecciation, and migration are produced during one episode of tectonic compression. Stress system is the same one shown in Figure 5–71.

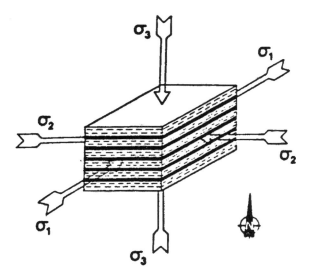

**OIL INJECTION BY NATURAL
HYDRAULIC FRACTURING
ALONG BED OR LAMINA PARTINGS**

**HORIZONTAL OR LOW-DIPPING,
DOLOMITIC, MONTEREY SHALE**

FIGURE 5–71 Oil injection by natural hydraulic fracturing along bed or lamina partings (after Redwine, 1981)

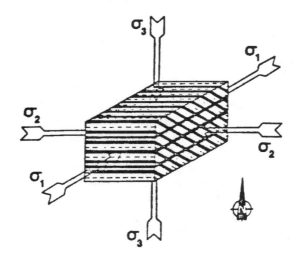

HORIZONTAL EXTENSION
FRACTURING, BRECCIATION,
AND OIL INJECTION BY NATURAL
HYDRAULIC FRACTURING

MODERATE TO STEEP-DIPPING,
DOLOMITIC MONTEREY SHALE

FIGURE 5–72   Horizontal extension fracturing, brecciation, and oil injection by natural hydraulic fracturing (after Redwine, 1981)

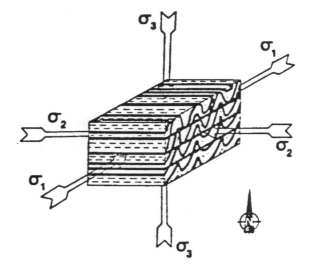

HORIZONTAL EXTENSION
FRACTURING, BRECCIATION,
AND OIL INJECTION BY NATURAL
HYDRAULIC FRACTURING

SLUMP-CONTORTED,
DOLOMITIC, MONTEREY SHALE

FIGURE 5–73   Horizontal extension fracturing, brecciation, and oil injection by natural hydraulic fracturing (after Redwine, 1981)

A summary of geology and physical properties of the Monterey has been presented by Isaacs (1984). Belfield (1983) has shown that there are better probabilities of success in the Monterey when deviated rather than vertical wells are drilled (see Figure 1–50).

Log interpretation can be attempted with a method developed by Cannon (1979) based on Equations 3–10 to 3–12 or a fracture plausibility log developed on the basis of Equation 3–21 (Grove and Whittakeer, 1985). A statistical approach seems reasonable due to the very thin layers of the Monterey that cannot be rigorously handled with currently available tools. Heflin et al (1979) have presented a different approach (uranium index) which is illustrated in Chapter 3. Knauss et al (1983) have used various logs for successfully selecting completion intervals in the Monterey Formation and overlying Reef Ridge Shale of the Lost Hill Field in California. Pickett plots combined with the P$^{\frac{1}{2}}$ statistical technique and Fracture Completion Logs (see Chapter 3) have also provided good results in some Monterey wells by allowing estimates of total, matrix and fracture porosity, fracture and matrix water saturation, and volume estimates of shale, opal A, opal CT, chert, silt, limestone and dolomite. Magnetic resonance image logging has in some instances forecasted oil, gas and/or water production, and the percentage of each phase. Imaging devices such as the FMS, FMI and CBIL have also been reported to provide good results in some cases.

Projections of reservoir performance can be developed utilizing a type curve published by Fetkovich et al (1984). It was developed using production history from Lompoc, Orcutt, West Cat Canyon, Santa Maria Valley and Zaca Fields.

The Monterey is difficult to analyze quantitatively. However, the interest in that formation was revitalized due to offshore discoveries of South Elwood oil field and Hondo within the Santa Ynez unit in 1969. Another discovery is the Point Arguello/Hueso field by Chevron which reportedly contains more than 500×10^6 stbo. These and other Monterey discoveries are important enough to make sure that interest remains focused in that area for years to come, the main concern being proper environmental protection.

Ezekwe et al (1991) have described the performance of a peripheral waterflood project in the main Body "B" Stevens reservoirs at the Elk Hills Oil field in California. The Stevens oil zone is the productive portion of the Elk Hills Shale member of the Monterey formation. According to Ezekwe et al (1991), at Elk Hills the clay shales and siliceous shales are not as substantially fractured as elsewhere in California, and sandstone reservoirs within the shale have provided the bulk of both production to date and proven reserves.

FRACTURED SHALES

Devonian fractured shales have been producing gas since the early 1900s along the western margin of the Appalachian Basin. Oil production has been obtained from the Mancos fracture shale, Rangely field, Colorado. Oil production has also been obtained from silicious shales of Ayukawa field in Japan and the Mississippian Bakken shale, Willison Basin (Montana and North Dakota). The Northeast Alden field in Caddo County (Oklahoma) and several other fields produce from the Woodford shale. The New Albani shale might account for as much as 90% of the known oil reserves in the Illinois Basin.

Horizontal wells have proved very important in fractured shales. For example, a horizontal drilling boom occurred recently in the fractured Bakken shale in North Dakota. Meridian Oil Company has been one of the most active operators trying to develop the thin Bakken shale (1–8 ft reservoir). As a result of horizontal drilling some of the wells have produced more than 500 bo/d. Cost for horizontal wells have decreased from 2.5 times to 1.2 times the cost of a vertical well as a result of advances in horizontal drilling experience (Moritis, 1990). A more complete discussion on fractured shales is presented in Chapter 4.

FRACTURED BASEMENT

The geologic aspects of the hydrocarbon reservoirs in fractured basement were discussed in Chapter 1. P'An (1982) has presented an excellent review of petroleum reservoirs in basement rocks. He indicates that most basement reservoirs occur either on platforms or in intermontane basins. In rare occasions they occur in foredeep basins.

Opportunity for entrapment of oil is provided by source beds that directly overlie the basement rocks. In general, basement reservoirs occur on highs or uplifts within the basin, and have been subjected to long periods of erosion and weathering. Most basement reservoirs occur below a regional unconformity. The so-called "second crops" are formed when the source bed lies below an unconformity. In most instances, however, the source bed lies above an unconformity.

I regard in this book as basement, igneous and metamorphic rocks. I do no consider sandstones and carbonates as basement rock even if they underlie unconformably oil-generating or oil-bearing formations.

There are many examples of naturally fractured basement rocks including La Paz and Mara (igneous and metamorphic) in Venezuela, Orth, Ringwald, Likica, Beaver, Trapp, Eveleigh, and Kraft-Prusa fields (metamorphic-quartzite) in Central Kansas, Hall-Gurney and Gorham fields (granite) in Central Kansas, Edison, El Segundo and Wilmington fields (schists) in California, Augila field (weathered granite and rhyolite) in Libya, Shaim oil field (paleozoic metamorphic rocks) in Russia, Yaerxia basement reservoir (caledonian metamorphic rocks) in Yumen, China, and Xinglongtai oil and gas field (Archean granite and Mesozoic volcanic rocks) in Liaoning, China. These reservoirs have been discussed in some detail by P'An (1982). Some of them have also been reviewed in Chapter 1.

My strong recommendation is not to stop drilling when basement rock is reached. On the contrary, keep on drilling 200 or 300 m (or more) especially if basement is overlain by an oil reservoir or a source rock.

TIGHT GAS RESERVOIRS

Tight (low permeability) gas reservoirs have been defined by the U.S. Federal Energy Regulatory commission (FERC) as those having an estimated in-situ gas permeability of 0.1 md or less and not exceeding certain prescribed production rates. In general tight reservoirs in the U.S. are grouped into tight gas sandstones and certain Devonian Shales.

Many of these tight gas reservoirs contain natural fractures of different origin. However, the intensity of fracturing is not enough to allow prolific production rates except in exceptional cases. An excellent source of information regarding this type of reservoir is presented in AAPG Studies in Geology #24, *Geology of Tight Gas Reservoirs*, edited by C.W. Spencer and R.F. Mast (1986). These types of reservoirs are also discussed in more detail in Chapter 4.

The importance of tight gas reservoirs lies in the potentially high recoverable resources. In fact, the U.S. Congress Office of Technology Assessment (OTA) has indicated that recoverable gas resources from tight sandstone formations are estimated to be between 100 and more than 400 trillion cubic feet. Recoverable gas resources from Devonian shales have been estimated between 20 and more than 100 trillion cubic feet. Devonian shales were described briefly in the section dealing with fractured shales and are discussed in more detail in Chapter 4.

Lawghrey and Harper (1986) have made comparisons between the Upper Devonian and Lower Silurian tight sandstones in Pennsylvania. They indicate that permeabilities and porosities, most of which are secondary, tend to be low. Permeabilities, however, are enhanced in some instances by natural fractures. As an example, a direct relationship has been established between the Tyrone-Mt. Union Lineament and gas production in Crawford County. Fractures have also been found in the Bradford field of McKean County.

Finley (1986) has discussed the Cotton Valley, Travis Peak, Cleveland and Olmos tight formations of Texas. In general permeabilities are significantly low (0.1 md or less). However, some zones have been reported to be naturally fractured to the point that fluid loss treatment materials have been required in some Cotton Valley wells. The contribution of natural fractures is not well know in the Travis Peak formation but is generally considered to be minimal. There is not direct evidence of natural fracturing in the Cleveland formation, and the extent of fracturing in the Olmos formation is unknown.

Coalbed methane and tight gas sands are thought to be interrelated (Rightmire and Choate, 1986). The environments of deposition leading to the formation of organic-reach swamps are the same as those conductive to deposition of repressive sand bodies presently identified as underlying many coal beds. Natural fractures (cleats) are a necessary requirement for commercial production of gas from coal beds.

Rose, Everett and Merin (1986) have discussed a potential basin-centered gas accumulation in the Cretaceous Trinidad Sandstone, Raton basin, Colorado. They indicate that the Raton basin is extensively fractured. Their postulate is based on geologic analogy, specific detailed geologic mapping, observed gas shows, and well log analysis. In this case, natural fractures are anticipated to have both negative and positive effects on fluid flow. The negative effect stems from fractures that have been injected with igneous material and thus become barriers to fluid flow. The positive effect stems from unfilled fractures that increase rock porosity and permeability, allowing greater sustained rates of production than unfractured reservoirs.

Pollastro and Scholle (1986) have discussed exploration and development of hydrocarbons from the Upper Cretaceous Niobrara formation. Biogenic gas is produced from the thermally immature, organic-rich Niobrara chalk in the eastern part of the Denver basin in eastern Colorado, northwestern Kansas, and southwestern Nebraska. These chalks have high porosity but very low permeability, and stimulation is required to achieve commercial production rates. Oil is produced from greater depth in the west of the basin where the Niobrara is naturally fractured.

The Wattenberger field, Denver basin, Colorado, produces gas from the J sandstone and the Codell sandstone (Weymer et al, 1986). The J (Muddy) sandstone is composed of two members: the Fort Collins and the Horsetooth. The porosity is mainly intergranular. There are also minor amounts of microporosity found in the matrix material and secondary interparticle porosity associated with leached feldspars and lithic rock fragments. A late stage natural fracturing of the Fort Collins member has been reported by Weimer et al (1986). Enhanced reservoir performance of the Codell sandstone is attributed to tensional natural fracturing associated with structural movement. The early production from the Codell sandstone was from a naturally fractured reservoir in the Boulder field. Oil discoveries are developed on 80-acre spacing and gas discoveries on 160-acre spacing.

The Upper Cretaceous Mesaverde group of the Piceance Creek basin, Western Colorado is also reported to contain vast amounts of gas-in-place in rocks which usually have very low permeability. Gas production is controlled by a network of open and partially mineralized fractures. These fractures are attributed to high pore fluid pressures that developed during hydrocarbon generation and to tectonic stresses resulting from periods of uplift and erosion during the Late Tertiary (Pitman and Sprunt, 1986).

The Corcoran, Cozzette and Rollins sandstones have primarily microporosity with extremely low permeability. Brown et al (1986) indicate that data from non-structural wells indicate little or no influence from natural fracturing. On the other hand, the two-well Buzzard Creek field is a structural accumulation where production rates are probably enhanced by tectonic natural fractures and possible structural closing.

The Natural Buttes field in the eastern part of the Uinta basin, Utah is reported to have large amounts of gas in Tertiary and Cretaceous non-marine sandstones (Pitman et al., 1986). Porosity and permeability have been significantly reduced due to clay mineral development and the formation of carbonate cement. Examination of Cretaceous cores indicate the presence of incipient small scale, closed, hairline fractures. An unusual feature is that the best fracture development is in shales. Fractures are rarely observed in sandstones.

NATURALLY FRACTURED RESERVOIRS STATISTICS

Naturally Fractured reservoirs are found all throughout the stratigraphic column as shown on Table 5–13. Many of the fields, areas and/or formations mentioned in that Table are discussed in this book.

Table 5–14 presents statistics associated with various naturally fractured oil reservoirs around the world. Table 5–15 shows statistics of gas reservoirs. Included in both tables are the field name (or general area), location, Age, dominant lithology, type of fractures, porosities, permeabilities, water saturations, reservoir area in acres, original oil or gas-in-place, percent recovery, and remarks associated with each reservoir.

The significantly different percent recoveries presented in Tables 5–14 and 5–15 indicate that it is very dangerous to apply rules of thumb when estimating recoveries from naturally fractured reservoirs. Rules of thumb and naturally fractured reservoirs do not mix well. My strong recommendation is to study each fractured reservoir in detail. In fact, I would go one step further and indicate, as did Dr. H.K. van Poollen many years ago, that each naturally fractured reservoir is a research project by itself.

Table 5–16 shows a list of basins that contain naturally fractured reservoirs and an indication of where horizontal drilling has taken place. Of the 39 basins listed for North America, 27 have seen horizontal drilling activity. This represents a very healthy 69.2%.

Of the 50 basins around the world (excluding North America) presented in Table 5–16, 10 have experienced horizontal drilling activity. This represents only 20% of the basins, and indicates that much horizontal drilling potential remains outside of North America.

Recovery factor statistics for the province of Alberta using data available from the Alberta Energy Resources Conservation board (ERCB) have been published recently in the Petroleum Society of CIM Monograph No. 1 "Determination of Oil and Gas Reserves" (1994). These statistics combine both fractured and unfractured reservoirs. However, there is an interesting comparison of oil recoveries from clastics (Figure 5–74) and carbonates (Figure 5–75). The total number of pools is 5918 divided into 3941 clastics and 1977 carbonates.

The clastics show an original oil-in-place of $4,865 \times 10^6$ m^3 (30.6×10^9 stbo). The carbonates $3,437 \times 10^6$ m^3 (21.6×10^9 stbo). Thus 41.4% of the original oil-in-place is stored in carbonates and 58.6% in clastics. The weighted mean recoveries from clastics is 11.33% as opposed to 29.09% from carbonates. Thus the carbonates contribute 999.8×10^6 m^3 (6.3×10^9 stbo) oil reserves and the clastics 551.2×10^6 m^3 (3.5×10^9 stbo). These statistics highlight the importance of carbonates in Alberta. Many of these carbonates have very efficient primary drives, mainly bottom water. The porosity of carbonate rocks is generally low (5% to 15%), but permeability can be very high, specially in dolomitized rock which is usually naturally fractured.

Figure 5–76 shows a very interesting crossplot of oil recovery vs. average porosity for 5915 Alberta pools. Included in this plot are clastics and carbonates. In general, the crossplot illustrates something very unconventional, i.e., the largest recoveries occur at the smallest average porosities. However, if the data is split into carbonates and clastics the reason is clear. These higher recoveries correspond to carbonate pools which have lower average porosities. Although I cannot claim that all carbonates are fractured it is important to remember from Figure 1–5A (Chapter 1) that for the same physical environment one of the biggest degrees of fracturing occurs in dolomites. Furthermore I indicated in Chapter 1 that other things being equal the largest intensity of fracturing occurs in those intervals with the lowest porosities.

The conclusion is clear. No where in the world should we overlook low porosity intervals!

TABLE 5–13 Stratigraphic Column Showing Examples of Naturally Fractured Reservoirs

ERA	PERIOD		
CENOZOIC	Quaternary		
	Tertiary	Pliocene	
		Miocene	Monterey, Elk Hills (Ca), Asmari (Iran), Gachsaran, Haft Kel, Asmari (Iran), Nido A & B (Philippines)
		Oligocene	Kirkuk, Ain Zalah (Iraq)
		Eocene	Altamont, Bluebell (Utah), Kirkuk (Iraq)
		Paleocene	Ekofisk (Norway)
MESOZOIC	Upper Cretaceous		Ekofisk (Norway), Reforma (Mexico), La Paz/Mara (Venezuela), Austin chalk (Texas)
	Lower Cretaceous		Alamein (Egypt)
	Jurassic		Lacq (France), Dukhan (Qatar), Edison, Mountain View, Playa del Rey, Wilmington (California)
	Triassic		Monkman Area, British Columbia
PALEOZOIC	Permian		Spraberry (Texas), Reeves (Texas)
	Pennsylvanian		Rangely Field (Colorado), Sichuan gas fields (China), Seminole Southeast (Texas)
	Mississippian		Mississippian Lime (Oklahoma), Savanna Creek (Alberta), Weyburn Unit (Saskatchewan), Northwest Lisbon (Utah), Little Knife (North Dakota), Bakken Shale (North Dakota)
	Devonian		Beaver River (British Columbia), Shales, Big Sandy (Appalachian Basin), Rainbow "A" (Alberta)
	Silurian		Howell Gas Field (Michigan)
	Ordovician		Shales (Quebec), Oklahoma City field, Healton Field (Oklahoma), Oriskany Sand fields (New York), Palm Valley (Australia), Cabin Creek (Montana), Killdeer (North Dakota), Pennel (Montana), Cottonwood Creek (Oklahoma)
	Cambrian		Amal (Libya)
PROTEROZOIC ARCHEOZOIC	Pre-Cambrian		Orth and Chase Fields (Kansas)

TABLE 5–14 Naturally Fractured Oil Reservoirs

Field/Area	Location	Age	Lithology	Natural Fractures	Type	Porosity %
Ain Zalah	Iraq	Cretaceous	LM	Abundant Upper Pay	Tectonic	NA
Alamein	Egypt	Cretaceous	DOL	Intense	Tectonic	11
Altamont Bluebell	Utah	Eocene/Paleoc.	SS	Hairline	Regional	3–7
Asmari Fields	Iran	Oligo/Miocene	LM	Abundant	Tectonic	8
Austin Chalk Fields	Texas	Upper Cretac	Chalk	Directional SW-NE Variable Intensity	Regional	NA
Bagre	Mexico		LM	Vertical, Caverns	Tectonic	18–19
Cabin Creek	Montana	Ordovician	DOL.	Vertical & Horizontal	Tectonic	1–25 13 Avg.
Cabin Creek	Montana	Silurian	DOL	Extension/ Breccia	Tectonic	6–23 15 Avg.
Cavone	Italy	Tertiary	LM	Vertical	Tectonic	5–9
Cottonwood Creek	Oklahoma	Ordovician	Arbuckle DOL	Vertical, Karst	Solution Collapse	3
Dukhan	Qatar	Jurassic	LM/DOL	Moderate	Tectonic	15–20
Elk Hills	California	Miocene	SH/CHERT	Moderate	Regional	20
Ekofisk	Norway	Cret/Paleoc	LM	Abundant	Tectonic	0–45 32 Avg.
Killdeer	N. Dakota	Ordovician	DOL	Present (minor)	Tectonic	12–15
Kirkuk	Iraq	Eocene/Olig.	LM	Abundant	Tectonic	0–30
La Paz/Mara	Venezuela	Cretaceous	LM	Abundant	Tectonic	2–3
Lacq Superieur	France	Upper Cretac.	LM	Abundant Some Karst	Tectonic	12 (matrix) 0.5 (fractures)
Little Knife	N. Dakota	Mississippian	DOL	Hairline Widely spread	Tectonic	8.5–27 14 Avg.
Nido A & B	Phillippines	Miocene	LM	Vertical Abundant	Tectonic	1–9 3 Avg.
Northwest Lisbon	Utah	Mississipian	DOL	Vertical/ Breccia	Collapse Brecciated	1–12 5.5 Avg.
Norman Wells	NW. Territories Canada	Devonian	LM	Vertical N 26°E	Regional	9.8
Oklahoma City	Oklahoma	Ordovician	Arbuckle DOL	Abundant, Cavities	Sollution Collapse	NA
Pennel	Montana	Ord/Sil/Mis	DOL	Vertical	Tectonic	2–22 11 Avg.
Rainbow "A"	Alberta	Middle Dev.	LM/DOL	Strong, Vertical	Contractional	3–15 10.1 Avg.
Reeves	Texas	Permian	DOL/LM	Vertical NE–SW	Tectonic	7.8–17.6 10.4 Avg.
Seminole Southeast	Texas	Pennsylvanian	LM	Vertical/Hairline/ solution	Tectonic	1–18 13 Avg.
Sitio Grande	Mexico	Cretaceous	DOL	Abundant/Caverns	Tectonic	6–11
Spraberry Trend	Texas	Permian	SS	Hairline	Regional	8–14
Zama-Keg River CC	Alberta	Devonian	LM	Moderate	Tectonic	7–10

TABLE 5–15 Naturally Fractured Gas Reservoirs

Field/Area	Location	Age	Lithology	Natural Fractures	Type	Porosity %
Beaver River	British Columbia	Devonian	LM	Abundant	Regional	2.7
Meillon	France	Jurassic	DOL	Moderate	Tectonic	1–8
Monkman Area	British Columbia	Triassic	LM/DOL	Thrust, Abundant	Tectonic	4.0
Palm Valley	Australia	Ordovician	SS	Abundant Tectonic	Tectonic	3–6
Savanna Creek	Alberta	Mississipian	LM	Abundant Breccia	Tectonic	4.0

Permeability md	Water Sat (%)	Area Acres	OOIP 10⁶ STBO	Reserves 10⁶ STBO	% Rec	Remarks
NA	NA	3840	NA	NA	NA	Types A and C Partitioning coefficient = 40%
2000	24	3100		120	> 40.0	Active water drive, Type A
0.01	30–50	NA	NA	> 300.0	30–40.0	Overpressured, gas expansion Type A
3–1200	17	NA	26300	10520.0	40.0	Type B, waterdrive, gas expansion Types A, B and C Successful horizontal wells
1020–1474	16–21				36–40.0	Strong water drive. Type A
0–142 7.9 Avg.	30	7620		224.0	33.5	Waterflood. Type A
5	Variable	8100	NA	NA	NA	Gas expansion. Water drive Type A
10–60 (matrix) 30-600 (fract)		4940	94.5	32.1	34.0	Type A. High matrix permeability
0.01 (matrix)	NA	1200	40.0	NA	NA	Waterflood, karst. Type A. Type B in Karst areas
30–250				2400.0		Type A, waterflood.
0–4570 32.3 Avg.	38–40	7700	NA	212	NA	Stevens Waterflood. Not substantially fractured. Type A
0.1–1000 1.0 Avg.	24	12071		5400.0	22.2	Gas inject., waterflood. Type A
N.A.	40–45	960		2.1	14.5	Probably Type A
0–1000	NA	30000	NA	1900.0	NA	Type A
< 0.1 (matrix) 1 (matrix) 10000 (fract)	100 (matrix) 60 (matrix) 0 (fract)	67000	NA	31.0 125	NA > 20.0	Type C Strong water drive. Steam injection. Type A
1–167	40	24000		195.0	NA	Solution gas. Limited water drive. Type A
0.01–3.3 1.0 Avg.	27	345	NA	NA	NA	Water drive, imbibition Types A and B
0.01–100 22 Avg.	39	5120		91.2	56.1	Gas cap and gravity drainage. Type A
4 (Horiz) 1 (vert)	10	48000	630.0	271.0	43	Waterflood, horizontal wells. Type A. Oil wet
NA	NA	2460	75.5	18.2	24.1	Oil wet, partial water drive. Type B & C where Karsted
0.1–35 9 Avg.	30–45	22300		279.0	20.8	Gas expansion, water drive, waterflood. Type A
184-570	10	625		90.3	87.8	Solvent flood, Type A and B
0.01–230 2.2 Avg.	42	5480		106.8	28.1	Sol. gas drive, waterflood. Type A
0.1–80 29 Avg.	20–80	1500		7.0	28.6	Waterflood. Type A
135	13–20	7680			19.0	Rapid decline of pressures. Waterflood. Types A, B and C
0.29–0.5	35	1.6 × 10⁶	3600	360	10.0	Underpressured, solution gas, waterflood. Type A.
0.5–5.8	14–19	—	3.2	1.0	31.0	One of many reefs. Types A and B

Permeability md	Water Sat (%)	Area Acres	OGIP 10⁹ SCF	Reserves 10⁹ SCF	% Rec	Remarks
20–200	25.0	10700	1633	245	< 15.0	Calculated OGIP might be too large
< 1 md	NA	NA	2300–3530	1950–2300	84.8–65.2	Active water drive
0.1 (matrix)	30.0	NA	NA	NA	NA	Thrust faulting
0.01–1	35.0		845	680	80.5	Arcuate anticline

TABLE 5–16 Some Basins that Contain Naturally Fractured Reservoirs (adapted from Fritz et al. 1991)

North America	Horizontal Drilling	North America	Horizontal Drilling
Alberta	X	North Slope	X
Anadarko	X	Palo Duro	X
Appalachian	X	Paradox	X
Ardmore	X	Permian	X
Arkoma		Piceance	X
Bear Lake	X	Powder River	X
Big Horn	X	Sabinas	
Black Warrior	X	Salinas (Mexico)	
Campeche		Salinas	
Cincinnati Arch	X	Salina/Forest City	X
Crazy Mountain	X	San Joaquin	X
Denver	X	San Juan	X
East Texas Salt Dome		Sand Wash	X
Fort Worth	X	Santa Maria	X
Green River	X	South Texas Salt Dome	X
Laramie		Uinta	
Los Angeles		Ventura	
Louisana Salt Dome	X	Williston	X
Michigan	X	Wind River	
Mississippi Salt Dome	X		

Central and South America	Horizontal Drilling
Maracaibo	X
Neuquen	X
Sergipe-Alagoas	

Africa	Horizontal Drilling
Cabinda	X
Ghadames	
Sirte	
Suez, Gulf of	
Western Desert	

Europe	Horizontal Drilling
Adriatic, North	X
Aquitaine	X
Caltanissetta	X
Carpathian	
Ebro Fan	
German Northwest	
Moesian	
Molasse	
North Sea, Northern	
North Sea, Southern	X

Europe	Horizontal Drilling
Pannonian	
Po	

Middle East	Horizontal Drilling
Arabian	
Zagros	

Asia and Oceania	Horizontal Drilling
Arita	
Amadeus	
East China	
Jiuquan	
Huabei	
Nigata	
Otway	
Sichuan	
Surat	
Taiwan	
Teshio	
Palawan, North	X
Sumatra, Central	
Sumatra, North	

TABLE 5–16 (Continued)

Asia and Oceania	Horizontal Drilling	USSR	Horizontal Drilling
Sumatra, South		Angara-Lena	
Java, East		Caucasus, North	
Java, West		Dnepr-Donets	
Zhungeer		Pechora	
		Pripyat	
		Sakhalin, North	
		Tunguska	
		Volga-Ural	
		West Siberia	X

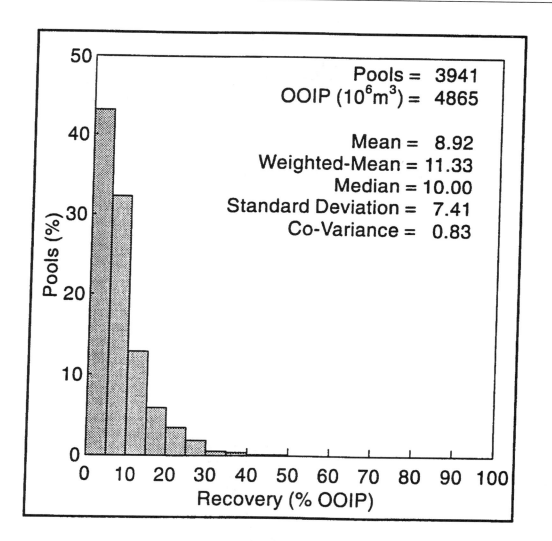

Clastic Oil Pools

FIGURE 5–74 Percent recoveries from 3941 clastic pools in Alberta. (Source: *Determination of Oil and Gas Reserves,* Petroleum Society of CIM Monograph No. 1, 1994)

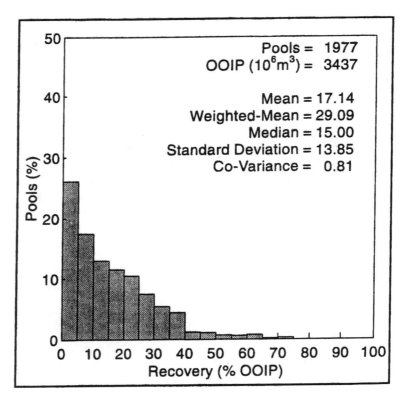

Carbonate Oil Pools

FIGURE 5–75 Percent recoveries from 1977 carbonate oil pools in Alberta. (Source: *Determination of Oil and Gas Reserves*, Petroleum Society of CIM Monograph No. 1, 1994)

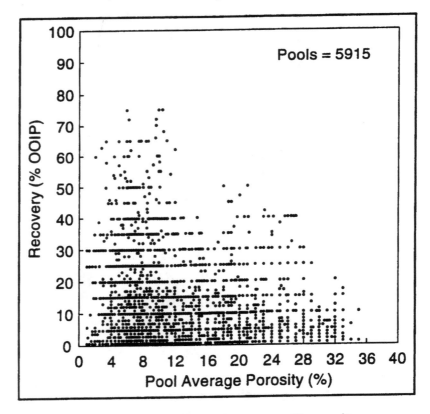

Oil Recovery vs. Porosity

FIGURE 5–76 Crossplot showing larger oil recoveries at smaller values of porosity. (Source: *Determination of Oil and Gas Reserves*, Petroleum Society of CIM Monograph No. 1, 1994)

REFERENCES

Aguilera, Roberto. "Evaluation of Fine-Grained Laminated Systems from Logs, Wasatch Formation, Utah." Ph.D. Dissertation T-1569, Colorado School of Mines (1973), 279 p.

Aguilera, R. and H.K. van Poollen. "Studies Show Occurrence of Fractured Reservoirs." *Oil and Gas Journal* (May 21, 1979).

Aguilera, R., and L. Acevedo. "FCL—A Computerized Well Log Interpretation Process for the Evaluation of Naturally Fractured Reservoirs." *Journal of Canadian Petroleum Technology*, Vol. 21, No. 1, pp 31–37 (Jan.–Feb. 1982).

Aguilera, R., and S.J. Song. "Well Test-NFR: A Computerized Process for Transient Pressure Analysis of Multiphase Reservoirs with Single, Dual or Triple Porosity Behaviour"; CIM paper 88-39-52, Annual Technical Meeting of the Petroleum Society of CIM, Calgary, Alberta (June 12–16, 1988).

Aguilera, R., L.N. Franks and A.D. Au. "Well Test Analysis of a Naturally Fractured Gas Reservoir—A Case History." *Journal of Canadian Petroleum Technology* (April 1992), 41–46.

Aguilera, R., A.D. Au, and L.N. Franks. "Isochronal Testing of a Naturally Fractured Gas Reservoir—A Case History." *Journal of Canadian Petroleum Technology*, V. 32, No. 8, (October 1993), 25–30.

Andresen, K.H., R.I. Baker, and J. Raoofi. "Development of Methods for Analysis of Iranian Asmari Reservoirs." Preceedings of the Sixth World Petroleum Congress, Section II (June 1963), 13–27.

Aronofsky, J.S., L. Masse, and S.G. Natanson. "A Model for the Mechanism of Oil Recovery from the Porous Matrix Due to Water Invasion in Fractured Reservoirs." *Transactions*. AIME 214 (1958), 17.

Au, A.D., L.N. Franks and R. Aguilera. "Simulation of a Naturally Fractured Gas Reservoir: A Case History." SPE paper 22920 presented at the 66th Annual Technical Conference and Exhibition of the Society of Petroleum Engineers, Dallas, Texas (October 6–9, 1991).

Bacon, J.R. and R.H. Kempthorne. "Waterflood Oil Recovery in Fractured Reservoirs with Directionally Drilled Wells." *Society of Petroleum Engineers Journal* (August 1984) 375–381.

Baker, D.A., and P.T. Lucas. "Strat Trap Production May Cover 280 Plus Square Miles." *World Oil* (April 1972), 65–68.

Barba, R.E. "Optimizing Hydraulic-Fracture Length in the Spraberry Trend." *SPE Formation Evaluation* (1989), 475–482.

Belfield, W.C. et al. "South Ellwood Oil Field, Santa Barbara Channel, California—A Monterey Formation Fractured Reservoir." In *Petroleum Generation and Occurrence in the Miocene Monterey Formation, California*, published by the Pacific Section, Society of Economic Paleontologists and Mineralogists, Los Angeles, California (May 20–22, 1983).

Berry, M.D., "Core Analysis of Palm Valley 7, Summary and Interpretation of Results; Magellan Petroleum (NT) Pty. Ltd., (Jan. 1991).

Beliveau, D., D.A. Payne, and M. Mundry. "Waterflood and CO_2 Flood of the Fractured Midale Field." *Journal of Petroleum Technology* (September 1993), 881–887.

Birks, J.A., "A Theoretical Investigation Into the Recovery of Oil from Fissured Limestone Formations by Water-Drive and Gas-Cap Drive." Proceeding of the Fourth World Petroleum Congress, Section II/F (1955), paper 2.

Blauer, R.E., I.N. Onat, and J. Lemieux. "Production Data Indicates Reservoir and Fracture Performance in the Spraberry." SPE paper 23995, Permian Basin Oil and Gas Recovery Conference held in Midland, Texas (March 18–20, 1992).

Brown, C.A., T.M. Smagala, and G.R. Haefele. "Southern Piceance Basin Model-Cozzette, Corcoran and Rollins Sandstones." *Geology of Tight Gas Reservoirs*, AAPG Studies in Geology #24 (1986), 207–219.

Brownscombe, E.R., and A.B. Dyes. "Water Imbibition Displacement—Can it Release Reluctant Spraberry Oil?" *Oil and Gas Journal* (November 1952), 264.

Brownscombe, E.R., and A.B. Dyes. "Water Imbibition Displacement, A Possibility for the Spraberry," *API Drilling and Production Practices* (1952), 383.

Cannon, D.E. "Log Evaluation of a Fractured Reservoir—Monterey Shale." SPWLA 20th Annual Logging Symposium (June 3–6, 1979).

Castelijns, J.H.P. and J. Hagoort. "Recovery of Retrograde Condensate from Naturally Fractured Gas-Condensate Reservoirs." *Society of Petroleum Engineers Journal* (December 1984) 707–717.

Chen, H.Y., S.W. Poston, and R. Raghavan. "Mathematical Development of Austin Chalk Type Curves." SPE paper 23527 (circa 1991).

Christie, R.S. and J.C. Blackword. "Characteristics and Production Performance of the Spraberry." API (1952).

Claycomb, E. "Pressure Buildup Characteristics in Austin Chalk Wells." *Austin Chalk Oil Recovery Conference*, Texas A&M, College Station, Texas (April 28, 1982).

Craft, M. "Evaluation of Austin Chalk from Cores." *Austin Chalk Oil Recovery Conference*, Texas A&M, College Station, Texas (April 28, 1982).

Crawford, G.E., A.R. Hagedorn, and A.E. Pierce. "Analysis of Pressure Buildup in a Naturally Fractured Reservoir." *Journal of Petroleum Technology* (November 1976), 1295–1300.

Darnborough, E. "The Buchan Field Development." SPE paper EUR 230 presented at the European Offshore Petroleum Conference and Exhibition held in London, England (October 21–24, 1980).

Davidson, D.A., and D.M. Snowdon. "Beaver River Middle Devonian Carbonate: Performance Review of a High Relief, Fractured Gas Reservoir with Water Influx." *Journal of Petroleum Technology* (December 1978), 1672–1678.

Del Orbe-Valdiviezo, J.F. "Drilling/Completion of Deep Reforma Wells." *Petroleum Engineer* (December 1975), 26–42.

Delgado, O.R., and E.G. Loreto. "Reforma's Cretaceous Reservoirs: An Engineering Challenge." *Petroleum Engineer* (December 1975), 56–66.

de Witt, W. "Devonian Gas-bearing Shales in the Appalachian Basin" *Geology of Tight Gas Reservoirs*, AAPG Studies in Geology #24 (1986), 1–8.

Determination of Oil and Gas Reserves, Petroleum Society of CIM Monograph No. 1, Calgary, Canada (1994), 362 p.

Do Rozario, R.F., and B.W. Baird. "The Detection and Significance of Fractures in the Palm Valley Gas Field." *The APEA Journal*, pp 264–280, (1987).

Dyes, A.B., and O.C. Johnston. "Spraberry Permeability from Buildup Curves," *Transactions*. AIME 198 (1953), 135.

Edlund, P.A. "Application of Recently Developed Medium-Curvature Horizontal Drilling Technology in the Spraberry Trend Area." SPE paper 16170 presented at the SPE/IADC Drilling Conference held in New Orleans, LA (March 15–18, 1987).

El Banbi, H.A. "Carbonate Field Developed to Recover 37% Primary Oil." *World Oil* (October 1974), 139–141.

Elkins, L.F., and A.M. Skov. "Determination of Fracture Orientation from Pressure Interference." *Transactions*. AIME 219 (1960), 301-304.

Elkins, L.F., and A.M. Skov. "Cycling Water Flooding the Spraberry Utilizes End Effects to Increase Oil Production Rates." *Journal of Petroleum Technology* (August 1963), 877–884.

Elkins, L.F. "Reservoir Performance and Well Spacing, Spraberry Trend Area Field of West Texas." *Transactions*. AIME 198, (1952), 301–304.

El Sayed, S.A., R. Baker, P.L. Churcher, and A.C. Edmunds. "Multidisciplinary Reservoir Characterization and Simulation Study of the Weyburn Unit." *Journal of Petroleum Technology* (October 1993), 930–973.

"Esso Expands Norman Wells Field in NWT." *Oil Week*, Canada (August 9, 1982) p.10.

Ezekwe, N., S. Smith, M. Wilson, M. Humphrey, and J. Murphey. "Performance of a Peripheral Waterflood Project in the Main Body 'B' Reservoirs (Stevens) at the Elk Hills Oil Field, California." SPE paper 21759 presented at the Western Regional Meeting held in Long Beach, California (March 21–22, 1991).

Fertl, W.H. and R.F. Holtz. "Logging Through the Austin Chalk." *Austin Chalk Oil Recovery Conference*, Texas A&M, College Station, Texas (April 28, 1982).

Fetkovich, M.J. et al. "Decline Curve analysis Using Type Curves: Case Histories." SPE paper 13169 presented at the 59th Annual Technical Conference and Exhibition held in Houston, Texas (Sept 16–19, 1984).

Findley, L.D. "Why Uinta Basin Drilling is Costly and Difficult." *World Oil* (April 1972), 77–91.

Finley, R.J. "An Overview of Selected Blanket-Geometry, Low-Permeability Gas Sandstones in Texas." *Geology of Tight Gas Reservoirs*, AAPG Studies in Geology #24 (1986), 69–85.

Freeman, H.A., and S.G. Natanson. "A Reservoir Begins Production: A Study on Planning and Cooperation." Sixth World Petroleum Congress, Section II (June 1963), 407–414.

Gholipour, A.M. "Pattern and Structural Position of Productive Fractures in the Asmari Reservoirs, South Western Iran." CIM paper 94–43 presented at the Canadian SPE/CIM/CANMET International Conference on Recent Advances in Horizontal Well Applications, Calgary, Canada (March 20–24, 1994).

Golaz, P., A.J. Sitbon, and J.G. Delisle. "Case History of the Meillion Gas Field." *Journal of Petroleum Technology* (August 1990), 1032–1036.

Graham, W.J., et al. "Effect of Production Restriction on Iranian Oil Reservoirs." Proceedings of the Fourth World Petroleum Congress (1952), paper 8, 395–446.

Gray, D.P., R.A. Morse and F.L. Oliver. "A Proposal for the Redevelopment of the Beaver River Gas Field, British Columbia, Canada." *Beaver River Resources, Ltd*, Vancouver, Dallas (circa 1992).

Gronseth, J.M. and P.R. Kry. "In-Situ Stresses and the Norman Wells Expansion Project." CIM paper 87–38–57 presented at the 38th Annual Technical Meeting of the Petroleum Society of CIM, Calgary, Canada (June 7–10, 1987).

Grove, G.P. and J.L. Wittaker. "Continuous Fracture Probability Determination as Applied to the Monterey Formation." SPE paper 13652 presented at the California Regional Meeting held in Bakersfield, California (March 27–29, 1985).

Guevara, E.H. "Geological Characterization of Permian Submarine Fan Reservoir of the Driver Waterflood Unit, Spraberry Trend, Midland Basin, Texas." Report of Investigations No. 172, Bureau of Economic Geology (1988).

Heacock, D.W. "An On-Site Method for Optimizing Oil Recovery During Drilling of Deviated and Off-Target Wells at Norman Wells." *Journal of Canadian Petroleum Technology* (May–June 1986), 60–65.

Heflin, J.D. et al. "Log Evaluation in the California Miocene Formations." SPE paper 6160 presented at the 51st Annual Fall Conference and Exhibition held in New Orleans, La. (October 3–6, 1979).

Hoel, M. "Crosslinked Gelled Water is Optimum Fracture Fluid for Spraberry Trend in West Texas." *Oil and Gas Journal* (August 15, 1988) 60–69.

Holditch, S.A. and D.E. Lancaster. "Economics of Austin Chalk Production." *Austin Chalk Oil Recovery Conference*, Texas A&M, College Station, Texas (April 28, 1982).

Holifield, R. "Austin Chalk Trend, Upper Gulf Coast, Texas." *Austin Chalk Oil Recovery Conference*, Texas A&M, College Station, Texas (April 28, 1982).

Isaacs, C.M. "Geology and Physical Properties of the Monterey Formation, California." SPE paper 12733 presented at the California Regional Meeting held in Long Beach, California (April 11–13, 1984).

Ives, George. "Development of Reforma Area Oil Fields." *Petroleum Engineer* (December 1975), 46–54.

Johnson, J.P. and D.W. Rhett. "Compaction Behavior of Ekofisk Chalk as a Function of Stress." SPE paper 15872 presented at the SPE European Petroleum Conference held in London, U.K. (October 20–22, 1986).

Julian, P.J. "Fracture Detection Techniques in the Georgetown and Austin Chalk Formations." SPE paper 11035 presented at the 57th Annual Fall Technical Conference and Exhibition held in New Orleans, LA. (Sept. 26–29, 1982).

Kazemi, H. "Pressure Transient Analysis of Naturally Fractured Reservoirs with Uniform Fracture Distribution." *Society Petroleum Engineers Journal* (December 1969), 451.

Kao J. "An Algorithm For Calculating Vapour-Liquid Equilibrium." Paper SPE 7605, presented at the SPE 53rd Annual Fall Meeting Houston (Oct. 1–3, 1978).

Kempthorn, R.H. and J.P.R. Irish. "Norman Wells—A Look at One of Canada's Largest Oil Fields." *Journal of Petroleum Technology* (June 1981) 985–991.

Knauss, M.E. et al. "Completion Interval Selection in the Lost Hills Field, Southeast Flank, Kern County, California." SPE paper 11699 presented at the California Regional Meeting held in Ventura, California (March 23–25, 1983).

Lane, C.M. and R.J. Watson. "Development of the Woodada Gas Field." *The APEA Journal* (1985), 316–328.

Laughrey, C.D., and J.A. Harper. "Comparisons of Upper Devonian and Lower Silurian Tight Formations in Pennsylvania-Geological and Engineering Characteristics." *Geology of Tight Gas Reservoirs*, AAPG Studies in Geology #24 (1986), 9–43.

Locke, C.D. and W.K. Sawyer. "Constant Pressure Injection Test in a Fractured Reservoir—History Match Using Numerical Simulator and Type Curve Analysis." SPE paper 5594 presented at the 50th Annual Fall Meeting, Dallas, Texas (September 28–October 1, 1975).

Lynch, M. and Z. Al-Shaieb. "Evidence of Paleokarstic Phenomena and Burial Diagenesis in the Ordovician Arbuckle Group of Oklahoma." In Johnson, S.K. (ed.), *Late Cambrian-Ordovician Geology of the Southern Midcontinent*, 1989 Symposium, Oklahoma Geological Survey Circular 92. (1989), 42–60.

Lyttle, W.J., and R.R. Ricke. "Well Logging in Spraberry." *Oil and Gas Journal* (December 13, 1951), 92.

Mardock, E.S., and J.P. Myers. "Radioactivity Logs Define Lithology in the Spraberry Formation." *Oil and Gas Journal* (November 29, 1951). 90.

Martinez, H.R.A. "Evaluación del Compartamiento de Yacimientos en Caliza Con Empuje de Agua Muy Activo." *Ingenieria Petrolera*, Mexico (June, 1975).

Mazzocchi, E.F., and K.M. Carter. "Pilot Application of a Blocking Agent—Weyburn Unit, Saskatchewan." *Journal of Petroleum Technology* (September 1974), 973–978.

McNaughton, D.A., Correspondence with Roberto Aguilera on Fracture Type Reservoir. (July 13, 1989).

McNaughton, D.A., and F.A. Garb. "Finding and Evaluating Petroleum Accumulations in Fractured Reservoir Rock." *Exploration and Economics of the Petroleum Industry*, vol. 13. Matthew Bender and Company Inc. (1975).

Milne, N.A., and D.C. Barr. "Sub-surface Fractures, Palm Valley Gas Field, Australia." *The APEA Journal*, pp. 321–341 (1990).

Moritis, G. "Horizontal Drilling Scores More Successes." *Oil and Gas Journal* (February 26, 1990) 53–64.

Nardon, S. et al. "Fractured Carbonate Reservoir Characterization and Modelling: A multidisciplinary Case Study from the Cavone Oil Field, Italy." *First Break*, vol. 9, No. 12 (December 12, 1991), 553–565.

Narr, W. and J.B. Currie. "Origin of Fracture Porosity—Example from Altamont Field, Utah." *Bulletin.* AAPG v. 66, No. 9 (September 1982), 1231–1246.

Novinpour, F., F.A. Sobbi, and A. Badakhshan. "Modelling the Performance of an Iranian Naturally Fractured Reservoir." paper No. 94–02 presented at the 45th Annual Technical Meeting of the Petroleum Society of CIM co-sponsored by AOSTRA, Calgary, Canada (June 12–15, 1994).

Parker, D. et al. "Austin Chalk Stimulation Techniques and Design." *Austin Chalk Oil Recovery Conference*, Texas A&M, College Station, Texas (April 28, 1982).

Oen, P.M., M. Engell-Jensen, and A.A. Barendregt. "Skojold Field, Danish North Sea; Early Evaluations of Oil Recovery Through Water Imbibition in a Fractured Reservoir." *SPE Reservoir Engineering* (February 1988), 17–22.

P'An Chung-Hsiang. "Petroleum in Basement Rocks." *Bulletin.* AAPG, v.66, No. 10 (October 1982), 1597–1643.

Peng, D.Y. and D.B. Robinson. "A New Two-Constant Equation of State." *Ind. Eng. Ehem. Fundam.* (1976), no. 1, 59–64.

Pisciotto, K.A. and R.E. Garrison. "Lithofacies and Depositional Environments of the Monterey Formation, California." *In The Monterey Formation and Related Siliceous Rocks of California*, Society of Economic Paleontologists and Mineralogists (May 29, 1981) 97–122.

Pitman, J.K., D.E. Anders, T.D. Fouch, and D.J. Nichols. "Hydrocarbon Potential of Nonmarine Upper Cretaceous and Lower Tertiary Rocks, Eastern Uinta Basin, Utah." *Geology of Tight Gas Reservoirs*, AAPG Studies in Geology #24 (1986), 235–251.

Pitman, J.K., and E.S. Sprunt. "Origin and Distribution of Fractures in Lower Tertiary and Upper Cretaceous Rocks, Piceance Basin, Colorado, and Their Relation to the Occurrence of Hydrocarbons." *Geology of Tight Gas Reservoirs*, AAPG Studies in Geology #24 (1986), 221–233.

Pollard, T. "Evaluation of Acid Treatments from Pressure Buildup Analysis." *Transactions.* AIME 38 (1959), 216.

Pollastro, R.M., and P.A. Scholle. "Exploration and Development of Hydrocarbons from Low-Permeability Chalks—an Example from the Upper Cretaceous Niobrara Formation, Rocky Mountain Region." *Geology of Tight Gas Reservoirs*, AAPG Studies in Geology #24 (1986), 129–141.

Poston, S.W. and H.Y. Chen. "Fitting Type Curves to Austin Chalk Wells." SPE paper 21653 presented at the Production Operation Symposium held in Oklahoma (April 7–9, 1991).

Read, D.L. and G.L. Richmond. "Geology and Reservoir Characteristics of the Arbuckle Brown Zone in the Cottonwood Creek Field, Carter County, Oklahoma." In Johnson, D.S. and J.A. Campbell (eds.), *Petroleum Reservoir Geology in the Southern Midcontinent*, Oklahoma Geological Survey Circular 95 (1993), 113–125.

Redwine, L. "Hypotheses Combining Dilation, Natural Hydraulic Fracturing, and Dolomitization to Explain Petroleum reservoirs in Monterey Shale, Santa Maria Area, California." *In The Monterey Formation and Related Siliceous Rocks of California*, Society of Economic Paleontologists and Mineralogists (May 29, 1981) 221–248.

Rightmire, C.T., and R. Choate. "Coal-Bed Methane and Tight Gas Sands Interrelationships." *Geology of Tight Gas Reservoirs*, AAPG Studies in Geology #24 (1986), 87–110.

Roehl, P.O. and P.W. Choquette. *Carbonate Petroleum Reservoirs*, Springer-Verlag New York Inc. (1985), 622 p.

Rose P.R., J.R. Everett, and I. S. Merin. "Potential-Centered Gas Accumulation in Cretaceous Trinidad Sandstone, Raton Basin, Colorado." *Geology of Tight Gas Reservoirs*, AAPG Studies in Geology #24 (1986), 111–128.

Sabet, M. and L. Franks. "Palm Valley—A Case for Interpretation, 1965–1984." *The APEA Journal*, Australia (1985) 329–343.

Sahuquet, B.C. and J.J. Ferrier. "Steam-Drive Pilot in a Fractured Carbonated Reservoir: Lacq Superieur Field." *Journal of Petroleum Technology* (April 1982) 873–880.

Schafer, J.N. "A Practical Method of Well Evaluation and Acreage Development for the Naturally Fractured Austin Chalk Formation." *The Log Analyst* (Jan.–Feb. 1980) 10–23.

Scott, John. "Pemex Trying to Double Reserves." *Petroleum Engineer* (December 1975), 20–24.

Smith, J.E. "The Cretaceous Limestone Producing Areas of the Mara and Maracaibo District—Venezuela." Proceedings of the Third World Petroleum Congress (1951), Section I, 56–71.

Smith, J.M. "Modularization Concepts for the Norman Wells Expansion Project." *The Journal of Canadian Petroleum Technology* (July–August, 1986) 52–54.

Sondergard, C.I. and R.S. Wu. "Norman Wells PVT Properties After Reservoir Repressurization." *Journal of Canadian Petroleum Technology* (November–December, 1991), 56–60.

Spencer, C.W. and R.F. Mast. *Geology of Tight Gas Reservoirs*, AAPG Studies in Geology #24 (1986), 299 p.

Stephenson, M. "The Cretaceous Limestone Producing Areas of the Mara and Maracaibo District—Venezuela—Reservoir and Production Engineering." Proceedings of the Third World Petroleum Congress (1951), Section II.

Streltsova, T.D. and R.M. McKinley. "Effect of Flow Time Duration on Buildup Pattern For Reservoirs with Heterogeneous Properties." *Society of Petroleum Engineers Journal* (March 1984) 294–306.

Strobel, C.J., M.S. Gulati, and H.J. Ramey. "Reservoir Limit Test in a Naturally Fractured Reservoir—A Field Case Study Using Type Curves." *Journal of Petroleum Technology* (Sept. 1976) 1097–1106.

Sulak, R.M. "Ekofisk Field: The First 20 Years." *Journal of Petroleum Technology* (October 1991), 1265–1271.

Thomas, L.K. et al. "Ekofisk Waterflood Pilot." *Journal of Petroleum Technology* (February 1987), 221–232.

Tyler, N. and J.C. Gholston. "Heterogeneous Deep-Sea Fan Reservoirs, Shaekelford and Preston Waterflood Units, Spraberry Trend, West Texas." Report of Investigations No. 171, Bureau of Economic Geology (1988).

van Wunnik, J.N.M. and K. Wit. "Improvement of Gravity Drainage by Steam Injection Into a Fractured Reservoir: An Analytical Evaluation." *SPE Reservoir Engineering* (February 1992), 59–66.

Warren, J.E., and P.J. Root. "The Behavior of Naturally Fractured Reservoirs." *Society Petroleum Engineers Journal* (September 1963). 245.

Weimer, R.J., S.A. Sonnenberg, and G.B.C. Young. "Wattenberg Field, Denver Basin, Colorado." *Geology of Tight Gas Reservoirs*, AAPG Studies in Geology #24 (1986), 143–164.

West, L.W. and F.M. Doyle-Read. "A Synergistic Evolution of the Norman Wells Reservoir Description." *Journal of Canadian Petroleum Technology* (March–April, 1988) 96–103.

White, C.W. "Drilling and Completion of a Horizontal Lower Spraberry Well Including Multiple Hydraulic Fracture Treatments." SPE paper 19721 presented at the 64th Annual Technical Conference and Exhibition of the Society of Petroleum Engineers held in San Antonio, Texas (October 8–11, 1989).

Wilkinson, W.M. "Fracturing in Spraberry Reservoir, West Texas." Bulletin American Association Petroleum Geologist 37 (February 1953), 250–265.

Chapter 6

Economic Evaluation and Reserves

Economic analysis and reserves have been discussed widely in the petroleum engineering literature and more recently in the Petroleum Society of CIM Monograph No. 1, "Determination of Oil and Gas Reserves," published in 1994. It has been shown in previous chapters, that other things being equal, a smaller number of wells is required to drain efficiently a naturally fractured reservoir as compared with a conventional "homogeneous" reservoir. Furthermore, the importance of drilling directional and horizontal wells for the purpose of intercepting the larger possible number of high inclination and vertical fractures has been stressed.

Following a basic introduction of key economic yardsticks for decision making, this chapter presents methods for handling economic aspects of acceleration projects such as those that occur in many naturally fractured reservoirs.

In general, fractured reservoirs should have larger spacing than homogeneous reservoirs. For example, Baker and Lucas (1972) have noted that practical experience and detailed economic evaluations have led to the conclusion that 640 acre is the optimum spacing for the Altamont trend. Closed spacing proved uneconomic.

Daniel (1954) indicated that the fractures are so closed at Kirkuk field (Iraq), that only a few wells located at the base of the highest dome (Baba) would be enough to drain the entire reservoir. A 2-mile spacing was expected to give adequate drainage (spacing of approximately 1280 acres).

In discussing the Ain Zalah field (Iraq), Daniel (1954) indicated that the degree of fracturing was so intense that drainage from the first and second pay could probably be achieved with only two or three wells. By contrast, he pointed out that the degree of fracturing at the Dukhan field (Qatar) was lower than that at Kirkuk and Ain Zalah and, consequently, the appropriate drainage of the reservoir required closer spacing.

These examples, described geologically in Chapter 1, indicate that the evaluation of naturally fractured reservoirs should involve finding an optimum point of equilibrium between spacing and economics.

If, for instance, one well is enough to drain a fractured reservoir uneconomically, it is likely that the drilling of additional wells (closer spacing) might make the project economic due to faster revenues. In this case we are in the presence of acceleration projects.

Following the discussion dealing with acceleration projects some key economic parameters related to directional and horizontal wells are presented.

The last part of this chapter presents reserves definitions and key concepts associated with the evaluation of proven, probable and possible reserves, and some guidelines for determination of reserves in naturally fractured reservoirs.

COMPOUND INTEREST

Compound interest is an interest rate which usually reflects the cost of borrowed money or the rate of return on invested capital. The principle behind compounding is simple. Assume, for example, that at present (time zero) you have a single sum of money PW (present worth) that you are going to invest at a compound interest rate (i) per period. You wish to know the future worth (FW) of this present single sum of money, n periods from now.

At the end of period 1, you would have drawn an interest

$$PW \times i$$

and the worth (FW_1) of your investment would be:

$$FW_1 = PW \times i + PW = PW (1 + i) \tag{6–1}$$

At the end of period 2, you would have drawn an interest:

$$PW (i + 1) \times i$$

and the worth of your investment would be:

$$FW_2 = PW (i + 1) \times i + PW (i + 1)$$
$$= PW (i + 1)^2 \tag{6–2}$$

At the end of the nth period you would have drawn an interest:

$$PW (i + 1)^{n-1} \times i$$

and the worth of your investment would be:

$$FW_n = PW (i + 1)^{n-1} \times i + PW (i + 1)^{n-1}$$
$$= PW (i + 1)^n \tag{6–3}$$

Equation 6–3 represents the foundation of economic evaluations. The factor $(1 + i)^n$ is referred to as single payment compound amount factor (Appendix).

Example 6–1. What is the future worth of $500 compounded annually at 6% at the end of four years?

$$FW = 500 (1 + 0.06)^4 = 631.24$$

Equation 6–3 can be rearranged to calculate present worth as follows:

$$PW = FW \left[\frac{1}{(1 + i)^n} \right] \tag{6–4}$$

The factor $1/(1 + i)^n$ allows the calculation of a present single sum of money equivalent to a future amount. This factor is referred to as single payment present worth factor (Appendix).

Example 6–2. What is the present worth of $631.24 received four years from now, if the compound interest is 6% per year?

$$PW = 631.24 \left[\frac{1}{(1 + 0.06)^4} \right] = 500$$

In some instances it is necessary to calculate the future worth of uniform series of equal investments (A). In these cases each investment generates compound interests for different periods. For example, the investment carried out at the end of period 1 will generate interest during all periods minus 1, while the investment carried out at n – 1 will generate interest for only one period, i.e., the period that goes from n – 1 to n.

The future worth in these situations can be represented by the equation:

$$FW = A (1 + i)^0 + A (1 + i)^1 + A (1 + i)^2 + \ldots + A(1 + i)^{n-1} \tag{6–5}$$

Example 6–3. What is the future worth of $500 investments made during four years at the end of each year if the compound interest is 6% per year?

$$FW = 500 (1 + i)^0 + 500 (1 + i)^1 + 500 (1 + i)^2 + 500 (1 + i)^3$$
$$= 500 (1 + 1.06 + 1.1236 + 1.1910) = 2,187.30$$

For simplicity, Equation 6–5 can be multiplied on both sides by $(1 + i)$ as follows:

$$FW (1 + i) = A (1 + i) + A (1 + i)^2 + A (1 + i)^3 + \ldots + A (1 + i)^n \qquad (6\text{–}6)$$

Subtracting Equation 6–5 from 6–6 yields:

$$FW (1 + i) - FW = A (1 + i)^n - A \qquad (6\text{–}7)$$

Solving for future worth, FW, results in:

$$FW = A \left[\frac{(1 + i)^n - 1}{i} \right] \qquad (6\text{–}8)$$

The factor $[(1 + i)^n - 1]/i$ is referred to as uniform series compound amount factor (Appendix). Equation 6–8 makes it possible to solve for future worth of uniform series of equal investments without the burden of calculating Equation 6–5.

Example 6–3 is solved as follows:

$$FW = 500 \left[\frac{(1 + .06)^4 - 1}{i} \right] = 2{,}187.30$$

Equation 6–8 can be rearranged to calculate a uniform series of equal payments at the end of each period as follows:

$$A = FW \left[\frac{i}{(1 + i)^n - 1} \right] \qquad (6\text{–}9)$$

The factor $i/[(1 + i)^n - 1]$ is referred to as sinking fund deposit factor (Appendix).

Example 6–4. What uniform series of equal payments must be made at the end of each year to have \$2,187.30 in four years if the compound interest rate is 6% per year?

$$A = 2187.30 \left[\frac{0.06}{(1 + 0.06)^4 - 1} \right] = 500$$

In some cases, it is necessary to calculate the present worth of a uniform series of equal investments. This can be accomplished by combining Equation 6–4 and 6–8 as follows:

$$PW = A \left[\frac{(1 + i)^n - 1}{i} \right] \left[\frac{1}{(1 + i)^n} \right] \qquad (6\text{-}10)$$

The combined factor $[(1 + i)^n - 1]/[i(1 + i)^n]$ is referred to as uniform series present worth factor (Appendix).

Example 6–5. What is the present value of a uniform series of \$500 equal payments made at the end of each year during four years if the compound interest rate is 6% per year?

$$PW = 500 \left[\frac{(1 + 0.06)^4 - 1}{0.06} \right] \left[\frac{1}{(1 + 0.06)^4} \right] = 1732.55$$

Equation 6–10 can be rearranged to calculate the uniform series of end of period payments (A) as follows:

$$A = PW \left[\frac{i (1 + i)^n}{(1 + i)^n - 1} \right] \qquad (6\text{–}11)$$

The above equation is used to calculate mortgage payments. The factor $i (1 + i)^n /[(1 + i)^n - 1]$ is referred to as capital recovery factor (Appendix).

Example 6–6. What is your mortgage payment if you borrow \$1,732.55 from a bank at a compound interest rate of 6% per year and you are going to pay your debt during four years at the end of each year?

$$A = 1{,}732.55 \left[\frac{0.06 (1 + 0.06)^4}{(1 + 0.06)^4 - 1} \right] = 500$$

PERIOD, EFFECTIVE, NOMINAL, AND CONTINUOUS INTEREST RATES

A period interest rate (i) is defined as the ratio between the nominal interest rate (r) and the number of compounding periods per year (m).

$$\text{Period interest rate} = i = \frac{r}{m} \tag{6–12}$$

If a financial agency pays 8% of nominal annual interest rate compounded quarterly, it means that the quarterly interest rate is 4/m = 8/4= 2%.

When the nominal annual interest rate is compounded for more than one period, the resulting effective interest (E) is larger than the nominal interest rate.

The future worth (FW) of a present amount (PW) at the end of year one is calculated from Equation 6–3 as:

$$FW = PW \, (1 + E)^1 \tag{6–13}$$

The future worth (FW) of the same present amount (PW) is calculated from Equation 6–3 for any desired number of periods, m, at the interest rate per period (i) as follows:

$$FW = PW \, (1 + i)^m \tag{6–14}$$

Combining Equations 6–13 and 14 results in an effective annual interest equal to:

$$E = (1 + i)^m - 1 \tag{6–15}$$

To calculate the continuous interest rate, the number of compounding periods per year (m) are made to approach infinity. Equation 6–15 can be combined with Equation 6–12 to yield:

$$E = (1 + r/m)^m - 1 \tag{6–16}$$

And the continuous interest rate is given by:

$$E_{continuous} = \lim_{m \to \infty} (1 + r/m)^m - 1 \tag{6–17}$$

$$= \lim_{m \to \infty} [(1 + r/m)^{m/r}]^r - 1$$

$$= \lim_{m \to \infty} [e^{m/r \ln (1 + r/m)}]^r - 1$$

$$E_{continuous} = e^r - 1 \tag{6–18}$$

Example 6–7. Calculate the effective interest rate which is equivalent to a nominal interest rate of 6% per year compounded semiannually and compounded continuously.

The equivalent interest rate compounded semiannually is calculated with the use of Equation 6–16 as follows:

$$E = (1 + .06/2)^2 - 1 = 6.09\%$$

The equivalent interest rate compounded continuously is:

$$E_{continuous} = e^{.06} - 1 = 6.18\%$$

DISCOUNTED CASH FLOWS

Discounted cash flows refer to the time value of money. In general, the most commonly used methods for economic evaluation in the Canadian and U.S. oil industry are (in order of preference) the rate of return, net present worth, and payout time (Mathur & Carey, 1974; Petroleum Society of CIM "Determination of Oil and Gas Reserves," 1994).

Present Worth

The present worth, also called present value, refers to a single sum of money at time zero which is equivalent to future income. Present worth is calculated using Equation 6–4 in the case of a single future sum of money. However, when there are two or more revenues, Equation 6–4 can be expanded as follows:

$$PW = FW_1 \left[\frac{1}{(1+i)^1}\right] + FW_2 \left[\frac{1}{(1+i)^2}\right] +$$

$$FW_3 \left[\frac{1}{(1+i)^3}\right] + \ldots + FW_n \left[\frac{1}{(1+i)^n}\right] \tag{6–19}$$

In the previous equation, the assumption is made that the incomes are received at the end of each year (in the remainder of this chapter, a period will be equivalent to one year unless otherwise specified). Some times it is assumed that the incomes occur at midyear and the exponents in the denominators of Equation 6–19 change as follows:

$$PW = FW_1 \left[\frac{1}{(1+i)^{0.5}}\right] + FW_2 \left[\frac{1}{(1+i)^{1.5}}\right] +$$

$$FW_3 \left[\frac{1}{(1+i)^{2.5}}\right] + \ldots + FW_n \left[\frac{1}{(1+i)^{n-0.5}}\right] \tag{6–20}$$

At other times, it is assumed that the cash flowbacks occur continuously. In this case the economic analysis is carried out with the use of the equation:

$$PW = FW_1 \left[\frac{e^r - 1}{re^{1r}}\right] + FW_2 \left[\frac{e^r - 1}{re^{2r}}\right]$$

$$+ FW_3 \left[\frac{e^r - 1}{re^{3r}}\right] + \ldots + FW_n \left[\frac{e^r - 1}{re^{nr}}\right] \tag{6–21}$$

Example 6–8. Assume that you are going to receive $150, $100, and $50 at the end of years one, two and three. What is the present value of these revenues if they are discounted at 6% per year?

The solution is carried out with the use of Equation 6–19, since the revenues are received at the end of each year.

$$PW = 150 \left[\frac{1}{(1+.06)}\right] + 100 \left[\frac{1}{(1+.06)^2}\right] + 50 \left[\frac{1}{(1+.06)^3}\right]$$

$$= 141.50 + 89.00 + 41.98 = 272.48$$

Usually Equation 6–19 is solved in tabular form as shown in Table 6–1. The single payment present worth factor $1/(1+i)^n$ is also referred to as discount factor. The interest rate (i) in this case refers to the company's discount rate.

The same problem has been solved in Table 6–2 by assuming that payments are received at the middle rather than at the end of the year. This case is solved using Equation 6–20. Finally, the problem has been solved assuming that the revenues are going to be received continuously. In this case the solution is carried out with the use of Equation 6–21 as shown in Table 6–3.

It is customary to calculate an overall present worth factor, usually referred to as deferment factor (D.F.) from the equation

$$D.F. = \frac{\text{total P.W.}}{\text{total undiscounted values}} \tag{6–22}$$

For example, in Table 6–3 the deferment factor is equal to 280.04/300 = 0.9334.

NET PRESENT WORTH

The net present worth is simply the difference between the present worth and the investment at time zero. If, for example, the revenues of Table 6–1 were obtained due to an investment of $100, the net present worth would be 272.49 – 100.00 = 172.49.

TABLE 6–1 Calculation of Present Worth for Example 6-8 Assuming Revenues at the End of the Year

Year	Revenue at end of year ($)	Discount factor	Present Worth ($)
1	150	$1/(1 + .06) = 0.9434$	$150 \times .9434 = 141.51$
2	100	$1/(1 + .06)^2 = 0.8900$	$100 \times .8900 = 89.00$
3	50	$1/(1 + .06)^3 = 0.8396$	$50 \times .8396 = 41.98$
	300		272.49

TABLE 6–2 Calculation of Present Worth for Example 6-8 Assuming Revenues at the Middle of the Year

Year	Revenue at middle of year ($)	Discount factor	Present Worth ($)
1	150	$1/(1 + .06)^{0.5} = 0.9713$	$150 \times .9713 = 145.70$
2	100	$1/(1 + .06)^{1.5} = 0.9163$	$100 \times .9163 = 91.63$
3	50	$1/(1 + .06)^{2.5} = 0.8644$	$50 \times .8644 = 43.22$
	300		280.55

TABLE 6–3 Calculation of Present Worth for Example 6-8 Assuming that Revenues Occur Continuously

Year	Revenue during the year ($)	Factor $= \dfrac{e^r - 1}{re^m}$	Present worth ($)
1	150	0.9706	145.59
2	100	0.9141	91.41
3	50	0.8608	43.04
	300		280.04

FUTURE WORTH

The future worth (FW) of a present amount can be calculated using Equation 6–14. Sometimes it is desirable to calculate the future worth of revenues which occur at the end of various years. In this situation the future worth can be calculated from the relationship:

$$FW = PW_1 (1 + i)^{n-1} + PW_2 (1 + i)^{n-2} + \ldots +$$

$$PW_{n-1} (1 + i)^{n-(n-1)} + PW_n (1 + i)^{n-n} \tag{6–23}$$

where $PW_1, PW_2, \ldots PW_n$ stands for revenues at the end of years 1, 2, ... n, respectively. The factors $(1 + i)^{n-1}$, $(1 + i)^{n-2}$, . . . , are compound factors.

Example 6–9. Calculate the future worth after three years of $150, $100, and $50 that will be received at the end of years 1, 2 and 3, if the discount rate is 6% per year.
  For the calculation, use Equation 6–23 as follows:

$$FW = 150 (1 + .06)^{3-1} + 100 (1 + .06)^{3-2} + 50 (1 + .06)^{3-3}$$

$$= 168.54 + 106.00 + 50 = 324.54$$

The computations are usually done in tabular form (Table 6–4)

RATE OF RETURN

The rate of return is usually calculated by trial and error from Equation 6–19 by assuming values of i until the present worth (PW) is equal to the investment, or:

$$NPW = -C + PW = 0 \tag{6–24}$$

where NPW is the net present worth and C is the invested capital at time zero.

TABLE 6–4 Calculation of Future Worth for Example 6-9

Year	Revenue at the end of year ($)	Compound factor	Future worth ($)
1	150	1.1236	168.54
2	100	1.0600	106.00
3	50	1.0000	50.00
			324.54

The statement has been usually made in petroleum engineering literature that reinvestment of incomes is implied by the rate of return calculation. This statement is based upon a misunderstanding of the rate of return calculation.

The rate of return indicates the return that an investor receives on the unamortized investment, no matter how the incomes are spent, wasted, or reinvested (Stermole, 1974; Dougherty, 1986; Aguilera, 1986). Mathematically, this concept can be written as:

First Year

$$C_1 = -C - (C \times i)$$

$$A_1 = C_1 + FW_1$$

Second Year

$$C_2 = A_1 \times (1 + i)$$

$$A_2 = C_2 + FW_2$$

Third Year

$$C_3 = A_2 \times (1 + i)$$

$$A_3 = C_3 + FW_3$$

nth Year

$$C_n = A_{n-1} \times (1 + i)$$

$$A_n = C_n + FW_n \tag{6–25}$$

where A_1, A_2, A_3,... A_n represent the unamortized investment upon which the rate of return is applied each year, C is the invested capital, and C_1, C_2, C_3,...C_n are the amounts of principal plus interest. When the value of A_n equals zero, the assumed i in Equation 6–25 equals the rate of return. A cumulative cash position diagram is most helpful to clarify the exact meaning of the rate of return (Stermole, 1974, Aguilera 1976, 1986).

Example 6–10. What is the rate of return of a project where we are going to invest $20,000, if we receive $6,309 at the end of each year during four years?

The solution to this problem can be carried out by assuming values of i in Equation 6–25 until an i is found such that A_n equals zero. If, for instance, we assume that i equals 10%, we obtain the following solution for Equation 6–25:

First Year

$$C_1 = -20,000 - (20,000 \times 0.10) = -22,000$$

$$A_1 = -22,000 + 6,309 = -15,691$$

Second Year

$$C_2 = -15,691 \times (1 + 0.10) = -17,260$$

$$A_2 = -17,260 + 6,309 = -10,951$$

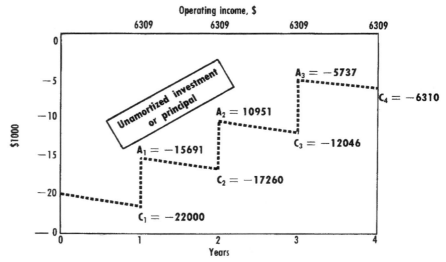

FIGURE 6–1 Cumulative cash position diagram for Example 6-10

Third Year

$C_3 = -10,951 \times (1 + 0.10) = -12,046$

$A_3 = -12,046 + 6,309 = -5,737$

Fourth Year

$C_4 = -5,737 \times (1 + 0.10) = -6,310$

$A_4 = -6,310 + 6.309 \approx 0$

The above results are plotted in the cumulative cash position diagram of Figure 6–1. Notice that the cumulative cash position is zero at the end of the fourth year. If a value of i different from 10% had been assumed, the cumulative cash position at the end of the 4th year would have been above or below the zero line.

It must be emphasized that the ROR of this single investment of $20,000 is 10%, no matter what is done with the income. If reinvestment of incomes is assumed, we are not talking anymore about a single project but about two or more projects. In this case consider the corporate growth rate rather than the ROR of a single project.

PAYOUT TIME

The payout time is probably the simplest yardstick for evaluating the desirability of an economic venture. It was heavily used in the early days to make decisions. However, it has given way to other methods due to the limitation that it does not give the ultimate cash flow productivity.

The payout time gives an indication of how long it would take to recover a given investment. This criterion can be used on a discounted or undiscounted basis.

Example 6–11. If the investment (C) in Problem 6–8 is $200, what would be the payout time and how many times would the project pay out?

The solution can be carried out as shown in Table 6–5. A plot of the number of times that the project has paid out vs. time illustrates this concept (Figure 6–2). The payout time occurs when the number of times paid out is equal to one (McCray, 1975; Petroleum Society of CIM "Determination of Oil and Gas Reserves," 1994).

TABLE 6–5 Payout Times for Example 6-11

	Undiscounted			Discounted		
Year	Revenue ($)	Cumulative ($)	Cum/C	P.W at 6%/yr ($)	Cumulative ($)	Cum/C
1	150	150	0.75	141.51	141.51	0.71
2	100	250	1.25	89.00	230.51	1.15
3	50	300	1.50	41.98	272.49	1.36

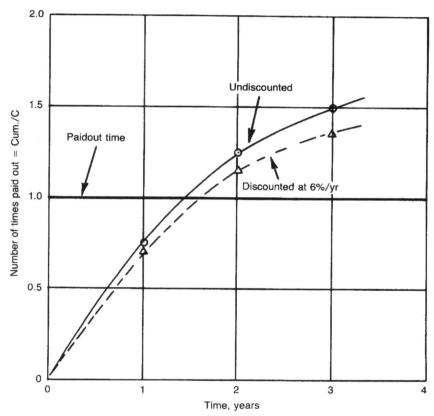

FIGURE 6–2 Number of times paid out vs. time, Example 6–11

PROFIT TO INVESTMENT RATIO

The profit–investment ratio relates the net cash flow to be obtained through the life of the project and the investment required to generate the cash flow. It is given by the equation:

$$\text{Profit to investment ratio} = \frac{PW - C}{C} = \frac{NPW}{C} \qquad (6\text{–}26)$$

For Example 6–11, the profit to investment ratio would be:

$$\frac{272.49 - 200.0}{200.00} = 0.36$$

This indicates that the profits are 36% of the investment.

Other yardsticks are used based on the company's preference. The ones discussed previously, however, are the most widely used (Mathur & Carey, 1974; Petroleum Society of CIM "Determination of Oil and Gas Reserves," 1994).

INFLATION

Inflation is a general rise in average prices which is not accompanied by an equivalent rise in productivity. Inflation decreases the quantity of goods and services that can be bought with one unit of money. A method widely used to predict inflation makes use of regression analysis to past data. This allows projections into the future.

Incorporating inflation into economic evaluations is a fuzzy subject. One approach consists of reducing inflated costs and incomes to a time zero by introducing inflation (f) into the single payment present worth factor $1/(1 + f)^n$. So, discounting is carried out with the factor

$$\frac{1}{(1 + f)^n(1 + i)^n}$$

This approach results in a rate of return (i) above the forecasted rate of inflation.

If the investor wants to make a profit of S percent per year, he can incorporate the rate of inflation into the following equation to obtain an apparent discount factor (i) (McCray, 1975):

$$i = S (1 + f) + f \tag{6–27}$$

The rate (i) obtained from Equation 6–27 is the one used in the conventional cash flow analysis.

Example 6–12. What actual profit can be obtained from a bank deposit if the bank pays an annual interest rate of 8% and the inflation rate is 6% per year.

The solution is obtained from Equation 6–27 as follows:

$$S = (0.08 – 0.06)/1.06 = 1.89\%$$

So the actual profit from keeping money in this bank is 1.89% per year.

Another approach to handling inflation is to carry out the economic analysis with the inclusion of the possible effects of inflation in both incomes and costs, since they do not always have washout effects. This is illustrated with an example presented by Stermole (1974).

Example 6–13. Assume an investment of $240,000 is going to give annual revenues of $67,000 and a salvage value of $70,000 at the end of five years. If the annual gross profit is $100,000 and operating costs are $33,000 annually, calculate the rate of return for the following cases:

a. There are no inflation effects in either gross income or operating costs.
b. Inflation in both gross income and operating costs is 5% per year.

Solution. Table 6–6 shows the solution for the case of no inflation effects. The ROR was calculated to be 18% per year on a trial and error basis.

Table 6–7 shows the solution for the case in which both operating costs and gross income are inflated by 5% per year. For this case the ROR was 21% per year. Note that an equal increase in inflation in both income and costs does not produce washout effects.

ACCELERATION PROJECTS

Many naturally fractured reservoirs can be drained efficiently with wide spacings. In some instances, however, closer spacings can lead to acceleration projects and more rewarding economics.

Dual Rates of Return

Whenever the net present worth at i = 0%/yr is negative for an investment, income, investment situation, a dual ROR exists if any real interest rate solutions exist for the NPW equation

TABLE 6–6 Cash Flow for Example 6-13 Without Effects of Inflation

Year	Investment $	Gross income $	Operating cost $	Net income $	Discount* factor at 18%/yr	Discounted cash flow $
0	240,000					−240,000
1		100,000	33,000	67,000	.8475	56,783
2		100,000	33,000	67,000	.7182	48,119
3		100,000	33,000	67,000	.6086	40,776
4		100,000	33,000	67,000	.5157	34,552
5		100,000	33,000	67,000	.4371	29,286
Salvage				70,000	.4371	30,597
						NPW = 319 ≅ 0

*The ROR ≅ 18%/year was determined by trial and error. Notice that if an ROR ≠ 18%/yr is used, the NPW is either larger or smaller than 0.

TABLE 6–7 Cash Flow for Example 6-13 with 5% per year Inflation in Gross Income and Operating Costs

Year	Investment $	Gross income $	Operating cost $	Net income $	Discount* factor at 18%/yr	Discounted cash flow $
0	240,000					−240,000
1		100,000	33,000	67,000	.8264	55,369
2		105,000	34,650	70,350	.6830	48,049
3		110,250	36,382	73,868	.5645	41,698
4		115,762	38,201	77,561	.4665	36,182
5		121,550	40,111	81,439	.3855	31,395
Salvage				70,000	.3855	26,988
						NPW = 319 ≅ 0

*The ROR ≅ 21% was determined by trial and error.

(Stermole, 1974). Dual (and multiple) rates of return are common in acceleration projects. This situation occurs when two (or multiple) values of i are found that satisfy Equation 6–19 or 6–25. A cumulative cash position diagram is most helpful to illustrate a case with dual rates of return.

Example 6–14. Table 6–8 shows the incomes for an example presented by Henry (1972). Dual rates of return (8.8 and 365.4%/yr) are found from analyzing the incremental project. In fact, 8.8 and 365.4% make the net present worth of the incremental project equal to zero. Figure 6–3 shows the cumulative cash position diagram for this example, which is generated by inserting the investment ($22), the ROR (8.8 and 365.4%/yr), and the future incomes in Equation 6–25.

Note in Figure 6–3 that the cumulative cash position does not always remain negative as in the case of Figure 6–1. The cash position shows positive values in Figure 6–3 for both rates of return at the end of the first and second years. When this situation arises, reinvestment is assumed in the positive portion of the cumulative cash position diagram at the same rate of return. Note that the negative operating incomes of the third and fourth years reduce the cumulative cash position to zero at the end of the fourth year.

The dual rate of return and reinvestment in the positive portion of the cumulative cash position diagram is not realistic; consequently, it should not be used for determining the proper level of investment in acceleration projects.

Henry (1972) has presented a review of various techniques available for handling rate of return calculations in acceleration projects.

TABLE 6–8 Incremental Cash Flow Due to Acceleration (after Henry, 1972)

Year	Base project	Acceleration project	Incremental
0		–22	–22
1	100	200	100
2	100	120	20
3	100	80	–20
4	100	0	–100

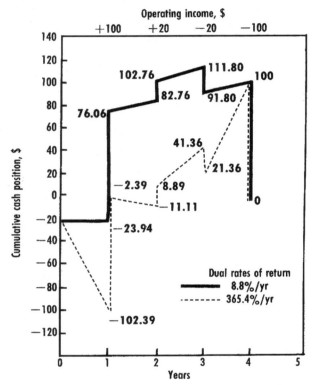

FIGURE 6–3 Cumulative cash position diagram for an incremental project with dual rates of return. Example 6–14 (after Aguilera, 1976)

Solomon (1956) has indicated that a unique ROR could be obtained in acceleration projects by reinvesting the revenues at the opportunity rate to the end of the base project. He indicated that an equivalent ROR could be determined by equating the terminal values of incremental cash flows to the terminal values of investments when the former was compounded at the opportunity rate.

Example 6–15. Calculate the ROR for the cash flow stream of Example 6–14 using Solomon's method.

The analysis is carried out as shown in Table 6–9. The investment, terminal values, and ROR are equated as follows:

$$ROR = \left(\frac{FW}{PW}\right)^{1/n} - 1 \tag{6–28}$$

$$ROR = \left(\frac{35.30}{22}\right)^{1/4} - 1 = 12.55\%$$

Stermole (1974) presented the same type of approach for decision-making and termed it "future worth modification for incremental ROR analysis." The same type of technique has been called "appreciation of equity" by Phillips (1965).

TABLE 6–9 Unique ROR Using Solomon Method

Year	Investment	Incremental cash flow	Compound FW factor at 10%/yr	Terminal Value
0	22			
1		100	$(1 + .10)^{4-1} = 1.3310$	133.10
2		20	$(1 + .10)^{4-2} = 1.2100$	24.20
3		−20	$(1 + .10)^{4-3} = 1.1000$	−22.00
4		−100	$(1 + .10)^{4-4} = 1.0000$	−100.00
				35.30

TABLE 6–10 Unique ROR Using Lefkovits et al. Modification

Year n	Δ Cash flow	Allocation of cash flow Interest	Allocation of cash flow Principal	Unpaid balance	Interest rate (%)
0				$22	18.0
1	$100	$4	$96	−74	10.0
2	20	−7	27	−101	10.0
3	−20	−10	−10	−91	10.0
4	−100	−9	−91	0	

Lefkovits et al (1965) provide another approach that leads to a unique rate of return in acceleration projects. In this technique the project ROR calculates interest on the unamortized balance as long as this balance is positive. When the balance becomes negative, the opportunity rate or the company's discount rate is used to compute interest.

Example 6–16. Calculate the ROR for the incremental cash flow shown in Table 6–10 using Lefkovits et al modification if an opportunity reinvestment rate of 10% is assumed.

The problem is solved by trial and error. For this case the ROR is calculated to be 18% per year (Table 6–10). In fact by using 18% when the unpaid balance is positive and 10% when it is negative, at the end of the fourth year the unpaid balance is zero.

Interest paid at the end of the first year is $22 \times 0.18 \simeq -4$ and the new unpaid balance is $22 + 4 - 100 = -74$

Since the unpaid balance is now negative, the interest is the opportunity reinvestment rate which has been assumed to be equal to 10%. Consequently, the interest for the second year is $-74 \times 0.10 \simeq -7$ and the new unpaid balance is $-74 - 7 -20 = -101$.

For the third year the interest is $-101 \times 0.10 \simeq -10$ and the new unpaid balance is $-101 - 10 + 20 = -91$.

Finally, for the last year the interest is $-91 - 9 + 100 = 0$, which indicates that the assumed ROR of 18% per year has been correct. Since the ROR is larger than the reinvestment opportunity rate (18 > 10), the acceleration project can be accepted as profitable. Table 6–10 summarizes the above calculations.

Henry (1972) discussed another proposed solution to the Lefkovits et al. method which consists of discounting each cash flow starting with the final period and working back to time zero or the beginning of the project. This solution is carried out using the equation:

$$V_n = P_n - C_{n+1} + (V_{n+1})/F \qquad (6-29)$$

where: V_n = accumulated value of all cash flows at the end of the year starting with the final period and working back to year n.

P_n = end of year incremental cash flow (other than investment)

C_{n+1} = capital investment at beginning of the year

F = $(1 + i)$ when V_{n+1} is negative and i is the opportunity rate, or $F = (1 + ROR)$ when V_{n+1} is positive and ROR is project rate of return.

Example 6–17. Calculate the ROR for the incremental cash flow shown in Table 6–11 using Henry's proposed solution to the Lefkovits et al. method. The $26 cash flow for the first year in Table 6–11 supports the $22 investment. Since the Value of V should be zero at time zero, Equation 6–29 can calculate the ROR as follows:

$$F_n = V_{n+1}/(V_n - P_n - C_{n+1}) \tag{6-30}$$

and

$$ROR = F - 1 \tag{6-31}$$

Consequently, the rate of return is

$$ROR = 26/(0 - 0 + 22) - 1 = 1.18 - 1 = 0.18$$

which is identical to the ROR calculated in the previous example. Table 6–11 shows a summary of the calculations and results.

Example 6–18. Calculate the rate of return for the incremental cash flow shown in Table 6–12. The investment is $400 and the reinvestment opportunity rate is 10% per year.

Conventional ROR calculations yield dual rates of return equal to 7.2 and 72.9% per year. A unique rate of return is 25.69% per year using the procedure from Example 6–17.

This example is different because the ROR has to be calculated using trial and error since the value of V_n changes from negative to positive in the third year. This calculation is carried out as follows:

a. A value of F is assumed (for instance F = 1.2569 or ROR = 25.69% per year) and V_n is calculated at the end of the first year with the use of Equation 6–29:

$$V_n = 500 - 0 + (319.39/1.2569) = 754.13$$

TABLE 6–11 Proposed solution Using Lefkovits et al. Modification

Year n	P_n	–	C_{n+1}	+	$(V_{n+1})/F$	=	V_n
0	0	–	22	+	(26)/1.18	=	0
1	100	–	0	+	(–81)/1.10	=	26
2	20	–	0	+	(–111)/1.10	=	–81
3	–20	–	0	+	(–100)/1.10	=	–111
4	–100	–	0	+	0	=	–100

TABLE 6–12 Solution Using the Lefkovits et al. Modification (after Henry, 1972)

Year n	ΔCash flow	Investment	ROR CALCULATION						
			P_n	–	C_{n+1}	+	$(V_{n+1})/F$	=	V_n
0		$600	0	–	600	+	(754.13)/1.2669	=	0
1	$500		500	–	0	+	(319.39)/1.2569	=	754.13
2	500		500	–	0	+	(– 198.67)/1.10	=	319.39
3	500		500	–	0	+	(– 768.54)/1.10	=	–198.67
4	500		500	–	0	+	(–1395.39)/1.10	=	–768.54
5	500		500	–	0	+	(–2084.93)/1.10	=	–1395.39
6	–500		–500	–	0	+	(–1743.43)/1.10	=	–2084.93
7	–500		–500	–	0	+	(–1367.77)/1.10	=	–1743.43
8	–500		–500	–	0	+	(– 954.55)/1.10	=	–1367.77
9	–500		–500	–	0	+	(– 500.00)/1.10	=	–954.55
10	–500		–500	–	0	+	(0.00)/1.10	=	–500.00

b. The ROR is calculated from Eqs. 6–30 and 31:

$$\text{ROR} = \frac{754.13}{500} - 1 = 0.2569$$

If the ROR value assumed in step (a) does not agree with the one calculated in step (b), the procedure must be repeated with a new value of ROR until agreement is reached.

Shoemaker (1963) has proposed another method which also allows the calculation of a unique rate of return. In this method the present worth (PW) of both base and accelerated projects are calculated at the investment opportunity rate with the use of Equation 6–19. Then, an adjusted cash flow is calculated with the use of the equation:

$$\text{ACF} = 1 - \frac{\text{PW}_{\text{base}}}{\text{PW}_{\text{accelerated}}} \times \text{CF}_{\text{accelerated}} \qquad (6\text{–}32)$$

where: ACF = adjusted cash flow
$\text{CF}_{\text{accelerated}}$ = accelerated cash flow

The rate of return is calculated conventionally by trial and error by assuming interest rates until the discounted adjusted cash flow is equal to the investment.

Example 6–19. Calculate the rate of return for the acceleration project shown in Table 6–13 if the investment opportunity rate is 10% per year.

The present worth of the base project cash flow is $316.99, using Equation 6–19. The present worth of the acceleration project is calculated to be $314.09. The adjusted cash flows are $14.14, 8.48, and 5.66, respectively (Equation 6–32).

Finally, the ROR is found to be 16.37% per year on a trial and error basis. Table 6–13 summarizes the results. Note that the PV at 16.37% ($22) is equal to the investment.

Kaitz (1967) proposed another method for calculated a unique rate of return in acceleration projects referred to as "percentage gain on investment" of PGI. In this approach the present worth of the cash flow stream is calculated at the company's discount rate, then an annuity having a present worth equal to the net present worth is calculated. Finally, the annuity is expressed as a percentage of the investment.

Example 6–20. Calculate the percentage gain on investment for the incremental cash flow shown in Table 6–14.

The annuity (A) is calculated using the equation:

$$A = \text{PW}\left[\frac{i(1+i)^n}{(1+i)^n - 1}\right] \qquad (6\text{–}33)$$

The above equation was also presented as equation 6–11.

$$A = 2.10\left[\frac{0.10(1+0.10)^4}{(1+0.10)^4 - 1}\right] = 0.6625$$

TABLE 6–13 Unique ROR Using Shoemaker Method (after Henry, 1972)

Year	Base project cash flow	Acceleration project cash flow	Adjusted* cash flow	PV at 16.37%
1	$100	$100	$14.14	$12.15
2	100	120	8.48	6.26
3	100	80	5.66	3.59
4	100			$22.00

PV Base Stream at 10 percent = $316.99
PV Acceleration Stream at 10 percent = $341.09
*(1 – 316.99/341.09) × Acceleration Cash Flow

TABLE 6–14 Kaitz Method (after Henry, 1972)

Year	Incremental net cash flow	PV at 10 %
0	–$22	–$22.00
1	100	90.01
2	20	16.54
3	–20	–15.03
4	–100	–68.30
	PV PROFIT	–$ 2.10

The percentage gain on investment, PGI, is calculated from:

$$PGI = \left(\frac{A}{C}\right) \times 100 \qquad (6\text{–}34)$$

$$PGI = \frac{0.6625}{22} \times 100 = 3.01\%$$

And the equivalent return on investment for this project is equal to the sum of PGI and the investment opportunity rate, or 3.01 + 10.00 = 13.01%.

In Henry's (1972) opinion, the Lefkovits (1959) method for evaluating acceleration projects yields a rate of return which is more comparable to the conventional ROR of simple projects. The other methods are also useful for accepting or rejecting projects provided that the analyst understands the concepts involved in each decision–making technique.

Present Value Profiles

The present worth (PW) or present value (PV) represents a present single sum of money at time zero and is calculated with the use of Equation 6–4. In this case, i is the company's discount rate. The net present worth (NPW) is the difference between the present worth and the investment at time zero.

The NPW is probably the best yardstick by which acceleration projects can be evaluated. A complete NPW analysis should include the individual analysis of the base, the accelerated and the incremental project at various discount rates. Present value profiles are most useful for evaluating acceleration projects, especially in the presence of dual rates of return (Wooddy, 1960).

Example 6–21. Table 6–15 shows the economic analysis of project A, which requires an investment of $100,000. The oil production is anticipated to decline during 10 years (Table 6–16). Assuming that the company's discount rate is 20% per year, the project can be accepted as profitable since the calculated rate of return is 27.32% per year and the NPW at 20% is $25,774.

Table 6–17 shows the economic analysis of an alternate project (B) that requires an investment of $168,000. This investment should permit the recovery of the same amount of oil in five rather than 10 years. The oil production with acceleration is presented in Table 6–16. This investment can also be accepted as feasible, since the calculated ROR (26.96%/yr) is greater than the company's discount rate (20%/yr) and the NPW at 20%/yr is $26,390 (Table 6–17).

Figure 6–4 shows present value profiles, i.e., plots of net present worth vs. various assumed discount rates for projects A and B. Notice that the two profiles intercept at discount rates of 16.41 and 23.98%/yr. This means that the net present worth of the incremental project (B-A) is zero at the interception points; thus, the incremental project has dual rates of return (16.41 and 23.98%/yr).

Figure 6–4 also shows a present value profile for the incremental project (B-A) that corroborates the dual rates of return. Analysis of Figure 6–4 reveals that for a company's

TABLE 6–15 Economic Evaluation
Project: Evaluation of Performance, Case A, Example 6-21
Basic Data
Oil Price = $10.000 per bbl
Rate of Discount = 20%/Year
First Year = 1
Number of Years to be Analyzed = 10
Solution

Year	Oil bbl/yr	Gross revenues $/yr	Operating expenses $/yr	Operating income $/yr	Investment $	Cash flow $/yr
1	4,000	40,000	10,000	30,000	100,000	−70,000
2	3,900	39,000	9,000	30,000	0	30,000
3	3,800	38,000	8,000	30,000	0	30,000
4	3,700	37,000	7,000	30,000	0	30,000
5	3,600	36,000	6,000	30,000	0	30,000
6	3,500	35,000	5,000	30,000	0	30,000
7	3,400	34,000	4,000	30,000	0	30,000
8	3,300	33,000	3,000	30,000	0	30,000
9	3,200	32,000	2,000	30,000	0	30,000
10	3,100	31,000	1,000	30,000	0	30,000
Totals	35,500	355,000	55,000	300,000	100,000	200,000

Discounted Profit to Investment Ratio = 0.26
Payout Time = 40.00 months
Payout = 300.00%
Present Worth at 20.00%/Year = $25774
ROR = 27.32%/Year

TABLE 6–16 Production Forecast for Example 6-21 With and Without Acceleration

	Production rate (bbl/year)	
Year	Without acceleration	With acceleration
1	4,000	8,000
2	3,900	7,400
3	3,800	6,800
4	3,700	6,700
5	3,600	6,600
6	3,500	0
7	3,400	0
8	3,300	0
9	3,200	0
10	3,100	0
	35,500	35,500

discount rate of 20%/yr, project B will yield a net present worth of $26,390 compared with $25,774 for project A.

Consequently, the investment in project B is the correct selection. The same selection is correct if the company's discount rate is between 16.41 and 23.98%/yr, as project B has larger net present worth than project A. Notice, however, that project A is the best selection for the following cases:

$$CDR < 16.41\%/yr$$

$$23.98\%/yr < CDR < 27.32\%/yr$$

TABLE 6–17 Economic Evaluation
Project: Evaluation of Performance, Case B, Example 6-21
Basic Data
Oil Price = $10.000 per bbl
Rate of Discount = 20%/Year
First Year = 1
Number of Years to be Analyzed

Year	Oil bbl/yr	Gross revenues $/yr	Operating expenses $/yr	Operating income $/yr	Investment $	Cash flow $/yr
1	8,000	80,000	15,000	65,000	168,000	–103,000
2	7,400	74,000	9,000	65,000	0	65,000
3	6,800	68,000	3,000	65,000	0	65,000
4	6,700	67,000	2,000	65,000	0	65,000
5	6,600	66,000	1,000	65,000	0	65,000
Totals	35,500	355,000	30,000	325,000	168,000	157,000

Discounted Profit to Investment Ratio = 0.16
Payout Time = 31.02 months
Payout = 193.45%
Present Worth at 20.00%/Year = $26390
ROR = 26.96%/Year

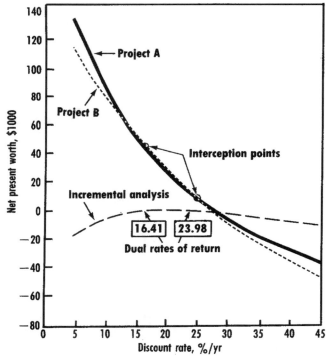

FIGURE 6–4 Present value profile for dual rates of return. Example 6–21 (after Aguilera, 1976)

where CDR is the company's discount rate. For a CDR greater than 27.32%/yr, project A is no longer feasible since the net present worth becomes negative.

Figure 6–5 shows a cumulative cash position diagram for the incremental project (B-A), which corroborates the dual rates of return. In fact, the cash position reduces to zero at the end of the tenth year with both rates of return.

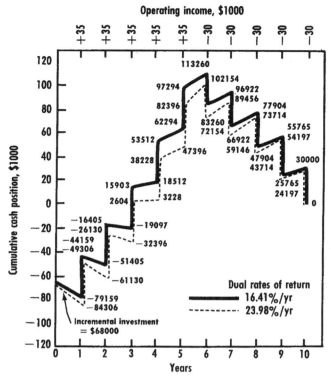

FIGURE 6–5 Cumulative cash position diagram for the incremental project. Example 6–21 (after Aguilera, 1976)

Notice in this example that both projects have the same starting point. If the acceleration project is to be carried out when a base project has already been initiated, the whole economic analysis must take as a starting point the initiation of the base project.

Future Worth Modification for Incremental ROR Analysis

Stermole (1974) presented a method referred to as future worth modification for incremental ROR analysis that allows the calculation of a pseudo-ROR. This does not give a measure of profitability, yet allows correct decisions in acceleration projects. The selection is made by comparing the pseudo-ROR with the company's discount rate.

The future worth of a project can be calculated from the relationship:

$$FW = PW_1 (1 + i)^{n-1} + PW_2 (1 + i)^{n-2} + \ldots + PW_{n-1} (1 + i)^{n-(n-1)} + PW_n (1 + i)^{n-n} \qquad (6\text{–}35)$$

where $(1 + i)^n$ is the single payment compound amount factor.

The future worth modification for incremental ROR analysis consists of calculating the future worth of both the base and the acceleration projects at the company's discount rate. Both projects are evaluated using the same life period, even if the accelerated project has a shorter life. Next, a pseudo rate of return is calculated from the equation

$$\text{Pseudo ROR} = \left(\frac{\Delta FW}{\Delta PW}\right)^{1/n} - 1 \qquad (6\text{–}36)$$

using the incremental investment (ΔPW) and the incremental future worth (ΔFW) determined from Equation 6–35. The acceleration project is the correct decision if the pseudo rate of return is larger than the company's discount rate.

Example 6–22. Table 6–18 shows the income data of an acceleration project presented by Wooddy and Capshaw (1960). The incremental analysis results in dual rates of return, 11 and 50%/yr (Figure 6–6).

TABLE 6–18   Income Acceleration Project (after Wooddy & Capshaw, 1960)

Year	Presently anticipated income, $	Accelerated income, $	Incremental cash flow $
0		−15	−15
1	10	20	10
2	10	20	10
3	10	20	10
4	10	20	10
5	10	20	10
6	10	0	−10
7	10	0	−10
8	10	0	−10
9	10	0	−10
10	10	0	−10

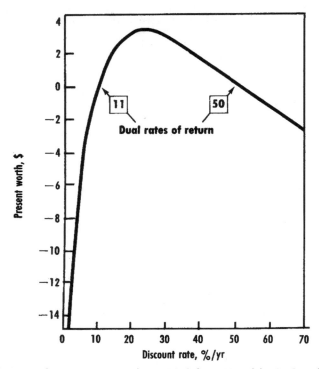

FIGURE 6–6   Dual rates of return. Example 6–22 (after Wooddy & Capshaw, 1960)

Equation 6–35 calculates the future worth of the base and accelerated projects (Table 6–19). Note that the single payment compound amount factors are the same for both cases in spite that the project lives are different (10 and 5 years). The pseudo–ROR has been calculated at 9.54%/yr from Equation 6–36. Since the pseudo–ROR is smaller than the company's discount rate (10%/yr), the correct decision is to reject the acceleration project. The same conclusion is reached by analyzing Figure 6–6. In fact, the future worth is negative at a discount rate of 10%/yr.

The same procedure has been repeated assuming a company's discount rate of 20%/yr (Table 6–20). For this case the calculated pseudo–ROR is larger than the discount rate (22.14 > 20%/yr). Consequently, the correct decision is to select the acceleration project. The same conclusion is reached by analyzing Figure 6–6. In fact, the future worth is positive at a discount rate of 20%/yr.

TABLE 6–19 Economic Analysis of an Acceleration Project by the Future Worth Modification for Incremental ROR Analysis
Discount Rate = 10%/yr
Incremental investment = $15

	Base Project			**Accelerated Project**		
Year	Operating Income	Compound Factor	FW	Operating Income	Compound Factor	FW
1	10	2.36	23.60	20	2.36	47.20
2	10	2.14	21.40	20	2.14	41.80
3	10	1.95	19.50	20	1.95	39.00
4	10	1.77	17.70	20	1.77	35.40
5	10	1.61	16.10	20	1.61	32.20
6	10	1.46	14.60	0	1.46	0
7	10	1.33	13.30	0	1.33	0
8	10	1.21	12.10	0	1.21	0
9	10	1.10	11.00	0	1.10	0
10	10	1.00	10.00	0	1.00	0
			159.30			196.60

$$\text{Pseudo} - \text{ROR} = \left(\frac{\Delta FW}{\Delta PW}\right)^{1/n} - 1$$

$$= \left(\frac{196.60 - 159.30}{15}\right)^{1/10} - 1 = 9.54\%/\text{yr}$$

Pseudo-ROR < Company's Discount Rate
∴ Reject acceleration project

TABLE 6–20 Economic Analysis of an Acceleration Project by the Future Worth Modification for Incremental ROR Analysis
Discount Rate = 20%/yr
Incremental investment = $15

	Base Project			**Accelerated Project**		
Year	Operating Income	Compount Factor	FW	Operating Income	Compount Factor	FW
1	10	5.16	51.60	20	5.16	103.20
2	10	4.30	43.00	20	4.30	86.00
3	10	3.58	35.80	20	3.58	71.60
4	10	2.99	29.90	20	2.99	59.80
5	10	2.49	24.90	20	2.49	49.80
6	10	2.07	20.70	0	2.07	0
7	10	1.73	17.30	0	1.73	0
8	10	1.44	14.40	0	1.44	0
9	10	1.20	12.00	0	1.20	0
10	10	1.00	10.00	0	1.00	0
			259.60			370.40

$$\text{Pseudo} - \text{ROR} = \left(\frac{\Delta FW}{\Delta PW}\right)^{1/n} - 1$$

$$= \left(\frac{370.40 - 259.60}{15}\right)^{1/10} - 1 = 22.14/\text{yr}$$

Pseudo-ROR > Company's Discount Rate
∴ Select acceleration project

Analyze the economics of naturally fractured reservoirs very carefully, because many of them represent actual acceleration projects which, in turn, require some special considerations.

COMPARING EVALUATION METHODS

The methods discussed previously can yield different results when ranking projects in order of priority according to potential profitability. Capen et al. (1976) analyzed 11 projects termed P to Z as shown in Table 6–21.

Assuming that the company's discount rate is 10%/yr, all projects would make some profit, as the ROR is greater than 10% in all cases. If the company discount rate is 15%/yr, projects R, S, V, and W would generate a negative present worth.

If there are no capital limitations, it is rather straight forward to determine which projects could be acceptable. In practice, this is not usually the case and companies have less capital to invest than projects. Correct ranking of projects for this situation becomes of paramount importance.

Assume, for example, that for the projects of Table 6–21 there is only $3,000 available (Capen et al, 1976). Since carrying out all the projects would require an investment of $11,000, the allocation of the $3,000 available must depend on some kind of project ranking.

Table 6–22 shows rankings for capital allocation for all projects presented in Table 6–21. If the investor prefers the "profit to investment ratio" criteria, he would probably select project Z as the one with top priority, followed by project W, project V, and project S.

An investor who prefers the ROR would probably select project U, followed by projects Q and P, respectively. An investor who gives equal weight to all evaluation techniques would probably be "democratic" and would give top priority to project U followed by projects Q and Z, respectively as shown in the last line of Table 6–22. It is evident, however, that there is only one selection criteria which would yield the most out of the $3,000 investment.

Capen et al (1976) have indicated that the best criteria for ranking investment efficiency are dictated by those projects that maximize the company's future worth. The yardstick proposed by them has been termed "growth rate" and is equivalent to Stermole's (1974) "future worth modification for incremental ROR analysis" and Solomon's (1956) approach for calculating a unique rate of return. The interest used in these calculations is the company's average opportunity rate.

Special care must be exercised when determining this rate because a small variation can produce large changes in the ranking of the projects. Note, for example, how the ranking changes when the growth rate is calculated at 5%, 10%, and 15%/yr, respectively.

In conclusion, it is clear that there is no agreement among the specialists on how to rank the investment efficiency of various projects. Consequently, the analyst who recommends some projects in a certain order of priority must have a full understanding of the yardstick by which he is basing his recommendations.

DEPRECIATION, DEPLETION AND AMORTIZATION

Depreciation

Depreciation is an annual reduction of income reflecting the loss in useful value of tangible equipment by reason of wear and tear (Megill, 1971).

The four depreciation methods more widely used are:

1. Straight line
2. Declining balance (usually double-declining balance)
3. Sum of the years-digits
4. Units of production

TABLE 6–21 Project Comparisons (after Capen et al., 1976)
After-tax net cash flow by year

						Project					
Year	P (dollars)	Q (dollars)	R (dollars)	S (dollars)	T (dollars)	U (dollars)	V (dollars)	W (dollars)	X (dollars)	Y (dollars)	Z (dollars)
0	–1,000	–1,000	–1,000	–1,000	–1,000	–1,000	–1,000	–500	–1,000	–1,000	–500
1	500	600	25	10	800	700	0	0	900	210	0
2	400	400	50	20	300	500	0	0	200	210	10
3	300	300	50	50	100	200	100	25	100	210	15
4	200	200	100	100	50	100	100	25	50	210	25
5	100	50	200	200	50	50	100	50	50	210	50
6	50	20	300	300	20	10	300	150	10	210	150
7	40	10	400	400	20	10	500	250	10	210	250
8	30	10	500	500	10	10	500	350	10	210	350
9	20	5	500	500	10	10	500	350	0	210	350
10	10	5	100	400	10	10	500	250	0	210	300

Criteria											
Profit-to-investment ratio	0.65	0.60	1.23	1.48	0.37	0.60	1.60	1.90	0.33	1.10	2.00
Payout years	2.3	2.0	6.7	6.8	1.7	1.6	6.8	7.0	1.5	4.8	7.0
Years 1 and 2 book profit (10 years straight-line write-off of initial investment), dollars	700	800	–125	–170	900	1,000	–200	–100	900	220	–90
Internal rate of return, percent	22.5	25.0	12.5	13.4	19.3	28.5	13.8	14.7	20.0	16.5	15.1
Present worth											
at 5%, dollars	446	432	582	724	249	448	794	487	227	622	518
at 10%, dollars	284	294	153	230	150	323	270	188	139	290	208
at 15%, dollars	153	179	–141	–102	66	217	–81	–10	64	54	3
Growth rate											
at 5%, percent	8.9	8.8	9.9	10.9	7.4	9.0	11.3	12.4	7.2	10.2	12.7
at 10%, percent	12.8	12.9	11.6	12.3	11.5	13.1	12.7	13.6	11.4	12.8	13.9
at 15%, percent	16.7	16.9	13.3	13.8	15.7	17.3	14.0	14.8	15.7	15.6	15.1
Investment efficiency (PW per dollar invested)											
at 5%	0.45	0.43	0.58	0.72	0.25	0.45	0.79	0.97	0.23	0.62	1.04
at 10%	0.28	0.29	0.15	0.23	0.15	0.32	0.27	0.38	0.14	0.29	0.42
at 15%	0.15	0.18	0.15	0.10	0.07	0.22	-0.08	-0.02	0.06	0.05	0.01

Straight Line

In this technique the cost of the property less its salvage value, if any, is deducted in equal annual amounts over the period of its depreciation life.

The straight line depreciation per year (SLD) is calculated from the equation:

$$SLD = (C - L)(1/n) \tag{6–37}$$

where C is the cost of the property and L is the salvage value.

Example 6–23. A machine purchased by $8,000 is expected to have an eight-year life and a salvage value of $2,000. Calculate the annual depreciation by the straight line method.

$$SLD = (8,000 - 2,000)(1/8) = \$750/yr$$

Declining Balance

In this technique a depreciation rate from 1/n to 2/n (double declining balance) is applied to a declining adjusted basis each year.

TABLE 6–22 Project Rankings for Capital Allocation (after Capen et al., 1976)

Measure	P	Q	R	S	T	U	V	W	X	Y	Z
Profit/investment	7	8	5	4	10	9	3	2	11	6	1
Payout	5	4	7	8	3	2	9	10	1	6	11
Book Profit	5	4	9	10	2	1	11	8	3	6	7
Internal R/R	3	2	11	10	5	1	9	8	4	6	7
Present worth											
at 5%	8	9	4	2	10	7	1	6	11	3	5
at 10%	4	2	9	6	10	1	5	8	11	3	7
at 15%	3	2	11	10	4	1	9	8	5	6	7
Growth rate											
at 5%	8	9	6	4	10	7	3	2	11	5	1
at 10%	6	4	9	8	10	3	7	2	11	5	1
at 15%	3	2	11	10	4	1	9	8	6	5	7
Investment efficiency											
at 5%	8	9	6	4	10	7	3	2	11	5	1
at 10%	6	4	9	8	10	3	7	2	11	5	1
at 15%	3	2	11	10	4	1	9	8	6	5	7
Democracy	5	2	11	8	9	1	7	6	10	4	3

The adjusted basis is calculated from the equation:

Adjusted basis = Cost or other basis − cumulative depreciation ± adjustments   (6–38)

The double-declining balance depreciation per year (DDB) is calculated using the equation:

$$\text{DDB} = \frac{2}{n} \text{ (adjusted basis)} \qquad (6\text{–}39)$$

Example 6–24. Calculate the annual depreciation for problem 6–23 using the double-declining balance method.

The solution is shown in Table 6–23. Note that the depreciation calculation is stopped at the end of year 5 because the Internal Revenue Service (IRS) does not permit depreciation of an asset past its salvage value.

Sum of the Year's Digits Method

In this technique a different depreciation rate is applied each year to the cost of the asset reduced by its estimated salvage value.

The sum of years-digits depreciation per year (SYD) is calculated using the equation.

$$\text{SYD} = (n - m + 1)(C - L)\Big/\left(\sum_{m=1}^{n} m\right) \qquad (6\text{–}40)$$

where n = asset depreciation life

m = depreciation year, 1, 2, ...n

$\displaystyle\sum_{m=1}^{n} m$ = sum of numbers 1 to m

C = cost or other basis

L = salvage value

Example 6–25. Calculate the annual depreciation for Problem 6–23 using the sum of the years-digits method.

TABLE 6–23 Double Declining Balance Depreciation

Year	Depreciation rate	Adjusted basis	DDB depreciation
1	2/n = 2/8 = 1/4	8,000	2,000
2	1/4	6,000	1,500
3	1/4	4,500	1,125
4	1/4	3,375	844
5	1/4	2,531	531
6–8	1/4	2,000	—

TABLE 6–24 Sum of the Years-Digits Depreciation

Year	Depreciation rate	C-L	SYD depreciation
1	8/36	6,000	1,333.36
2	7/36	6,000	1,166.69
3	6/36	6,000	1,000.02
4	5/36	6,000	833.35
5	4/36	6,000	666.68
6	3/36	6,000	500.01
7	2/36	6,000	333.34
8	1/36	6,000	166.67

The solution is done as in Table 6–24. Note that the incremental SYD remains constant at $166.67.

Units of Production

This method depreciates an asset which is controlled mainly by depletion of reserves. The units of production depreciation (UP) are calculated using the equation:

$$UP = (C - D)\ P/R \qquad (6\text{–}41)$$

where C =   cost of equipment
            D =   cumulative depreciation
            P =   production per year
            R =   remaining reserves

Example 6–26. Calculate annual depreciation by the units of production method if an equipment costs $100,000. The well is expected to produce 4000 volume units during the first year and 3000 unit volumes during the second year. The ultimate recovery is anticipated at 10,000 volume units.

The solution is obtained as in Table 6–25.

Depletion

Depletion is a reduction in income reflecting the exhaustion of a mineral deposit (Megill, 1971).

There are two methods of calculating depletion: cost depletion and percentage depletion. Depletion is computed at the end of the year by both methods and the one that gives the larger tax deduction is the one to use.

TABLE 6–25 Units of Production Depreciation

Year	Depreciable balance C-D	Remaining reserves R	Production P	Annual depreciation UP
1	100,000	10,000	4,000	40,000
2	60,000	6,000	3,000	30,000

Cost Depletion

This method can be calculated using the equation:

$$\text{cost depletion} = (\text{adjusted basis})\,(P/R) \tag{6–42}$$

where:
adjusted basis = cost basis ± adjustments – cumulative depletion
P = mineral units (in this case oil, gas, or liquid volume units) produced and sold during the year.
R = mineral units (or hydrocarbon reserves) recoverable at the beginning of the year.

The capitalized costs that go into the analysis are usually the mineral right acquisition and/or the leasing costs.

Example 6–27. Calculate cost depletion if you have paid $200,000 as lease cost for a property which has reserves equal to 1,000,000 stbo at the beginning of the first year. Reserves increase by 500,000 stbo at the beginning of the second year. Yearly production during the first three years is expected at 50,000 stbo.
The solution is done as in Table 6–26.

Percentage Depletion

Percentage depletion is "a specified percentage of gross income (sales revenues dollars) from the sales of minerals removed from the mineral property during the tax year, but the deduction for depletion under this method cannot exceed 50% of taxable income from the property after all deductions except the deduction for depletion" (Stermole, 1974). The percentage depletion for both oil and gas is 22% of gross income.

Example 6–28. You have paid $200,000 for a lease from which reserves are estimated at 1 MMstbo. Yearly production is 70,000 stbo. Price of oil is $12/bbl. Allowable depreciation is $30,000. Operating and overhead expenses per year are $100,000. Calculate cost depletion and percentage depletion. Discuss what type of depletion must be selected.
The calculations are presented in Table 6–27. The calculated percentage depletion ($184,800) for both years is less than 50% of the taxable income before depletion ($710,000 × 0.50 = $355,000). Consequently, the full calculated depletion is allowed.
Since the percentage depletion is greater than the cost depletion ($184,800 > $14,000), percentage depletion is selected.

Amortization

Yearly amortization is calculated during a five-year period using the equation:

$$\text{Yearly amortization} = (C\text{-}L)\,(1/5) \tag{6–43}$$

where C is the capital expenditure and L is the salvage value.
Some permissible expenditures that can be amortized each year are: research and development expenses; the cost of acquiring a lease for business purpose; organizational expenses of the company; and payments in connection with trademark and trade name acquisition.

TABLE 6–26 Cost Depletion Calculation

Year	R stbo	P stbo	Adjusted basis cost	Cost depletion = Adjusted basis × (P/R)
1	1,000,000	50,000	200,000	10,000
2	1,450,000	50,000	190,000	6,552
3	1,400,000	50,000	183,448	6,552

TABLE 6–27 Cost and Percentage Depletions Calculations

	Cost Depletion				Percentage Depletion				
Year	Adjusted basis $	P stbo	R stbo	Cost depletion $	Gross income $	Operating costs $	Depreciation $	Taxable income before depletion $	Percentage depletion (0.2)(gross inc.) $
1	200,000	70,000	1,000,000	14,000	840,000	100,000	30,000	710,000	184,800
2	186,000	70,000	930,000	14,000	840,000	100,000	30,000	710,000	184,800

Example 6–29. A company spends $50,000 in research and development. Calculate the allowable amortization per year assuming that the salvage value is zero.

The solution is carried out using Equation 6–43 as follows:

$$\text{Amortization} = (50,000 - 0)\,(1/5) = \$10,000/\text{year}$$

TAX TREATMENT

Income taxes can be considered as disbursements in the same way that material costs and labour expenses are disbursements. Figure 6–7 shows a general cash flow diagram for a corporation.

Once the taxable income has been determined, the tax is applied to determine the income tax and the net profit. For corporations, the first $25,000 is taxed at a rate of 22%. Taxable income above the first $25,000 is taxed at 48%.

Income tax considerations can change substantially from one case to the other. Consequently, carry out all economic comparisons on an after-tax basis.

Capitalized investments, intangible drilling and development costs, and depreciation deductions must be considered for proper tax treatment of depletable petroleum operations (Stermole, 1974).

Capitalized Investments. Leasing or acquisition costs can be capitalized and recovered by means of depletion allowance or abandonment loss.

Intangible Drilling and Development Costs. Intangible costs refer to expenditures which do not create tangible depreciable assets and, in general, do not have a salvage value. Typical intangible costs include site preparation, rig transportation and operations, drilling mud, formation evaluation, cement, completion tools, perforating jobs, squeeze cementing, and stimulation jobs (Megill, 1971).

Typical tangible costs include casing, tubing, and well equipment. Intangible costs can be deducted as expenses in the year in which they occur.

Depreciation. Depreciation includes deductions over the life of assets such as compressors, pipe, rigs, and pumps.

Example 6–30, adapted from Stermole's (1974) book, illustrates project evaluation in an after-tax basis.

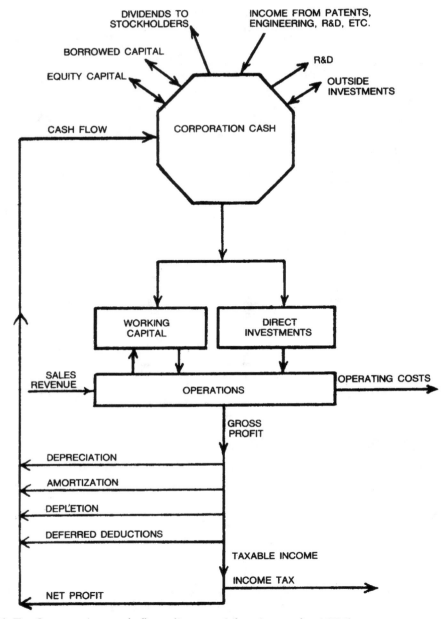

FIGURE 6–7 Corporation cash flow diagram (after Stermole, 1974)

Example 6–30. Calculate the after tax net present worth and the rate of return of a natural gas development project using the following basic data:

Investment at time zero, C_o	$1,000,000
Intangible drilling costs during year 1 to be written-off for tax deduction purposes against other income at the end of year 1, C_1	$2,500,000
Depreciable well costs at end of year 1, depreciated straight line over the next 10 years (years 2 to 11) with zero salvage value, C_2	$1,000,000
Working capital requirements	Negligible
Project life	11 yr
Effective income tax return	50%
Depletion rate	22%
Discount factor	12%

TABLE 6–28 After Tax Economic Evaluation—Example 6-30

(1) Year	(2) Gas prod. MMscf/yr	(3) Gas price $/Mscf	(4) Gross revenue M$	(5) Operating cost M$	(6) Depreciation	(7) Taxable before depletion (4)-(5)-(6)
0						
1						
2	1429	1.75	2,500	700	100	1,700
3	1528	1.80	2,750	720	100	1,930
4	1622	1.85	3,000	740	100	2,160
5	1711	1.90	3,250	760	100	2,390
6	1795	1.95	3,500	780	100	2,620
7	1875	2.00	3,750	800	100	2,850
8	1561	2.05	3,200	820	100	2,280
9	1214	2.10	2,550	840	100	1,610
10	837	2.15	1,800	860	100	840
11	432	2.20	950	880	100	–30

(8) Depletion 22% x (4) M$	(9) Taxable after depletion (7) – (8) M$	(10) Taxes 50% × (9) M$	(11) Net profit (9) – (10) M$	(12) Cash flow (11) + (6) + (8) $M	(13) P.W. factor at 12% (end of the year)	(14) Cash flow discounted at 12% M$
				–1,000*	1.0000	–1000
				–2,500**	0.8929	–2232
550	1,150	575	575	1,225	0.7972	977
605	1,325	662	663	1,368	0.7118	974
660	1,500	750	750	1,510	0.6355	960
715	1,675	837	838	1,653	0.5674	938
770	1,850	925	925	1,795	0.5066	909
825	2,025	1,012	1,013	1,938	0.4523	877
704	1,575	788	781	1,591	0.4039	643
561	1,049	524	525	1,186	0.3603	428
396	444	222	222	718	0.3220	231
0	–30	–15	–15	85	0.2875	24
						3,629

*Investment at time zero
** 50%(Tax) of intangibles plus depreciable well costs.
NPW at 12% = $3,629,000
ROR = 35%.

Estimated production rates are shown in column 2 of Table 6–28. Gas price is assumed at $1.75/Mscf in year 2, followed by a constant yearly increase of $0.05/Mscf.

Operating costs are assumed at $700,000 for year 2, followed by a constant increase of $20,000/yr which accounts for the effects of inflation. The solution to this problem is presented in Table 6–28. The net present worth is $3,629,000 and the rate of return is 35%.

WELL SPACING

In general, naturally fractured reservoirs should have larger spacing than non-fractured reservoirs. Under conditions of acceleration, the optimum number of wells for a given reservoir can be determined using the future worth modification for incremental ROR analysis (Example 6–22). An important characteristic of some of these reservoirs is their gravity segregation mechanism.

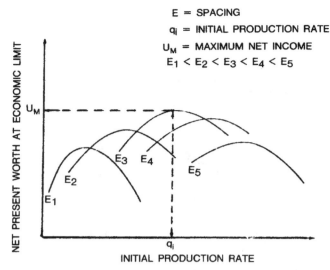

FIGURE 6–8 Determination of optimum spacing for a reservoir with gravity segregation drive (after Garaicochea & Acuna, 1978)

Garaicochea and Acuña's (1978) approach for determining optimum spacing consists of predicting performance of the reservoir under various spacing schemes. Economic evaluations are carried out to calculate the economic recoverable reserves and the net present worth of each project. By carrying out various calculations, it is possible to generate a plot of initial production rate vs. net present worth (Figure 6–8). Optimum spacing and optimum production rate can be determined from this type of analysis.

For a given spacing, the reservoir life decreases as production increases and, consequently, the net present worth is larger. However, there is a production limit at which the pressure gradients are so large that the effects of gravity segregation become negligible, leading to lower recoveries and thus lower net present worths.

Another possibility for calculating the optimum number of wells is to prepare a cross-plot of net present value vs. number of wells using as a parameter oil price as illustrated in Figure 5–55. For that case, five is the optimum number of wells.

RISK EVALUATION

All investments involve a level of uncertainty. The uncertainty arises from many causes including cost estimates, demand, sales, and taxes. Mechanical uncertainty can arise from drilling and completion problems. Nature uncertainty can arise from weather conditions and production capabilities of the reservoir.

Whenever possible, formalize the evaluation risk rather than merely give a general description such as fair, reasonably good, more less, or quite risky. An excellent discussion on risk analysis is presented in the Petroleum Society of CIM Monograph No. 1 "Determination of Oil and Gas Reserves" (1994).

Binomial Theorem

If there is enough history in a given region, it is possible to calculate the probability of any combination of future events from the binomial theorem (Campbell, 1970).

$$(a + b)^n = a^n + na^{n-1}b + \frac{n(n-1)}{1 \times 2} a^{n-2}b^2 + \ldots$$

$$+ \frac{n(n-1)(n-2)\ldots(n-r+2)}{1 \times 2 \times \ldots \times (r-1)} a^{n-r+1}b^{r-1} + b^n \qquad (6\text{–}44)$$

where:

a	= probability of one result occurring = 1 – b
b	= probability of the alternate result occurring = 1 – a
r	= the successive number of the event

Applying the previous equation assumes that all attempts in an exploratory program have an equal geologic chance (Campbell & Schuh, 1970). In some cases this assumption is reasonable. In exploration programs with unequal geologic chances, the approach is similar, but the mathematics are more complicated. Equation 6–44 has also been used successfully to estimate subsurface distance between vertical parallel natural fractures based on core data (Aguilera, 1988. See also Chapter 1).

The following rules apply to probability evaluation using the binomial theorem (Campbell, 1970):

1. "a" must be larger than "b" for the series to be convergent
2. The exponent of "a" is equal to "n" minus the exponent of "b"
3. The exponent of "b" is less than the number of the event by one
4. The last factor in the denominator is the same as the exponent of "b"
5. The last factor in the numerator of the coefficient is greater than the exponent of "a" by one

Example 6–31. Use the binomial theorem to calculate the probability of outcomes if you drill (a) 5 wells and (b) 10 wells and the exploration history of the area indicates that the chance of getting a commercial well is 10%.

a. For this case a = 0.90 and b = 0.10. The solution can be done as in Table 6–29. Note that there is a 59.05% chance of getting all five wells dry. And there is only 32.81% chance of getting one producing well out of the five wells drilled.
b. Following an approach identical to the one indicated in Table 6–29, it can be shown that if ten wells are drilled, the chance of getting one producing well 38.7%.

In other words, there is an increase in success probability if the number of exploratory attempts increases.

Example 6–32. This example is adapted from Grayson (1960). It indicates that drilling of a well is expected to result in any of the five outcomes shown in the first column of Table 6–30. The second column shows the probability of the event occurring. The last five columns show different possible alternatives of the investor. Drilling of the well costs $500,000. An oil price of $20/bbl is assumed. Miscellaneous expenses would amount to $200,000. The expected monetary value is shown in the last row. It is calculated by

TABLE 6–29 Example of the Application of the Binomial Theorem in Probability Studies (after Campell, 1970)

Combination of Events	Probability of event occurring	
5 dry holes	$(0.9)^5 =$	0.5905
4 dry holes – 1 success	$(5)\,(.10)\,(.90)^4 =$	0.3281
3 dry holes – 2 successes	$\dfrac{5 \times 4}{1 \times 2}\,(.10^2)\,(.90)^3 =$	0.0729
2 dry holes – 3 successes	$\dfrac{5 \times 4 \times 3}{1 \times 2 \times 3}\,(.10)^3\,(.90)^2 =$	0.0081
1 dry hole – 4 successes	$\dfrac{5 \times 4 \times 3 \times 2}{1 \times 2 \times 3 \times 4}\,(.10)^4\,(.90) =$	0.0045
5 successes	$(.10)^5 =$	0.00001

TABLE 6–30 One Example of Using Probability Numbers

		Possible acts				
Possible events	**prob. of event occurring**	**Don't drill**	**Drill with 100% interest**	**Drill with 50% partner**	**Farm-out, keep 1/8 override**	**Farm-out, come back in for 50% interest after payout**
Dry hole	.60	$0	$–500,000	$–250,000	$ 0	$ 0
25,000 bbl	.10	0	–200,000	–100,000	62,500	0
50,000 bbl	.15	0	300,000	150,000	125,000	150,000
250,000 bbl	.10	0	4,300,000	2,150,000	625,000	2,150,000
500,000 bbl	.05	0	9,300,000	4,650,000	1,250,000	4,650,000
Expected monetary value		$0	$ 620,000	$ 310,000	$ 150,000	$ 470,000

multiplying the probability of the event occurring by the appropriate amount of money shown in the last five columns.

For example, in the case where drilling is carried out with 100% interest, the calculation is carried out as follows:

$$(0.60)\,(-500,000) + (0.10)\,(-200,000) + (0.15)\,(300,000) +$$

$$(0.10)\,(4,300,000) + (0.05)\,(9,300,000) = \$620,000$$

Analysis of Table 6–30 indicates that the alternative that yields the least expected monetary value is to farm-out, keeping 1/8 override. The most favorable option appears to drill with 100% interest. However, this should be done only if the investor can afford to lose $500,000 in case of a dry hole.

SIZE OF FIELDS FOUND

Lahee (1956) has indicated that the size of oil fields found range from a few very large fields to many small fields (Figure 6–9). Lahee's (1956) data indicate that the occurrence of oil fields can be described by a log–normal distribution.

Campbell and Schuh (1970) derived additional distributions from Lahee's (1956) data for groups of fields (Figure 6–9). These curves are very useful for evaluating the chances of making a profit from an exploratory program when used in conjunction with the binomial theorem.

According to Campbell and Schuh (1970) the minimum average field size needed to yield a certain rate of return can be estimated from the equation:

$$F_a = \frac{nC}{rD} \tag{6–45}$$

where: F_a = average size of the "r" discoveries, b/field
         n = number of exploration attempts
         C = cost of each attempt
         r = number of fields that may be discovered
         D = present worth of oil reserves at the time of discovery, $/bbl

The statistical chance (S) of making a profit is given by the equation (Campbell & Shuh, 1970):

$$S = \sum_{r=1}^{r=n} P_b P_a \tag{6–46}$$

where: P_b = binomial probability of discovering r oil fields in n attempts
         P_a = chance of making at least the desired profit from exactly r discoveries

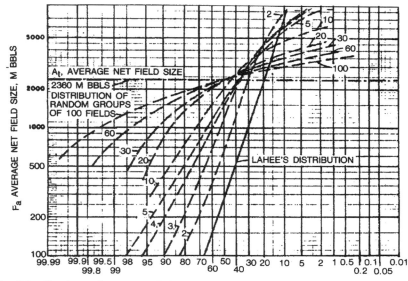

FIGURE 6–9 Distribution of average field size for random groups of fields (after Campbell & Schuh, 1970)

TABLE 6–31 Calculation of Chance of Making at Least a 10% Rate of Return for n = 10, Example 6-33 (after Campbell & Schuh, 1970)

Possible number of discoveries (r)	Probability of getting r discoveries $P_b(r)$	Avg. field size needed to make 10% rate of return F_a Mbbls	Probability of getting F_a or more $P_{(a)}$
0	0.1969	—	0
1	0.3474	3500	0.16
2	0.2759	1750	0.35
3	0.1298	1165	0.50
4	0.0401	875	0.64
5	0.0085	700	0.73
6	0.0012	583	0.80
7	0.0001	500	0.84
8	0.0000		
9	0.0000		
10	0.0000		

$$S_{10} = \sum_{r=1}^{r=n} P_a P_b = .250$$

$$S_{10} = 25\%$$

Example 6–33. Calculate the chance of making at least a 10% rate of return in 10 exploration attempts if the chance of making a commercial discovery is 15%. The cost of each exploration attempt is $2,500,000. The present worth of oil reserves is $7.10/bbl.

The calculation is carried out as shown in Table 6–31. The probability of getting r discoveries (2nd column) is calculated using Equation 6–44 in the same fashion discussed in Example 6–31. For this case the chance of success from past experience is assumed at 15%.

So the probability of having no discoveries is given by:

$$P_{b(o)} = (a)^{10} = 0.85^{10} = 0.1969$$

The probability of having one discovery is given by

$$P_{b(1)} = n \ a^{m-1} \ b = 10 \times 0.85^{10-1} \times 0.15 = 0.3474$$

Other probabilities are calculated in the same fashion.

The third column of Table 6–31 is calculated using Equation 6–45. For example, the average field size needed to make a 10% rate of return, if there is only one discovery, is given by:

$$F_a = (10 \times 2,500,000) / (1 \times 7.10) = 3521 \ \text{Mbbl} \simeq 3500 \ \text{Mbbl}$$

If there are two discoveries, the average field size needed to make a 10% rate of return is given by:

$$F_a = (10 \times 2,500,000) / (2 \times 7.10) = 1761 \ \text{Mbbl} \simeq 1750 \ \text{Mbbl}$$

The probability of getting the required field size (F_a) is read from Figure 6–9 for each value of r. Note that r is indicated in Figure 6–9 as "distribution of random groups" for different amounts of fields.

For example, for r = 1 and F_a = 3500 Mbbl, Lahee's (1956) distribution indicates that there is a 16% probability of getting a field with 3500 Mbbl or more.

For r = 2 and F_a = 1750 Mbbl, the probability of getting 1750 Mbbl or more is 35%.

Finally, the chance of making at least 10% rate of return is calculated from Equation 6–46 as follows:

$$S_{10} = \sum_{r=1}^{r=n} P_a P_b = (0.3474 \times 0.16) + (0.2759 \times 0.35)$$
$$+ (0.1298 \times 0.50) + (0.0401 \times 0.64) + (0.0085 \times 0.73)$$
$$+ (0.0012 \times 0.80) + (0.0001 \times 0.84) = 0.25$$

Thus, there is a 25% chance of making at least a 10% rate of return.

MONTE CARLO SIMULATION

The same type of approach can be used to calculate probability distribution of rates of return as for the uncertainty of calculating oil–in–place presented in Chapter 3. McCray (1975) presents lucid treatments of the theory behind the Monte Carlo method and gives a computer program to calculate probability outcomes of rates of return. Aguilera (1978) has presented an approach for estimating the uncertainty of calculating original oil-in-place in naturally fractured reservoirs using a Monte Carlo Simulation. The subject is also covered in the Petroleum Society of CIM Monograph No. 1 "Determination of Oil and Gas Reserves" (1994).

DIRECTIONAL AND HORIZONTAL WELLS

Throughout this book I have placed some emphasis on the importance of drilling directional or horizontal wells for improving the probabilities of success in naturally fractured reservoirs. Drilling costs per well are of course higher. However, we need fewer horizontal than vertical wells to drain efficiently a naturally fractured reservoir. The bottom line is that we might end up with a more economic project if we drill directional and horizontal wells rather than vertical wells.

The following cost comparison is extracted from a publication by Horwell (1984). It is based on the cost analysis performed on the first four horizontal wells (Lacq 90, Lacq 91, Rospo Mare 6D and Castera Lou 110) drilled by Elf Aquitaine (FORHOR project). All of these wells were drilled using the low angle method. The reader is cautioned that costs are changing continuously and dramatically and estimates presented here should be used only as indicators.

Data available are not sufficient to perform a complete analysis of cost evolution with parameters such as depth, trajectory, horizontal length, diameter of hole, etc. What experience to date indicates, however, is that the more we learn about horizontal wells the smaller is the cost increase as compared with vertical wells. Furthermore, the larger the number of wells in a given area the bigger is the reduction in cost for each successive well.

The method of ratios has been selected to facilitate preliminary evaluations. It consists of comparing horizontal well cost to a vertical well cost. Both wells being drilled from the same location in the same reservoir.

All ratios assume that the drilling operations are performed using the know–how and field practices acquired during Horwell's FORHOR research and development project.

Direct Cost Ratio

Standard project:
—Vertical depth   5000 ft
—Horizontal displacement   4000 ft
—Total length drilled   8000 ft
—Horizontal section   1300 ft
—Well program   17½" × 12¼" × 8½"

Comparing a horizontal well to a vertical or slightly slanted well (inclination less than 35°), the direct cost ratio varies from 2.00 to 2.75.

Within this margin, the factors influencing the variations are:

—length of the horizontal hole
—accuracy of trajectory control
—intensity of coring
—importance of logging program

A long, intensively cored and logged horizontal well will cost 2.75 the cost of a standard vertical well. A horizontal well with 1300 ft of horizontal section, modest coring (1 or 2 cores) and a light logging program would cost two times the cost of a standard deviated well (less than 35° deviation).

Total Cost Ratio

When adding to the direct cost constant expenses such as civil works on location, moving, and pumping equipment, the ratio decreases from 2.00/2.75 to 1.70/2.75, that is a decreasing factor of about 0.85, for onshore wells. For offshore wells, the ratios can be lower once the share of platform cost is incorporated.

Sensitivity of Cost to Well Geometry

Depth. The cost ratio should decrease with depth, due to a reduction of the ratio of length drilled.

Diameter of Hole. If a horizontal hole drilled with a horizontal section of 8½" diameter is compared to a vertical well of 6" diameter, the direct cost ratio is to be increased by 50% to 65%.

Comparison With Highly Deviated Wells

This concerns deviated wells with inclination over 35°. From Elf Aquitaine experience, the cost ratio between a horizontal well and a directional well tends towards the ratio of drilled length.

Forecasted Evolution

On a medium term basis, direct cost ratio should tend towards the ratio of drilled length because of (1) more experience in horizontal drilling, and (2) generalization of MWD practices which give better cost/efficiency of these tools.

Other Experiences

Based on Sohio's experience in the Prudhoe Bay Field, Alaska, a typical development well costs $2.7 to 3 million (1986 dollars). Completed cost of the first horizontal well was 30–40% higher, but the company reduced that with experience (Petzet, 1986). The direct cost ratio for this case is in the order of 2.7.

More recently Broman et al (1990) presented drilling and completion costs for 16 non-conventional (horizontal) wells in Prudhoe Bay, Alaska. The costs are presented in Figure 6–10. This Figure shows that there were drastic decreases in drilling and completion costs as experience was gained during 1986. From there on the drilling and completion costs have remained relatively stable.

Gorody (1984) has indicated that the cost of horizontal drain-holes using the high angle method is primarily a function of time. The time parameter varies with the complexity of the completion and field operating conditions. Number of drainholes, degree of precision in their orientation, formation hardness, cased target zones, and deep target zones, all contribute to the total cost of the job. With the previous information, it is possible to make reliable cost-estimates for a single or a multiple drainhole program.

In general, the cost per horizontal drainhole using the high angle method (excluding the vertical section) will range between $50,000 and $150,000 (1984 dollars) including rig and associated equipment (Gorodi, 1984).

RESERVES

The following definitions have been extracted from Petroleum Society of CIM Monograph No. 1 "Determination of Oil and Gas Reserves" (1994). They were developed by the Standing Committee on Reserves Definitions of the Petroleum Society of the Canadian Institute of Mining, Metallurgy and Petroleum. For a more complete treatment of the subject the reader is referred to the above mentioned Monograph No. 1.

Resources

Resources are the total quantities of oil and gas and related substances that are estimated, at a particular time, to be contained in, or that have been produced from, known accumulations, plus those estimated quantities in accumulations yet to be discovered.

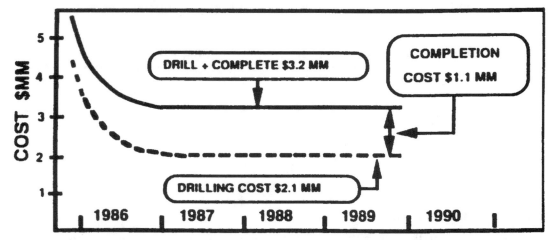

FIGURE 6–10 Non-conventional (horizontal) well costs at Prudhoe Bay (after Broman et al., 1990)

Initial Reserves

Initial reserves are those quantities of oil and gas and related substances that are estimated, at a particular time, to be recoverable from known accumulations. They include cumulative production plus those quantities that are estimated to be recoverable in the future by known technology under specified economic conditions that are generally accepted as being a reasonable outlook for the future. Figure 6–11 shows how initial reserves are classified.

Specified Economic Conditions

In order for oil and gas and related substances to be classified as reserves, they must be economic to recover at specified economic conditions. The estimator should use, as the specified economic conditions, a price forecast and other economic parameters that are generally accepted as being a reasonable outlook for the future. The revenue, appropriately discounted, must be sufficient to cover the future capital and operating costs that would be required to produce, process, and transport the products to the marketplace. A more detailed discussion of discounting future cash flow has been presented earlier in this chapter and in Chapter 21, *Cash Flow Analysis*, and Chapter 26, *Uses of Reserves Evaluations* of the Petroleum Society of CIM Monograph No. 1 "Determination of Oil and Gas Reserves."

If required by securities commissions or other agencies, current prices and costs may also be used. In either case, the economic conditions used in the evaluations should be clearly stated. Occasionally, the estimator also may wish to determine the impact of higher or lower price forecasts on estimates of reserves as compared to the most reasonable forecast. These cases (current, higher or lower prices) should not be reported as the most reasonable reserves estimates, but should be identified as sensitivity cases with the assumptions clearly stated. They illustrate the impact of different specified economic conditions on estimates of reserves.

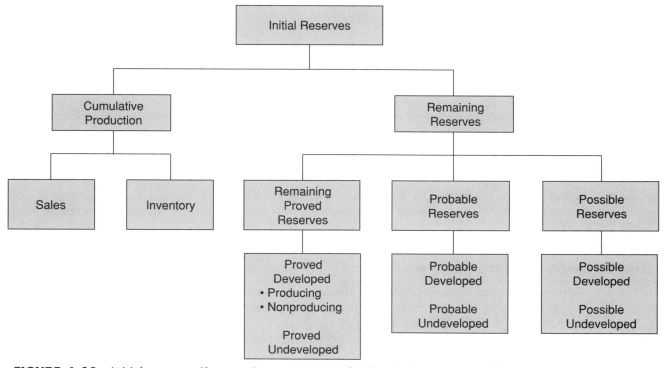

FIGURE 6–11   Initial reserves (Source: Determination of Oil and Gas Reserves, 1994)

Remaining Reserves

Remaining reserves (Figure 6–11) are estimated quantities of oil and natural gas and related substances anticipated to be recoverable from known accumulations, from a given date forward, by known technology under specified economic conditions that are generally accepted as being a reasonable outlook for the future.

Remaining Proved Reserves

Remaining proved reserves are those remaining reserves that can be estimated with a high degree of certainty, which for purposes of reserves classification means that there is generally an 80% or greater probability that at least the estimated quantity will be recovered. These reserves may be divided into proved developed and proved undeveloped to identify the status of development. The proved developed may be further divided into producing and nonproducing categories.

Probable Reserves

Probable reserves are those remaining reserves that are less certain to be recovered than proved reserves, which for purposes of reserves classification means that generally there is a 40% to 80% probability that the estimated quantity will be recovered. Both the estimated quantity and the risk-weighted portion reflecting the respective probability should be reported. These reserves can be divided into probable developed and probable undeveloped to identify the status of development.

Possible Reserves

Possible reserves are those remaining reserves that are less certain to be recovered than probable reserves, which for purposes of reserves classification means that generally there is 10% to 40% probability that the estimated quantity will be recovered. Both the estimated quantity and the risk-weighted portion reflecting the probability should be reported. These reserves can be divided into possible developed and possible undeveloped to identify the status of development.

Development and Production Status

Each of the three reserves classifications, remaining proved, probable, and possible, may be divided into developed and undeveloped categories (Figure 6–11). The developed category for proved reserves is often divided into producing and nonproducing.

Developed Reserves

Developed reserves are those reserves that are expected to be recovered from existing wells and installed facilities or, if facilities have not been installed, that would involve a low expenditure to put the reserves on production (i.e., when compared to the cost of drilling a well).

Developed Producing Reserves

Developed producing reserves are those reserves that are expected to be recovered from completion intervals open at the time of the estimate. These reserves may be currently producing or, if shut in, they must have previously been on production, and the date of resumption of production must be known with reasonable certainty.

Developed Nonproducing Reserves

Developed nonproducing reserves are those reserves that either have not been on production, or have previously been on production, but are shut in, and the date of resumption of production is unknown.

Undeveloped Reserves

Undeveloped reserves are those reserves expected to be recovered from known accumulations where a significant expenditure (i.e., when compared to the cost of drilling a well) is required to render them capable of production.

In multi-well pools, it may be appropriate to allocate the total reserves for the pool between the developed and undeveloped categories or to subdivide the developed reserves for the pool between developed producing and developed nonproducing. This allocation should be based on the evaluator's assessment as to the reserves that will be recovered from specific wells, the facilities and completion intervals in the pool, and their respective development and production status.

OTHER DEFINITIONS

The definitions presented above and Monograph No. 1 "Determination of Oil and Gas Reserves" were developed with the hope that they will help to simplify and standardize the science and art of estimating oil and gas reserves throughout the world. There are, however, many other definitions that the reader should be aware of. These ones have been summarized by Lang (1989) as follows:

1. US. Department of Energy

Proved Reserves are those reserves of oil and gas which geological and engineering data demonstrate with reasonable certainty to be recoverable in the future under existing economic and operating conditions.

Source: 1977 Annual Report of U.S. Crude Oil and Natural Gas Reserves (February 1980).

2. Canada

Established Reserves are those reserves recoverable under current technology and present and anticipated economic conditions, specifically proved by drilling, testing or production, plus that judgement portion of contiguous recoverable reserves that are interpreted to exist, from geological, geophysical or similar information, with reasonable certainty.

Dec. 1978—Joint Task Force Under the auspices of the Interprovincial Advisory
          Committee on Energy.

          —adopted by:
          Canadian Petroleum Association
          Independent Petroleum Assoc. of Canada
          National Energy Board
          Energy Resources Conservation Board
          Province of British Columbia
          Province of Saskatchewan
          Province of Manitoba
          Other provinces in Canada

3. Securities & Exchange Commission and Financial Accounting Standards Board

Proved oil and gas reserves are the estimated quantities of crude oil, natural gas, and natural gas liquids which geological and engineering data demonstrate with reasonable certainty to be recoverable in future years from known reservoirs under existing economic and operating conditions (i.e., prices and costs as of the date the estimate is made). Prices include consideration of changes in existing prices provided only by contractual arrangements, but not on escalations based upon future considerations.

—Proved developed reserves
—Proved undeveloped reserves.

Sources: (1) SEC Accounting Rules (1982).
Rule 4-10 Financial accounting and reporting for oil and gas producing activities pursuant to the federal securities laws and the Energy Policy and Conservation Act of 1975.

(2) Statement of Financial Accounting Standards No. 25, February 1979. Financial Accounting Standards Board (FASB).

4. National Policy No. 2-B

Proved Reserves are those reserves estimated as recoverable under current technology and existing economic conditions, from that portion of a reservoir which can be reasonably evaluated as economically productive on the basis of analysis of drilling, geological, geophysical and engineering data, including the reserves to be obtained by enhanced recovery processes demonstrated to be economic and technically successful in the subject reservoir.

—Proved Producing Reserves
—Proved Non-Producing Reserves

Probable Additional Reserves are those reserves which analysis of drilling, geological, geophysical and engineering data does not demonstrate to be proved under current technology and existing economic conditions, but where such analysis suggests the likelihood of their existence and future recovery . . .

Source: Ontario Securities Act and Regulation with Policy Statements, 1984.

5. Canadian Institute of Chartered Accountants

Proved Reserves are defined as those quantities of crude oil, including condensates and natural gas liquids, natural gas, and sulphur which upon analysis of geological and engineering data, appear with reasonable certainty to be recoverable in the future from know reservoirs under existing economic and operating conditions. Proved reserves are limited to those quantities which can be expected, with little doubt, to be recoverable commercially at current or contractual prices and currently prevailing costs under existing regulatory practices and with existing conventional equipment and operating methods . . .

—Proved developed reserves
—Proved undeveloped reserves

When reserve quantities are disclosed on a proved basis, with separate identification of that part which has been developed, it may also be desirable to disclose quantities of Probable Reserves as at the end of the period. When quantities of probable reserves are disclosed, it is necessary to state how probable reserves are defined and to provide general narrative comments regarding the uncertainties involved in such estimates and uncertainties regarding the means and timing of developing the reserves.

Source: CICA Handbook, Page 3528
September 1986—Accounting Recommendations

6. Society of Petroleum Evaluation Engineers

Proved reserves of crude oil, condensate, natural gas, and natural gas liquids are estimated quantities that geological and engineering data demonstrate with reasonable certainty to be commercially recoverable in the future from known reservoirs under existing economic conditions using established operating procedures and under current governmental regulations.

Probable reserves are the estimated quantities of commercially recoverable hydrocarbons associated with known accumulations, which are based on engineering and geological data similar to those used in the estimates of proved reserves, but for various reasons, these data lack the certainty required to classify the reserves as proved. In some cases, economic or regulatory uncertainties may dictate the probable classification. Probable reserves are less certain to be recovered than proved reserves.

Possible reserves are the estimated quantities of commercially recoverable hydrocarbons associated with known accumulations, which are based on engineering and geological data which are less complete and less conclusive than the data used in estimates of probable reserves. In some cases, economic or regulatory uncertainties may dictate the possible classification. Possible reserves are less certain to be recovered than proved or probable reserves.

Developed
—Producing
—Non Producing
Undeveloped

December 1985.

7. Society of Petroleum Engineers

Reserves are estimated volumes of crude oil, condensate, natural gas, natural gas liquids, and associated substances anticipated to be commercially recoverable from known accumulation from a given date forward, under existing economic conditions, by established operating practices, and under current government regulations. Reserve estimates are based on interpretation of geologic and/or engineering data available at the time of the estimate.

Proved Reserves can be estimated with reasonable certainty to be recoverable under current economic conditions. Current economic conditions include prices and costs prevailing at the time of the estimate. Proved reserves may be developed or undeveloped.

Unproved Reserves are based on geologic and/or engineering data similar to that used in estimates of proved reserves but technical, contractual, economic, or regulatory uncertainties preclude such reserves being classified as proved. They may be estimated assuming future economic conditions different from those prevailing at the time of the estimate.

—Probable Reserves are less certain than proved reserves and can be estimated with a degree of certainty sufficient to indicate they are more likely to be recovered than not.
—Possible Reserves are less certain than probable reserves and can be estimated with a low degree of certainty, insufficient to indicate whether they are more likely to be recovered than not.

Developed
—Producing
—Non Producing
Undeveloped

Approved February 27, 1987

8. American Gas Association

Proved Reserves are the volume of natural gas estimated from analysis of geological and engineering data to be recoverable from a known oil or gas reservoir, assuming current operating and anticipated economic conditions.

Source: 1986 Report on Natural Gas Reserves

9. World Petroleum Congresses

Proved Reserves of petroleum are the estimated quantities, as at a specific date, which analysis of geological and engineering data demonstrates, with reasonable certainty, to be recoverable in the future from known reservoirs under the economic and operating conditions at the same date.

> —Proved Developed Reserves
> —Proved Undeveloped Reserves

> Unproved Reserves of petroleum are the estimated quantities, as at a specific date, which analysis of geological and engineering data indicates might be economically recoverable from already discovered deposits, with a sufficient degree of probability to suggest their existence . . .

—Probable Reserves of petroleum are the estimated quantities, at a specific date, which analysis of geological and engineering data indicates might be economically recoverable from already discovered deposits with a reasonably high degree of probability, which suggests the likelihood of their existence, but not sufficient to be classified as proved . . .

—Possible Reserves of petroleum are the estimated quantities, as at a specific date, which analysis of geological and engineering data indicates might be economically recoverable from already discovered deposits with only a moderate degree of probability which suggests the chance of their existence, but not sufficient to be classified as probable.

Source: 1987 Report of Special Study Group on Classification and Nomenclature Systems for Petroleum and Petroleum Reserves.

GUIDELINES FOR ESTIMATING OIL AND GAS RESERVES

The reader is referred to Chapter 3 of the Petroleum Society of CIM Monograph No. 1 "Determination of Oil and Gas Reserves" for an excellent treatment on guidelines for estimation of oil and gas reserves. The discussion that I present in this section, however, is based exclusively on my experience with naturally fractured reservoirs. I recommend to use statistical procedures to quantify the uncertainty associated with hydrocarbons-in-place and reserves.

Volumetric Estimates

Most naturally fractured reservoirs I am familiar with are characterized by low matrix porosities (much lower than 10%) and low matrix permeabilities (much lower than 1 md).

For these reservoir characteristics it is difficult to place a reasonable certainty on volumetric estimates of original-hydrocarbons-in-place, recoveries and hence reserves. As a consequence, as a general rule, I recommend to place reserves from volumetric estimates in the possible category which for purposes of the reserves classification recommended in this book means that generally there is a 10 to 40% probability that the estimated quantity will be recovered.

Material Balance Estimates

Material balance techniques are very useful in conventional unfractured reservoirs but can easily lead to gross errors in naturally fractured reservoirs. A tank material balance cannot properly handle permeability anisotropy, permeability contrast between matrix and fractures, matrix and fracture porosities, skin, net pay, structural position of each well, gas or water influx via natural fractures, etc. . . As a consequence, as a general rule, I recommend to place reserves from material balance estimates in the probable category which for purposes of the reserve classification recommended in this book means that generally there is a 40 to 80% probability that the estimated quantity will be recovered.

Material balance estimates can be upgraded to the proved category when the production history is long and the quality of the pressure data and the oil, gas and water production data is good. Beware of possible changes in fracture communication and possible fracture closing as the reservoir is being depleted.

Production Decline Estimates

For short production histories reserves estimates from production decline curves should be placed in an unproved category. Long production histories should lead to reasonable estimates of proved reserves. In general, I do not recommend decline curves for estimating reserves of gas reservoirs unless the wells are at a late stage of production where a constant surface compression pressure is being utilized.

Reservoir Simulation Estimates

Although imperfect, this is the tool that in my opinion provides the most reliable source of information for estimating recoveries and proved reserves. A significant amount of high quality data is required. The longer the production history the more reliable are the forecasting results. I must emphasize that although reservoir simulators are good in general, the quality of the results will be only as good as the quality of the input data. "Trash in-trash out."

What to Do When Production Histories Are Short or Non-Existent?

In my opinion the best way for estimating preliminary reserves in these situations is by running a well-designed, well-supervised interference test. The larger the number of wells involved in the test the better. This test will provide useful information such as level of anisotropy, fracture hydrocarbon storage and, if the test is long enough, matrix hydrocarbon storage. An estimate of fracture-matrix interaction can also be obtained from this test.

What to Do When There Is Only One Well?

Following collection of a good initial pressure, a long flow period is a "must" followed by a long buildup test. An estimate of the radius of investigation leads to a volume of hydrocarbons-in-place within the investigated area. This assumes a good knowledge of net pay, matrix and fracture porosity, and matrix and fracture water saturation. Keep in mind that the radius of investigation equation is the "conventional" one only if a pressure equilibrium has been reached between matrix and fractures. Otherwise a different equation including matrix and fractures has to be used (Aguilera, 1987).

The long flow period also allows to make an estimate of hydrocarbons-in-place via equations developed for Reservoir Limit Tests. Careful! Effective compressibility plays an important role in these calculations. In Reservoirs of Type A (refer to Chapter 1) the effective compressibility of the naturally fractured reservoir in approximately equal to the effective compressibility of a conventional unfractured reservoir. However, in naturally fractured reservoirs of Type B and specially Type C the effective compressibilities are quite different from those of conventional unfractured reservoirs. This is the result of the fractures being more compressible than the matrix.

The use of a conventional unfractured effective compressibility in a naturally fractured reservoir of Type B or C will lead to very large values of hydrocarbons-in-place and hence very large and unrealistic reserves.

In my experience calculations of OOIP and reserves from a well designed reservoir limit test in an undersaturated reservoir using an effective compressibility equal to 10 times the effective compressibility of a conventional unfractured reservoir provides reasonable estimates. Reserves from these estimates usually increase as the reservoir is developed.

Does a Dry Hole Between Two Producing Wells Mean That the Proved Reserves Have to be Decreased?

Not necessarily. Most naturally fractured reservoir I am familiar with produce from vertical and subvertical natural fractures which cut a matrix that usually has a permeability equal to a fraction of millidarcy. This tight matrix cannot flow at commercial rates into a wellbore thus leading to a "dry hole." However, the tight matrix might flow efficiently into natural fractures that communicate with another wellbore.

In this case the "dry hole" might have penetrated a hydrocarbon-saturated matrix but has not intersected the vertical and subvertical natural fractures. If the "dry hole" is between two producing wells that have been previously shown to be in communication via an interference test, my recommendation is not to delete the proved reserves that have been booked between those two wells. The in-between hydrocarbons will be recovered by those two wells. In fact, if the "dry hole" had been successful, all it would have done is to accelerate the recoveries of the reservoir without adding any reserves.

Wells that do not produce hydrocarbons at commercial rates have been used successfully as observation wells in many naturally fractured reservoirs.

REFERENCES

Aguilera, Roberto. "Economic Analyses of Acceleration Projects." Paper SPE 6086 presented at the 51st annual fall meeting of SPE of AIME, New Orleans (October 3–6, 1976).

Aguilera, R. "The Uncertainty of Evaluating Original-Oil-in-Place in Naturally Fractured Reservoirs." SPWLA, Paper A (June 13–16, 1978).

Aguilera, R. and H.K. van Poollen. "Naturally Fractured Reservoirs—How to Analyze Reservoir Economics. *Oil and Gas Journal* (June 4, 1979).

Aguilera, R. "Discussion of What Discounted Cash Flow Rate of Return Never Did Require." *Journal of Petroleum Technology* (June 1986) 676–678.

Aguilera, R. "Well Test Analysis of Naturally Fractured Reservoirs." *SPE Formation Evaluation* (September 1987) 239–252.

Aguilera, Roberto. "Determination of Subsurface Distance Between Vertical Parallel Natural Fractures Based on Core Data." *Bulletin*. AAPG (July 1988), 845–851.

Aldwell, R.H. and D.I. Heather. "How to Evaluate Hard–to–Evaluate Reserves." SPE paper 22025 presented at the SPE Hydrocarbon Economics and Evaluation Symposium held in Dallas, Texas (April 11–12, 1991).

Baker, D.A., and P.T. Lucas. "Strat Trap Production May Cover 280 Plus Square Miles." *World Oil* (April 1972), 65–68.

Broman, W.H., T.O. Stagg, and J.J. Rosenzweig. "Horizontal Well Performance Evaluation at Prudhoe Bay." CIM/SPE paper No. 90-124 (1990).

Campbell, J.M. Oil Property Evaluation. Prentice–Hall Inc. (1959).

Campbell, J.M. "Optimization of Capital Expenditures in Petroleum Investments." *Oil and Gas Property Evaluation and Reserve Estimates*, SPE of AIME Petroleum Reprint Series No. 3 (1970), 168–175.

Campbell, W.M. and F.J. Schuh. "Risk Analysis: Over-all Chance of Success Related to Number of Ventures." *Oil and Gas Property Evaluation and Reserve Estimates*, SPE of AIME Petroleum Reprint Series No. 3 (1970) 168–175.

Capen, E.C., R.V. Clapp, and W.W. Phelps. "Growth Rate - A Rate of Return Measure of Investment Efficiency." *Journal of Petroleum Technology* (May 1976) 531–543.

Cronquist, C. "Reserves and Probabilities—Synergism or Anachronism?" *Journal of Petroleum Technology* (October 1991), 1258–1264.

Daniel, E.J. "Fractured Reservoirs of Middle East." *Bulletin*. AAPG (May 1954), 774–815.

Davidson, L.B. "Investment Evaluation Under Conditions of Inflation." Paper SPE 5013 presented at the 49th annual meeting of SPE of AIME, Houston (October 6–9, 1974).

Delgado, O.R., and E.G. Loreto. "Reforma's Cretaceous Reservoirs. An Engineering Challenge." *Petroleum Engineer* (December 1975).

DeSorcy, G.J. "Classification and Nomenclature Systems for Petroleum Reserves." Paper No. 89–40–97 presented at the 40th Annual Technical Meeting of the Petroleum Society of CIM held in Banff, Canada (May 28, to 31, 1989).

Determination of Oil and Gas Reserves, Petroleum Society of CIM Monograph No. 1, Calgary, Canada (1994).

Dougherty, E.L. "What Discounted Cash Flow Rate of Return Never Did Require." *Journal of Petroleum Technology* (June 1986), 85–87.

Garaicochea, F., and A. Acuña. "Espaciamiento Optimo de Pozos en Yacimientos con Segregación de Gas Liberado." *Revista del Instituto Mexicano del Petróleo* (April 1978), 31–41.

Gorodi, A.W. "TEDSI Develops Horizontal Drilling Technology." *Oil and Gas Journal*, (October 1, 1984) 118–126.

Grayson, C.J. *Drilling Decisions Under Uncertainty*. Harvard University Press (1960).

Henry, A.J. "Appraisal of Income Acceleration Projects." *Journal of Petroleum Technology* (April 1972), 393–398.

Horwell. "Horizontal Well Technology." Rueil–Malmaison, France (1984).

Kaitz, M. "Percentage Gain on Investment—An Investment Decision Yardstick." *Journal of Petroleum Technology* (May 1967), 679–687.

Lahee, F.H. "How Many Fields Really Pay Off?" *Oil and Gas Journal* (September 17, 1956), 369–371.

Lang, R.V. "Reserves Definitions—Current and Future Considerations." Paper 89–40–98 presented at the 40th Annual Technical Meeting of the Petroleum Society of CIM, Banff, Canada (May 28–31, 1989).

Lefkovits, H.D., H. Kanner, and R.B. Harbottle. "On Multiple Rates of Return." Proceedings Fifth World Petroleum Congress, Section IX, paper E, (1959).

Mathur, S.B., and O.L. Carey. "Economic Decision–Making Practices in the U.S. Petroleum Industry." Paper SPE 5011 presented at the 49th annual meeting of SPE of AIME, Houston (October 6–9, 1974).

McCray, A.W. *Petroleum Evaluations and Economic Decisions*. Prentice–Hall, Inc. (1975).

Megill, R.E. *An Introduction to Exploration Economics*. Petroleum Publishing Company (1971).

Megill, R.E. "Reserve Estimates Require More than a Single Answer." *AAPG Explorer* (November 1990).

Newendorp, P.D. *Decision Analysis for Petroleum Exploration*. Petroleum Publishing Company (1975).

Oil and Gas Property Evaluation and Reserves Estimates. SPE of AIME Petroleum Reprint Series No. 3 (1970).

Ovreberg, O., E. Damsleth, and H.H. Haldorsen. "Putting Error Bars on Reservoir Engineering Forecasts." *Journal of Petroleum Technology* (June 1992), 732–738.

Petzet, G.A. "Prudhoe Bay Horizontal Well Yields Hefty Flow." *Oil and Gas Journal* (February 17, 1986), 42–43.

Phillips, C.E. "The Appreciation of Equity Concept and its Relationship to Multiple Rates of Return." *Journal of Petroleum Technology* (February 1965), 159–163.

Robinson, J.G. "Definition of Reserves and Values and Application of Risk." Paper No 89–40–99 presented at the 40th Annual Technical Meeting of the Petroleum Society of CIM held in Banff, Canada (May 28 to 31, 1989).

Stermole, F.J. *Economic Evaluation and Investment Decision Methods*. Investments Evaluation Corporation, Golden, Colorado (1974).

Schoemaker, R.P. "A Graphical Short-Cut for Rate of Return Determinations-Part 2: Special Applications." *World Oil* (August 1963), 157,2.

Wooddy, L.D. and T.D. Capshaw. "Investment Evaluation by Present Value Profile." *Journal of Petroleum Technology* (June 1960), 15–18.

Appendix (After Sternole)

i = 5%

Period	Single payment compound amount factor	Single payment present worth factor	Uniform series compound amount factor	Sinking fund deposit factor	Capital recovery factor	Uniform series present worth factor
1	1.050	0.9524	1.000	1.00000	1.05000	0.952
2	1.103	0.9070	2.050	0.48780	0.53780	1.859
3	1.158	0.8638	3.153	0.31721	0.36721	2.723
4	1.216	0.8227	4.310	0.23201	0.28201	3.546
5	1.276	0.7835	5.526	0.18097	0.23097	4.329
6	1.340	0.7462	6.802	0.14702	0.19702	5.076
7	1.407	0.7107	8.142	0.12282	0.17282	5.786
8	1.477	0.6768	9.549	0.10472	0.15472	6.463
9	1.551	0.6446	11.027	0.09069	0.14069	7.108
10	1.629	0.6139	12.578	0.07950	0.12950	7.722
11	1.710	0.5847	14.207	0.07039	0.12039	8.306
12	1.796	0.5568	15.917	0.06283	0.11283	8.863
13	1.886	0.5303	17.713	0.05646	0.10646	9.394
14	1.980	0.5051	19.599	0.05102	0.10102	9.899
15	2.079	0.4810	21.579	0.04634	0.09634	10.380
16	2.183	0.4581	23.657	0.04227	0.09227	10.838
17	2.292	0.4363	25.840	0.03870	0.08870	11.274
18	2.407	0.4155	28.132	0.03555	0.08555	11.690
19	2.527	0.3957	30.539	0.03275	0.08275	12.085
20	2.653	0.3769	33.066	0.03024	0.08024	12.462
21	2.786	0.3589	35.719	0.02800	0.07800	12.821
22	2.925	0.3418	38.505	0.02597	0.07597	13.163
23	3.072	0.3256	41.430	0.02414	0.07414	13.489
24	3.225	0.3101	44.502	0.02247	0.07247	13.799
25	3.386	0.2953	47.727	0.02095	0.07095	14.094
26	3.556	0.2812	51.113	0.01956	0.06956	14.375
27	3.733	0.2678	54.669	0.01829	0.06829	14.643
28	3.920	0.2551	58.403	0.01712	0.06712	14.898
29	4.116	0.2429	62.323	0.01605	0.06605	15.141
30	4.322	0.2314	66.439	0.01505	0.06505	15.372
35	5.516	0.1813	90.320	0.01107	0.06107	16.374
40	7.040	0.1420	120.800	0.00828	0.05828	17.159
45	8.985	0.1113	159.700	0.00626	0.05626	17.774
50	11.467	0.0872	209.348	0.00478	0.05478	18.256
55	14.636	0.0683	272.713	0.00367	0.05367	18.633
60	18.679	0.0535	353.584	0.00283	0.05283	18.929
65	23.840	0.0419	456.798	0.00219	0.05219	19.161
70	30.426	0.0329	588.529	0.00170	0.05170	19.343
75	38.833	0.0258	756.654	0.00132	0.05132	19.485
80	49.561	0.0202	971.229	0.00103	0.05103	19.596
85	63.254	0.0158	1245.087	0.00080	0.05080	19.684
90	80.730	0.0124	1594.607	0.00063	0.05063	19.752
95	103.035	0.0097	2040.694	0.00049	0.05049	19.806
100	131.501	0.0076	2610.025	0.00038	0.05038	19.848

i = 10%

Period	Single payment compound amount factor	Single payment present worth factor	Uniform series compound amount factor	Sinking fund deposit factor	Capital recovery factor	Uniform series present worth factor
1	1.100	0.9091	1.000	1.00000	1.10000	0.909
2	1.210	0.8264	2.100	0.47619	0.57619	1.736
3	1.331	0.7513	3.310	0.30211	0.40211	2.487
4	1.464	0.6830	4.641	0.21547	0.31547	3.170
5	1.611	0.6209	6.105	0.16380	0.26380	3.791
6	1.772	0.5645	7.716	0.12961	0.22961	4.355
7	1.949	0.5132	9.487	0.10541	0.20541	4.868
8	2.144	0.4665	11.436	0.08744	0.18744	5.335
9	2.358	0.4241	13.579	0.07364	0.17364	5.759
10	2.594	0.3855	15.937	0.06275	0.16275	6.144
11	2.853	0.3505	18.531	0.05396	0.15396	6.495
12	3.138	0.3186	21.384	0.04676	0.14676	6.814
13	3.452	0.2897	24.523	0.04078	0.14078	7.103
14	3.797	0.2633	27.975	0.03575	0.13575	7.367
15	4.177	0.2394	31.772	0.03147	0.13147	7.606
16	4.595	0.2176	35.950	0.02782	0.12782	7.824
17	5.054	0.1978	40.545	0.02466	0.12466	8.022
18	5.560	0.1799	45.599	0.02193	0.12193	8.201
19	6.116	0.1635	51.159	0.01955	0.11955	8.365
20	6.727	0.1486	57.275	0.01746	0.11746	8.514
21	7.400	0.1351	64.002	0.01562	0.11562	8.649
22	8.140	0.1228	71.403	0.01401	0.11401	8.772
23	8.954	0.1117	79.543	0.01257	0.11257	8.883
24	9.850	0.1015	88.497	0.01130	0.11130	8.985
25	10.835	0.0923	98.347	0.01017	0.11017	9.077
26	11.918	0.0839	109.182	0.00916	0.10916	9.161
27	13.110	0.0763	121.100	0.00826	0.10826	9.237
28	14.421	0.0693	134.210	0.00745	0.10745	9.307
29	15.863	0.0630	148.631	0.00673	0.10673	9.370
30	17.449	0.0573	164.494	0.00608	0.10608	9.427
35	28.102	0.0356	271.024	0.00369	0.10369	9.644
40	45.259	0.0221	442.593	0.00226	0.10226	9.779
45	72.890	0.0137	718.905	0.00139	0.10139	9.863
50	117.391	0.0085	1163.909	0.00086	0.10086	9.915
55	189.059	0.0053	1880.591	0.00053	0.10053	9.947
60	304.482	0.0033	3034.816	0.00033	0.10033	9.967
65	490.371	0.0020	4893.707	0.00020	0.10020	9.980
70	789.747	0.0013	7887.470	0.00013	0.10013	9.987
75	1271.895	0.0008	12708.954	0.00008	0.10008	9.992
80	2048.400	0.0005	20474.002	0.00005	0.10005	9.995
85	3298.969	0.0003	32979.690	0.00003	0.10003	9.997
90	5313.023	0.0002	53120.226	0.00002	0.10002	9.998
95	8556.676	0.0001	85556.760	0.00001	0.10001	9.999

	i = 15%					
Period	Single payment compound amount factor	Single payment present worth factor	Uniform series compound amount factor	Sinking fund deposit factor	Capital recovery factor	Uniform series present worth factor
1	1.150	0.8696	1.000	1.00000	1.15000	0.870
2	1.322	0.7561	2.150	0.46512	0.61512	1.626
3	1.521	0.6575	3.472	0.28798	0.43798	2.283
4	1.749	0.5718	4.993	0.20027	0.35027	2.855
5	2.011	0.4972	6.742	0.14832	0.29832	3.352
6	2.313	0.4323	8.754	0.11424	0.26424	3.784
7	2.660	0.3759	11.067	0.09036	0.24036	4.160
8	3.059	0.3269	13.727	0.07285	0.22285	4.487
9	3.518	0.2843	16.786	0.05957	0.20957	4.772
10	4.046	0.2472	20.304	0.04925	0.19925	5.019
11	4.652	0.2149	24.349	0.04107	0.19107	5.234
12	5.350	0.1869	29.002	0.03448	0.18448	5.421
13	6.153	0.1625	34.352	0.02911	0.17911	5.583
14	7.076	0.1413	40.505	0.02469	0.17469	5.724
15	8.137	0.1229	47.580	0.02102	0.17102	5.847
16	9.358	0.1069	55.717	0.01795	0.16795	5.954
17	10.761	0.0929	65.075	0.01537	0.16537	6.047
18	12.375	0.0808	75.836	0.01319	0.16319	6.128
19	14.232	0.0703	88.212	0.01134	0.16134	6.198
20	16.367	0.0611	102.444	0.00976	0.15976	6.259
21	18.822	0.0531	118.810	0.00842	0.15842	6.312
22	21.745	0.0462	137.632	0.00727	0.15727	6.359
23	24.891	0.0402	159.276	0.00628	0.15628	6.399
24	28.625	0.0349	184.168	0.00543	0.15543	6.434
25	32.919	0.0304	212.793	0.00470	0.15470	6.464
26	37.857	0.0264	245.712	0.00407	0.15407	6.491
27	43.535	0.0230	283.569	0.00353	0.15353	6.514
28	50.066	0.0200	327.104	0.00306	0.15306	6.534
29	57.575	0.0174	377.170	0.00265	0.15265	6.551
30	66.212	0.0151	434.745	0.00230	0.15230	6.566
35	133.176	0.0075	881.170	0.00113	0.15113	6.617
40	267.864	0.0037	1779.090	0.00056	0.15056	6.642
45	538.769	0.0019	3585.128	0.00028	0.15028	6.654
50	1083.657	0.0009	7217.716	0.00014	0.15014	6.661
55	2179.622	0.0005	14524.148	0.00007	0.15007	6.664
60	4383.999	0.0002	29219.992	0.00003	0.15003	6.665
65	8817.787	0.0001	58778.583	0.00002	0.15002	6.666

	i = 20%					
Period	Single payment compound amount factor	Single payment present worth factor	Uniform series compound amount factor	Sinking fund deposit factor	Capital recovery factor	Uniform series present worth factor
1	1.200	0.8333	1.000	1.00000	1.20000	0.833
2	1.440	0.6944	2.200	0.45455	0.65455	1.528
3	1.728	0.5787	3.640	0.27473	0.47473	2.106
4	2.074	0.4823	5.368	0.18629	0.38629	2.589
5	2.488	0.4019	7.442	0.13438	0.33438	2.991
6	2.986	0.3349	9.930	0.10071	0.30071	3.326
7	3.583	0.2791	12.916	0.07742	0.27742	3.605
8	4.300	0.2326	16.499	0.06061	0.26061	3.837
9	5.160	0.1938	20.799	0.04808	0.24808	4.031
10	6.192	0.1615	25.959	0.03852	0.23852	4.192
11	7.430	0.1346	32.150	0.03110	0.23110	4.327
12	8.916	0.1122	39.581	0.02526	0.22526	4.439
13	10.699	0.0935	48.497	0.02062	0.22062	4.533
14	12.839	0.0779	59.196	0.01689	0.21689	4.611
15	15.407	0.0649	72.035	0.01388	0.21388	4.675
16	18.488	0.0541	87.442	0.01144	0.21144	4.730
17	22.186	0.0451	105.931	0.00944	0.20944	4.775
18	26.623	0.0376	128.117	0.00781	0.20781	4.812
19	31.948	0.0313	154.740	0.00646	0.20646	4.843
20	38.338	0.0261	186.688	0.00536	0.20536	4.870
21	46.005	0.0217	225.026	0.00444	0.20444	4.891
22	55.206	0.0181	271.031	0.00369	0.20369	4.909
23	66.247	0.0151	326.237	0.00307	0.20307	4.925
24	79.497	0.0126	392.484	0.00255	0.20255	4.937
25	95.396	0.0105	471.981	0.00212	0.20212	4.948
26	114.475	0.0087	567.377	0.00176	0.20176	4.956
27	137.371	0.0073	681.853	0.00147	0.20147	4.964
28	164.845	0.0061	819.223	0.00122	0.20122	4.970
29	197.814	0.0051	984.068	0.00102	0.20102	4.975
30	237.376	0.0042	1181.882	0.00085	0.20085	4.979
35	590.668	0.0017	2948.341	0.00034	0.20034	4.992
40	1469.772	0.0007	7343.858	0.00014	0.20014	4.997
45	3657.262	0.0003	18281.310	0.00005	0.20005	4.999
50	9100.438	0.0001	45497.191	0.00002	0.20002	4.999

Period	Single payment compound amount factor	Single payment present worth factor	Uniform series compound amount factor	Sinking fund deposit factor	Capital recovery factor	Uniform series present worth factor
i = 30%						
1	1.300	0.7692	1.000	1.00000	1.30000	0.769
2	1.690	0.5917	2.300	0.43478	0.73478	1.361
3	2.197	0.4552	3.990	0.25063	0.55063	1.816
4	2.856	0.3501	6.187	0.16163	0.46163	2.166
5	3.713	0.2693	9.043	0.11058	0.41058	2.436
6	4.827	0.2072	12.756	0.07839	0.37839	2.643
7	6.275	0.1594	17.583	0.05687	0.35687	2.802
8	8.157	0.1226	23.858	0.04192	0.34192	2.925
9	10.604	0.0943	32.015	0.03124	0.33124	3.019
10	13.786	0.0725	42.619	0.02346	0.32346	3.092
11	17.922	0.0558	56.405	0.01773	0.31773	3.147
12	23.298	0.0429	74.327	0.01345	0.31345	3.190
13	30.288	0.0330	97.625	0.01024	0.31024	3.223
14	39.374	0.0254	127.913	0.00782	0.30782	3.249
15	51.186	0.0195	167.286	0.00598	0.30598	3.268
16	66.542	0.0150	218.472	0.00458	0.30458	3.283
17	86.504	0.0116	285.014	0.00351	0.30351	3.295
18	112.455	0.0089	371.518	0.00269	0.30269	3.304
19	146.192	0.0068	483.973	0.00207	0.30207	3.311
20	190.050	0.0053	630.165	0.00159	0.30159	3.316
21	247.065	0.0040	820.215	0.00122	0.30122	3.320
22	321.184	0.0031	1067.280	0.00094	0.30094	3.323
23	417.539	0.0024	1388.464	0.00072	0.30072	3.325
24	542.801	0.0018	1806.003	0.00055	0.30055	3.327
25	705.641	0.0014	2348.803	0.00043	0.30043	3.329
26	917.333	0.0011	3054.444	0.00033	0.30033	3.330
27	1192.533	0.0008	3971.778	0.00025	0.30025	3.331
28	1550.293	0.0006	5164.311	0.00019	0.30019	3.331
29	2015.381	0.0005	6714.604	0.00015	0.30015	3.332
30	2619.996	0.0004	8729.985	0.00011	0.30011	3.332
35	9727.860	0.0001	32422.868	0.00003	0.30003	3.333

i = 40%						
Period	Single payment compound amount factor	Single payment present worth factor	Uniform series compound amount factor	Sinking fund deposit factor	Capital recovery factor	Uniform series present worth factor
1	1.400	0.7143	1.000	1.00000	1.40000	0.714
2	1.960	0.5102	2.400	0.41667	0.81667	1.224
3	2.744	0.3644	4.360	0.22936	0.62936	1.589
4	3.842	0.2603	7.104	0.14077	0.54077	1.849
5	5.378	0.1859	10.946	0.09136	0.49136	2.035
6	7.530	0.1328	16.324	0.06126	0.46126	2.168
7	10.541	0.0949	23.853	0.04192	0.44192	2.263
8	14.758	0.0678	34.395	0.02907	0.42907	2.331
9	20.661	0.0484	49.153	0.02034	0.42034	2.379
10	28.925	0.0346	69.814	0.01432	0.41432	2.414
11	40.496	0.0247	98.739	0.01013	0.41013	2.438
12	56.694	0.0176	139.235	0.00718	0.40718	2.456
13	79.371	0.0126	195.929	0.00510	0.40510	2.469
14	111.120	0.0090	275.300	0.00363	0.40363	2.478
15	155.568	0.0064	386.420	0.00259	0.40259	2.484
16	217.795	0.0046	541.988	0.00185	0.40185	2.489
17	304.913	0.0033	759.784	0.00132	0.40132	2.492
18	426.879	0.0023	1064.697	0.00094	0.40094	2.494
19	597.630	0.0017	1491.576	0.00067	0.40067	2.496
20	836.683	0.0012	2089.206	0.00048	0.40048	2.497
21	1171.356	0.0009	2925.889	0.00034	0.40034	2.498
22	1639.898	0.0006	4097.245	0.00024	0.40024	2.498
23	2295.857	0.0004	5737.142	0.00017	0.40017	2.499
24	3214.200	0.0003	8032.999	0.00012	0.40012	2.499
25	4499.880	0.0002	11247.199	0.00009	0.40009	2.499

i = 50%

Period	Single payment compound amount factor	Single payment present worth factor	Uniform series compound amount factor	Sinking fund deposit factor	Capital recovery factor	Uniform series present worth factor
1	1.500	0.6667	1.000	1.00000	1.50000	0.667
2	2.250	0.4444	2.500	0.40000	0.90000	1.111
3	3.375	0.2963	4.750	0.21053	0.71053	1.407
4	5.062	0.1975	8.125	0.12308	0.62308	1.605
5	7.594	0.1317	13.188	0.07583	0.57583	1.737
6	11.391	0.0878	20.781	0.04812	0.54812	1.824
7	17.086	0.0585	32.172	0.03108	0.53108	1.883
8	25.629	0.0390	49.258	0.02030	0.52030	1.922
9	38.443	0.0260	74.887	0.01335	0.51335	1.948
10	57.665	0.0173	113.330	0.00882	0.50882	1.965
11	86.498	0.0116	170.995	0.00585	0.50585	1.977
12	129.746	0.0077	257.493	0.00388	0.50388	1.985
13	194.620	0.0051	387.239	0.00258	0.50258	1.990
14	291.929	0.0034	581.859	0.00172	0.50172	1.993
15	437.894	0.0023	873.788	0.00114	0.50114	1.995
16	656.841	0.0015	1311.682	0.00076	0.50076	1.997
17	985.261	0.0010	1968.523	0.00051	0.50051	1.998
18	1477.892	0.0007	2953.784	0.00034	0.50034	1.999
19	2216.838	0.0005	4431.676	0.00023	0.50023	1.999
20	3325.257	0.0003	6648.513	0.00015	0.50015	1.999

Index